Relentless Evolution

JOHN N. THOMPSON

THE UNIVERSITY OF CHICAGO PRESS Chicago and London

JOHN N. THOMPSON *is the Distinguished Professor of Ecology and Evolutionary Biology at the University of California, Santa Cruz. Among his previous books are* The Coevolutionary Process *and* The Geographic Mosaic of Coevolution, *both published by the University of Chicago Press.*

The University of Chicago Press, Chicago 60637
The University of Chicago Press, Ltd., London

Printed in the United States of America

22 21 20 19 18 17 16 15 14 13   1 2 3 4 5

ISBN-13: 978-0-226-01861-4 (cloth)
ISBN-13: 978-0-226-01875-1 (paper)
ISBN-13: 978-0-226-01889-8 (e-book)

Library of Congress Cataloging-in-Publication Data

Thompson, John N.
Relentless evolution / John N. Thompson.
pages ; cm
Includes bibliographical references and index.
ISBN 978-0-226-01861-4 (cloth : alkaline paper)—ISBN 978-0-226-01889-8 (e-book) 1. Evolution (Biology) 2. Coevolution.
3. Adaptation (Biology). I. Title.
QH366.2.T54 2013
576.8—dc23

2012034475

∞ This paper meets the requirements of ANSI/NISO Z39.48–1992 (Permanence of Paper).

Contents

Preface

We now know that species are always evolving. Natural selection constantly adjusts the traits of populations generation after generation. In this book I explore what we have learned about why evolution is so relentless. Amid the flood of exquisite new results, I wanted to step back and ask what we now know, and still need to know, about the processes that drive adaptive evolution and shape the entangled web of life. I especially wanted to consider, from multiple perspectives, what we have learned from some of the best-studied examples of adaptive evolution in species and their interactions with each other. My hope is that these chapters show we have made great progress in recent years in our understanding of why evolution is unrelenting.

The book begins with what we know about the pace of adaptation and the intensity of natural selection (part 1) and the genetics of adaptation and coadaptation (part 2). Using those first two sections as background, I then explore the ecological forces that drive ongoing adaptation (parts 3 and 4) and diversification (part 5). The book ends with two synthetic chapters (part 6).

While I was writing this book, I sometimes felt as if I were racing against evolution itself in synthesizing the results of recent studies. Nothing captured that feeling more than an exchange I had one day with Peter and Rosemary Grant. I had written to them to ask if I could use in this book a modified version of a graph they had published a few years earlier. The graph showed evolution of beak size in one of Darwin's finches over more than thirty years. They wrote back to say that using the graph was fine with them, but had I seen their recently published updated graph showing evidence, from the past several years, that the birds had undergone another evolutionary shift?

This book continues an arc of argument about evolution of the web of life that I have been probing over the past four decades, with parts of the work summarized in three earlier books. In *Interaction and Coevolution* (Thompson 1982), I considered how natural selection acts on different forms of interaction among species—parasitism, predation, grazing, competition, and various forms of mutualism. In *The Coevolutionary Process* (Thompson 1994), I suggested how specialization, the local adaptation of populations, and environmentally varying selection on interacting species continually reshapes coevolution among species. I also discussed how the rift between ecology and evolutionary biology held back for decades our general understanding of the evolutionary process. Then, in *The Geographic Mosaic of Coevolution* (Thompson 2005), I broadened the discussion and considered how geographic variation in species shapes their interactions and the coevolutionary process. This book extends the discussion by considering how selection imposed by physical and biotic environments together drive continuing evolution in populations.

With so much research on evolution and coevolution being published each year, it is impossible to synthesize the entire history of past work. Therefore, this book draws primarily on research published since 2005, when *The Geographic Mosaic of Coevolution* was published. Even so, these chapters capture only some of the great breadth and depth of new research on adaptive evolutionary change.

I am grateful to the many colleagues and students who have shared their results and views on evolution, read chapters or parts of chapters, sent figures or photos, or provided insights during discussions at various stages of work on this book. I particularly thank the following colleagues: Cerisse Allen, Bruce Anderson, Jordi Bascompte, Robert Beardmore, Craig Benkman, May Berenbaum, Michael Boots, Seth Bordenstein, Paul Brakefield, Edmund Brodie, Jr., Edmund Brodie III, Michael Brockhurst, Judie Bronstein, Jeremy Burdon, Brendan Bohannan, Angus Buckling, Lynda Delph, Marcel Dicke, Rodolfo Dirzo, Camila Donatti, Paul Ehrlich, Johan Ehrlén, Niles Eldredge, Samantha Forde, Magne Friberg, Douglas Futuyma, Sergey Gavrilets, George Gilchrist, Charles Godfray, Richard Gomulkiewicz, Deborah Gordon, Peter Grant, Rosemary Grant, Ivana Gudelj, Paulo Guimarães, Jr., Ilkka Hanski, Michael Hassell, David Hembry, Jason Hoeksema, Robert Holt, Laurence Hurst, David Jablonski, John Jaenike, Steven Johnson, Pedro Jordano, Atsushi Kawakita, Joel Kingsolver, Britt Koskella, Anna-Liisa Laine, Bruce Lieberman, Curtis Lively, Jonathan Losos, Richard Lenski, Cristina Lorenzi, Bruce Lyon, Marc Mangel, Robert Marquis, Mark McPeek, Rodrigo Medel, William Miller, Nancy Moran, David Nash, Scott Nuismer, Takayuki Ohgushi, Anton Pauw, Galen Pelzmann, Grant Pogson, Peter Price, Trevor Price, Elizabeth Pringle, Robert Raguso, Paul Rainey, Robert Ricklefs, Tomas Roslin, Victor Rico-Gray, Loren Rieseberg, Samuel Scheiner, Johanna Schmitt, Christopher Schwind, Dolph Schluter, J. Mark Scriber, Adam Siepielski, Christopher Smith, Douglas Soltis, Pamela Soltis, Ayco Tack, Jeremy Thomas, Peter Thrall, Hirokazu Toju, Philip Ward, Gerrit Velema, Christer Wiklund, and Arthur Zangerl.

I am especially grateful to Paul Ehrlich for a crucial conversation as I was beginning to formulate my ideas for this new book.

I thank the following colleagues who made helpful comments on multiple chapters of the book and, in some cases, on an entire draft of the book: Magne Friberg, Paulo Guimarães, Jr., Pedro Jordano, Kristen Ruegg, Christopher Schwind, and Hirokazu Toju. Three anonymous reviewers for the University of Chicago Press made multiple insightful comments for which I am grateful, as did Kathryn Gohl. Her careful copyediting improved every chapter.

I am indebted to my editor, Christie Henry, who has helped support this book in many ways throughout its development, as she has done with the two previous books I have published with the University of Chicago Press. She and the others at the Press have made the process wonderfully enjoyable.

As with all my previous books, my wife Jill has made important suggestions on all the chapters and helped in multiple ways. I am deeply grateful for her assistance.

I am also indebted to the organizations that provided me with the time and settings to work for extended periods on the book. The National Science Foundation generously provided an OPUS (Opportunity for Promoting Understanding through Synthesis) award. The Whiteley Center at the Friday Harbor Laboratories of the University of Washington provided an idyllic setting for extended periods of writing. I am grateful to Arthur Whiteley for creating the center and to Kathleen Cowell for the many ways in which she helped to make day-to-day life at the center so productive. I am also grateful to the University of California, Santa Cruz, for providing, as it has for many years, a collegial environment in which to discuss nascent ideas with colleagues.

Part 1

The Process of Adaptation

1

Adaptive Evolution

If you read this book at the rate of about a chapter a day, by the time you finish it some species will have evolved. They will have been microbial species, and the populations will have evolved in ways almost imperceptible. If you have the right experimental tools, though, you can document the evolution. If you wait a little longer, say a year, the same will have had happened to some plants, insects, and other species with short generation times. We did not realize until recently the relentlessness of evolution, because we lacked the tools, and we often looked for it only where we expected we would most likely find it—principally in environments we have greatly changed.

In the old days of research on evolution, just a few decades ago, we hoped at best to catch glimpses of evolution in action. Scientists and nonscientists alike thought that evolutionary processes acted over long periods of time. We thought that any chance of seeing evolution occur would be due to luck or to extremely unusual circumstances. It was common for biologists to talk about "ecological time" as compared with "evolutionary time." Ecological processes happened quickly; evolutionary processes happened slowly.

Most of us biologists therefore felt we could ignore rapid evolution as a potential explanation for the changing patterns we often find in populations and biological communities. When asked for examples of evolution occurring over short timescales, we would rely on a few well-studied cases. We would point to the increases in dark-winged forms of peppered moths in regions of high industrial pollution, the rapid evolution of resistance to pesticides in some insects, or the continuing evolution of human influenza virus during the past century. There were some other examples from which we could choose, but few had been analyzed in detail. They were collectively viewed as the fortunate exceptions we could study.

Those days are over. Well-studied examples of ongoing evolution within our lifetimes are being published in professional journals at such a fast rate that it is hard to keep up with them. Even those of us who have studied the ongoing evolution of populations have become increasingly impressed by the speed at which some populations are evolving in nature. The examples come from studies in the fields of ecology, epidemiology, medicine, microbiology, agriculture, forestry, wildlife management, marine biology, fisheries biology, population genetics, and molecular biology. We have now come to expect that insects and weeds will evolve resistance to pesticides, influenza viruses will evolve at speeds that will keep epidemiologists nervous, and new strains of antibiotic resistant bacteria will continue to proliferate and cause concern within the medical community. We now know that even the simple act of harvesting fish populations has led to marked evolutionary changes in some species.

As we have come to realize the sometimes rapid pace of evolution, many biologists and some policy makers and resource managers have increasingly turned to the problem of how to manage it. How do we slow the rate at which insects evolve resistance to pesticides and bacteria evolve resistance to antibiotics? How do we conserve and restore biological communities amid global change that is driving evolutionary change in some species? How do we control invasive species that are evolving as they spread across new continents and oceans? Amid our growing appreciation of the pervasiveness of evolutionary change, just about every possible view has now been expressed on how human activities may alter the future evolution of species.

That discussion, though, only highlights the more long-standing debate in evolutionary biology of what drives ongoing evolutionary change—sometimes quickly, sometimes more slowly, but ongoing nevertheless. We can point to particular cases and their causes: rising or falling temperatures, changing patterns of rainfall, sexual selection within species, competition, predation and trophic cascades, parasitism, mutualism, the balance between mutation and random loss of genes, and the occasional odd asteroid. These ad hoc explanations simply underscore the fact that almost all species live in a constantly changing world that demands evolutionary change in populations.

If we are to interpret how our world is changing through climate change, habitat modification, and the wholesale movement of species among continents, we need to understand much better the background chatter of endless year-to-year evolution and its causes. We need to know the extent to which continual evolutionary change is truly important in shaping and maintaining the web of life at every timescale and across every spatial scale.

This book explores the pace, genetics, and ecological drivers of adaptive evolutionary change. It is about why natural selection is generally stronger and adaptive evolution more dynamic than, until recently, we have thought.

The early chapters focus on adaptive evolutionary change in populations over tens, hundreds, and thousands of years rather than millions of years. This is the part of evolutionary change that is most directly and immediately important to the ecological dynamics of biological communities, to the conservation of species, and to human society as species all around us continue to adapt amid environmental change. The later chapters explore the consequences of ongoing evolution for ecological speciation, adaptive radiation, and the continual reformulation of the web of life.

THE PROBLEM TO SOLVE

The great problem to solve about life on earth has gradually shifted over the past century and a half since Darwin's *Origin of Species*. We began with the problem of whether species evolve. The problem has been solved so completely that we are now faced with a problem at the opposite extreme. Why is evolution so relentless, altering populations generation after generation? After all, species generally seem well adapted to the environments in which they live, yet they continue to evolve even in environments that have not undergone major recent changes. Most of these evolutionary changes occur through modification of genes and traits that already have been subject to selection for many thousands of generations.

Superficially, these small changes seem like aimless evolutionary meanderings. Slowly, though, we have come to realize that these continual adjustments in adaptation are often surprisingly important to the persistence of populations. These small changes capture the ecology of evolution. That appreciation has made us realize that we need as deep an understanding of the ecological drivers of evolutionary change as we have tried to develop for the genetic and molecular processes that translate ecological selection into evolving traits. These chapters explore how our understanding is progressing.

We know the component parts of the process of adaptive evolution. It begins with differences among physical environments that impose selection on populations to adapt to local temperatures and the availability of water, light, and nutrients. Without major environmental change, populations become well adapted to their local physical conditions. Populations and species, though, do not live in a vacuum. They adapt, speciate, and go extinct as parts of continually changing webs of interacting species. Much of the ongoing evolution of each species is about exploiting other species and avoiding exploitation. The result is a process of reciprocal evolutionary change—coevolution—that shapes the web of life in different ways in different environments. Occasionally those webs are torn apart by huge physical upheavals that lead to mass extinctions, creating new opportunities for diversification. Overall, the physical environments provide the basic templates for adaptation and diversification, but interactions among species multiply and modify, in myriad ways, how selection acts within and among those templates.

That much now seems obvious to us after many decades of hard-won paleontological, evolutionary, and ecological data. We are, though, still struggling with fundamental questions about the ecological structure and dynamics of evolutionary change. How can natural selection on species be so unrelenting without constantly reorganizing much of the web of life? If natural selection is so strong on populations, why are species not constantly undergoing directional evolutionary change? If natural selection often does not lead to directional change, then what forms of selection are most important in driving much of the generation-to-generation evolutionary change we see in populations? Is selection imposed by species on each other inherently different from selection imposed by physical environments? These questions are increasingly important at a time when we are altering the earth's physical environments and the web of life itself.

THE CENTRAL ARGUMENTS

This book weaves together two arguments on why evolution is so relentless. The central argument is that evolution is as much an ecological process as it is a genetic process. At a superficial level, we all know that, but the drive to understand the molecular mechanisms of evolution can make it seem at times that evolution can be understood mostly by a deeper understanding of molecular mechanisms alone. It cannot. Much of the dynamics of evolution is about the interplay between genes and environments (genotype-by-environment interactions) and about the ever-changing coevolution among species in different environments (genotype-by-genotype-by-environment interactions). Adaptation and adaptive diversification are, at their core, the result of the intermingling of molecular and ecological processes. Species are constantly adapting and re-adapting because they are forced to do so by the ever-changing web of life. Much of adaptive evolution, then, is about the continual redeployment of standing genetic variation in different ways in constantly changing physical and biotic environments. The pacesetters of day-to-day evolution seem to be at least as much, and maybe more, ecological rather than genetic.

The second argument is that much of adaptive evolution does not lead anywhere, yet these small changes are crucially important. These continual microevolutionary changes keep populations in the evolutionary game as they interact with other species that are themselves constantly evolving. These seemingly aimless meanderings are the essential dynamics of evolution, with directional change and speciation as occasional outcomes. Species do not fail to undergo sustained directional change because natural selection has been asleep on the job; species fail to undergo sustained directional change because natural selection time and again comes up with slightly variant ways of jury-rigging species to keep them as viable evolutionary products even as the world continues to change around them.

Together, these two arguments constitute a view of evolution that is unrelenting because selection on populations constantly changes as genes are expressed in different ways in different environments and as interactions among species vary in their effects among environments. Almost every major study of selection in nature has found differences either in the form or the strength of selection among years and among populations. Much of evolution is about selection that favors modest changes in degrees of novelty within populations and relatively modest divergence among populations. Most adaptive radiations of species are about variations on an ecological theme. This book, then, examines why evolution can appear to be either frenetic or sluggish, depending on the lens we use to examine it. Some chapters focus more on the pace and dynamics of evolutionary change, and others focus more on the drivers that fuel ongoing adaptive change.

THE EXPANDING STUDIES OF ADAPTIVE EVOLUTION

We have progressed in recent years in our understanding of adaptive evolution through work by ecologists painstakingly studying which individuals survive and reproduce in different environments and different years; coevolutionary biologists studying how species coadapt to each other across complex environments; microbiologists, population geneticists, physiologists, and developmental biologists examining experimental evolution and coevolution in the laboratory; structural biologists analyzing how selection remolds the shapes of organisms in different ecological contexts; molecular biologists exploring the mechanisms of evolutionary change; paleobiologists examining patterns of change over longer timescales; and theoreticians probing mathematical models of the rates of evolution. Rather than simply trying to infer how evolutionary processes might have created the patterns we see in nature, we now have the tools to study these processes directly in the laboratory and in nature. We can compare evolution in populations that have been manipulated directly by human activities with those only indirectly affected by our activities or with those in the few remaining environments still mostly free of human activities.

These studies have documented hundreds of cases of ongoing evolution of species over the past century (table 1.1). They are examples of what is sometimes called contemporary evolution (Hendry and Kinnison 1999). It is evolution on timescales that can affect the dynamics of populations, communities, and ecosystems. These studies in eco-evolutionary dynamics represent part of a renewed attempt to link the fields of ecology and evolutionary biology. The connections between these fields have waxed and waned over the past century, but these disciplines are now coming together again in new ways (Abrams and Matsuda 1997; Thompson 1998; Hairston et al. 2005; Whitham et al. 2006; Fussman et al. 2007; Kinnison and Hairston 2007; Wade 2007; Strauss et al. 2008; Bailey et al. 2009; Pelletier et al. 2009; Ellner et al. 2011; Schoener 2011).

Compiled lists and summaries of rapid evolution such as those in table 1.1 include taxa, traits, and trends as diverse as life itself (e.g., Thompson 1998; Hendry and Kinnison 1999; Kinnison and Hendry 2001; Reznick and Ghalambor 2001; Hairston et al. 2005; Carroll et al. 2007; Hendry et al. 2007; Ellner et al. 2011). Any such list would rise rapidly into the thousands if it included every case of a population evolving pesticide resistance or antibiotic resistance and every case of evolution in a local population caused by human activities. Examples of rapid evolution are limited much more by the number of biologists studying it than by the number of actual examples in nature.

The examples now include just about every kind of trait biologists have studied: morphology, physiological pathways, life histories, behaviors, and

Table 1.1 *Examples of ecologically important characteristics of species that have evolved in nature over the past two centuries*

Traits	*Examples*	*References*
Morphology	Beak size and shape in birds	Grant and Grant 2008a
	Mouthparts of insects	Dingle et al. 2009
	Body size in harvested fish	Darimont et al. 2009
	Latitudinal clines in insect size	Huey et al. 2000; Gilchrist et al. 2004
	Wing and body color in insects	Saccheri et al. 2008; Forsman et al. 2011
	Shell color in snails	Cook and Pettitt 2008; Ozgo 2011
	Sexually selected morphs	Sinervo and Lively 1996; E. Svensson et al. 2005
	Range of variation in traits	Olsen et al. 2009

Table 1.1 (*continued*)

Traits	*Examples*	*References*
Physiology	Detoxification enzymes in insects	Berenbaum and Zangerl 1998
	Resistance to pesticides	Powles and Yu 2010
	Resistance to transgenic crops	Tabashnik et al. 2008
	Resistance to animal parasites	Duffy and Sivars-Becker 2007
	Plant resistance to herbivory	Oduor et al. 2011
	Plant chemical defenses	Zangerl and Berenbaum 2005
	Invertebrate resistance to toxic prey	Hairston et al. 1999
	Antibiotic resistance in microbes	Goetz 2010
	Changed patterns of viral virulence	Nelson and Holmes 2007; Bhatt et al. 2011
	Maintenance of different enzyme forms	Niitepõld et al. 2009
Behavior	Migratory patterns in birds	Pulido and Berthold 2010
	Dispersal distances in toads	Alford et al. 2009
	Anti-predator behavior in fish	O'Steen et al. 2002
Life history	Diapause in insects	Feder et al. 2010
	Flowering time in plants	Franks et al. 2007; Dlugosch and Parker 2008
	Growth rates in fish and toads	Edeline et al. 2007; Phillips 2009
	Resident vs. dispersive traits in fish	Pearse et al. 2009
	Age of maturation in fish	Walsh and Reznick 2011
Species interactions	Shifts of insects onto novel hosts	Singer et al. 2008; Forbes et al. 2009
	Shifts of pathogens onto novel hosts	Lowder et al. 2009
	Mutualism from antagonism	Weeks et al. 2007
	Coevolution of hosts and parasites	Decaestecker et al. 2007; Lively 2010a

Notes: The list includes only a small sample of known cases of rapid evolution.

interactions with other species. They include native species living in their normal environments, native species living in environments greatly altered by humans, and introduced species living in environments either similar or different from where they lived in their native ranges. They include vertebrates, invertebrates, plants, fungi, and microbes. They involve not only examples of how populations adapt to their physical environments but also how they adapt to each other and to other species. Considered together, these examples have told us that rapid evolution is occurring in all major taxa in most environments. It is not limited to fast-growing microbes or insects or small plants, and it is not limited to highly modified environments that impose novel selection pressures on populations.

If we lengthen the timeline to thousands of years, evolutionary change becomes even more evident. Changes in climate, habitats, and the geographic distributions of species in the past 10,000–12,000 years since the end of the Pleistocene have resulted in many populations that have diverged from each other as they have adapted to different environments. Deer mice (*Peromyscus maniculatus*) in the Sand Hills of Nebraska have evolved a light coat color rather than the normal dark color during the past 8,000 years (Linnen et al. 2009), and some crossbill populations have diverged to specialize on different conifers since the Pleistocene (Benkman 2010). The fastest observed rates have been in species that we are trying to manipulate for our own ends, but that is also where we most often look for rapid evolution. We know we are directly fueling the evolutionary process through our manipulation of other species, but we are coming to realize that our manipulations are often just a highly efficient and specialized form of what species everywhere impose on each other.

TRACKING RAPID EVOLUTION THROUGH QUIRKS IN LIFE HISTORIES

In some cases, quirks in the biology of species make it possible to track the genetic signatures of rapid evolution directly by comparing the current generation with ancestral generations. This approach, sometimes called resurrection ecology (Kerfoot and Weider 2004), has been used to study species with dormant stages, that can remain alive for many years and brought back later to an active state. These include, for example, invertebrates in which dormant eggs or other resting stages become buried in lake sediments. This approach has also long been the mainstay of studies of adaptive evolution in laboratory experiments on microorganisms that can be frozen alive and then thawed later, still alive.

Among studies in nature, this approach has been especially successful in studies on the evolution of water fleas (*Daphnia*). These small crustaceans are abundant in many lakes worldwide, and they have been used in many studies of evolutionary and coevolutionary dynamics (Little et al. 2006; Decaestecker et al. 2007; Duffy et al. 2008; Ebert 2008; Walsh and Post 2011). *Daphnia* produce eggs capable of remaining unhatched but viable in lake sediments for many decades. Lake bottoms therefore contain a record of genetic change in water flea populations, and each layer of the sediment captures the genetic composition of each species in that particular year. By coring sediments, it is possible to collect water fleas from each layer: the most recent populations are at the top, and the oldest populations are at the bottom. The populations from the different layers are then analyzed for the evolution of traits. DNA can also be extracted from eggs collected in each layer, making it possible to directly observe genetic changes in these populations over time.

By sampling resting eggs from sediments, researchers have now observed the signature of rapid evolution of water fleas in multiple studies. These studies show that some of this rapid evolution is driven by interactions with other species. A study of *Daphnia galeata* in the sediments of Lake Constance in central Europe showed that, as eutrophication increased in this lake, so did the abundance of nutritionally poor or toxic cyanobacteria, and, in turn, the resistance of *Daphnia* to these cyanobacteria (Hairston et al. 1999). Elsewhere, a study of sediments from a small pond in Belgium showed rapid evolution of interactions between one of the most commonly studied water fleas, *Daphnia major*, and one of its bacterial endoparasites, *Pasteuria ramosa* (Decaestecker et al. 2007). By hatching dormant *Daphnia* eggs from eight layers in the sediment, researchers could compare populations from eight points in time over the past thirty-nine years. Bacterial populations could also be resurrected from each layer. Each *Daphnia* population could then be challenged with bacteria from the next layer down, from the layer in which the *Daphnia* eggs and bacteria were collected, and from the next layer up.

This experiment, in effect, exposed each water flea population to a past, a present, and a future parasite population from the same lake. This experimental design makes it possible to track changes over time in resistance in the water fleas and virulence in the parasites. The changes were fast. During the thirty-nine-year period, the parasite population repeatedly evolved to track evolutionary changes in resistance in the *Daphnia* population.

Other studies using *Daphnia* eggs from sediments have shown a signature of evolution involving not only adaptation within species but also hybridization among species. For example, Lake Constance, on the border of Austria, Germany, and Switzerland, and Lake Greifensee, in Switzerland, have experienced great variation in phosphorus levels over the past century as a result of human activities. The variation has, in turn, affected the ecological food web in these lakes (Brede et al. 2009). Eutrophication of the lakes reached its peak in the 1970s and 1980s and decreased in subsequent decades. These environmental changes resulted in genetic change in water flea species that matched the history of eutrophication. During the first half of the twentieth century, the lakes were inhabited by one species, *D. hyalina*. A second species, *D. galeata*, invaded both lakes during the 1940s and 1950s and mated with *D. hyalina*, producing hybrids. Pure forms of *D. hyalina* disappeared from these lakes. As of 2004, the last year of the study, the water fleas in these lakes still retained the genetic signature of this evolutionary event.

Field studies using other *Daphnia* and their parasites have taken advantage of other techniques or environmental conditions to confirm rapid evolution at these timescales. Artificial populations set up in the field in Finland have shown detectable evolution in populations over two years, which is about fifteen generations (Zbinden et al. 2008). In Bristol Lake, Michigan, a population of *Daphnia dentifera* underwent detectable evolution after an epidemic of parasitic yeasts in 2004. The water flea population did not increase in average resistance to the parasite, but it showed much greater genetic variance after the epidemic, indicating that the parasite population had exerted disruptive selection on the *Daphnia* population (Duffy et al. 2008). This is a much more subtle form of evolution than directional selection. It is, though, just as ecologically important, because it alters the range of genetic forms of *Daphnia* species within the community in which it lives. Mathematical modeling of this interaction has suggested that epidemics of yeast infection may end due at least partially to rapid evolution in the water flea population (Duffy et al. 2009). Overall, the many studies of *Daphnia* have shown that not only do these populations commonly evolve over just a few decades but they are a capable of evolving in many different ways.

TRACKING THE MOLECULAR SIGNATURE OF RAPID EVOLUTION

In other cases, it is possible to use molecular substitutions to track the signature of rapid evolution going back decades. Some of the best data are on human influenza viruses, which evolve at astonishingly fast rates. Many of the changes occur in the virus's glycoprotein coat called hemagglutinin, and those changes are often associated with pandemics of the disease. The changes occur in an almost clocklike fashion, with observable change occurring every decade over much of the past century (Suzuki and Nei 2002). In recent years, adaptive evolution in influenza A subtype H3N2 has con-

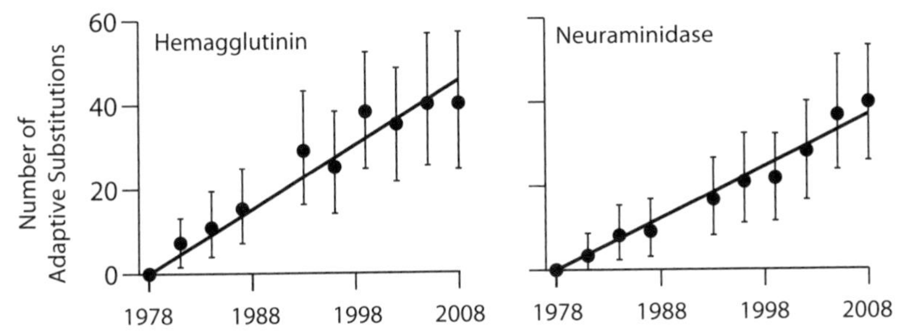

Fig. 1.1 The number of adaptive substitutions in hemagglutinin and neuraminidase genes of human influenza virus subtype H3N2 during two decades. Values are means and 95 percent bootstrap percentiles; the line is a linear regression through the data. Adaptive evolution was determined using a modification of methods that evaluate the ratio of nonsynonymous to synonymous mutations. After Bhatt et al. (2011).

tinued to occur at a remarkably constant rate in hemagglutinin and also in neuraminidase, which is another envelope glycoprotein (fig. 1.1).

The same clocklike pattern of change has occurred in the pre-2009 version of subtype H1N1. Although these genes evolve at a slightly slower rate in H1N1 than in H3N2, the pattern of change over time is just as linear (Bhatt et al. 2011). These genes, though, are not just evolving through chance molecular substitutions. Analyses of the ratio of the rate of nucleotide substitutions that affect amino acids to the rate of substitutions that do not affect amino acids suggest that these genes are under strong selection (box 1.1). In contrast to these two genes, some other influenza virus genes in both subtypes have evolved at a much slower rate, and some have undergone few adaptive substitutions. Even within hemagglutinin and neuraminidase, one part of each of these genes is evolving at a fast rate while the other part remains mostly unchanged (Bhatt et al. 2011).

These genomic analyses show that each influenza virus subtype possesses genes evolving at markedly different rates. At any moment, there are multiple strains of human influenza virus circulating in populations, with evolution ticking away in different ways in each viral subtype and each viral type. The problem for public health officials each year is to guess which of the current strains are the most likely to occur in numbers high enough to be included in this year's flu shots.

MEASURING EVOLUTIONARY RATES

Quantifying the rate of evolutionary change would seem to be a relatively easy task, if changes in traits can be measured over time. It has, though, often

BOX 1.1 Detecting Selection through Rates of Molecular Substitution

Molecular analyses evaluating evidence for natural selection often use the ratio of the rate of nucleotide substitutions that affect amino acids (dN) to the rate of substitutions that do not affect amino acids (dS). This ratio expresses the proportion of substitutions likely to have had some real effect on organisms. Substitutions with no effect on the structure of amino acids, which are often called silent substitutions or synonymous mutations, are assumed to be generally neutral with respect to natural selection. Substitutions affecting amino acids (nonsynonymous mutations) are assumed to be under positive selection and adaptive, if they are retained in a population over time. Nonsynonymous mutations that are maladaptive are assumed to be removed by purifying selection from a population.

The dN/dS ratio has become a tool for identifying genes under selection among divergent populations. It is, though, useful only as a general signature of selection, because multiple molecular and selective factors can influence the ratio (Kimura 1977; Holt et al. 2008; Stoletzki and Eyre-Walker 2011). A dN/dS ratio greater than 1 suggests that natural selection favors changes in the protein, and a ratio less than 1 suggests that selection disfavors changes in the protein. This ratio works best for comparisons among populations. When applied to evolution within populations—or, as in influenza, within subtypes—the interpretation of the ratio is more complicated, because repeated sampling of populations over time makes it more difficult to assess the relationship between this ratio and the strength of selection (Kryazhimskiy and Plotkin 2008). Even so, new statistical methods have made it possible to improve the interpretation of dN/dS ratios for selection within single lineages. Application of these methods to influenza viruses has reinforced the view that parts of the hemagglutinin and neuraminidase protein molecules are undergoing high rates of adaptive evolutionary change (Kryazhimskiy et al. 2008; Bhatt et al. 2011).

turned out to be difficult, because evolution can take multiple forms, and different traits in populations evolve at different rates. The simplest measures of evolutionary rates are of species that have undergone directional change over time in one trait or a composite of traits, such as a decrease in average adult size of individuals or an increase in average tooth size. Body or tooth size is a result of selection acting on multiple traits that are correlated with each other to varying degrees. Measurement of these complex traits is where much of the early work on rates of evolution began, and it required standardized ways of measuring the rate of evolutionary change.

Population geneticists in the 1920s and 1930s had already developed formal mathematical theory to track the rates at which gene frequencies would evolve under different genetic conditions (Fisher 1930; Wright 1931; Haldane 1932). The problem was how to transfer that knowledge to the study of the evolution of traits governed by multiple genes that interact with each other in complicated ways. In trying to show what paleontology could contribute to our understanding of the evolutionary process, George Gaylord Simpson (1944) adopted a statistical approach focusing on the rate of change in the morphology of organisms in the fossil record. In the first sentence of the first chapter of his 1944 volume, he writes, "How fast, as a matter of fact, do animals evolve in nature?" This book, titled *Tempo and Mode in Evolution*, helped make the analyses of evolutionary rates one of the central problems in evolutionary biology.

Simpson approached the problem by studying the rates of origination and extinction of genera. Because taxa are identified in the fossil record through their differences in morphological traits, these studies were, in effect, analyses of the net rate of evolution of complex sets of traits. Simpson suggested that the rates of origination and extinction of taxa are a special case of the rates of morphological evolution. The rest of his thought-provoking book is an exploration of possible ways of evaluating, quantitatively rather than subjectively, the rates of evolutionary change and the implications of those rates for our understanding of the tempos and modes of evolution.

Simpson's studies prompted J. B. S. Haldane (1949) to go further and consider how to measure the rate of evolutionary change in a particular trait such as tooth size or bone length. Haldane wanted to know how fast traits evolve on average per year or per generation. He proposed two possible standardized units of evolutionary change: one that he called a darwin and another that he did not name, but later Philip Gingerich (1993, 2009) called it a haldane. Those two units remain the two major ways in which rates of directional change are determined and compared among populations and species.

A darwin is the simpler but less informative unit. All that is needed is the average value of a trait of a species at one point in time, the average of that trait at another point in time, and the time interval between those two measurements (box 1.2). In a broad sense, a darwin is easy to understand: it

BOX 1.2 Two ways of measuring the rate of evolution, where the question to be answered is how much a measured trait in a population has changed in a particular direction over time

Darwin. The measure darwin uses the means of traits to quantify the rate of change in a trait over time. It requires three values: the mean of a collection of measurements of a trait of a species at one point in time (x_1), the mean of a collection of measurements of the same trait at another point in time (x_2), and the time interval between those two measurements ($t_2 - t_1 = \Delta t$).

The measured traits are converted to natural logarithms (ln) to keep the changes proportional to each other. The ln-transformed values create a log-normal distribution of measurements that makes it possible to compare rates for different traits (Gingerich 2000). A darwin scales the amount of change in a trait to the time interval over which that change has been measured (Δt) and measures the rate as the change in e, the base of the natural logarithm, per million years.

$$d = (\ln x_2 - \ln x_1)/\Delta t$$

Haldane. The measure haldane uses the means and standard deviations of traits to quantify the rate of change in a trait expressed in standard deviations per generation (g). It requires the same three values as a darwin but also requires the pooled standard deviations (s_p) of the two samples.

$$h = [(\ln x_2 - \ln x_1)/S_p]/g$$

The expression used here for haldanes uses the natural logarithms of the means to show the relationship between haldanes and darwins. In practice, haldanes are sometimes calculated using the means themselves rather than the logarithms of the means, because the difference in calculations is sometimes slight.

See further details and variations on the details of calculations in Haldane (1949), Hendry et al. (2008), and Gingerich (2009).

is the amount of change that has accumulated over time. How that change is calculated, though, is less intuitive to most people. It is the change in value of a trait in natural logarithms per million years. Haldane suggested that darwins be measured as change per million years, because he was thinking at that time about change in the fossil record, and he thought that evolution was generally slow. He acknowledged that evolution sometimes could be fast, as in cases of human-driven evolution, and noted that, by his definition of a darwin, domesticated animals and plants "have changed in rates measured in kilodarwins."

Using darwins as the unit of measurement, Haldane calculated the rate of change in fossil horses based on the assumption that the rate was the same on most genes responsible for the observed changes in morphology. His calculations suggested that, if natural selection were largely responsible for the evolution of horses, then it acts only weakly and would rarely be observable over short time spans. He went on to argue that it is therefore not surprising that progressive changes in gene frequencies have rarely been observed and that, when they have been observed at all, they are probably due to our alteration of natural environments (Haldane 1949). Haldane was right about many things, but not on this point.

In the same paper, Haldane suggested an alternative unit of evolutionary change based on change over time in the distribution of variation in a trait (box 1.2). This idea has an intuitive appeal. As Haldane wrote, variation within a population is the raw material available for evolution. He therefore based his alternative measure on change in standard deviations per generation. Calculation of an evolutionary rate in haldanes therefore requires a solid assessment of variation in a trait in addition to the average value. Haldanes are also more intuitive when thinking about the relentlessness of evolution, because they measure change per generation, whereas darwins are measured in change per million years. In recent years, evolutionary rates have become commonly measured in haldanes, although studies also commonly report rates in darwins.

Simpson's (1953) reaction to Haldane's measures was positive but cautious. He wrote that he could see the utility of using natural logarithms of the measurement rather than the original measurements if the genetics of size lead to proportional changes in characters, but he predicted that a measure based on changes in natural logarithms would be an "unnecessary complication." He argued that Haldane's suggestion of basing rates on changes in the standard deviation per generation was interesting, and Simpson hoped it would be tried out for various groups. Nevertheless, he doubted that it would be widely used. As it turned out, he was wrong, but it took decades and the ease of modern computing power for haldanes to become a standard measure.

WHAT AND HOW TO MEASURE

The most accurate determinations of evolutionary rates are those made by monitoring populations across multiple generations for change in the means and variation of traits. This close monitoring determines the generation-to-generation dynamics of change and often allows simultaneously an evaluation of the potential causes of change. In practice, darwins and haldanes are often used to measure the net change in a trait over many generations based on samples at just two points in time: a beginning point and an ending point. Those two points are sometimes fossils separated by millions of years (i.e., Haldane's original use) and sometimes extant populations separated by a hundred years. All such measures of change are only indicators that some change has occurred in one or more populations due to any of many causes (Hendry and Kinnison 1999).

In some cases, the measures have been used to calculate the accumulated divergence between two or more populations over time, rather than change in a population over time. These are not the same thing. Divergence among populations is the sum of the net changes that have occurred in each population. Each population could have evolved at a different rate. At one extreme, all the change could have occurred in only one of the populations. At the other extreme, each population could have diverged at the same rate.

Even after controlling for these problems as much as possible, compilations of rates show a wide range of values (e.g., Bone and Farres 2001; Kinnison and Hendry 2001; Westley 2011). Even fast rates are barely perceptible from one generation to the next, because they generally involve small shifts in the average value of a trait. Most estimates of haldanes are well below 1, which is one standard deviation per generation. Haldanes that are estimated from short-term field sites are often near 0.3 haldanes, although some values are higher and many are lower (Gingerich 2009). A change of almost a third of a standard deviation in a generation is still fast, suggesting that natural selection is sometimes strong on populations. Rates measured using fossils give much lower estimates, often in the range of 0.1–0.2 haldanes. Overall, the longer the timescale over which the rate is estimated, the lower the estimate of the net rate of change (Gingerich 1983, 2009). Longer timescales damp the shorter-term fluctuations.

When Haldane proposed these measures in 1949, he was fully aware that these indices captured only some information about evolutionary rates. He wrote, "it is likely that better indices of evolutionary rate can be made than any which I have suggested." Although we now have multiple useful ways of studying evolutionary rates, discussions over alternative measures continue. We know that the strength and sometimes the direction of evolution vary over time within time within populations. The value of measurements in darwins or haldanes is that they provide a standardized starting point for evaluating why some populations or species change faster than others. Both measures are simply descriptions of the net change in populations over time.

Whether the changes are the result of evolution driven by natural selection requires additional information from observational or experimental studies. Without additional information, the changes cannot be attributed with certainty to evolution. Any change could be due either to evolution or simply to differences in how genes are expressed as environments change. For example, the mean and variation in body size in a population living in a stressful environment could change if the environment became more benign and nutrition of individuals improved.

More generally, darwins and haldanes almost certainly understate the actual rate and dynamics of evolutionary change, because they measure only net change. They mask any reversals in the direction of evolutionary change that occurred between the times when the measurements were made. If natural selection varies erratically over time as environments change, or if natural selection oscillates over time—favoring first one trait, then another, then the first one again—then populations could have low values for darwins and haldanes despite much ongoing and ecologically important evolution.

This point is important for our understanding of the evolutionary process. Ultimately we want to understand the extent to which evolution truly is unrelenting and strong enough to shape variation within species and the dynamics of populations, communities, and ecosystems. The seemingly small evolutionary shifts in the traits of populations can be as ecologically and evolutionarily important as the long-term directional changes. I return to this point repeatedly and explore it from multiple perspectives in the upcoming chapters.

THE CHALLENGES AHEAD

Rapid evolution is now being found in nature in such a wide range of taxa that it must be one of the working hypotheses for the dynamics of populations and communities over even short periods of time. Studies of eco-evolutionary dynamics have increasingly shown that evolutionary and ecological changes can influence each other, fostering yet more change (Hairston et al. 2005; Carroll et al. 2007; Palkovacs et al. 2009; Post and

Palkovacs 2009; Bassar et al. 2010; Hanski 2011; Schoener 2011). Examples of strong selection and rapid evolution have accumulated to such an extent in recent years that we should expect that the ecological dynamics observed in populations over timescales of just a few decades are often driven in part by rapid evolution.

Among the most ecologically important evolutionary changes are changes in the timing and location of life history events within populations—for example, shifts in the ages at which individuals reproduce, shifts in dispersal or migratory patterns, or shifts in the time during a year when individuals search for prey or hosts. Most small shifts in life history events undoubtedly reflect plastic responses of species to environmental cues, but some observed changes seem to have involved adaptive evolutionary change as well (Gienapp et al. 2007a; Franks and Weis 2008; Montague et al. 2008). Any shift in the timing or location of a life history event has the potential to ripple throughout a community, because interactions between species begin with the simple act of encountering each other. Much of evolution is about staying in sync with prey, hosts, or mutualists and staying out of sync with predators, parasites, or competitors. As environments change, each species responds independently to the altered environmental cues. That inevitably leads to species becoming more in sync with some species and more out of sync with other species (e.g., Møller et al. 2011).

Life history shifts therefore have the potential to ripple in their effects on selection throughout communities at the local level, the regional level, and even at the transcontinental level. Changes in flowering time over the past two hundred years have been reported for many communities worldwide, and those changes can affect interactions with pollinators within and among ecosystems (Elzinga et al. 2007). At larger geographic scales, the migratory habits of species sometimes depend on the availability of particular prey species or mutualistic species at their breeding grounds, wintering grounds, or along their migration routes. Shifts in the timing of migration or shifts in the timing of events in prey or mutualistic species have the potential to alter selection on multiple species among the communities that the migrants normally visit on their yearly cycle.

One of the important current challenges, then, in evolutionary biology is to understand how rapid evolutionary changes in life histories ripple throughout webs of interacting species, fostering yet more evolutionary change within and among ecosystems. The potential rate of change in these and other traits depends in part on the strength of natural selection. In the next chapter, we turn to the assessment of selection in nature.

2

Natural Selection

One of the major reasons that rates of evolution measured on populations today are generally faster than rates measured on fossils is surely that evolution in the fossil record detects net changes over long periods of time. Selection on populations from one generation to another is more dynamic than can be assessed by long-term averages. This point has been made repeatedly by evolutionary biologists, but it needs continued emphasis. Lack of sustained directional selection in populations or species is not, in itself, evidence that natural selection is weak.

This chapter considers what we currently know about the forms and intensity of selection in populations in nature, including populations under direct or indirect selection by humans.

BACKGROUND: HARVEST SELECTION AND RAPID EVOLUTION BY STRONG FILTERING

Rates measured as darwins and haldanes capture evolutionary change best when populations are under sustained directional selection. That is exactly what we do well as humans in our manipulations of other species: we often select consistently for traits at one extreme of a distribution of values. Historically and now, the most common way has been through harvest selection. The effects of these actions on evolution provide an indication of how fast populations can evolve when subjected to strong selection.

The relatively simple act of harvesting wild populations for our own use has caused rapid evolutionary change in multiple species (Edeline et al. 2007; Allendorf and Hard 2009). Large-scale commercial fishing with nets of particular sizes, trophy-hunting for large wildlife species, and selective culling of plants have all resulted in measurable changes in the characteristics of animal and plant species over the past century. These changes are not restricted to the year in which the harvest is done. Some are true evolutionary changes that have altered the traits of populations in subsequent generations.

We have become one of the strongest agents of directional selection on other species through our consistent selection on their traits (Ehrlich 2001; Palumbi 2001a, 2001b; Ehrlich and Ehrlich 2008). We are now, in the words of Paul and Anne Ehrlich (2008), the dominant animals. The speed at which our activities have caused evolutionary change in many species shows that populations often harbor a great ability to evolve in response to environmental change. There are now enough studies of our effects that we can ask whether humans are accelerating the rate of evolution in populations over what normally occurs in nature.

One major study of 29 species has suggested that harvested populations have undergone observable change in morphological and life history traits 300 percent faster than that found in unharvested populations (Darimont et al. 2009). It is a pace of evolution observable within human lifetimes. The study included populations of multiple fish species such as cod, flounder, multiple sockeye salmon, Atlantic salmon, herring, and pike. The analysis also included trophy-hunted populations of bighorn sheep and caribou and populations of ginseng and snow lotus, which are harvested for culinary, medicinal, or other purposes. All these species show strong evidence of change in the mean values of traits over time. Some of these changes could have been due to phenotypic plasticity of traits in changing environments, but plasticity is unlikely to explain all these sustained changes.

Some harvested populations have changed in multiple ways. Most have declined in sizes of morphological traits such as body size or horn size, and the average decrease has been almost 20 percent. Most populations have shown alterations in life histories, producing individuals that reproduce at earlier ages or smaller sizes. Commercially harvested species show greater changes than recreationally harvested species. Presumably, this difference occurs because commercially harvested species are subject to consistent natural selection year after year, and a high proportion of the population is harvested each year. Harvested populations are accumulating changes 50 percent faster than populations undergoing other forms of selection driven by human activities such as disturbance of habitats.

Discussions continue about whether changes over the past century in this species or that species are due to strong natural selection or a combination of natural selection and other causes (Conover et al. 2005; Brown et al. 2008). Nevertheless, the number of careful studies showing selection and evolution continues to grow, leaving little doubt that harvest selection is altering the traits of multiple species that we are manipulating in many ecosystems. That does not mean all harvested species will eventually be the size of minnows or mice. It all depends on how we choose to manipulate other species. A study reversed evolution in experimental populations of silverside fish (*Menidia menidia*) within twelve generations by imposing intense selection for larger sizes on these populations (Conover et al. 2009).

FROM HARVEST SELECTION TO DOMESTICATION AND PEST MANAGEMENT

Human-induced evolution generally has been most rapid when we have selectively killed whole classes of individuals generation after generation. The extreme is truncation selection, which is the traditional tool of plant and animal breeders. Before the age of genetic engineering, it was how we developed improved crops and domestic livestock. We chose individuals with one group of extreme traits—larger seeds or fruits; more docile or larger animals—and either killed all the rest or allowed only the favored group to breed.

We have favored the evolution of major pests and pathogens in much the same way, but often more indirectly. We have used artificial selection to breed a small number of plants and animals for our own purposes, and in the process we have often made these crops and domesticated animals more susceptible to pest species. We tried to compensate with wholesale use of pesticides and made the problem worse as pest after pest rapidly evolved resistance. In some cases it has taken only a few decades of intense selection for pest populations to evolve to a point at which they are made up of almost exclusively of resistant individuals. Organic pesticides began to be applied extensively and intensively to crops after World War II, and the International Survey of Herbicide Resistant Weeds now includes 200 species in which resistance has been reported (Heap 2011).

Today about 700 pesticides are used to control plants, fungi, insects, and other organisms, and these pesticides selectively act on about 95 biochemical binding sites or biochemical lesions in pest species (Casida 2009). The large number of pesticides in use and the specificity of their action continue to make them important agents of selection on crops in many parts of the world. We have known for decades that when we broadcast pesticides across environments, we often kill all but the small proportion of pests that happen to be genetically resistant to that particular pesticide. Those resistant individuals become the parents for the next generations of pests.

The evolution of pesticide resistance has become so common that major pesticide producers and agribusiness owners increasingly have had to assimilate knowledge of evolutionary biology as they have developed strategies for deploying pesticides in ways less likely to impose strong, consistent selection on pest populations. The Insecticide Resistance Action Committee was formed in 1984 and includes among its sponsors most of the major producers of pesticides. The efforts have expanded more recently to include the Herbicide Resistance Action Committee and the Fungicide Resistance Action Committee. These are just three of the growing number of efforts devoted to confronting the problem of rapid evolution in pests and pathogens.

It is a huge and complex evolutionary and applied problem, because each of our crop plants has multiple potential pests. Over 1,000 arthropod species are associated with tea (Hazarika et al. 2009), with different combinations of these species occurring in different geographic regions. Imposing strong evolutionary change on any one of these species has the potential to create a domino effect on the other species.

There continue to be fierce debates about which strategies of pest management are most effective at slowing the rate at which pests evolve resistance to control measures. When the European Parliament voted in 2009 to approve a set of regulations banning almost a quarter of the pesticides available in Europe, reactions within the farming and scientific communities were highly variable and strongly voiced on all sides (Coelho 2009). Here is where evolutionary biology, policy, environmental science, and social welfare come together in ways that make clear the need to understand rapid evolution for the future of our societies.

The problem continues to grow. The Arthropod Pesticide Resistance Database project directed by Mark Whalon at Michigan State University makes quarterly updates to its list of cases of pesticide resistance in insects, mites, and other arthropods based on new reports in the scientific literature (Whalon et al. 2011). The database, which is available online, includes not only crop pests but also other species, such as mosquitoes, that have been the targets of control programs using pesticides. The list currently includes over 1,900 records distributed over more than 500 species: aphids, moths, leafhoppers, scale insects, whiteflies, true bugs, thrips, beetles, flies, parasitic wasps, cockroaches, mites, ticks, midges, lice, fleas, and others, including a long list of mosquito species. Resistance in many of these arthropods has been reported from a single locality, but resistance is more widespread in other species.

The number of resistant populations found in each species, and the number of pesticides to which each species is resistant, also continues to expand. Rat and mouse populations were controlled using warfarin beginning in the 1950s, but some populations had already evolved resistance by the late 1950s. By the 1980s resistant populations had been reported in Europe, North America, Asia, and Japan (Ishizuka et al. 2008). That led to the development of a new class of rodenticides, sometimes called superwarfarins, but some rodent populations in Europe may already be evolving resistance against these new chemical compounds (Kohn et al. 2000). One of the extremes in this kind of pesticide-driven evolutionary arms race has occurred in Colorado potato beetles. Since the 1950s, these beetles have evolved resistance to 52 different compounds, including all the major classes of insecticide (Alyokhin et al. 2008).

SELECTION AND DRUG RESISTANCE

The growing medical crisis resulting from the rapid evolution of antibiotic resistant pathogens has followed the same pattern of ongoing evolutionary change as in agriculture. Widespread distribution of antibiotics has favored evolution of resistant forms of pathogens (Luciani et al. 2009; Rice 2009; Gullberg et al. 2011). Multidrug treatments have increased the selection pressures on microbial populations (Hegreness et al. 2008). Arguably, application of evolutionary approaches in medicine has been slower than in agriculture. Since at least the 1940s, plant breeders have taken, directly or indirectly, a coevolutionary approach to the rapid evolution of pests (Flor 1942, 1955), and the first mathematical model of rapid coevolution between pathogens and plants was published half a century ago (Mode 1958). It took a couple more decades for researchers to begin to apply similar kinds of thinking to medicine and animal epidemiology. In a series of papers beginning in 1979, Roy Anderson and Robert May began to explore ecological and evolutionary approaches to epidemiology (Anderson and May 1979; May and Anderson 1983). Soon thereafter, Paul Ewald (1994) and others pushed this approach further by suggesting multiple ways in which differences in the life histories of pathogens and parasites could influence their level of virulence. These studies provided a basis for thinking about how human activities influence the evolution of pathogens.

Despite decades of subsequent sophisticated models of how parasites may evolve rapidly during epidemics or in response to antibiotic treatment, the gap between evolutionary biology and medicine remains. Few medical schools or veterinary schools provide advanced training in evolutionary biology. The perceived need, though, is growing as researchers have realized that "evolutionary biology . . . is still lacking the attention it deserves from the medical community" (Restif 2009), that "the canyon between evolutionary biology and medicine is still wide" (Nesse and Stearns 2008), and that there is still "avoidance of the e-word" in studies of antibiotic resistance (Antonovics et al. 2007). Fortunately, for the sake of human health, the canyon is narrowing each year as researchers worldwide have come to appreciate that evolution truly is happening in many species over timescales as short as a few decades.

That appreciation is accompanied by a growing sense of urgency as we run out of antibiotics and antimalarial drugs to combat newly evolved resistant forms of our parasites and those of our domestic animals (Buckling and Brockhurst 2005). The first generation of antibiotics was produced by purifying forms of natural antibiotics. *Penicillium* mould produced penicillin, and *Streptomyces* bacteria produced streptomycin. These antibiotics were initially effective, because bacteria resistant to these compounds were generally rare in populations. Resistance to antibiotics often comes at a cost such as slower growth rates (Gagneux et al. 2006). Consequently, resistant forms tend to be favored in natural populations only when the bacterial population is being subjected to these compounds. As the use of the first generation of antibiotics grew, so did the number of resistant bacterial populations. That, in turn, fostered the search for a new generation of antibiotics, to which the bacteria also quickly evolved resistance. The result has been the spread of bacterial populations now resistant to multiple forms of antibiotics, and we are running out of simple options.

Antibiotic use is so widespread that it may even be affecting the evolution of nontarget species. Antibiotic resistance genes are often located on plasmids, which are extrachromosomal genetic elements that are sometimes transferred between bacteria. Plasmids affecting antibiotic resistance occur naturally in bacteria and may have other biological functions in natural environments (Martinez 2008). Widespread use of antibiotics in medicine and veterinary practices, however, could be favoring the spread of genetic elements that happen to confer antibiotic resistance, as antibiotics make their way into water and soil. That may have occurred, for instance, as the use of penicillin increased. Mass production of penicillin began in the 1940s, and more widespread distribution of antibiotics rose quickly from 1950s onward. Soil samples collected at five sites in the Netherlands at various times between 1940 and 2008 have shown a large and significant increase in eighteen antibiotic resistance determinants (Knapp et al. 2010). In these samples, antibiotic resistance genes have risen in abundance by more than fifteen times since the 1970s.

Since the early 1960s only a few new classes of antibiotics have been introduced, and between 1962 and 2000 none of these was a fundamentally new kind of antibiotic built on a novel chemical structure (Fischbach and Walsh 2009). Multiple new natural, synthetic, and semisynthetic antibiotics have been in development during the past decade, but as the search for new candidates continues, so has the search for ways of slowing down the evolution of resistance. It has meant paying increased attention to the results coming from studies in the fields of ecology, population biology, and coevolutionary biology. A new generation of mathematical models and empirical studies has challenged some of the earlier assumptions about the evolution of resistance.

For example, although resistant bacteria are often out-competed by nonresistant forms in the absence of antibiotics when a new antibiotic is first introduced, the disadvantage does not always remain. As natural selection continues to act, it favors compensatory evolution in the resistant forms. Any new mutation that decreases the cost of resistance is favored. After multiple generations of selection, the resistant forms may survive and grow almost as well as nonresistant forms. This result was first shown in laboratory studies several decades ago (Lenski 1988). It has now been found to hold in cases such as tuberculosis, in which prolonged treatment of patients with antibiotics can result in the evolution of multi-drug-resistant strains of tuberculosis bacteria with a competitive fitness similar to susceptible forms (Gagneux et al. 2006).

Once compensatory evolution has occurred in resistant forms, they do not necessarily disappear from bacterial populations quickly just because we stop using a particular antibiotic. Loss of resistance can be slowed by persistence of antibiotics in the environment for some time, or by their continued use in some places even after their use has been discontinued in other places. We end up with the situation we have now in which bacteria are resistant to multiple forms of antibiotics.

After decades of extensive use of antibiotics in the treatment of humans and domesticated animals, it is likely now to take decades to reduce some antibiotic-resistant forms of bacteria to relatively low levels. For example, in the 1990s Norway and Denmark stopped using avoparcin in poultry because some studies suggested it confers cross-resistance to vancomycin, an antibiotic that has been used in humans. Within a few years, resistant forms dropped quickly to about 10 percent, but they persisted at low levels for almost a decade. By 2007, about 1–2 percent of the bacteria were resistant (Johnsen et al. 2009).

The current situation with malaria is even more complicated because it involves at least four species of the malarial parasite *Plasmodium*, multiple species of mosquitoes, bacteria in the gut of mosquitoes, and genetic differences among human populations in susceptibility to malaria. There are coevolutionary interactions at every stage of the process. Humans vary in their susceptibility to malaria, mosquitoes vary in their resistance to malaria parasites, and malarial parasites vary in their ability to infect particular mosquitoes (Niarè et al. 2002; Lambrechts et al. 2005). There is even significant variation within populations of mosquito species in their resistance to particular genetic forms of the malarial parasites, and also variation within populations of malarial parasites to infect particular genetic forms within a mosquito population (Lambrechts et al. 2005). Resistance is determined in part by yet other interacting species. *Enterobacter* bacteria in the midgut of mosquitoes affect the resistance of *Anopheles* mosquitoes to infection by *Plasmodium falciparum* (Cirrimotich et al. 2011). Adding to the complexity, different combinations of *Plasmodium* and mosquito species occur in different geographic areas. Hence, there is a great deal of variation on which natural selection can act on every component of this interaction, and the intensity of natural selection can be high on each species (Bongfen et al. 2009).

Hundreds of millions of people are infected with malarial parasites, and each year over a million die. For many decades the fight against malaria relied on a small number of chemicals that inevitably became less effective as *Plasmodium* populations evolved resistance. Resistance to antimalarial drugs has evolved multiple times and has spread as resistant parasites have dispersed to other populations by way of their mosquito vectors. Resistance to chloroquine in *Plasmodium falciparum*, which is the most virulent malarial parasite, has evolved at least three times and spread among continents (Mita et al. 2009). Evolution of resistance to pyrimethamine has an even more complicated history of repeated origins around the world (Mita et al. 2009). Studies of the mutational steps involved in resistance suggest that resistance to this class of compounds has likely involved three mutational pathways (Lozovsky et al. 2009). Bed nets impregnated with pyrethroids have been a mainstay of antimalarial programs, but these too are losing their effectiveness as resistance rises in some populations of each of the mosquito species that act as vectors of the disease (Ranson et al. 2011).

Although new chemical compounds are under development, artemisinin and its derivatives are currently the most important. Artemisinin is highly effective against today's *Plasmodium* populations and has become a vital tool at a time when resistance to antimalarial drugs continues to evolve in *Plasmodium* populations. The compound is derived from a species of wormwood (*Artemisia annua*), but it is costly to purify and make into a drug that could be used throughout the many parts of the world in which incomes are low. It has prompted attempts to develop semisynthetic versions of the compound, while other scientists and policy makers work to contain the spread of artemisinin-tolerant parasites (Van Noorden 2010; Sutherland et al. 2011).

Even as research and development of artemisinin-based therapies move forward quickly, there is a growing appreciation among medical researchers that ongoing evolution of pathogens imposes limits on any attempt to rely on one new drug or class of drugs (Fidock et al. 2008; Greenwood et al. 2008). Researchers are looking for the spread of resistance to artemisinin as the first cases begin to appear (Noedl et al. 2008). This reflects a growing acceptance of the inevitability of rapid evolution of any large population subjected to strong and consistent natural selection.

SELECTION IN HUMAN-ALTERED ENVIRONMENTS

Do all these examples of evolution fueled by human activities imply that we are particularly good as agents of selection? If we were to assume that human influence results in much faster rates of evolution, then we would also be assuming that, in the absence of human influence, populations are at or near some kind of equilibrium that is disrupted by human activities. That view is implicit in many mathematical models of evolution, which commonly assume that selection is often weak and acts only slowly to shift traits within populations. A major reason for this assumption of weak selection in many models is that, in some cases, it more readily allows the models to be solved analytically. Models of weak selection have played a major role in evolutionary biology and continue to inform many important questions about the interaction of evolutionary processes (O'Fallon et al. 2010; Charlesworth 2012). Nonetheless, the insufficiency of those models for understanding many cases of contemporary evolution is becoming evident as multiyear field studies worldwide show at least occasional strong selection on populations. What remains unclear is whether evolutionary change in human-altered environments is faster than in less-altered environments, despite the extreme results on some heavily harvested species.

Even analyses using the same data on human impacts on evolution have led to different conclusions. In a large meta-analysis of published studies analyzing changes in the continuous phenotypic traits of animals, Hendry et al. (2008) found that continuously distributed traits of animals (e.g., body size) changed faster in human-altered environments than in more natural situations. Human alteration included harvesting, pollution, environmental acidification, and introduction of populations into new regions, among other influences. Natural situations included those changes in which no human influences were obvious. They summarized the data using darwins or haldanes, analyzed it in multiple ways, and concluded that human activities raise the rates of phenotypic change beyond the baseline found in natural situations. Using the same data but analyzing it in a different way, Gingerich (2009) could find no difference in the rates of change in human-altered environments as compared with less disturbed environments. He concluded that selection "is equally potent with or without us."

In another analysis, using more than 5,500 estimates of directional change based on darwins or haldanes in 90 species, Peter Westley (2011) found no clear differences in evolutionary rates in introduced invasive and native species. Both groups undergo periods of abrupt change interspersed with periods of less change. Interpretation of any current meta-analyses is complicated, as Hendry et al. (2008) note, by the likelihood that not all the suggested changes are evolutionary. Some may be due to phenotypic plasticity in expression of traits, because not all these studies were done in ways that allowed separation of evolutionary change from plasticity in traits. As studies continue to sort out actual evolution from plastic responses, we may be able to resolve more sharply the role of human activities on the rates and patterns of evolution.

It seems unlikely, though, that we can completely separate the effects of humans from other factors imposing selection on populations. Almost all major ecosystems worldwide have now been altered to some degree by human activities. Introduced species are evolving rapidly on continents far away from where they originated (Huey et al. 2005), but so are native species in environments that have been altered much less (Grant and Grant 2008b). Examples of ongoing, sometimes rapid, evolution occur in populations in every kind of environment in which it has been studied.

STRENGTHS AND FORMS OF SELECTION

Even as we accumulate more examples of ongoing evolution in nature, there are still surprisingly few formal analyses of the strength, forms, and dynamics of selection in the wild spanning multiple years and generations. When John Endler (1986) undertook an analysis of the published studies, he was able to find estimates on 25 species published through 1983. Fifteen years later, Kingsolver et al. (2001) and Hoekstra et al. (2001) were able to include 62 species that had been studied between 1984 and 1997. Most estimates were of a single population evaluated once for an average of four traits, usually using a single measure of Darwinian fitness. Of these estimates, 40 percent were for plant species, 31 percent for invertebrate species, and 29 percent for vertebrate species. A decade later, Siepielski et al. (2009) searched 1,569 studies published between 1986 and 2008 and were able to

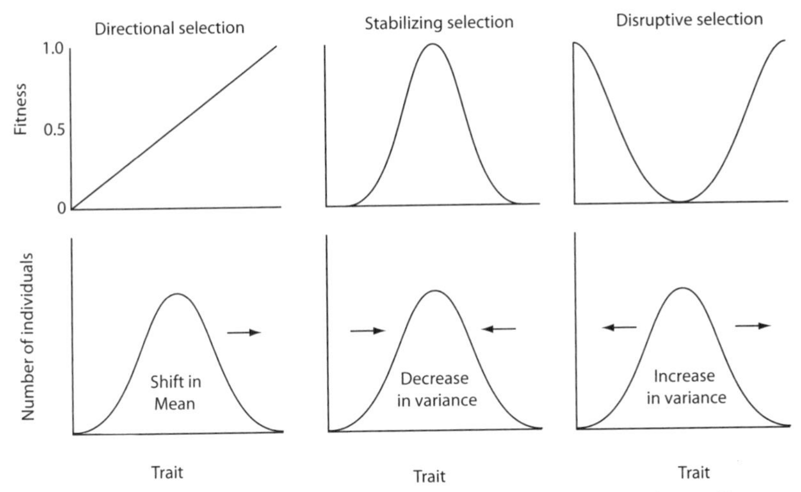

Fig. 2.1 Three forms of natural selection on a trait within a population. The top three panels show the relative fitness of individuals having particular values of a trait. The bottom three panels show how each form of selection affects either the mean or the variance of the trait in subsequent generations.

find studies of 73 species in which selection had been analyzed for 2 or more years. The studies ranged in length from 2 to 45 years, with a median of 3 years and a mean of 7.6 years. Most estimates were of plants or vertebrates. Although the world is made up mostly of invertebrates, there were multiyear estimates of only seven invertebrates. There were no published multiyear estimates for a fungal species or a microbial species.

These meta-analyses have imposed a strict set of criteria on the studies they have included. They have all been on selection measured on continuous traits in unmanipulated populations. The lists of species would have been longer if the analyses had included species subjected to selection in experiments. The more recent analyses have used the technique introduced by Russell Lande and Stevan Arnold (1983; Arnold and Wade 1984a, 1984b). This technique allows a quantitative assessment of selection on traits that vary continuously such as size or shape, and it provides ways of assessing the extent to which fitness of a trait varies linearly or nonlinearly with increasing or decreasing values of the trait (fig. 2.1). The Lande-Arnold method was a major advance in the study of selection, although it takes a bit of time to get the jargon straight (box 2.1).

Although studies of the forms and strength of selection continue to expand, fewer than one hundred of the millions of species on earth have been

BOX 2.1 The vocabulary of selection on quantitative traits: Selection gradients, selection differentials, and the G-matrix

The Lande-Arnold method allows selection on a trait to be partitioned into directional selection (called a standardized linear selection gradient) and nonlinear selection (called a standardized quadratic selection gradient). These measures estimate the extent to which selection acts directly on a particular trait. A linear selection gradient measures directional selection by determining the extent to which fitness increases or decreases linearly among individuals with more or less of that trait—for example, fitness of individuals at different positions along a gradient of increasing height or weight (fig. 2.1). As in the measurement of haldanes, a linear selection gradient often standardizes selection by expressing it in units of standard deviations. These calculations of selection can also be standardized relative to the mean, which may be more appropriate for some analyses of the strength of selection (Hereford et al. 2004; Stinchcombe 2005).

The nonlinear component of selection gradients measures more complex relationships between traits and fitness (fig. 2.1). At the extremes, fitness could be high in individuals with intermediate values of a trait and low in individuals with either high or low values of the trait. This is the first step in demonstrating stabilizing selection on a trait, but a full demonstration of stabilizing selection requires additional analyses (Mitchell-Olds and Shaw 1987). Alternatively, fitness could be low in individuals with intermediate values and high in individuals with either high or low extreme values, suggesting the potential for disruptive selection on the trait.

The Lande-Arnold method also makes it possible to assess total selection on a trait. The estimate includes direct selection and also the indirect selection acting on other traits correlated with the trait under study. These are called standardized linear and quadratic selection differentials. If selection is mostly linear, then linear selection gradients and differentials are effective at capturing the evolutionary changes estimated by darwins or haldanes.

More generally, we often want to know how evolution occurs among multiple traits that are genetically correlated and therefore do not evolve

BOX 2.1 ***(continued)***

independently. The analysis requires measurement of the genetic variation in each trait, the genetic correlations (covariances) among traits, and a combined value of the selection gradients. Formally, the overall evolutionary change among these traits (ΔZ) is equal to the matrix of genetic variances and covariances (G) times the vector of the selection gradients (b). If the traits are uncorrelated, then each trait will evolve independently. If the traits are highly correlated, then evolution in some directions will be more constrained than evolution in other directions. The G-matrix therefore determines how difficult it is for traits to evolve in one direction or another even when subject to strong selection.

studied in any detail. Some species, though, have been studied repeatedly in multiple populations or life history stages, resulting in more than 3,000 published estimates of selection (Endler 1986; Kingsolver et al. 2001; Bell 2008; Siepielski et al. 2009; Siepielski et al. 2011a). Most estimates are for morphological or life history traits (e.g., growth rates). There have been relatively few evaluations of selection on behavior or physiology, and it is unclear how that bias affects our current impressions of the strength and forms of selection.

Collectively, the current estimates suggest that selection acting on reproductive traits (mating success and fecundity) is stronger than selection acting on survival (Siepielski et al. 2011a). That conclusion, however, is tentative. It could depend in part on the traits most commonly measured in these studies, the taxa (mostly vertebrates and plants) that have been studied, or the relatively short timescales over which measurements have been taken on most populations.

Studies have tended to find weak selection more commonly than strong selection during these short lengths of time (fig. 2.2). These analyses show distributions with peak values less than 0.3 (Siepielski et al. 2011a). The values tend to be bunched at the low end of the potential range of values, regardless of whether selection acts through mating success, fecundity, or survival.

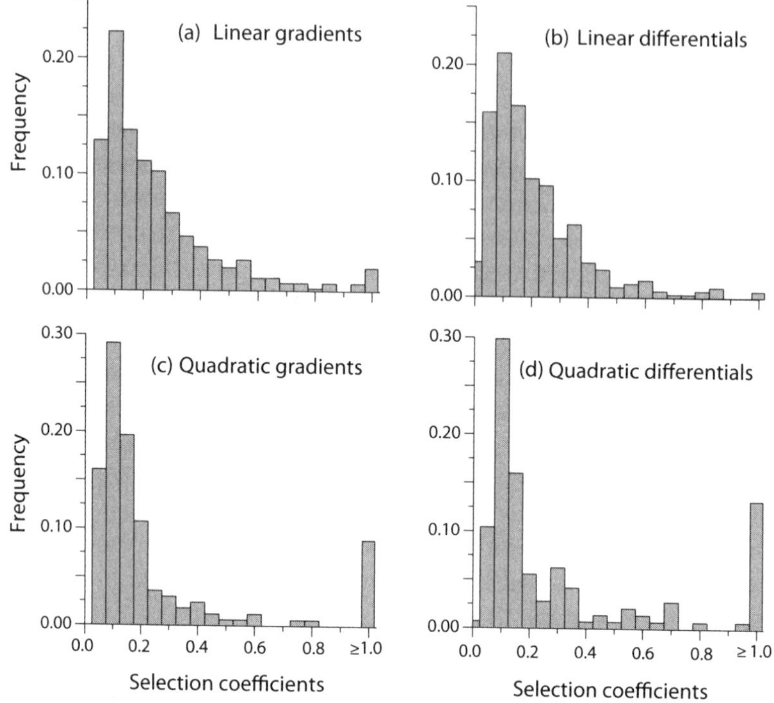

Fig. 2.2 Distribution of estimates of the strength of selection acting on traits of populations in the wild. The estimates use either selection gradients or selection differentials. Graphs (*a*) and (*b*) present values for linear (i.e., directional) selection, whereas (*c*) and (d) present values for nonlinear (i.e., quadratic) selection. Selection coefficients are shown as the means of the absolute values. After Siepielski et al. (2009).

INTERPRETING CURRENT RESULTS ON THE STRENGTHS AND FORMS OF SELECTION

These distributions of the strength of selection should be viewed with caution as we consider how natural selection continues to fuel evolutionary change. Most studies evaluate selection on only one trait or a few composite traits. Individuals are made up of many traits, and ultimately we want to know the overall strength of selection on a population, not the strength of selection on a few traits. Also, all but a few studies are less than a decade in length. If selection generally acts strongly on a population only once in ten or twenty years, then studies assessing populations for a median of three years are often likely to miss the years of strong selection.

It seems that the longer we study a population beyond a single year, the more likely it is that we will find occasional years of strong selection

(Grant and Grant 2008b; Ehrlèn and Münzbergova 2009; Siepielski et al. 2009; Sletvold and Ågren 2010). This observation suggests that a reasonable working hypothesis for current research is that selection is usually episodic, with occasional bouts of strong selection over timescales of decades. Equally important, most studies show some evidence of at least weak selection acting commonly on populations.

The published studies suggest that directional selection may be more common than stabilizing or disruptive selection, but, again, these patterns must be viewed cautiously. It is easier to get clear results showing directional selection, frequency-dependent selection, or sometimes disruptive selection, than stabilizing selection, because much of the analysis can focus on how selection acts on the most common individuals within a population. (Arnold and Wade 1984a; Mitchell-Olds and Shaw 1987; Schluter 1988; Shaw and Geyer 2010). To demonstrate stabilizing or disruptive selection, we need to be able to show that selection either favors or disfavors individuals with intermediate rather than extreme traits in a population. Simply showing a nonlinear relationship between a trait value and fitness is not good enough, because we must also know if the nonlinear part of the curve affects many or few individuals. If the relationship between traits and fitness is nonlinear only at trait values rarely found in a population, then selection will be mostly directional. If, though, an important part of the curve includes individuals in the actual population, then the curve does matter. The problem is that it is often difficult to get an accurate measure of how selection acts on extreme individuals, because they are, by definition, rare. That makes it especially difficult to get accurate measures of stabilizing selection in some populations.

Adding to the problem, populations are surely often subject to two or more forms of selection simultaneously. Some alternative methods have been proposed to handle the complex structure of fitness (Shaw et al. 2008; Shaw and Geyer 2010), and there are promising new methods to evaluate how selection shapes populations when both selection gradients and demography vary over time (Horvitz et al. 2010). These approaches offer the opportunity to evaluate the structure and strength of selection under different assumptions, but they have not yet been tested in a wide range of populations.

Even so, we now have enough studies of contemporary selection in nature to know that linear and nonlinear selection on populations are common. These studies suggest that stabilizing selection seems to be not much more common than disruptive selection (Kingsolver and Pfennig 2008). Our interpretations from field studies, though, are from short-term analyses

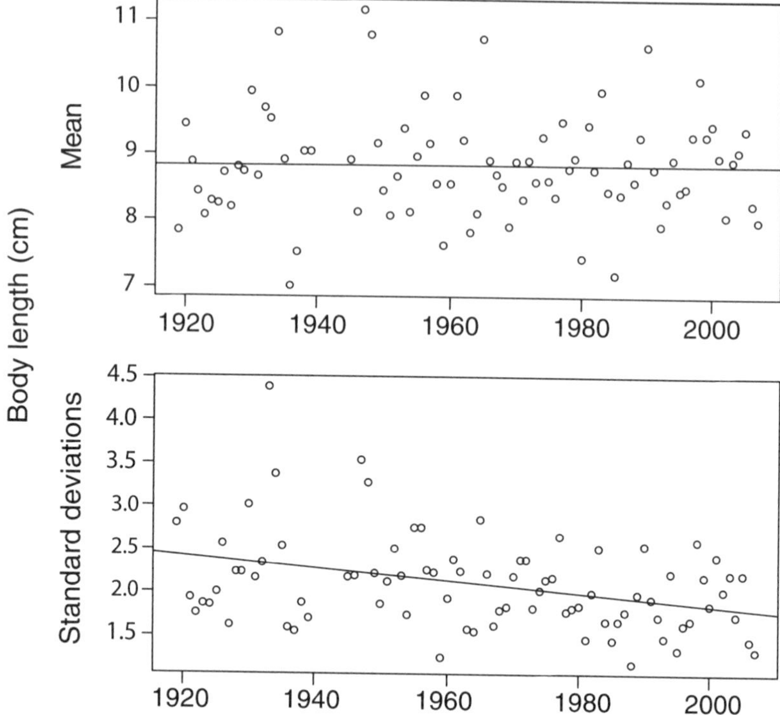

Fig. 2.3 Change in the mean and standard deviation of body length of juvenile Atlantic cod along the Skagerrak coast of Norway between 1919 and 2007. After Olsen et al. (2009).

of selection on species that may persist for millions of years. Many species show considerable stasis in morphology in the fossil record (Eldredge and Gould 1972; Eldredge et al. 2005), and lineages often show strong evidence of niche conservatism (Wiens et al. 2010). Stabilizing selection could therefore act either among populations at larger spatial scales (i.e., metapopulations and higher levels) or at longer temporal scales rather than at the shorter-term and local-population scale of most studies.

How stabilizing selection contributes to the morphological and niche conservatism of many species remains one of the difficult problems in evolutionary biology (Eldredge et al. 2005; Estes and Arnold 2007; Futuyma 2010; Wiens et al. 2010). The most convincing studies are the few that have quantified evolutionary change in the variation of traits in populations over time relative to changes in average values. Among the clearest examples are the changes in body size of Atlantic cod (*Gadus morhua*) along the Norwegian coast over the past ninety years (Olsen et al. 2009). Juveniles of this nonmigratory coastal population have been surveyed annually at multiple stations using a standardized seine. The only exception was during World War II, when only two stations were sampled and therefore excluded from the analysis. Average body length has not changed over the past century, but variation in body length, as indicated by the standard deviation, has decreased significantly (fig. 2.3).

These data have been supplemented in more recent decades with mark-recapture studies, which can assess the strength of current natural selection on body size. These studies show that selection over the past century has acted against fish with extreme traits: large and fast-growing juvenile cod at the one extreme, and small and slow-growing juveniles at the other extreme. Together, the survey and mark-recapture studies suggest strong selection that restricts variation in juvenile cod size and growth rate to a narrow range near the average for the population. These results alone, though, do not show unequivocally that cod in this region have been evolving recently through stabilizing selection. Some of the change could have been due to a narrowing of the range of environmental factors that affect body length and growth rate. Nevertheless, growth rate is known to be a heritable trait in Atlantic cod (Gjerde et al. 2004), and it seems likely that at least some of the observed change is the result of stabilizing selection.

MULTIPLE RATES OF EVOLUTION ON THE SAME TRAITS

Why evolution is not even more unrestrained remains one of the major problems to solve in evolutionary biology (Futuyma 2010). At the broadest scale, evolution is constrained by fundamental physical and biochemical trade-offs, forcing species to adapt and diverge much more in some directions than in others (Brown and West 2000; Hechinger et al. 2011; Tilman 2011). The effectiveness of selection varies with the direction in which it is pushing a population.

Evolutionary rates are therefore not a fundamental property of the traits themselves. Rates also depend on forms and directions of selection imposed on traits. A trait that is not evolving noticeably, or is evolving slowly, in one environment, can sometimes evolve quickly in another environment. A trait that is not evolving under one form or direction of selection can sometimes evolve quickly under a different form or direction of selection. Hence, response to selection on a trait in one environment is not a reliable index of the potential rate of evolution of a trait.

Among the most elegant experiments, of recent decades, showing different rates of evolution on the same trait are those on eyespots of butterflies. These eyespots have become one of the models for studies of the genetics of traits and their modification by natural selection acting on the induction of genes during development. In nature, eyespots are often involved in avoidance of predators, and also sometimes in sexual selection (Lyytinen et al. 2004; Robertson and Monteiro 2005; Vallin et al. 2007; Olofsson et al. 2010; Prudic et al. 2011). They occur commonly not only on the wings of butterflies but also on the bodies of caterpillars (Janzen et al. 2010). Eyespots on adult wings vary within and among species in their number and size and in the color and shape of the concentric rings of scales from which they are composed. In the laboratory, these characteristics of eyespots can evolve through artificial selection. They can also be manipulated during development by surgically altering the presence or position of the precursor cells, called eyespot foci, found in the pupae (Beldade et al. 2002, 2008).

Wing development is controlled by a set of conserved genes that also appear to control the formation of the eyespot foci and rings (Keys et al. 1999). Some genes contribute to eyespot variation in very specific ways, and spontaneous mutations found in populations have been shown to have repeatable specific effects on eyespot color or size (Beldade et al. 2002; Monteiro et al. 2007; Saenko et al. 2010). These genes affect developmental pathways and signaling by regulating, for example, diffusion of signals that affect eyespot size and the concentration signal gradients that affect color composition (Allen et al. 2008).

The interaction of these genes results in a greater likelihood of evolution of eyespot patterns in some directions than in others. Within *Bicyclus* and *Mycalesis* butterflies, which are Old World tropical species, the two eyespots on the dorsal side of the forewing tend to evolve in tandem, as can be seen in comparisons among species (fig. 2.4), although there is a great deal of variation among species. Some species have no or barely visible eyespots; others have moderately sized anterior and posterior eyespots; and still others have large anterior and posterior eyespots. Other traits in these butterflies also show coordinated evolution in one direction or another (Brakefield 2010).

It is possible, however, to produce butterflies outside the normal patterns through artificial selection that reduces only, or mostly, either the anterior or posterior eyespot (fig. 2.4). Selection to reduce only the anterior or posterior eyespot takes more generations than selection along the path of parallel increase or decrease of both eyespots (fig. 2.5). Experiments on hindwing eyespots show that selection to change color patterns is also much more effective along some directions than others. Hence, genetics and developmental pathways impose relative, rather than absolute, constraints on the evolution of these traits.

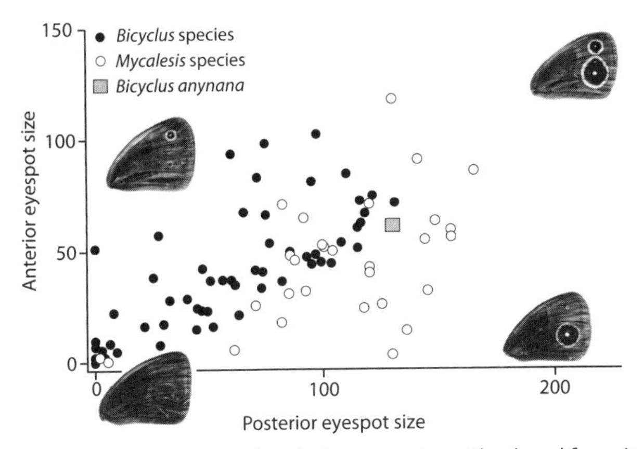

Fig. 2.4 Relative sizes of anterior and posterior eyespots on the dorsal forewings of *Bicyclus* (*filled circles*) and *Mycalesis* (*open circles*) butterflies. The shaded square in the middle indicates the average position of eyespots in *B. anynana*. This species has been used in multiple experiments that have selected for changes in eyespots in this lineage. The four wings in the figure are representative *B. anynana* wings after 25 generations of selection in each of the four directions. These four wings are placed in approximately the correct position on the graph. After Brakefield and Roskam (2006).

In other species, the tendency for traits to evolve further in one direction than in another is driven more by counteracting selection pressures than by genetic or developmental processes. One of the major current challenges in evolutionary biology is to understand the extent to which restrictions on the rates of evolutionary change are governed by conflicting selection pressures rather than by genetic or developmental constraints. Evolving in one direction could, for example, increase competition with other species, while evolving in another direction could increase the chances of attack by predators or parasites or lessen the effectiveness of interactions with mutualists. The options for sustained major evolutionary changes in traits are probably often few, because the web of life is so entangled that it is difficult for natural selection to cut a straight path through it.

The opportunities for major directional change are further restricted by the ways in which genes interact with each other as selection acts on multiple traits. In population genetics, the inheritance of multiple phenotypic traits is summarized in a matrix of genetic relationships called the G-matrix (box 2.1). Some studies have suggested that parts of this matrix often remain unchanged under different forms of selection (Arnold et al. 2008). These relatively unchangeable parts of the matrix could restrict the trajectories of directional evolutionary change in complex environments.

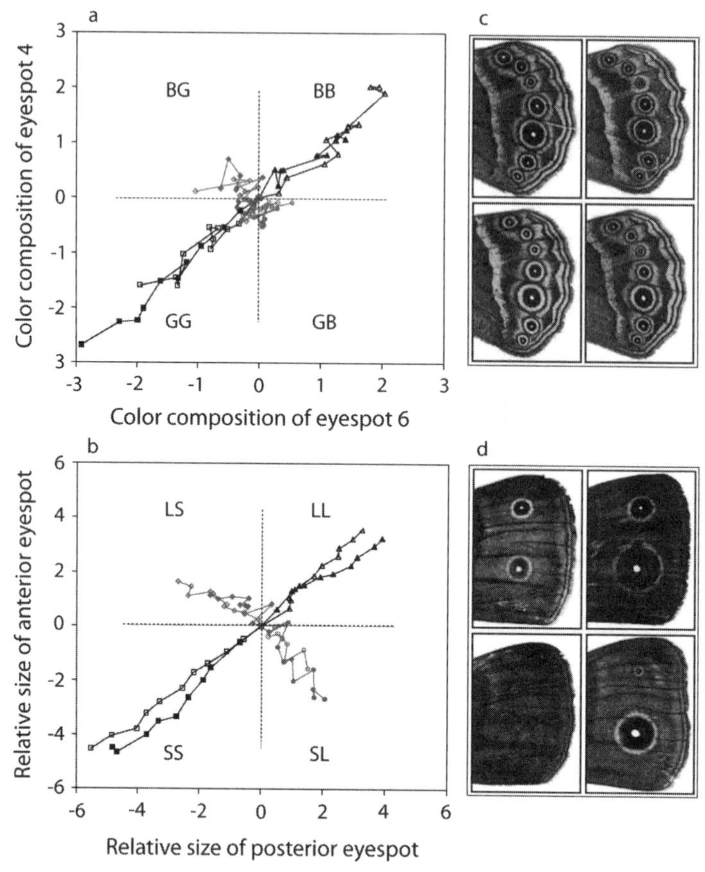

Fig. 2.5 Evolutionary response to artificial selection on (*a*) the color composition of the fourth and sixth eyespots on the ventral hindwings of *Bicyclus anynana* butterflies or (*b*) the size of the anterior and posterior eyespots on the dorsal forewings relative to wing size. The starting point for each experiment is the mean of the population at the beginning of the experiment, which is at the center of each graph. The values track the changes over time in the experimental populations relative to the control populations, measured in standard deviations. Each experiment was conducted twice from the same starting population, and an unselected control population maintained under the same environmental conditions was also measured each generation. Black lines show concerted changes in size of color composition of the anterior and posterior eyespots (e.g., gold on eyespot 4 and gold on eyespot 6, or large anterior eyespot and small posterior eyespot); gray lines show antagonistic changes (e.g., black on eyespot 4 but gold on eyespot 6, or large anterior eyespot but small posterior eyespot). Panels (*c*) and (*d*) show representative phenotypes in each of the directions of selection after ten generations. After Allen et al. (2008).

Fruit flies (*Drosophila*), which can evolve fast in the lab, illustrate the problem of the evolution of multiple ecologically important traits in complex environments. These flies often feed on highly ephemeral resources such as rotting fruits. As more flies discover and lay their eggs on a fruit, competition increases among developing larvae. That can favor flies with traits that allow larvae to compete better against other larvae. As the flies develop, however, they must defend themselves against parasitic wasps. The parasitic wasps pierce the integument of larvae and lay their eggs. The developing wasps eat the larva from the inside, pupate, and then cut their way out as they emerge as adults. The flies are able to evolve resistance against some of these parasitoids by encapsulating the wasps' eggs and smothering them. The flies are also attacked by microsporidians and other vertically inherited pathogens that affect host reproduction (Werren et al. 2008; Vijendravarma et al. 2009).

The ability of *Drosophila* to evolve resistance against these enemies comes at a cost. When *Drosophila* are selected to be more resistant to attack by parasitoids, they simultaneously become less competitive at acquiring food (Kraaijeveld and Godfray 1997; Kraaijeveld et al. 2001). Devoting energy to one activity comes at the expense at the other. Natural selection on a population could either favor more resistant individuals or more competitive individuals, but not both, at least in the short term. There is a similar cost of resistance against microsporidians and other pathogens (Modak et al. 2009; Vijendravarma et al. 2009). Experiments on some other species suggest that the costs to fitness against some enemies can be reduced over time as natural selection favors individuals with combinations of genes that confer resistance but at a lower cost to survival and reproduction (Weeks et al. 2007). In general, though, it seems that trade-offs from conflicting selection pressures are common in nature and vary among populations (Rose 2005; Futuyma 2010), preventing unbridled selection in particular directions within species.

THE CHALLENGES AHEAD

We still have an imprecise understanding of the strength and forms of selection in nature, how variation in selection and conflicting pressures are distributed among populations and across time, and how human-imposed selection is altering the process of selection. We also still have an incomplete understanding of the efficacy of selection at different developmental stages. Traits downstream along a developmental pathway may evolve quickly, but traits upstream are sometimes more constrained (Rausher et al. 2008). At any developmental stage, gene interactions affect the potential rate of evolution in various directions. The rate of evolution therefore depends only in part on the strength of selection. Fast rates of evolution depend on strong selection, but strong selection does not always lead to fast evolutionary rates.

We need a better understanding of which kinds of traits are most likely to evolve most quickly and which kinds of selection are mostly likely to act on those traits (Thompson 2009b). We also need more refined ways of measuring the intensity of stabilizing selection. That would help us understand whether our perception of no evolutionary change in some populations has sometimes been due to our failure to perceive ongoing selection and evolution that reduces variance but leaves average values unchanged.

Multiple alternative ways have been proposed to assess the strength and forms of selection, patterns of fitness, and evolutionary trajectories in populations (e.g., Lynch 1990; Hunt et al. 2008; Gingerich 2009; Shaw and Geyer 2010). For now, the Lande-Arnold methods remain the standard measures for selection, and darwins and haldanes remain the standard measures for rates of directional change. The fact that there is no single way to measure the strength and structure of selection on a population reflects not only the reality that selection acts on multiple traits but also that selection and evolution are hierarchical. Genes differ in the rates at which they evolve, as do chromosomes, nuclear and organelle genomes (e.g., mitochondria, chloroplasts), populations, species, and higher lineages. Hence, analyses of whether one group of organisms evolves faster or is subject to more intense selection than another group of organisms apply only to that particular set of molecular or phenotypic traits in that environmental setting.

Among the greatest current challenges is a better understanding of the underlying genetics of rapidly evolving traits. Some, but not all, examples of rapid evolution involve quantitative traits that are likely controlled by multiple genes. Some instances of rapid evolution depend on novel mutations. More generally, studies from evolutionary developmental biology ("evo-devo") suggest that the evolution of novel forms is often about changes in the regulatory mechanisms of ancient genes rather than the origination of new genes (Brakefield 2011), what Sean Carroll (2005) has called "teaching old genes new tricks." Increasingly, the problem is not just about what drives evolution but about what shapes evolvability itself (Hendrikse et al. 2007; Pigliucci 2008; Wagner and Zhang 2011). It is to the genetic underpinnings of adaptive evolution that we now turn.

Part 2

The Ecological Genetics of Adaptation

3

Genes

The studies of selection gradients in chapter 2 focus on traits controlled by multiple genes that often produce a near continuous distribution of values (e.g., body sizes, height). But many traits important to adaptation are controlled in other ways that often create less continuous distributions: single genes, a small group of genes (oligogenic traits), an inversion of a section of chromosome, a repeated section of a chromosome, duplication of the entire genome (polyploidy), or insertion of transposons. We do not yet know much about the dynamics of adaptive evolution driven by these kinds of genetic change rather than by the evolution of traits controlled by the interaction of many genes. It is, though, an important problem to solve for our understanding of evolution, because the genetics of adaptive traits could affect the dynamics of adaptive change.

This chapter and the next two explore what we have been learning about the genetics of adaptation at three levels—genes, genomes, coevolving genomes—and how it affects our views of the evolutionary process and the relentlessness of evolution.

BACKGROUND: ENVISIONING AND REENVISIONING THE PROBLEM

Our views of the pace of physical and biological change have always depended on our views of the mechanisms of change. In the mid-1800s Charles Lyell developed a theory of geology based on slow and continuous changes caused by slow-acting physical processes acting over long periods of time (Lyell 1830, 1832, 1833; Rudwick 1970). Creeks slowly cut channels that deepened and widened over eons, eventually forming canyons. This uniformitarian school of thought won out over the catastrophist school that focused on major cataclysmic events.

Lyell's arguments had a strong effect on Charles Darwin, who, in developing the theory of natural selection, focused on gradual change in traits over long periods of geological time. That view was reinforced in the late 1800s by biometricians such as Karl Pearson and others who showed mathematically how slow, continuous changes could accumulate large effects (Provine 1971; Magnello 2009). With the rediscovery of Mendel's work in the early twentieth century, though, the pendulum swung back hard and fast. Mendel's studies showed that mutations could produce discretely different traits—wrinkled peas rather than smooth peas in Mendel's experiments—and biologists began to consider that evolution may sometimes, perhaps even often, proceed through macromutations. In that new view, natural selection became a secondary, fine-tuning force in evolution. It became a process that weeds out maladapted mutations rather than one that molds evolution of the trait itself across many generations.

The New Synthesis of the 1920s and 1930s resolved these conflicting views by showing that, although mutation was the ultimate source of variation, it was random. It is natural selection that produces adaptation by favoring beneficial mutations and disfavoring detrimental mutations in successive generations over time. That realization, though, begged the next questions: were traits determined by mutation and selection acting over many genes or a few genes?

Ronald Fisher (1930) argued for a view of evolution in which traits are often governed by many genes, each with a small effect. Traits evolve gradually by weak selection distributed across genes. This process is often called Fisher's infinitesimal model of evolution. By that model, some variant forms (alleles) of a gene may be dominant over other alleles, but traits could mostly be viewed in what we would now consider a digital way. An allele adds either one increment to a trait or it adds no increment. Alleles, then, are additive, and expression of a trait depends on many genes, each contributing in a small way.

In contrast, Sewall Wright (1931) argued for more complex relationships among gene interactions and more complex roles for natural selection. In his view, genes commonly affect multiple traits (pleiotropy), and some genes mask the phenotypic effects of other genes (epistasis). By this view, genes often did not interact in simple ways, and they sometimes produce discontinuous distributions of traits even when multiple genes contribute to a trait. The genetic complexity of evolving traits was highlighted further by Thomas Hunt Morgan (1926) and others, who showed that genes are carried on chromosomes. That meant that groups of neighboring genes on chromosomes could be transmitted together to offspring rather than independently. This so-called linkage disequilibrium among genes showed the possibility that some combinations of traits could evolve in tandem.

Nevertheless, Fisher's infinitesimal model of evolution dominated the development of theoretical population genetics over the next half century. There were pointed criticisms of that approach (e.g., Mayr 1963), prompting Haldane (1964) to write a spirited defense of what some biologists had dubbed bean bag genetics. There were good arguments for using the infinitesimal model as a logical starting point for modeling many traits involving size or shape. Many traits are distributed continuously and approach a normal distribution in large populations. In addition, it is mathematically convenient to model evolution using the assumption of weak natural selection acting on many genes, each with a small effect, and it therefore provides a good first approximation of how a trait might evolve if the gene interactions are not complex.

Studies during the decades after 1930, however, showed that many traits do not fit the infinitesimal model. Beginning in the 1940s, studies of gene-for-gene interactions between plants and pathogens showed that plant defenses often involve major genes (Flor 1942, 1956). In the middle decades of the twentieth century, field studies of polymorphic traits, such as the color or banding patterns of snails and the cryptic coloration of moths, showed examples of selection acting on discrete rather than continuous traits (Cain and Sheppard 1950; Kettlewell 1959, 1973; Clarke 1962; Ford 1964). Breeding experiments with brightly colored butterflies suggested that the complex warning coloration on their wings may evolve through gene interactions, with a few genes having large phenotypic effect and other genes having smaller, modifying effects (Clarke and Sheppard 1959, 1963). By the 1980s and 1990s, laboratory studies of microbial evolution showed that selection on major genes often drives the early stages of adaptation, and subsequent studies showed that changes in single genes are important at multiple stages in the evolution of populations (Wichman et al. 2005; Blount et al. 2008; Scanlan et al. 2011).

New molecular tools and experimental methods have continued to reveal that some stages of adaptive evolution often involve a small to moderate number of genes, some of which have large phenotypic effects (Orr 2005). Yet other studies have shown that species have often co-opted whole genomes of other species during evolution (upcoming in chap. 5). We are, then, still trying to understand the extent to which adaptive evolution is shaped by many rather than few alleles and when even more complex genetic and genomic processes drive evolutionary change (Turelli and Barton 2006; Stinchcombe et al. 2009).

In the late 1980s, it became increasingly possible to map many quantitative trait loci (QTLs). These are regions of the genome that contain one or more loci contributing to phenotypic variation in a quantitative trait such as size or height (Tanksley 1993; Lynch and Walsh 1998). A QTL can have a large effect on a phenotypic trait (i.e., large effect size) or a small effect. Studies of QTLs showed that genetic control of some traits was restricted to one or a few chromosomal regions rather than many regions. The new generation of genomic and bioinformatic techniques has provided further evidence that some traits are controlled by genes localized to one or a few positions on chromosomes, but other traits are controlled by genes distributed across the genome (Nadeau and Jiggins 2010; Cubillos et al. 2011). Major genes, minor genes, and larger genomic processes are clearly all involved in adaptive evolution, and we are continuing to get closer to understanding when evolution proceeds through different genetic mechanisms.

DEMONSTRATING THE EFFECTS OF MAJOR GENES ON EVOLUTION

The evolution of adaptation through a combination of major genes and minor genes is especially clear for plants. Plant defenses against pathogens and herbivores sometimes involve major genes that affect a plant's ability to detect attack and other genes that generate the actual defenses (Schneider and Collmer 2010). The first set of genes often evolves through gene-for-gene interactions. Attack by a pathogen population favors mutant plants capable of detecting a gene product (e.g., a protein) produced by the pathogen. It does not matter which gene product it is. All that matters is that the plant can use a particular gene product produced by the pathogen as a signal that it is under attack. Recognition of the signal then induces the plant to ramp up its chemical or other defenses, which may involve multiple other genes.

As the recognition gene spreads through the plant population, it favors mutant pathogens that either do not produce that particular gene product or produce it in a way that masks its presence. This first cycle of gene-for-gene coevolution can be followed by many similar cycles. Over time, host populations can build up multiple genes that recognize different genetic forms of the pathogen, and pathogen populations can build up multiple genetic forms capable of escaping detection by all the host's recognition genes.

For decades this form of coevolution provided the basis by which crop breeders selected for new crop varieties resistant to particular pathogens. Gene-for-gene coevolution is found mostly in interactions between plants and pathogens (Thompson and Burdon 1992; Thrall and Burdon 2002), but there are suggested examples of similar kinds of major gene effects in other taxa. These include various forms of gene-for-gene-like effects in *Drosophila*

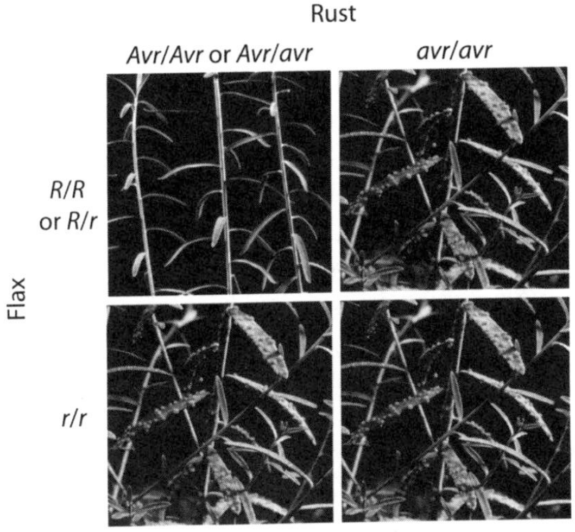

Fig. 3.1 Gene-for-gene interaction between Australian wild flax (*Linum marginale*) and flax rust (*Melampsora lini*). Flax plants are resistant to flax rust when they have an *R*-gene capable of detecting a gene product produced by rusts with the corresponding *Avr* gene. Otherwise, they are susceptible to attack. Photographs courtesy of Jeremy Burdon.

flies and their parasitoids and in interactions between microbial species (Dubuffet et al. 2007; Forde et al. 2008a).

The best-studied example of gene-for-gene evolution in natural populations is in Australian wild flax, *Linum marginale*, and its interactions with flax rust, *Melampsora lini* (Barrett et al. 2009; Burdon and Thrall 2009; Dodds and Thrall 2009; Antonovics et al. 2011b). Susceptible and resistant plants can be readily scored (fig. 3.1), and the range of resistant genes found within populations can be determined by challenging plants grown from seed with different genotypes of the pathogen. These studies have shown that the plants and the rusts are evolving at fast rates, with different combinations of host and pathogen genes rising and falling in different ways in different populations over the course of a decade. Both selection and random genetic drift appear to contribute to the genetic composition of resistance in local populations (Jarosz and Burdon 1992; Burdon and Thompson 1995; Thrall et al. 2002).

Because Australian wild flax is closely related to cultivated flax (*Linum usitatissimum*), it has been possible to enlist the molecular tools used in studying the genetics of resistance in this crop species to study coevolution in natural populations of wild flax. Australian wild flax has at least fourteen genes or alleles involved in gene-for-gene interactions, and cultivated flax has at least thirty alleles distributed over five loci (Burdon 1994; Ravensdale et al. 2011). Local wild flax populations differ in the number and combination of resistance genes. Some populations have as many as eighteen resistance phenotypes, resulting from various gene combinations, and others have a single resistance phenotype (Burdon and Thompson 1995; L. Barrett et al. 2008). In some populations, some host individuals are susceptible to attack by all local genetic forms of the rust, while others are susceptible to only some forms, and some may be resistant to all local forms. Across Australia, there is considerable variation in the combinations of genes found locally and regionally in the plant and pathogen populations (Burdon and Thrall 2000; Thrall and Burdon 2003; Barrett et al. 2009; Laine et al. 2011).

More generally, studies of plant species show that a mutation in a single gene can have a large effect on fitness in natural populations, even when the entire remainder of the genome remains the same. An elegant set of experiments using monkey flowers (*Mimulus* spp.) in western North America has demonstrated that changes in a single gene can have large effects on which pollinator species visit the flowers. The *YUP* gene, or more formally the *YUP* quantitative trait locus, controls the amount of carotenoid pigment produced in the petals. When the dominant *YUP* allele is present, the plant does not deposit carotenoids in the flowers, and the petals appear pink because they contain only anthocyanins. When the recessive *YUP* allele is present in a homozygous state, the flowers contain carotenoids as well as anthocyanins, and they appear red or orange-red (Bradshaw and Schemske 2003). Absence of that gene prevents carotenoid deposition in the floral petals, which can reduce visitation by bees by 80 percent (Schemske and Bradshaw 1999).

Repeated genetic crosses make it possible to create lines that differ almost solely in the alleles at the *YUP* locus and then present those plants to pollinators in the wild. Those experiments show that the attraction of hummingbirds or bees can be altered greatly by changes in this one gene (Bradshaw and Schemske 2003). Another allele that increases nectar production can double the rate of visitation by hummingbirds (Schemske and Bradshaw 1999). Single alleles of large effect therefore have the potential to drive rapid adaptive evolution of a population.

To be sure, many plant traits are controlled by multiple genes of small phenotypic effect. Flowering time, for example, is a genetically complex trait that can strongly influence plant fitness, because it can directly affect the probability that a flower will be pollinated and donate pollen to nearby plants. Studies of flowering time in maize using analyses of quantitative trait loci have shown that this trait is controlled by the additive effects of many genes (Buckler et al. 2009). There are no major effects due to single genes, and the effects of particular alleles differ among genetic lines. In *Arabidopsis*, flowering time is controlled by more than sixty quantitative trait loci under some field conditions, although a few genes have major effects under some greenhouse conditions (Brachi et al. 2010).

It is not known, though, if most other plant species have similarly complex genetic control of flowering time. Most analyses of the genetic control of flowering time have been on crop species or a few model plant species, and the genetic architecture of flowering varies even among these species (Cockram et al. 2007; Rhone et al. 2010). Experimental studies combined with molecular studies have identified in some plant populations QTLs or families of QTLs that contribute importantly to the pattern of seasonal response in multiple plant species (Alonso-Blanco et al. 2009; Huang et al. 2010; Wilczek et al. 2010). It seems likely, then, that the genetic complexity of flowering time varies among species.

GENOME-WIDE STUDIES

As genome-wide studies of species continue to expand, they are suggesting some patterns in the number of genes involved in the evolution of traits. These patterns, however, must be regarded for now as useful working hypotheses until more genomic analyses are complete. A comparative analysis of QTLs in plants has suggested that each QTL under biotic selection (e.g., herbivores, pathogens, pollinators) has a larger effect on phenotypic traits than do QTLs under abiotic selection (Louthan and Kay 2011). For traits under both abiotic and biotic selection, almost 11 percent of QTLs have effects so large that they account for more than 20 percent of the variance found among individuals in those traits. In this analysis, abiotic traits were controlled on average by 3.8 QTLs, whereas biotic traits were controlled by 5.3 QTLs.

Among animals, rapid expansion of genome-wide studies has changed our views of the genetic architecture of traits (Flint and Mackay 2009). Most of the work until recently has been on model organisms such as laboratory mice (*Mus musculus*), *Drosophila melanogaster* flies, and humans. Early studies of quantitative trait loci in the 1990s indicated that many quantitative traits are controlled by a few QTL regions of large effect. Each of these QTLs can include multiple genes, but early studies of QTLs suggested that they might often be localized rather than spread across the genome. As genetic mapping studies have incorporated more QTLs, more individuals, and greater computing power, it has become possible to detect more QTLs of small effect. These studies now suggest that some animal traits may be influenced by more genes than once thought. Genome-wide association studies of over 180,000 humans have shown, for example, that adult height is influenced by at least 180 loci (Allen et al. 2010).

These studies have found that more traits are affected by many genes—or, more accurately, many QTLs—of small effect than previously thought (Hindorff et al. 2009; Manolio 2010). In human disease risk studies, which include the most extensive genome-wide data, the effects are often evaluated as values called odds ratio effect sizes. By this method the effects are scaled such that an effect of two doubles the risk of developing the disease. Most effect sizes of genes influencing human diseases are less than 2 and have an average near 1.3 (Hindorff et al. 2009; Manolio 2010). Moreover, some surveys of the human genome suggest that the genetics of disease in animals is similar to the genetics of other quantitative traits (Altshuler et al. 2008; Flint and Mackay 2009),

That does not mean, though, that the infinitesimal model of the evolution of adaptation is sufficient for our general understanding of adaptive evolution in animals. There is still much to learn about the evolution of traits that are not determined by many genes each with a small effect on phenotypes. Even in species that have been studied in detail, such as laboratory mice, *Drosophila*, and humans, the genetics of selection has been studied in detail for only a small number of traits.

One of the most difficult genetic problems to solve is how gene interactions influence the effects of genes on phenotypes. These interactions have turned out to be difficult to detect in many studies. If the main phenotypic effect of a gene is small, then the effects of the interaction of that gene with other genes of small effect will often be even smaller (Flint and Mackay 2009). Assessing the importance of these small effects has become an increasingly important issue not only for evolutionary biology but also for genetically based medicine (Crespi 2010). The risk of some infectious diseases has been linked to the major effects of particular alleles, such as the effect of the sickle-cell allele on risk of malaria (Penman et al. 2009). Examples like this one have fostered medical genomic research devoted to identifying genes with large effects on the risk of infectious diseases, metabolic imbalances, or cardiovascular disorders. There remains, though, disagreement about how often the risk of developing diseases or disorders is due to the occurrence of a few genes of large effect rather than many genes of small effect.

More broadly, the genetics of many traits are likely to differ among organisms with different body plans. Plants and some animal taxa such as corals are built on a design of highly repeated modules (e.g., cutting off a branch of a tree is very different from cutting off a mammalian leg). Modular design creates opportunities for phenotypic traits different from those found in less modular organisms (Herrera 2009). Clearly, we have a long way to go in figuring out how the genetics of traits varies across the web of life.

MOVING FROM PATTERN TO PROCESS

Describing the current genetic architecture of traits, though, can get us only so far in our understanding of the evolution of adaptation. If traits are often shaped by multiple genes that differ in the magnitude of their effects on traits, then we need to know how, and in what order, those various genes tend to accumulate during the process of adaptation. Studies of the genetics of domesticated plants and animals suggest that genetic architecture and evolution can be a two-way street. The genetics of traits can affect rates and patterns of evolution, but the evolutionary process can reshape the genetics of traits.

The evolution of domestic dogs is an extreme but insightful example. Although morphological traits in mammals are often controlled by many genes, that pattern does not hold for domesticated dogs (Boyko et al. 2010). Genome-wide studies indicate that dogs went through a strong genetic bottleneck early in their domestication, and additional genetic bottlenecks occurred as specific breeds were developed. These genome-wide studies also suggest that multiple regions of dog genomes have been through recent strong selection (Akey et al. 2010). Many modern breeds originated during Victorian times, when novelty was the focus of selection by breeders. This emphasis favored animals that looked very different from other breeds (Boyko et al. 2010). As a result of that selective process, only two to six QTLs are needed to explain over 70 percent of the variation found in many dog morphological traits (Boyko et al. 2010). Strong artificial selection and genetic bottlenecks have therefore changed the genetic architecture of traits in dogs relative to some other mammals.

In natural populations, even if genome-wide studies suggest that a complex trait is controlled by many genes, particular alleles can become the focus of strong selection in some environments. The number of genes controlling a trait is not the same as the number of genes controlling the process of adaptation. For example, more than 150 genes are known to affect body pigmentation in mammals (Hofreiter and Schoeneberg 2010). Even so, some major evolutionary changes in fur color in mammals have resulted from selection on one or a few genes.

Geographic differences in the coat color of deer mice (*Peromyscus maniculatus*) are a clear example. Deer mice are generally a cryptic shade of brown that matches to varying degrees the color of the soil in their environment. Individuals that match the soil color have a selective advantage over poorly matched individuals in environments where they are subject to predators that search for prey by sight (Vignieri et al. 2010). Differences in coat color can occur even among nearby populations living on contrasting soils. Deer mice in the Sand Hills of Nevada have an unusually light color that closely matches the light background of the soil, whereas those in the surrounding habitats, which have darker soils, have darker coat colors (Linnen et al. 2009). The light color in the Sand Hills mice is due to a novel banding pattern that results from a single mutation that arose within the past 8,000 years in one gene. Analysis of the strength of selection has suggested that natural selection has been strong enough to drive this population to its light coat color in that length of time (Linnen et al. 2009).

Similar examples of geographic differences in coat color linked to a single mutation have been found in other species such as beach mice (*Peromyscus polionotus*). These mice have a light coat color and live on the primary dunes and barrier islands of Florida's Gulf and Atlantic coasts (Mullen et al. 2009). A single gene (*Mc1r*), which results in light color, explains between 10 and 36 percent of the variation in color pattern found among the mouse populations. That is only part of the story (Hoekstra et al. 2006). The *Mc1r* allele is present in the Gulf Coast populations but not in the Atlantic coastal populations, suggesting that these populations differ in the genetic mechanisms by which they have evolved light coat colors. The barrier islands on the Gulf Coast are less than 6,000 years old. It is therefore likely that selection has favored the rapid spread of light coloration over that short length of time (Hoekstra et al. 2006).

Body armor is another trait that would seem to be genetically complex, but, as with coat color in mammals, careful geographic and molecular studies accompanied by field experiments have shown that selection on single genes may drive much adaptive evolution. Marine populations of threespine stickleback, *Gasterosteus aculeatus*, have lateral plates, pelvic spines, and dorsal spines that have been lost to varying degrees in freshwater populations of this species (fig. 3.2). Reductions in these traits have occurred repeatedly in some coastal freshwater populations in British Columbia in less than 12,000 years. During that time, the Pleistocene glaciers retreated, the land rebounded, and some stickleback populations became isolated in coastal lakes from their marine ancestors (Walker and Bell 2000; Chan et al. 2010; Schluter et al. 2010). Within those lake environments, the fish underwent reduction in their spines and in their lateral plates that act as body armor.

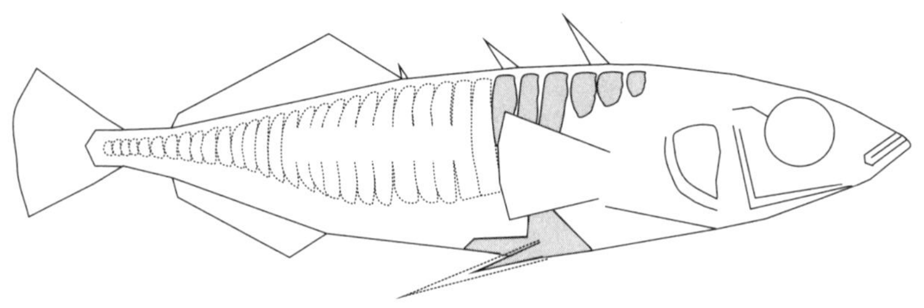

Fig. 3.2 Distribution of body armor (shaded regions) on threespine sticklebacks. Dotted regions show lateral plates and pelvic spines found in marine ancestors that are absent or reduced in most freshwater populations. Multiple other morphological traits have also changed in the transition to freshwater. After Schluter et al. (2010).

These traits are controlled mostly be two genes. The production of the pelvic apparatus, which consists of a pelvic girdle and a set of spines, is controlled to a major degree by *Pitx1* (*pituitary homeobox transcription factor 1*) (Chan et al. 2010). The production of lateral plates is controlled to a major degree by the gene *Eda* (*ectodysplasin*) (Colosimo et al. 2000). Individuals carrying only the full-armor allele for lateral plates have a full set of plates along both sides of the body, and individuals carrying only the low-armor allele usually have few plates. Heterozygous individuals carrying both alleles have an intermediate number of plates.

Field experiments on sticklebacks show that natural selection on body armor is driven by predation. The causes of this developmental change in selection are not yet fully understood. In marine environments the plates probably aid in defense against predators. Body armor makes it difficult for predators to ingest sticklebacks, and it decreases damage to sticklebacks if they are caught by a predator but escape (Reimchen 2000). In freshwater, juvenile sticklebacks with the low-armor allele grow faster, which would move them more quickly through the early stages of life when fish are highly vulnerable to predatory insects.

The direction of selection on body armor, though, changes as a fish grows. At the earliest ages before development of lateral plates, the proportion of individuals with the low-armor allele in experimental field populations decreases. Only later does the proportion of fish with low armor increase with age. Hence, early in life there are opposing, but not yet identified, selective forces acting on this gene or some other genes (R. Barrett et al. 2008). Some other traits in sticklebacks, such as body shape, are controlled by multiple genes that vary in the degrees to which they affect phenotypes (Albert et al. 2008). The results for spines and body armor, however, suggest that selection on a single loci can have major effects on fitness.

Major loci can even control some ecologically crucial aspects of body size and shape, as has been found in molecular studies of the evolution of the beaks of Darwin's finches in the Galápagos Islands (Grant and Grant 2008a, 2008b). The rapid evolution of beak size in some of these finches over the past half century in response to changes in available resources (e.g., seed sizes and shapes) has become the premier example of how complex traits can evolve rapidly in relatively long-lived organisms. Beak development is controlled by only a small number of signaling molecules, one of which, *Bmp4*, is important in controlling the development of beak shape (Abzhanov et al. 2004). A second pathway, controlled by the gene *calmodulin*, affects beak length (Abzhanov et al. 2006). Hence, potentially simple changes in one or a few developmental pathways may account for some of the major differences in beak morphology found among these bird species (Grant et al. 2006).

THE EARLIEST STAGES OF ADAPTATION

If strong selection on major genes is often part of the process of adaptation, then the early stages of adaptation to a new environment could be so rapid that the importance of these genes would be missed unless a population had been under study from the moment it entered the new environment. Otherwise, much of the process of selection on those genes could be over by the time a population is studied a few years or decades later. A researcher analyzing the process of adaptation after those initial decades would have the impression that selection is mostly a slow process acting on many minor genes.

One of the best-studied examples of rapid evolution driven by a major gene during a period of major environmental change is the evolution of melanic-winged forms of peppered moths (*Biston betularia*) in industrialized areas of Europe and North America. This example has also been the most widely cited instance of rapid evolution over the past century (Kettlewell 1973; Majerus 1998; Brakefield and Liebert 2000; Cook and Turner 2008). It is so well known that we often take it for granted, but it continues to provide new and insightful evidence of how rapidly populations can evolve in response to some forms of environmental change.

As industrial pollution increased in Europe in the 1800s, the frequency of rare black moths increased and that of the salt-and-pepper winged forms decreased. The major cause was clear: as pollution killed the lichens on the

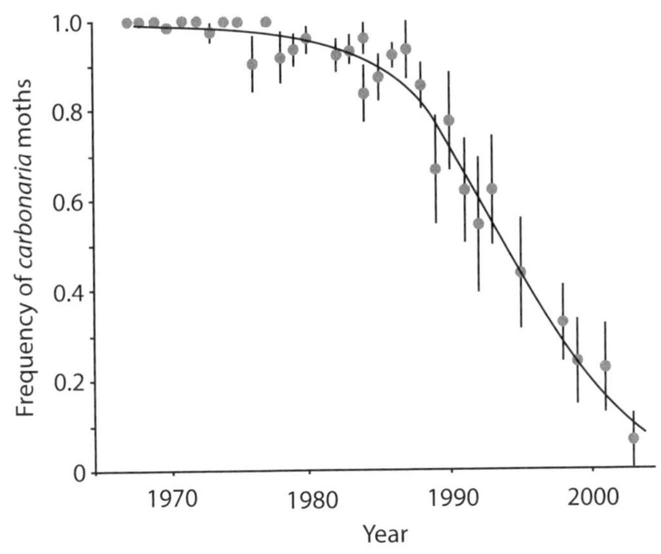

Fig. 3.3 Rapid decrease of the melanic *carbonaria* morph of peppered moths, *Biston betularia*, following the decrease in industrial pollution in north Leeds, England. Values are the means of samples ±1 SE. The curve is calculated from estimates of selection favoring peppered moths scaled forward and backward from the year 1991. After Cook and Turner (2008).

increasingly sooty trees where the moths rested during the day, the peppered form became more visible to predators, and melanic moths became less visible. The wing colors of these moths have a fairly simple genetic basis (van't Hof et al. 2011), and selection drove rapid evolution toward melanic moths in multiple industrialized areas.

As pollution controls were introduced during the latter part of the twentieth century, selection began to favor the peppered form. Today, melanic moths are rare in regions where, just half a century ago, they were by far the more common form. In Leeds, England, where populations have been continuously monitored for decades, the decline of melanic moths was slow during the 1970s and early 1980s but then much faster since then (fig. 3.3). It took less than thirty years for those populations to evolve in color pattern from one extreme to the other.

The causes of this rapid evolutionary change have been subject to intense scrutiny over many years, and the evidence has stood up well. The rate of change has been too fast to be accounted for only by movement of individuals among regions, by mutation, or by random genetic drift (Cook and Turner 2008). The changing values for Leeds closely fit predictions of a rate of change due solely to natural selection (fig. 3.3). Although gene flow occurs among regions, which can affect the rate of change due to selection in local populations, selection continues to drive the decline in melanic morphs in regions with improved air quality.

Analyses of industrial melanism in this moth species now span more than a century (Cook 2003; Saccheri et al. 2008; van't Hof et al. 2011). The importance of these observational and experimental studies, produced by a small number of dedicated researchers over many years, cannot be overemphasized. This body of research has shown, with remarkable clarity, how quickly natural selection focused on major genes can alter populations, first in one direction and then in another, as environmental conditions change.

The development of QTL techniques has made it possible to probe more deeply and more quickly into the earliest stages of adaptation. One approach is to generate a set of recombinant inbred populations that are genetically variable only for a particular set of traits. Individuals from those populations are then placed into the field and evaluated for how natural selection acts on survival and reproduction at every life history stage. These studies make it possible to understand how selection on a trait is distributed across the lifetimes of individuals. It can also help in assessing why selection on any particular genes, or set of genes, may have such strong effects on fitness.

This approach is becoming useful in addressing questions on the genetics of adaptation in populations colonizing new regions. The colonists are adapting to truly new environments, and the experiments make it possible to evaluate the effects of major genes on the early stages of adaptation. The plant species *Arabidopsis thaliana* has been the most commonly used plant species so far in these studies (Huang et al. 2010; Wilczek et al. 2010) because it is the standard model species for studies on the genetics, developmental biology, and physiology of plants. In one particularly informative set of studies, recombinant inbred populations were established in old fields in Rhode Island and Kentucky (Huang et al. 2010). Old fields are the usual habitats for this weedy species in North America since its introduction from Europe. The populations planted into these fields had no past history at those sites, and it was therefore possible to evaluate how these populations evolved from the moment at which the seeds were planted in these fields.

During the early stages of adaptation, single QTLs of this species had large effects on fitness at each developmental stage, including the ability of seeds to germinate in each environment, survival of seedlings following germination, and reproduction once a plant reached maturity. These effects on fitness were driven by genes that affected the timing of events in individuals from germination to flowering (i.e., phenology and dormancy). Allele frequencies of some QTLs changed by as much as 90 percent in a single generation, suggesting strong selection on these populations. Fitness, though, was driven not only by direct selection on individual major genes. Some pairs of genes on different chromosomes had higher fitness than other gene combinations (Huang et al. 2010).

RAPID EVOLUTION OF BEHAVIORS

Most of the data on the genetics of adaptive evolution are on morphological traits. We still do not know if adaptive evolution of physiology and behavior is generally more complex or simpler than morphological evolution. It is crucial, though, that we understand the genetics of behavior, which is often mediated through physiological mechanisms. Behaviors often determine the selection pressures to which a local population is subject (Gordon 2011). Slight changes in behaviors—movement patterns, the choice of other species with which to interact—can alter selection on a wide range of other traits.

The evolution of even some seemingly complex behaviors may be under simple genetic control. Data are still few, however, because rapid evolution of behavioral traits in natural populations is difficult to study. Unlike morphologies, behaviors generally do not leave a permanent record unless they are associated with building structures (e.g., nests, burrows). Each measurement is of a transient behavior. Many behaviors are also highly plastic, requiring careful experiments to understand the genetic underpinnings.

The study of one group of behaviors has suggested that at least some behaviors can evolve quickly and do so in response to selection on a major gene or localized group of genes. These are the behavioral traits associated with preference for host or prey populations. Preferences for food are likely under strong selection in most species, because small differences in host or prey choice can have large effects on subsequent survival and reproduction. Examples of genetically based preference and specialization have now been studied in detail in multiple insect and microbial species (Janz 2003; Nylin et al. 2005; Tanaka et al. 2007; Thrall et al. 2007; Abrahamson and Blair 2008; Feder and Forbes 2008; Scriber et al. 2008; Singer et al. 2008; Thompson 2009a). There are also some studies of the genetics of prey selection by vertebrates and marine invertebrates (Arnold 1981; Sanford et al. 2003; Sotka et al. 2003; Ghedin et al. 2005) and specificity in interactions between plants and fungi (Shefferson et al. 2007).

We can see traces of continuing evolution of behavior by evaluating variation within and among populations. Swallowtail butterflies in the *Papilio machaon* group feed throughout Eurasia and North America on plants in the carrot family (Apiaceae) and also some plants in the citrus (Rutaceae) and sunflower (Asteraceae) families. In some cases, they have colonized plant species that have been introduced from other continents, and the problem to solve is whether this colonization has involved evolutionary change or whether the butterflies can incorporate these new hosts without any change in their behaviors.

These butterflies have genetically determined preference hierarchies that can be studied experimentally through choice experiments. When offered multiple plant species in a controlled environment, a female might lay, say, 50 percent of her eggs on one plant species, 25 percent on a second, 15 percent on a third, and 10 percent on a fourth. By testing multiple offspring of a single female, it is possible to determine whether these preferences are repeatable, suggesting a possible genetic basis to this preference hierarchy. By testing multiple families within a population, it is possible to ask whether there is variation in preference hierarchies within a population. By testing multiple populations of a species and also multiple species, it becomes possible to build a hierarchical view of how preference has evolved during the diversification of a group of species. Finally, by crossing populations or species with different preference hierarchies, it becomes possible to begin to sort out the genetic basis of these behavioral differences.

Swallowtail butterflies show genetic variation in preference hierarchies at all these phylogenetic levels. Individuals vary within populations (Wiklund 1981; Thompson 1988a); populations differ more from each other than do individuals within populations (Tiritilli and Thompson 1988; Thompson 1993; Bossart and Scriber 1995; Wehling and Thompson 1997; Wiklund and Friberg 2008); and species differ more than populations (Thompson 1995, 1998, 2009a; Mercader and Scriber 2008). Among the North American species of the *Papilio machaon* group, there is almost a smooth progression in preference hierarchies among species, when individuals are tested against a standard set of potential host plants (fig. 3.4). Yet these butterflies are also highly constrained in the range of species they exploit. Even though these butterflies use plants in three families, the particular plant species that they use in these families all contain a similar set of chemical compounds sought by females when searching for hosts on which to lay their eggs (Murphy and Feeny 2006).

Small shifts in preference hierarchies may be possible through fairly simple genetic changes. Differences among species in how females rank

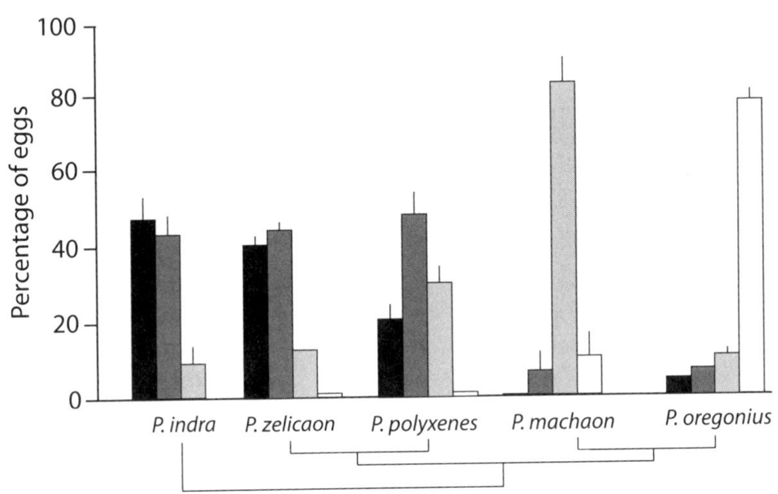

Fig. 3.4 Variation in relative oviposition preference for four plant species within the *Papilio machaon* clade of swallowtail butterflies in North America. The test plants include *Lomatium grayi* (*black*) and *Cymopterus terebinthinus* (*dark gray*), which are native species within the Apiaceae; *Foeniculum vulgare* (*light gray*), which is an introduced species within the Apiaceae; and *Artemisia dracunculus* (*white*), which is a native species within the Asteraceae. Values are means (±1 SE). One population from each species is included here to show the full range of variation in relative preference found within this lineage of butterflies. Additional variation occurs within each species but is less than the major interspecific differences. Bars under the names indicate the phylogenetic relationships among these species. Data on preference from Thompson (1993, 1998, 2009a) and Wehling and Thompson (1997).

hosts are controlled primarily by one or more sex-linked genes (Thompson 1988b). The ability of larvae, which can feed on a wider range of plants than found in each local community, appears to be controlled by other genes (Thompson et al. 1990). Experimental trials using a standardized set of potential host plants show that females differ among populations in their preference hierarchies. These populational differences generally fall within a narrower range than those found among species (fig. 3.5). Some of the differences may result from selection imposed by the regional differences in the availability of host plants, but that is just one of the potential sources of selection. Studies in North America and Europe have suggested that host use in some swallowtail species may be controlled as much, or even more, by selection imposed by predators, which are more likely to attack larvae on some plant species or in some environments than in others (Murphy 2004; Wiklund and Friberg 2009).

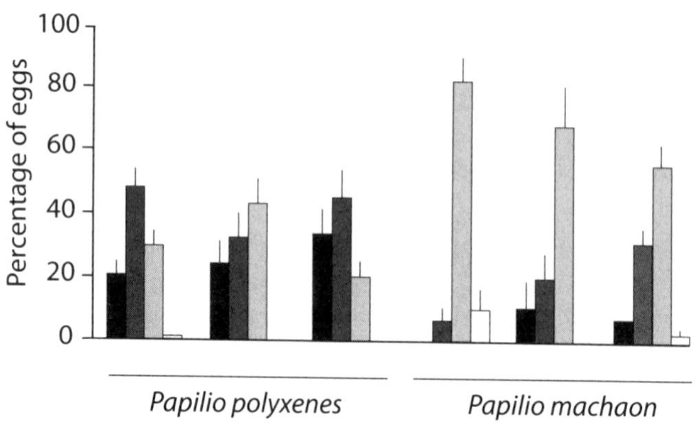

Fig. 3.5 Differences among populations of black swallowtail (*Papilio polyxenes*) and Old World swallowtail (*P. machaon*) butterflies in their relative preferences for four host plants in North America. The test plants include *Lomatium grayi* (*black*), *Cymopterus terebinthinus* (*dark gray*), *Foeniculum vulgare* (*light gray*), and *Artemisia dracunculus* (*white*). See fig. 3.4 legend for more details. The three populations of *P. polyxenes* are from (*left to right*) Colorado, Nebraska, and New York, and the three populations of *P. machaon* are from different parts of Alaska. Data from Thompson (1998).

In the wild, these populations feed mostly on native plant species, but some populations of *P. machaon* in Alaska have a relatively high preference for introduced fennel, even though they never encounter it (fig. 3.5). Farther south in Washington State through California, some populations of *P. zelicaon* have colonized fennel, where it is a dominant, invasive weed. Some of those individuals show a higher preference for fennel than do many populations of this species throughout western North America that feed only on native species (Wehling and Thompson 1997; Thompson 1998). Without knowledge of the history of evolution of preference hierarchies in the fennel-feeding *P. zelicaon* populations, it is difficult to know if populations are evolving an increased preference for fennel in places where native hosts are decreasing in abundance. It is, though, clear that this butterfly clade has an evolutionary history of subtle shifts among populations in relative preference for host plants, and that they have the capacity to incorporate new species when available.

Direct evidence of rapid evolution of oviposition preference comes from studies that have been conducted for multiple years before the introduction or significant spread of a new species. For example, a population of Edith's checkerspot butterfly (*Euphydryas editha*) near Carson City, Nevada, appears to have evolved a strong preference for plantain (*Plantago lanceo-*

lata), which is a common Eurasian weed that is now widespread in North America (Singer et al. 2008). Studies of this butterfly population in the late 1970s showed that females strongly preferred their major native host. By the 1990s most field-caught females preferred the introduced plant to their major native host plant, and the same was true for females that have been grown from eggs in the lab under controlled conditions (Singer et al. 1993).

Studies of checkerspot butterflies reinforce the view that butterflies are continually evolving, and major genes can be the focus of selection (Ehrlich and Hanski 2004). The metapopulation of Granville fritillary butterflies on the islands off southwestern Finland is the best-studied natural metapopulation of any animal species in the world. The hundreds of local populations within this metapopulation have undergone annual monitoring since the early 1990s. Each local population occurs in a meadow that is separated by forests, woodlands, and agricultural fields from other populations (Hanski and Meyke 2005). The populations show a pattern of repeated colonization and extinction of meadows that is characteristic of metapopulations.

Local populations are variable for the gene locus phosphoglucose isomerase (*Pgi*), and the frequency of *Pgi* varies among populations (Hanski 2011). This enzyme is known to affect the temperatures at which some butterflies are able to fly, and alleles at this locus differ in the optimum temperature at which they function (Watt et al. 2003; Saastamoinen and Hanski 2008). Consequently, butterflies with different alleles, or different allelic combinations, differ in their ability to fly at different temperatures, and they differ in their flight metabolic rates. Consequently, some of these butterflies are likely better at dispersing to new habitats, and it should be possible to pick up the signal of that difference by sampling butterflies in newer and older local populations.

As predicted, the frequency of one allele, which results in high flight metabolic rate, is greater in new populations than old populations (Haag et al. 2005). Moreover, individuals heterozygous at this locus are better able to fly at low temperatures, and they are better able to fly longer distances than are homozygous individuals (Niitepõld et al. 2009). Differences among local populations in the frequency of particular *Pgi* alleles also contribute importantly to the differences among local populations in other life history traits and in growth rates (Hanski and Saccheri 2006; Klemme and Hanski 2009; Wheat et al. 2011). These results, together with those showing that local populations are continually being established while others go extinct, indicate that the dynamics of this metapopulation is as much an evolutionary process as it is an ecological process.

The careful and detailed study of Glanville fritillary butterflies has been crucial for our understanding of the continual interplay of ecological and evolutionary processes. These studies show how the evolutionary dynamics of behavior—as well as ecological factors such as habitat size or distance to neighboring suitable habitats—affect the dynamics of metapopulations. The results have been used as one of the major models for the development of the mathematical theory of metapopulation dynamics. Early studies left unexplained some puzzling variation in the dynamics that could not be explained by ecological factors alone. By incorporating rapid evolution of traits as one of the working hypotheses to explain metapopulation dynamics, these studies have become a major model of how ongoing evolutionary and ecological processes together shape metapopulations (Hanski et al. 1994; Kuussaari et al. 1996; Ehrlich and Hanski 2004; Orsini et al. 2008; Wheat et al. 2010).

THE CHALLENGES AHEAD

Our early scientific knowledge on rapid evolution was built on a famous example, the evolution of melanism in peppered moths, which was clearly the result of human alteration of environments. That example involved major genes, as did examples of gene-for-gene interactions in agriculture. Through these and other examples, it became common to associate major gene effects with selection imposed by unusual situations created, directly or indirectly, by human activities. Careful study of multiple species in the wild in recent decades, however, has made it clear that evolution through selection on major genes is common in many kinds of adaptation that are not the direct result of human intervention. Some seemingly complex morphological, physiological, and behavioral traits are controlled to a large extent by one or a few genes.

Even when the underlying genetic architecture of a trait is complex, selection can sometimes act strongly on particular genes. We are, though, still trying to sort out how much of evolution is driven at any stage of adaptation by strong selection on one gene, a few genes, many genes, or even whole genomes. In the next chapter we turn to adaptive evolution governed by more complex gene interactions.

4

Genomes

If major genes or major genomic processes are commonly involved in the process of adaptation, then we need a broader perspective on the process of adaptive change. As Allen Orr (2005) has noted, we have a rich theory of evolution through natural selection based on the infinitesimal model, which applies to varying degrees to some traits in some species. Those models, however, do not fit neatly and completely with the rich body of data on adaptation based on major genes, a combination of major genes and minor genes, or broader genomic interactions.

To be sure, there is solid mathematical theory on selection based on many different forms of genetic architecture. Nevertheless, we are still in the midst of developing a solid theoretical basis for understanding when the process of adaptation is likely to involve few genes, many genes, or broader genomic processes such as duplication of whole genomes (polyploidy) or selection on hybrids between divergent populations or species. As the theory and data continue to develop in tandem, some studies suggest that the genetics of adaptation depend on how big a change is under way and how far along evolution has proceeded toward an adaptive peak. It may also depend on whether a population is adapting to, say, a gradually changing physical environment rather than to a parasite that matches each evolutionary change with a counter-evolutionary change.

In this chapter I consider how complex genomic interactions can affect the structure and dynamics of evolutionary change at multiple stages of the adaptive process.

BACKGROUND: EVOLUTION WITHIN CHANGING ADAPTIVE LANDSCAPES

The infinitesimal model of evolution remains useful because the predictions hold over a short number of generations for traits controlled by many genes, regardless of the details of the genetics (Turelli and Barton 1994).

As studies in adaptive evolution have increased, though, it has become important to understand the process of adaptive change across many timescales. If the earliest stage of adaptation to changing environments involves selection on major genes that have pleiotropic or epistatic effects on other genes, then those advantageous alleles can become fixed rapidly and alter the subsequent rates and trajectories of adaptive evolution (Agrawal et al. 2001; Stinchcombe et al. 2009).

If, instead, selection early on in the evolution of new adaptation is distributed across many genes, each with a small effect on fitness, then evolution may not proceed by selective sweeps. Instead, populations may maintain considerable genetic variation for those traits for many generations. That has been the inference from genome-wide analyses of selection on development time in *Drosophila melanogaster* analyzed after about 600 generations of laboratory selection on the rate of development (Burke et al. 2010).

If we envision, as did Sewall Wright (1932), a fitness landscape with adaptive peaks and valleys of gene combinations, then we can ask what kinds of genetic and genomic architecture of traits are most likely to move a population through an adaptive valley from one adaptive peak to another. We can also ask how evolution proceeds if adaptive peaks are distant and separated by deep valleys, rather than close and separated only by shallow saddles. Adaptive peaks, valleys, and saddles are an imperfect, some would say poor, metaphor for the evolutionary process (Reiss 2009). But they capture in a general heuristic way the evolutionary problem of how the genetics of adaptive change may mold the range of possibilities in the adaptive evolution of populations.

One way to address the problem of the genetics of adaptive change is to explore how major and minor genes contribute to adaptation when populations are large and natural selection is the only evolutionary process moving populations around the adaptive landscape. Selection distributed over many minor genes will move a population slowly and gradually. In contrast, selection on major genes may move a population quickly, but those genes could have detrimental pleiotropic effects on other genes unless those effects are mitigated by minor genes.

If a population is adapting to a gradually changing environment—that is, a slowly moving adaptive peak—some models suggest that selection on polygenic traits may often be more effective than selection on major genes (Lande 1983). There is a greater chance that common, minor mutations, rather than rare, major mutations, will be beneficial. This result, however, depends on the distribution of mutation sizes and the strength of natural selection relative to the rate of environmental change. More drastic environmental change may favor selection mediated by genes with more major effects on phenotypes.

For example, selection on major genes has often occurred during artificial selection on domesticated species or on populations introduced into new environments. Studies on the evolution of flowering time in experimental populations of wheat planted in three contrasting environments in France have shown rapid evolutionary change in flowering time during the first seven generations (Rhone et al. 2008). Eighty percent of the observed genetic variation is due to a single gene, *VRN-1*. That gene affects vernalization, the process by which plants use the onset and duration of cold temperatures to time the onset of flowering.

Lande (1983), however, argued some decades ago that evolution in species subject to artificial selection may not mimic nature. Fluctuations in natural environments may not be as extreme as those that occur under artificial selection on domesticated species or on artificially disturbed populations. The reason is that previous adaptation and coevolution among species will already have created an underlying basis of adaptation to that environment.

But environments are always changing, affecting the roles of major and minor genes during selection in subsequent generations. Selection could act at any stage of adaptation to alter the genetic basis of adaptive traits. Arnold et al. (2008) have argued that the central problem to understand is how populations evolve and diverge across adaptive landscapes as selection shapes patterns of inheritance among multiple phenotypic traits (i.e., the G-matrix of population genetics). That is, there is continuing interplay between the genetics of evolution and the evolution of genetics.

THE MULTIPLE GENETIC MODELS OF EVOLUTION

It is worth exploring this process of genetic change in more detail. Even though Fisher was the major force in the development of the infinitesimal model of evolution, he also explored a geometric model of adaptation that allowed random mutations to differ in the sizes of their phenotypic effects (Fisher 1930). He allowed each mutation to have multiple phenotypic effects (pleiotropy), some of which may be beneficial and some detrimental. When he analyzed the probability that a mutation of a particular phenotypic size would be beneficial, he found that an infinitesimal mutation had a 50 percent chance of being beneficial but that mutations of larger size rapidly approached a zero percent chance. He therefore concluded that natural selection shapes adaptation through many small genetic changes.

That conclusion, however, has undergone considerable modification since then. Motoo Kimura (1983) showed that, although mutations of small phenotypic size are more likely to be beneficial, they are also more likely to be lost soon after they appear while they are still rare, because they are under only weak selection. He concluded that adaptation is driven more often by selection on mutations of intermediate size. Some studies of experimental evolution have provided support for that view (e.g., Barrett et al. 2006).

Kimura's conclusion underwent yet more modification when Allen Orr (1998, 1999) realized that Kimura had focused on the distribution of mutation sizes at a particular stage in adaptation. The real problem, though, is to understand the distribution of mutation sizes across all the stages of an adaptive change. When analyzed in that way, adaptation is most likely to involve a combination of mutations of different phenotypic size, with larger mutations more common at early stages and smaller mutations more common later in the process. The closer a population approaches its optimal adaptive state, the greater the chance that a large mutation will have a detrimental effect. In Orr's (2005) words, adaptation is "characterized by a pattern of diminishing returns."

Another way of approaching the problem is to ask what kinds of genetic architecture are most likely to allow a population to evolve fast enough in a novel environment to begin growing. When first introduced into a new environment, a population may experience natural selection so strong that individuals are eliminated from the population faster than they can be replaced, driving the population down to unsustainable levels and eventual extinction. This situation could happen, for example, when a population is introduced into a very different physical landscape or when it is confronted with a newly introduced enemy such as a parasite. The population will persist only if it evolves fast enough to overcome the rate of local population decline (Gomulkiewicz and Holt 1995; Orr and Unckless 2008; Gomulkiewicz and Houle 2009; Holt 2009).

Some theoretical results suggest that making it through that population bottleneck must often involve selection on a combination of major genes and minor genes. By exploring multiple forms of genetic inheritance, Richard Gomulkiewicz and colleagues (2010) found that if many genes with small phenotypic effects each contribute to fitness, then mean population fitness can increase and populations can at least begin to expand in a new environment. The increase, however, is often too small and slow to prevent extinction. At the other extreme, some models suggest that reliance on current genetic variation and new mutations at one or a few loci will also often be ineffective at slowing population decline fast enough to prevent extinction, unless population sizes are initially very large or population declines are small (Orr and Unckless 2008).

Current models therefore suggest that the evolutionary solution is between these two extremes: a population in a new environment is more likely to persist if selection acts on major genes that are modified by minor genes (Gomulkiewicz et al. 2010). This conclusion applies to the special problem of evolution in a declining population, following, for example, invasion of a new habitat or attack by a newly introduced invader that increases deaths. The models explore the genetic architecture most likely to allow the time needed to attain a mean fitness high enough for a population to grow. At that point, selection will still continue, because the population will likely still be far from its final state of genetic equilibrium in the new environment.

These models ignore the effects of other genetic effects such as dominance and epistasis and other evolutionary processes such as random genetic drift, but the simulations suggest that the results are robust under a wide range of conditions (Gomulkiewicz et al. 2010). These and other models therefore suggest that successful adaptation to significantly changed environments may often involve a moderate number of genes of varying effect rather than many genes of small effect (Orr 2005; Bell 2008, 2010).

GENOME DUPLICATION AND PULSED EVOLUTION

Part of the problem of evolving from one adaptive peak to another is achieving some kind of coordinated evolution among the multiple genes and traits under selection. Doubling the entire genome (polyploidy) is one solution that appears to be part of the evolutionary history of many taxa, and there is a growing appreciation of the role of whole genome duplication in evolution. It can produce large, but coordinated, effects on many phenotypic traits, potentially leading to pulses of rapid evolutionary change across adaptive landscapes. In some cases the doubling of chromosome numbers occurs within species (autopolyploidy) and in other cases through hybridization with other species (allopolyploidy) (Abbott and Lowe 2004; Soltis et al. 2004, 2007). Polyploidy is common in some taxa and rare in others, but many lineages of plants, ferns, mosses, fungi, invertebrates, and vertebrates have polyploidy in their ancestry.

Whole genome duplication has been implicated in the early diversification of seed plants and the early diversification of angiosperms (Jiao et al. 2011). Major genome duplication events took place about 319 million years ago before seed plants diversified, and then again 192 million years ago before angiosperms diversified. These duplication events resulted in the diversification of regulatory genes involved in flower and seed development. These are processes at the core of the ecological diversification of plants.

Whole genome duplication has continued to occur during the radiation of angiosperms. About 35 percent of vascular plant genera include one or more polyploid species, and, by one estimate, 15 percent of angiosperm speciation events and 31 percent of fern speciation events have been accompanied by an increase in ploidy (Wood et al. 2009). These estimates have been revised upward in recent years, as more species have been studied using cytogenetic and molecular tools. Even these estimates, however, may underestimate the rates of polyploidization, because they do not include instances in which multiple tetraploid populations have been formed from the same diploid populations, as is known to occur (Soltis and Soltis 1999). These estimates also do not include unnamed cryptic polyploid species that remain embedded within named species. Such cryptic species appear to be more common than previously suspected, at least in plants (Soltis et al. 2007).

The reason that whole genome duplication is potentially so important as a driver of evolutionary change is that it can affect a broad range of important phenotypic traits. Examples include the types of habitats used by plant populations (Lumaret et al. 1987; Thompson and Lumaret 1992), interactions between bacteria and fungi (McBride et al. 2008), resistance of snails to parasites (Osnas and Lively 2006), survival of crustaceans under various physical conditions (Jose and Dufresne 2010), and the mating calls of tree frogs (Keller and Gexhardt 2001). The list of phenotypic traits affected by polyploidy continues to grow, but formal analyses of the effects of polyploidy on evolutionary rates and dynamics at short timescales are still few.

The potential for rapid evolution following genome duplication is high, because it could lead rapidly to the establishment of new populations and further genetic or chromosomal change within the genome (Oswald and Nuismer 2007; Leitch and Leitch 2008; Lim et al. 2008). In some ways, it is not a surprise that polyploidization should have the potential to be a powerful form of major adaptive evolutionary change. Humans have been selecting for polyploid plants for thousands of years during the process of artificial selection on crop species. Polyploid individuals often have traits beneficial to farmers such as larger seed size, higher oil content, or greater resistance to pathogens (Salamini et al. 2002; Matsuoka 2011; Sicard and Legras 2011; X. Wang et al. 2011). Many agricultural plants are polyploid species, including alfalfa, wheat, oats, sugar cane, coffee, potatoes, cotton, peanuts, brassicas, and strawberries among others. All crop plants probably have polyploidization as part of their ancestry either before or after domestication (Udall and Wendel 2006). Polyploidization has also been employed in attempts to achieve higher artemisinin content from *Artemisia annua* plants, for use in malaria control programs (Lin et al. 2011).

Through studies over the past two decades, we now know that polyploidy can have a large effect on ecological interactions with other species in nature. Most studies have been of plant species (Thompson et al. 1997; Husband and Schemske 2000; Kennedy et al. 2006; Arvanitis et al. 2007, 2010; Münzbergov· 2007; Halverson et al. 2008; McBride et al. 2008). The species that has been studied in the greatest detail for these ecological effects is the perennial herb *Heuchera grossulariifolia* (Saxifragaceae), which occurs in the northern Rocky Mountains of the United States as diploid and tetraploid populations (Thompson et al. 1997, 2004; Nuismer and Ridenhour 2008; Thompson and Merg 2008). The tetraploid populations arose through a doubling of the complete chromosome sets found in ancestral populations (autotetraploidy) (Wolf et al. 1989). Moreover, the doubling occurred independently at least three times (Segraves et al. 1999), resulting in tetraploids of different evolutionary origin in different river systems. The tetraploid populations usually occur near diploid populations, and local populations of mixed polyploidy occur in the rocky canyons of some rivers. The current complex distribution of diploid and tetraploid populations is probably a result of Pleistocene events that separated populations in the various deep river systems of these mountains.

Where *H. grossulariifolia* populations of different ploidy level overlap, they differ in their insect herbivores and pollinators. The three major herbivores are all moth species. The flower-feeding *Greya politella* moths and the flower-and-stem-feeding *Eupithecia misturata* moths attack mostly tetraploid plants, whereas the stem-boring *Greya piperella* moths attack mostly diploid plants (Thompson et al. 1997; Nuismer and Thompson 2001; Janz and Thompson 2002). The major pollinators include several bee species and *Greya politella* moths (Segraves and Thompson 1999; Thompson and Merg 2008). Among the bees, *Lasioglossum* species preferentially visit diploid plants at one of the sites but not at the other (fig. 4.1). Queens of the bumblebee species *Bombus centralis* preferentially visit tetraploids at both sites, and workers of that species differ between sites in the cytotype they prefer. These insect species readily distinguish between neighboring plants of different ploidy as they approach the flowers.

Populations differ in which pollinators are most responsible for seed set. Taking into account frequency of visits, number of visits, and efficiency

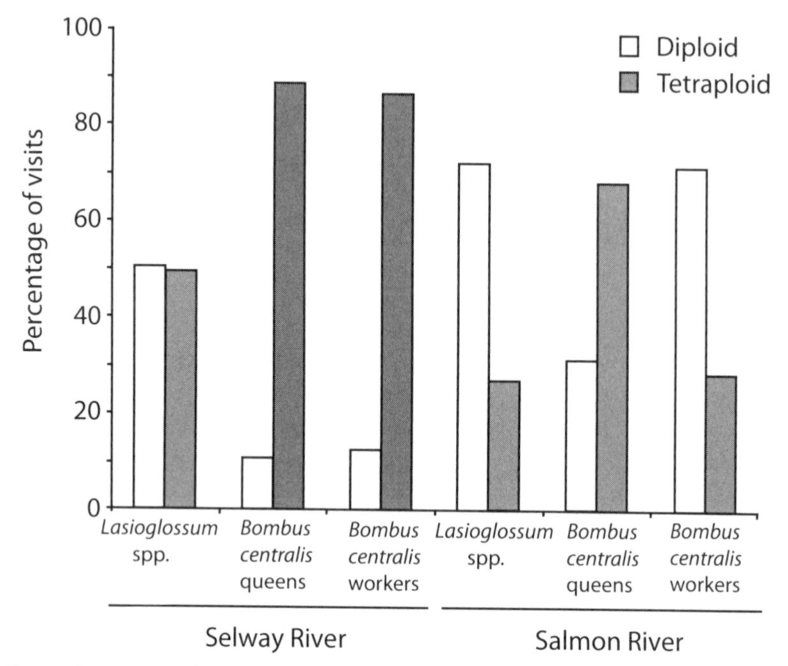

Fig. 4.1 Percentage of visits by bees that are the major floral visitors to diploid and autotetraploid *Heuchera grossulariifolia* plants along the Selway and Salmon rivers, Idaho. Values combine data from four years of study. Data from Thompson and Merg (2008).

of pollination during each visit, we find that bumblebee queens of *Bombus centralis* are responsible for two-thirds or more of the seed set of tetraploid plants on both the Selway River and the Salmon River. The smaller *Lasioglossum* bees are the major pollinators of diploid plants on the Selway River, and bumblebee workers are the major pollinators of diploid plants on the Salmon River (Thompson and Merg 2008). Hence, chromosomal doubling has affected the pollinators and the herbivores attracted to *H. grossulariifolia* plants, and the effects of these insects on plant fitness.

It is likely that differences in morphology, chemistry, and flowering times all contribute to the differences in how diploid and tetraploid plants interact with insects. The plant populations differ in morphology (Segraves and Thompson 1999), and current selection favors adaptive divergence of the flowering times of diploids and tetraploids (Nuismer and Cunningham 2005; Nuismer and Ridenhour 2008). Selection is driven in part by the low seed set in hybrid offspring, which are usually sterile triploids.

Experimental studies that have artificially induced polyploidization in *H. grossulariifolia* suggest that some of the differences between cytotypes may have occurred immediately after polyploidization. When diploid seedlings are treated with colchicine to induce polyploid formation, the resulting tetraploid plants differ in multiple phenotypic traits from the untreated diploid plants (Oswald and Nuismer 2011). It is unknown whether colchicine-induced seedlings produce polyploid phenotypes similar to those formed naturally, but these experiments suggest that tetraploids may have begun adaptive divergence from their diploid parents soon after polyploidization.

HYBRIDIZATION AND HYBRID ZONES

Hybridization is the other whole genome process that has the potential to foster rapid evolutionary change by altering the genetics of evolving traits. How hybridization affects the dynamics of adaptation in populations depends on how the fitness of hybrids compares with the fitness of parental species in each environment in which hybrids occur. Where the fitness of hybrids is less than that of their parental species, selection acts against them and hybrid populations are not self-sustaining. In these cases, hybrids persist in a region where two species overlap through a balance between two factors: dispersal of individuals of the two parental species into a habitat, and the intensity of selection against hybrids within that habitat (Endler 1977; Barton and Hewitt 1985; Harrison 1993; Price 2008). In contrast, where hybrids are locally more fit than their parents, hybrids may persist locally as relatively stable populations once they are established.

In either case, the result is a strong interplay between evolutionary and ecological processes because selection acts on survival and reproduction of hybrids and the parental species. In between these two extremes are habitats in which hybrids have neither higher nor lower fitness than one or both parental species. These situations allow ready diffusion of genes from one species to the other (i.e., introgression) and contribute to the continuing evolution of populations (Endler 1977).

We now know that hybridization and introgression are common in many taxa. Hybrids between species have been reported within all the major lineages of life, and they seem especially common in plants, fungi, and insects (Schardl and Craven 2003; Soltis and Soltis 2009). There may, though, be some bias in our perception that these groups are particularly prone to hybridization. These are the most species-rich taxa, and they therefore offer more opportunities for hybridization than do less diverse taxa such as vertebrates. Even so, plants seem especially prone to form hybrid populations. By some estimates, up to 25 percent of plant species form hybrids, in contrast to 16 percent of butterfly species and between 6 and 9 percent of mammal and bird species (Mallet 2005, 2008).

In some cases, hybrids are common enough to form identifiable hybrid zones. Each hybrid zone may spread out over many kilometers or only tens of meters (Mullen et al. 2008; Ruegg 2008; Hermansen et al. 2011; Scriber 2011). At local scales, hybrid zones can appear as miniatures of the patterns observed over larger scales, sometimes forming patches of hybrids intermixed with parental populations and sometimes forming smooth clines of introgressed genes (Mallet et al. 2009; Ross and Harrison 2002; Harrison 2010). Where hybrids are as fit as either parent, genes will readily diffuse between the species, often creating wide hybrid zones. Where hybrids are less fit than either parental species, the width of hybrid zones depends on the balance between dispersal of individuals of each species into the hybrid zone and selection against hybrids. Where hybrids are more fit than the parental species, the width of the hybrid zone will depend on the width of that environment (Endler 1977; Barton and Hewitt 1985; Price 2008).

The mobility of individuals sometimes provides surprisingly little information on the size and shape of hybrid zones. Steep clines can occur over fairly short distances even in a species that seems to be highly mobile, such as Swainson's thrush (fig. 4.2). Individuals from some populations of these birds migrate thousands of kilometers between their breeding sites and their overwintering sites, but populations of the coastal and inland subspecies remain separated during the breeding season except in a hybrid zone in southwestern British Columbia.

Swainson's thrushes are not unusual among vertebrates in forming hybrid zones. At least 200 hybrid zones have been reported in birds (Price 2008), and hybrids between species or highly divergent populations occur within many other vertebrate lineages (e.g., Chatfield et al. 2010; Duvaux et al. 2011; Carson et al. 2012). On large continents, these and other hybrid populations often cluster into regions that are major geological transition zones, sometimes associated with Pleistocene glacial processes (Swenson and Howard 2005; Hewitt 2011). On smaller land masses, hybrid zones are less likely to be clustered into regions, but they are common. Within New Zealand, for example, molecular analyses have documented hybridization among nineteen native species, including birds, fish, insects, and plants, and hybrids also occur among introduced taxa, which now make up a large proportion of New Zealand species (Morgan-Richards et al. 2009). Hybrid zones may play an increasingly important role in the ongoing dynamics of evolution worldwide, as we continue to move species among continents and oceans.

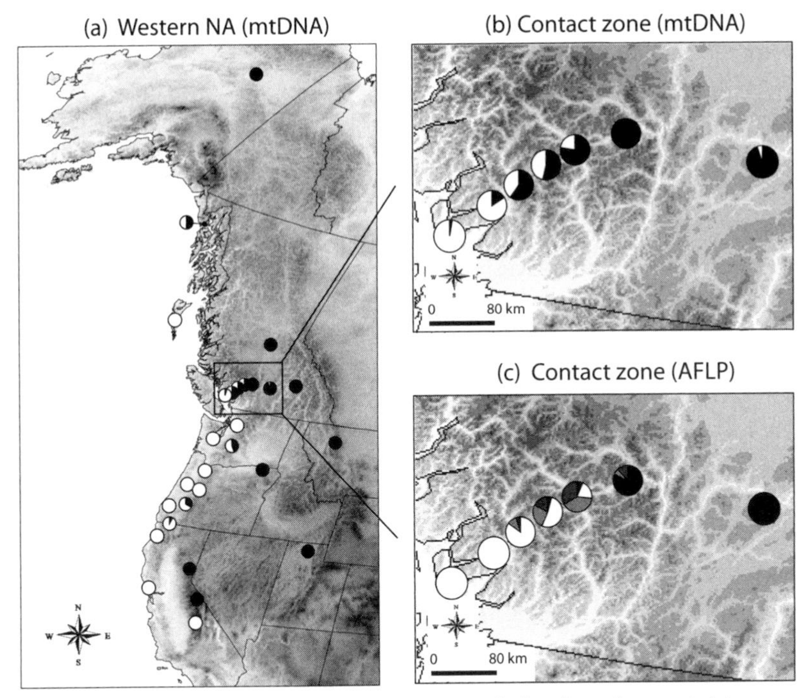

Fig. 4.2 Hybrid zone between subspecies of Swainson's thrush, *Catharus ustulatus*, (*a*) across far western North America, and (*b*) and (*c*) within the contact zone. Panels (*b*) and (*c*) indicate the hybrid zone using two methods: analysis of mitochondrial DNA (mtDNA) and genome-wide analysis of amplified fragment length polymorphisms (AFLP). Pie diagrams indicate the proportion of individuals belonging to the coastal (*white*) or inland (*black*) populations. Hybrid individuals are indicated by light gray and unassignable individuals by dark gray. Horizontal scale represents 80 km. After Ruegg (2008).

There have been fewer studies of hybrid zones in the oceans than in terrestrial or freshwater environments, but evidence for hybridization in marine environments is accumulating. Examples include reef corals (Willis et al. 2006), fish (Ouanes et al. 2011), and mussels (Shields et al. 2010). Eleven coral reef fish species have been reported to hybridize at Christmas and Cocos Islands in the eastern Indian Ocean (Hobbs et al. 2009). In some marine lineages, hybridization is limited to relatively young species, and the introgression of genes from one species to the other is sometimes asymmetric (e.g., Addison and Pogson 2009). In general, though, as more species are studied using molecular tools, it increasingly seems that hybridization has been underestimated in oceanic environments as well as in other environments.

SHIFTING HYBRID ZONES

Once formed, the genetic composition of populations within hybrid zones can continue to evolve rapidly in local communities and across large geographic scales. A hybrid with relatively high fitness in one environment may have relatively low fitness in another environment (Campbell and Waser 2007; Grant and Grant 2008a, 2008b), because genes are expressed in different ways in different environments, and gene combinations that fare well in one environment can fare poorly in another environment. As environments change, so does local selection favoring or disfavoring hybrids.

During a ten-year study of *Sinularia* soft corals in the Pacific Ocean near Guam, hybrids initially covered less than 15 percent of the habitat but eventually increased to cover more than 30 percent (fig. 4.3). During that time, the parental species became almost locally extinct. This change appears to have been due to multiple causes, including sedimentation, competition between hybrid and parents through release of deterrent chemicals, and differences in predation, which was lower on hybrids than on the parental species (Slattery et al. 2008). This kind of rapid evolutionary change in hybrid zones is probably common, but few studies have tracked the relative proportions of hybrids and their parental species for a decade or more and simultaneously tracked the potential ecological causes of differential selection on the genotypes.

Changes in hybrid zones have now been observed in a wide range of taxa. In a review of studies of shifting hybrid zones, Buggs (2007) found twenty-three studies showing observational evidence that hybrid zones have moved over time. Sixteen other studies using molecular markers suggested that the sites of introgression between species have shifted over the past century. These studies include hybrid zones between species of plants, crickets, katydids, ticks, ants, water striders, butterflies, crayfish, fish, salamanders, lizards, birds, and mammals. They include distances of less than 1 km to more than 200 km. In California, for example, introduced barred tiger salamanders (*Ambystoma tigrinum*) have hybridized with threatened native California tiger salamanders (*A. californiense*) over the past sixty years. Some, but not all, genes of the introduced species have introgressed 90 km into the range of the native species during that time (Kays et al. 2010). The introgression of only some genes into the native species over this short length of time suggests that natural selection is driving this evolutionary change.

As they move across landscapes, hybrid zones have the potential to change adaptive evolution in the parental species, if ecologically important genes from one species introgress into some populations of the other spe-

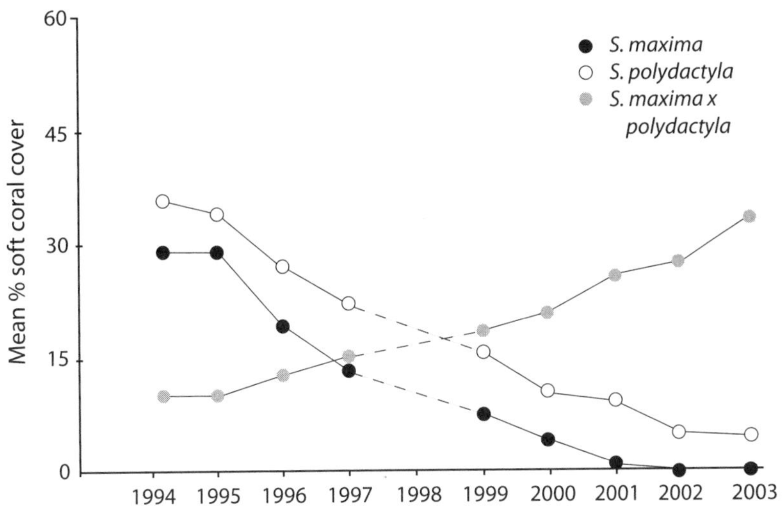

Fig. 4.3 Change over a decade in the mean percentage of sea floor covered by two species of *Sinularia* soft coral species and hybrids between these species in the Pacific Ocean near Guam. No samples were collected in 1998. After Slattery et al. (2008).

cies (Grant et al. 2004, 2010b; Mallet et al. 2007). As long as hybrids are not completely sterile or infertile, there is the potential for some of them to mate with individuals from the parental species. These matings allow diffusion of genes between the species and formation of novel adaptive gene combinations.

Evolutionary change through introgressive hybridization has been proposed for an increasingly wide range of taxa as the techniques for detecting introgression have improved. Most studies are based on a correlation between recent introgression and recent ecological changes in species, which then sets the stage for more detailed studies. For example, as coyotes have rapidly spread from western North America into eastern North America over the past century, they have hybridized in a few places with wolves, based on observed changes in their mitochondrial DNA and skull morphology. One potential explanation for the tendency of some eastern coyotes to attack deer in addition to smaller prey is that hybridization with wolves resulted in larger coyotes capable of attacking deer (Kays et al. 2010).

HYBRID ZONES AND SPECIES INTERACTIONS

Hybridization and introgression also have the potential to shape the evolution of mutualistic interactions among species by producing novel combinations of adaptive traits. *Heliconius* butterflies have become one of the major models for these studies. These brightly colored neotropical butterflies are distasteful to predators and advertise that distastefulness with bright patterns on their wings. Within a geographic locality, two or more *Heliconius* species often converge on the same wing colors and patterns. Across Central and South America, groups of *Heliconius* species have converged on different colors and patterns in different places. More than 25 percent of *Heliconius* species are known to form hybrids with other species (Mallet et al. 2007). Lawrence Gilbert (2003) has suggested that the geographically varying color patterns found in the mimicry rings of *Heliconius* butterflies have resulted from introgression of color-pattern genes across narrow hybrid zones.

Two of the most widespread species, *H. erato* and *H. melpomene* co-occur throughout this huge geographic area and have repeatedly diverged together in their color patterns from other populations. *Heliconius erato* arose in western South America and *H. melpomene* arose in eastern South America (Quek et al. 2010). As these species expanded and came into contact, each major color pattern arose repeatedly. Molecular analyses of divergence among *Heliconius* species and populations have shown that these species have had complex histories of divergence and hybridization among races rather than simple patterns of expansion and contact after the Pleistocene (Dasmahapatra et al. 2010; Quek et al. 2010). Hence, hybridization could have had a direct role in the formation of new color patterns in local mimicry complexes, as *Heliconius* species have spread over the past several million years. As molecular results continue to inform the historical patterns of contact among these species and populations, these butterflies remain one of the most useful groups for studying how hybrid zones affect adaptive evolution in geographically diverging populations.

Hybrid zones can also generate new assemblages of interacting species that impose different selection pressures on the hybrid and parental populations. Among the clearest and most detailed are the long-term studies by Thomas Whitham and colleagues, who have studied natural selection across hybrid zones of cottonwoods (*Populus*) in the southwestern United States and in the herbivores of these plants. Two cottonwood species, narrowleaf cottonwood (*P. angustifolia*) and Fremont's cottonwood (*P. fremontii*), occur along elevational gradients in this region, with hybrids between the species occurring in some habitats at intermediate elevations. Where hybrids occur, backcrosses between them and the parental species also occur. In one intensively studied thirteen-kilometer hybrid zone along the Weber River near Ogden, Utah, the hybrids backcross only to narrowleaf cottonwoods. There is therefore a complex genetic structure to these cottonwood species and the hybrid zones between them.

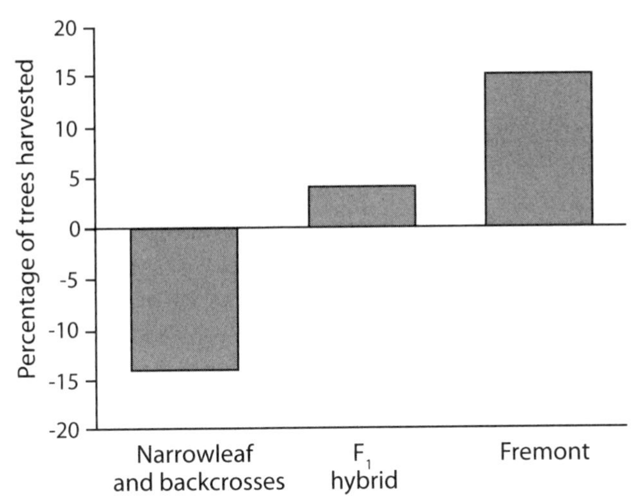

Fig. 4.4 Selective attack by beavers among narrowleaf cottonwoods (*P. angustifolia*), Fremont cottonwoods (*P. fremontii*), and F1 hybrids between these species. Attack on narrowleaf cottonwoods includes hybrid plants that have backcrossed to narrowleaf cottonwoods. Values show the extent of preference for each plant genotype relative to its local availability. Values above the horizontal line are greater than expected; values below the line are less than expected. After Bailey et al. (2004).

The animal and fungal community associated with these tree populations follows directly from the effects of hybridization. The effects travel up through the population, community, and ecosystem levels (Whitham et al. 2006; Schweitzer et al. 2008). At the population level, mite populations on cottonwoods are distributed nonrandomly, feeding on narrowleaf cottonwoods and on hybrids but not on populations of pure Fremont cottonwoods (Whitham et al. 1999). Additionally, mite populations on the hybrid plants are genetically distinct from those on narrowleaf cottonwoods and fare better on those plants when mites are experimentally reciprocally transferred among these hosts (Evans et al. 2008). Molecular and ecological evidence suggests that the mites on the parental and hybrid plant populations are two cryptic species rather than one species (Evans et al. 2008). Beavers also discriminate among these cottonwood populations. They tend to avoid narrowleaf cottonwoods (*Populus angustifolia*) and backcross hybrids, which have relatively high tannin levels in comparison to Fremont cottonwoods (fig. 4.4). Through their selective harvesting of trees with low tannin, beavers can alter the composition of cottonwoods in an area, shifting the stand composition to narrowleaf and backcross trees (Bailey et al. 2004; Whitham et al. 2006).

At the community level, the web of life surrounding cottonwoods differs on hybrids and their parents. Hybridization affects interactions not only with mites and beavers but also with the broader community of arthropods associated with these trees. It even affects the pattern of infection by endophytic fungi, interactions between leaf galls and birds, rates of decomposition in the surrounding litter, and rates of mineralization of nitrogen (Schweitzer et al. 2008). At even a finer scale, individual cottonwood genotypes differ in their interactions with other species and in their effects on ecosystem processes. Collectively these effects generate a strong influence of plant genotype on population, community, and ecosystem processes (Keith et al. 2010; Smith et al. 2011).

Studies on how hybrid individuals interact with other species show that each genome forms a unique web of interactions with other species (Johnson 2008; Mooney and Agrawal 2008; Schaedler et al. 2010; Cook-Patton et al. 2011; Genung et al. 2012). These genomic effects are most obvious when the cases involve hybridization or polyploidy, but they occur in all populations that have been studied in detail. In some cases the differences can be due to single genes, but in other cases they are due to the combined effects of many genes. Each individual or clone attracts its own subset of antagonistic and mutualistic interactions from the larger pool of species that interact with that species, guaranteeing that a population is constantly tested with multiple forms of natural selection.

THE CHALLENGES AHEAD

It has been common to view adaptation involving hybridization, genome duplication, or major genes as mechanisms of evolutionary change that apply to stressful environments, novel environments, or colonizing populations—that is, evolution under conditions that are, in some way, unusual. It is based on the view that these are mechanisms that can generate an evolutionary response to a major qualitative change in the environment (e.g., introduction of a new enemy) rather than a smaller quantitative change (e.g., a few degrees change in mean temperature). The smaller changes can be accomplished readily through weak selection on quantitative traits. Increasingly, though, selection acting on major genetic and genomic differences among individuals seems to occur in many environments. Whether major events in adaptive evolution occur more commonly by these means, rather than by accumulation of small changes over many genetic loci, remains unresolved.

If we are to solve how the genetics of evolution depend in part on the environments in which selection is acting, we require an understanding of two fundamental aspects of adaptive landscapes that remain elusive simply because they are difficult to analyze in natural populations. We need better knowledge of the ruggedness of adaptive landscapes (Orr 2009). How deep are the adaptive valleys between adaptive peaks, how different are the heights of the peaks above those valleys, and how are adaptive ridges distributed among the peaks? Do adaptive landscapes form adaptive cordilleras (Armbruster 1990) or do they favor plateaus full of adaptive holes (Gavrilets 1997)? How do adaptive landscapes appear in nature when fitness is controlled by many traits, rather than a few?

We also need to decipher how the shapes of the adaptive landscapes change during different kinds of environmental change. It is one thing to think about qualitative or quantitative environmental change. It is another to translate that simple dichotomy into an understanding of how different forms of environmental change alter adaptive landscapes. If enemies are rapidly evolving in response to evolutionary changes in a host population, then the adaptive landscapes for the host populations are themselves constantly being reconfigured. The adaptive landscape may look more like turbulent water than a mountain range.

Evolutionary changes brought about by polyploidy or hybridization can introduce new genetic variation into a population quickly, thereby altering fitnesses, evolutionary rates, and ecological dynamics. Introgressive hybridization has even been proposed as a mechanism to help rescue genetically impoverished species that may have little ability to respond to environmental change, and there is a growing discussion on when and where such management procedures are appropriate (Araguas et al. 2009; Baskett and Gomulkiewicz 2011). It is part of the growing broader discussion of how to use our understanding of rapid evolutionary dynamics as a tool in the conservation of species and the web of life. We now turn to how the web of life itself provides major mechanisms for evolutionary change.

5

Coevolving Genomes

Soon after life began, the genetics of species evolving in isolation from each other quickly ceased even as a possibility. The genomes of most species living today are insufficient to allow individuals to survive and reproduce in real ecosystems. The vast majority of living organisms have evolved in ways that require them to use a combination of their own genetic machinery and that of one or more other species. The interactions between species are therefore not something added onto species. They are not rococo embellishments to the evolution of life.

Instead, interactions are at the core of the adaptation of species and the diversification of life. They are part of what makes each species unique, and they seem to be the primary reason that the earth is inhabited by tens of millions of species rather than by just a few. Much of evolution is about selection that constantly rebalances the use of genes inherited from parents and the use of genes co-opted each generation from other species. The interplay of inherited genes and co-opted genomes appears to be the fundamental evolutionary engine of continually changing biodiversity.

In this chapter I explore our increasing appreciation that much of evolution is about the coevolution of disparate genomes.

BACKGROUND: WHY HAS IT TAKEN SO LONG?

When we think about evolutionary rates, we often begin with a classical population genetic perspective: mutation, selection, random genetic drift, and gene flow—all acting directly on the core genome that individuals inherit from their parents. Evolutionary rates, though, are just as much about the pace at which interactions among species evolve as they are about the rates of genome evolution of each species. The web of life itself is just as much a product of evolution as are the component species. The combined rate of evolution of species and their interactions would therefore provide just as important a measure of the rate of evolution as the rate of genome evolution in any individual species, because the genome of each species include only some of the genes needed for survival and reproduction. It is the sum total of the evolving genes needed by a species that matters.

Interactions among species have become so important in evolution because they provide opportunities well beyond those encoded in the genes of a species. There are millions of species on earth, but they are involved in tens or hundreds of millions of interactions in what Darwin called the entangled bank of life. Each species is a highly coordinated set of genomes that can be exploited by other species. When Darwin described the entangled bank, he understood clearly that the history of the earth was the history of manipulation and defense, creating interactions ranging from antagonistic to mutualistic. Darwin drew many of his examples of adaptation from traits that have evolved through interactions with other species.

In the decades immediately following publication of *The Origin of Species*, most tests of the theory of natural selection were on how selection has favored the manipulation of other species or defenses against other species. Within three years of the publication of *The Origin*, Henry Bates (1862) was interpreting the similar patterns of bright colors found among coexisting unrelated species, such as some tropical butterflies, as products of natural selection. He argued that selection has favored individuals of palatable species that mimic unpalatable species. We now take Batesian mimicry for granted, but it was a remarkably insightful early application of the theory of natural of selection. It implies that selection can favor similarity in two species that never interact directly. They interact only indirectly through a third species, in this case a predator that avoids individuals of the palatable species because they closely resemble the warningly colored individuals of the unpalatable species.

Fritz Müller (1879), who was one of Darwin's favorite correspondents, extended the idea to cases in which convergence of color patterns occurs in two or more unpalatable species. Natural selection favors a shared warning signal that is recognized by predators. Müller's studies produced detailed observations on how natural selection has favored species that manipulate or otherwise use the traits of other species (West 2003). Together the evolution of Batesian and Müllerian mimicry became, as R. A. Fisher (1930), wrote, "the greatest post-Darwinian application of Natural Selection, [playing] an especially important role towards the end of the nineteenth century and the beginning of the twentieth century." Meanwhile, Fritz Müller's brother Hermann explored similar questions on the ways in which animals and plants manipulate each other through his studies on how natural selection shapes coevolution between plants and pollinators. Building on

Darwin's (1862) book on orchid pollination, Müller's studies appeared in a series of papers in the journal *Nature* (e.g., Müller 1873), culminating in a large book called *The Fertilization of Flowers* (Müller 1883).

These studies of how evolution often proceeds through selection acting on species to manipulate each other came almost to a complete stop in the early twentieth century. With the growing appreciation of Gregor Mendel's work on the genetic basis of traits, evolution became the study of genetics without ecology, and ecology became the study of species without genetics. By the time Charles Elton's *Animal Ecology* (1927) was published, the rift between ecology and evolution was so wide that he began his chapter on evolution by writing, "It may at first sight seem out of place to devote one chapter of a book on ecology to the subject of evolution." (See Thompson 1994 for an extended discussion of the history of the rift between evolution and ecology.) Almost a century later, we are still struggling to reassert the role of evolution through manipulation of other species as a major mechanism of evolutionary change. There are still some textbooks on evolution that do not include even a single chapter on the evolution of interactions between species, and some beginning biology texts hardly discuss the coevolutionary process. Fortunately, that has been changing quickly in recent years.

Gradually, during the second half of the twentieth century, biologists began again to appreciate the importance of coevolution. It started as multiple threads that gradually became intertwined (Thompson 1994). Harold Flor's studies (1942) showed that plants battled pathogens using gene against gene. David Lack's studies (e.g., Lack 1947) suggested how competitors may coevolve, and Paul Ehrlich and Peter Raven's important paper (1964) suggested how plants and insects may drive adaptive radiation in each other. Daniel Janzen (1966) showed, through innovative and detailed field studies, the remarkably intricate ways in which the disparate genomes of acacias and ants have coevolved to form mutualisms (fig. 5.1). In the 1980s, the importance of parasites and symbiotic mutualists as major drivers of evolution started to become more evident (Hamilton 1980; Price 1980; Thompson 1982; Smith and Douglas 1987). More recently, new techniques in microbiology have allowed more systematic evaluations of the pervasive roles of microbial species and their genomes in evolution.

As these studies have continued, two observations in particular have helped us realize the central importance of species interactions in evolution. One occurred through the development of molecular tools, which have shown us that microbial symbioses are pervasive throughout the web of life. The other observation came from the rise of evolutionary ecology, which has shown that fitness in most species depends importantly on interactions

Fig. 5.1 *Pseudomyrmex* ant foraging on an acacia that provides the ants with nectar from extrafloral nectaries, protein and lipids from Beltian bodies, and housing in its swollen thorns. These three plant structures, and the year-round production of leaves, have all been modified by natural selection to co-opt the behaviors of ants to defend the plants against herbivores and vines. Following the landmark studies of these interactions by Daniel Janzen (1966), other studies have indicated much variation in how interactions between acacias and ants have been modified among environments as the plants evolve in response to interactions with other species (Brody et al. 2010; González-Teuber and Heil 2010; Palmer et al. 2010; Stanton and Palmer 2011).

with other species. In recent decades, these results increasingly have been captured in discussions of evolution through phrases such as "the extended phenotype" (Dawkins 1982), "symbiosis as a source of evolutionary innovation" (Margulis and Fester 1991), "acquiring genomes" (Margulis and Sagan 2002), and "much of evolution is coevolution" (Thompson 2005). In reviewing studies of symbiotic interactions between animals and bacteria a little over a decade ago, Margaret McFall-Ngai (2002) concluded that our understanding is in its infancy and that the magnitude of the unknown is daunting. It is still daunting even now, but we are learning fast.

We are still discovering the many ways in which microbial species shape the evolution of animals, plants, fungi, and other microbial species. Exploring the genetic diversity of microbes has demanded work by consortia of researchers. These studies have shown us the great diversity of microbial interactions at the macro-ecosystem level of oceans, soil, and air, and at the micro-ecosystem level of the human gut and the myriad of symbioses found among most living species (e.g., Turnbaugh et al. 2007; Ehrlich and Nelson 2010). New results on symbioses are coming in so fast that major researchers are writing reviews that convey a sense of being almost overwhelmed. Nancy Moran (2007) has noted that studies of symbioses are now "vast and growing quickly." Angela Douglas, in *The Symbiotic Habit* (2010), has written that recent technical and conceptual developments are revolutionizing the field of symbiosis. Edward DeLong (2009) thinks that metagenomic data on microbial species are proliferating so fast that the functional and ecological data needed for the interpretation of how these genomes interact cannot keep pace.

The advances continue on all fronts—technical, analytical, and conceptual. The technical and analytical advances include not only methods for gathering metagenomic and proteomic data, but also statistical algorithms that allow deeper interpretation of the parts of genomes devoted directly to interactions with other species. Innovations in analyses of fossil data have fostered studies that have shown the maintenance of interactions among taxa over tens or even hundreds of millions of years and the role of interspecific interactions in driving adaptation and diversification (Aberhan et al. 2006; Jablonski 2008; Labandeira 2010; Vermeij and Grosberg 2010; Dietl and Flessa 2011). The conceptual ecological advances have included a change in coevolutionary research in recent years, shifting from approaches that are more typological—this species coevolves with that species—to approaches that are more populational—this species is a collection of populations that differ in how they coevolve with other species. Coevolutionary studies are showing that interacting genomes coevolve not only through slow and stately change over long periods of geologic time, but often also in highly dynamic ways that can quickly result in geographic mosaics of adaptation and counteradaptation among populations (Thompson 2005; Wade 2007; Gandon and Nuismer 2009). These developing approaches are indicating that interactions among species affect the pace and dynamics of evolution in all species.

THE GENETIC LANDSCAPE AS IT APPEARS NOW

We now know that symbiosis, living in intimate association in or on other species, is the most common lifestyle on earth (Price 1980; Moran 2002; Schmid-Hempel 2008; Douglas 2010). Symbiont species make up the majority of living species. Some of these symbionts are mutualistic with their hosts and others are parasitic, but each has a genome that requires the genetic machinery of a host species, and often the host requires some of the genetic machinery of the symbiont.

As life has diversified into three major groups—archaea, bacteria, and eukaryotes—the possibilities for coevolution among highly disparate genomes have also diversified. Although once grouped with bacteria, the archaea are highly distinct and have genomes that are a mosaic of genes found in bacteria and eukaryotes. Another way of saying this is that eukaryotes are a mosaic of genes found in bacteria and archaea (Yutin et al. 2008). Some researchers view these three groups as constituting three equally distinct domains of life, whereas others view archaea and eukaryotes as more closely related to each other than to bacteria (Gribaldo et al. 2010). By that latter view, there are two domains of life, Archaea and Bacteria, with eukaryotes arising from the archaea. It remains common in many studies, though, to contrast eukaryotes with prokaryotes (archaea and bacteria taken together). Some discussions in the upcoming chapters do so for convenience in discussing some major evolutionary differences in the structure and dynamics of ongoing evolution.

The origins of genes in these different groups have become increasingly important, because they provide clues as to how symbiotic interactions have originated, how genes have been transferred between symbionts and hosts, how often major events of horizontal transfer of genes have occurred among microbial species, and how, collectively, these processes have shaped adaptation and diversification of eukaryotes (Cavalier-Smith 2009; Koonin 2010). In the process of diversification, eukaryotes developed an endomembrane system, a cytoskeleton, and, crucially, a coevolved interaction with bacteria that eventually became organelles we now call mitochondria. The symbiosis with mitochondria was crucial to the diversification of eukaryotes, although exactly when and how that interaction developed in early eukaryotes remain unclear.

More generally, functional and ecological studies are showing in ever-greater detail the extent to which eukaryotic organisms rely upon symbionts (Moran 2007; Douglas 2010). A key part of the radiation of eukaryotes has involved choosing among the many possible prokaryotic symbionts or phylogenetically distant eukaryote symbionts, and then using those symbioses to interact with yet other species. There is a broad range of potential choices. Among insects alone, bacterial symbionts have been drawn from across the phylogenetic diversity of bacteria (Moran et al. 2008). Although

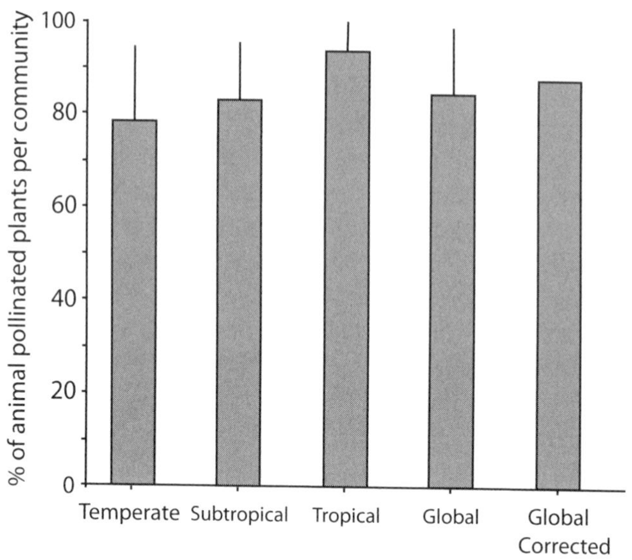

Fig. 5.2 Estimated percentage of plant species pollinated by animals in temperate, subtropical, and tropical habitats. Values are means ±1 SD. The global estimate combines the estimates for all habitats, and the corrected global mean value adjusts the estimate for differences among habitats in the percentage of plant species in each habitat type. After Ollerton et al. (2011).

many of these symbionts are mutualistic, others are parasitic under some or all environmental conditions.

Strip away the intracellular symbionts that most eukaryotes harbor—such as mitochondria for respiration or chloroplasts for photosynthesis—and most of life as we know it would simply cease to exist. These symbioses were central events in the diversification of life (Keeling 2010). Strip away the gut microflora and microfauna of most animals and they could not digest their food. Strip away the mutualistic associations called lichens, rhizobia, and mycorrhizae, and the structure of terrestrial communities would collapse. Do the same with the mutualistic associations between plants and pollinators, and most plants would cease to produce seeds (fig. 5.2). By current estimates, 88 percent of the earth's more than 350,000 plant species rely on animals for pollination (Ollerton et al. 2011), and most plants rely on multiple other mutualistic interactions (Bronstein et al. 2009). Whether the DNA sequence is a mitochondrion, a chloroplast, a gut bacterium, a mycorrhizal fungus, or a pollinator, these co-opted genomes are a fundamental part of the genomes of the species that use them.

Even the simplest forms of eukaryotes engage in complex interactions with other species. *Dictyostelium discoideum* amoebas usually feed as solitary predators of bacteria when food is abundant. When food becomes scarce, they aggregate to form multicellular dispersing bodies (Kessin 2001; Bourke 2011). Before dispersal, most of the cells differentiate into spores and fruiting bodies, but about 20 percent die to form a sterile stalk that aids dispersal of the multicellular body. As the cells aggregate, some of the spores carry bacteria, which they use to seed a new site and then harvest it (Brock et al. 2011). There is even variation within local populations in the tendency to farm. Some local populations include farmer and nonfarmer clones, favoring one or the other lifestyle depending on the local abundance of bacteria (Brock et al. 2011). Unlike in more sophisticated agricultural systems found in species such as attine ants, these amoebae do not actively cultivate the bacteria. They illustrate, however, that active co-opting of the genomes of other species occurs even among the simplest of eukaryotes.

Among more complex eukaryotes, host development and symbiotic interactions proceed hand in hand. Some invertebrate taxa pass symbionts between generations directly within or on the surface of eggs (Kaltenpoth et al. 2009; Kaiwa et al. 2010). Other species acquire symbionts anew each generation, often using complicated, multistep genetic signaling, which allows hosts to distinguish their normal mutualistic symbionts from other species (Simms and Taylor 2002; Ercolin and Reinhardt 2011). Once the relationship is established, hosts then must regulate these symbiont populations using an array of biochemical, physiological, and sometimes physical mechanisms. Meanwhile, the symbionts continue to evolve to manipulate hosts, guaranteeing that there is never an end to the evolution of these relationships.

Symbioses are so deeply embedded in the history of eukaryotes that species commonly have specialized structures to accommodate symbiont genomes. Reef-building corals, sponges, and some molluscs have evolved ways of harboring photosynthetic algae and cyanobacteria (Venn et al. 2008). *Nautilus*, whose history extends back more than 450 million years, have a specialized organ that houses a symbiont involved in host excretion, and yet other potential symbioses are still being discovered in these ancient cephalopods (Pernice et al. 2007).

Many insects have specialized tissues to harbor particular bacterial symbionts; plants form rhizobial nodules or mycorrhizae for their root symbionts; and fungi encapsulate their algal symbionts to form lichens (Allen et al. 2003; Moran et al. 2008; Douglas 2010). Leaf-cutting ants create gardens to cultivate the fungi on which they feed (Mueller et al. 2005), and some adult wood-boring insects carry in a special organ the fungi on which their larvae feed (Scott et al. 2008). Among vertebrates, half or more of the cells associated with the immune system may be devoted to the epithelial mucosa that interacts with the microbial assemblages in the digestive tract (McFall-Ngai 2011). The process of passing beneficial symbionts to offspring and other close relatives is such a central part of the ecology and evolution of many species that it may even have played a role in the evolution of multiple types of social behavior (Troyer 1984; Lombardo 2008).

Maintaining these symbiont genomes and the accompanying symbiont-housing tissues, organs, and structures demands processes encoded early in the developmental program of many eukaryotes, beginning at fertilization (McFall-Ngai 2002). If a developing individual is to acquire one of these symbionts from the surrounding environment, the tissues to harbor the symbiont population must often begin to form early in the development of the embryo. Later, an individual must discriminate between beneficial symbionts and potential pathogens, provide living conditions and nutrients that allow the symbiont population to grow but not too much, control invasion by other microbes, and sometimes transfer the symbionts to offspring directly by feeding them to the offspring.

How Hawaiian bobtail squid (*Euprymna scolopes*) maintain their symbionts has become a model for much research on symbiosis (McFall-Ngai 2008; McFall-Ngai et al. 2010). These squid cultivate their light-producing *Vibrio* bacterial symbionts in a special light organ that requires the presence of the bacteria to develop properly. The light decreases the shadow produced by the squid when foraging in moonlight, thereby making them less visible to predators. Juvenile squid acquire their bacteria from the surrounding environment by drawing water into the mantle cavity. Once inside, the bacteria become attached to mucous on the outer edge of the light organ and then migrate into the organ. The bacteria live in the epithelium of this organ, where they multiply and influence further development of the organ by mediating host cell death and the proliferation of epithelium within the organ.

By evening, the bacterial population has grown sufficiently to emit collectively a strong and continuous light, which is controlled by the squid using tissues that regulate the amount of light emitted. Each day at dawn, the squid bury themselves in sand and expel up to 95 percent of their light-producing *Vibrio* bacterial symbionts (Nyholm and McFall-Ngai 2004; McFall-Ngai 2008). The remaining cells then grow throughout the day in the protected, nutrient-rich environment of the specialized light organ in which they are housed. This active daily process captures at its most graphic the importance of co-opted whole genomes and the complexity of their maintenance in eukaryotes.

COPING WITH ASSEMBLAGES

Vertebrates and invertebrates may differ in the number of microbial taxa they harbor. This difference could reflect fundamental differences in the ways they cope with microbes or, to some extent, the simple fact that vertebrates are usually much larger than invertebrates and therefore provide a greater range of microbial niches. The current view is that vertebrates form large microbial assemblages specific to particular organs (fig. 5.3), whereas most invertebrates form mutualistic symbioses with smaller assemblages of microbial taxa and sometimes even with single microbial species to assist with a particular physiological need. Our views may be skewed, though, because there has been much more work on the microbial assemblages of vertebrates than invertebrates. Nevertheless, at the extremes, the problems of microbial assemblage management seem to differ for vertebrates and invertebrates.

The major problem for invertebrates that manage only a few symbionts is to recognize self from nonself, whereas the problem for vertebrates is to control diverse microbial assemblages (Cooper and Alder 2006). Adding to the complexity, each part of the vertebrate body harbors a microbial assemblage that includes some species found in other parts of the body and some species unique to that body part. In addition, all taxa must defend against a wide range of potential pathogens. Over 1,400 infectious species have been documented as pathogenic in humans alone (Taylor et al. 2001).

The innate immune system of invertebrates and the combined innate and adaptive immune system of vertebrates evolved to handle these problems (Rolff and Reynolds 2009; Iwasaki and Medzhitov 2010). Within vertebrates, the two systems interact (Slack et al. 2009). The innate system is effective at responding to the relatively predictable features of microbial species by producing generic receptors, whereas the adaptive immune system produces tailor-made lymphocytes specific to the infections that occur within a particular individual (Cooper and Alder 2006). The adaptive immune system, however, appears to be the more flexible at coping with a large microbial assemblage, and it may have been shaped as much by selection to control mutualistic symbionts as by selection to fight against pathogens (McFall-Ngai 2007; Lee and Mazmanian 2010).

With up to thousands of microbial species inhabiting the bodies of vertebrates, it is inevitable that some of these populations will evolve within individual hosts during the host's lifetime. Some estimates suggest that the

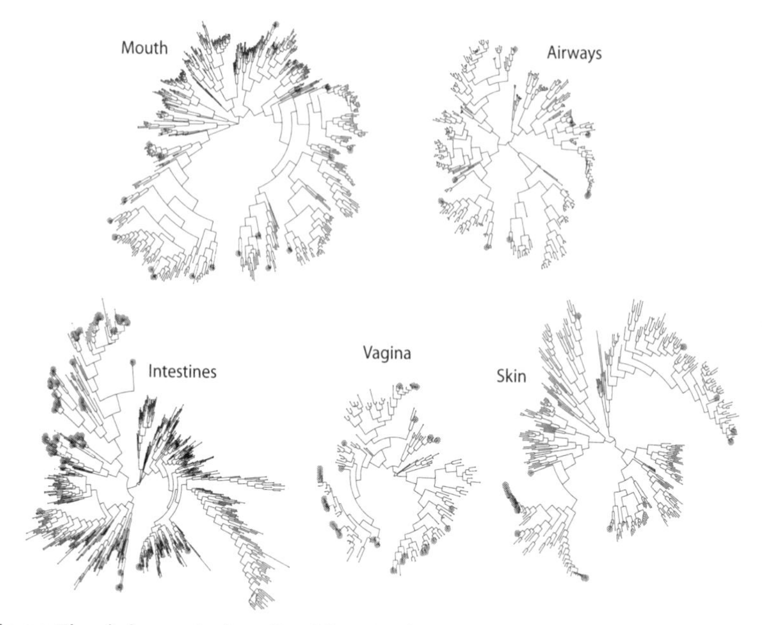

Fig. 5.3 The phylogenetic diversity of the microbiota inhabiting different parts of the human body. Gray circles indicate species whose sequences are known. After Lee and Mazmanian (2010).

number of genes in the microbiota of the human digestive system is more than a hundred times the number of genes found in the human genome, and those genes are distributed over hundreds of species that collectively include up to one hundred trillion cells (Bäckhed et al. 2005). Those numbers guarantee that mutation, selection, and evolution and coevolution are occurring within the microbial assemblages of every individual human during a lifetime.

The normal assemblage of microbes within vertebrates seems to have a core group of interacting symbiont species that undergo succession during development. Some of this successional process may be as much evolutionary as it is ecological, following changes in the host development. There are multiple hypotheses on how these interactions are maintained in ways that do not lead to a breakdown of this mutualism, including the maintenance of "controller" symbionts that regulate other symbionts (Thompson 2005). Regulation, though, could also work in more complicated ways. Some studies have suggested that the normal intestinal microbiota of humans form products, such a fatty acids, that may protect against inflammatory diseases by influencing immune responses. The vertebrate immune system may therefore rely on coevolved relationships with gut symbionts to function properly (Maslowski et al. 2009). Hence, a piece of the system is encoded in our nuclear DNA, and the rest may be in the genomes of our symbionts.

Understanding how coevolution regulates these and similar large assemblages of interacting species is one of the great current challenges in evolutionary biology. These appear to be organized assemblages. Reciprocal transplants of the gut microbiota of zebrafish and mice result in the development of a microbial assemblage whose composition reflects the host from which it came, but also reflects, in relative abundance, the recipient host (Rawls et al. 2006). Within zebrafish, domesticated populations harbor an assemblage of microbiota similar to wild populations (Roeselers et al. 2011), suggesting that the core of the microbial assemblage may function as a coevolved group of species. The formation of microbial assemblages within vertebrates must therefore involve a combination of coevolution among the microbial species and also coevolution between the microbial assemblage and the host population (Van den Abbeele et al. 2011). The host population provides the template for coevolution among the microbial species but also coevolves with the assemblage. The process undoubtedly results in some redundancy in which some microbial taxa with similar roles are interchangeable under different environmental and developmental conditions.

Given the complexity and ubiquity of mutualistic and parasitic interactions, it is not surprising that the genes involved with immune function are generally the fastest evolving genes within vertebrates and invertebrates. Favorable immunity genes in *Drosophila* have been estimated to evolve to fixation at twice the rate of other genes in the genome (Obbard et al. 2009). Genes involved in pathogen detection and immune response are among the regions of the human genome that show the strongest signal of adaptive evolution during human history (Williamson et al. 2007).

Long-lived plants are in some ways more like large vertebrates than small invertebrates in their need to control a diversity of continually evolving symbionts and pathogens (Whitham 1981; Petit and Hampe 2006). The roots harbor complex arrays of fungal and bacterial species involved in mycorrhizal and sometimes rhizobial associations (Heath 2010; Hoeksema 2010). The floral nectar of many species includes yeasts whose interactions with plants and plant reproduction have been mostly unexplored (Herrera et al. 2009), and the leaves and other vegetative tissues of many plants are packed with anywhere from a single fungal species to scores of species (Clay et al. 2005; Faeth 2009; Rudgers et al. 2009; Saikkonen et al. 2010). Trees and other long-lived woody plants must confront local populations of microbial and insect enemies capable of evolving within the lifetime of an individual tree.

The diverse array of fungal endophytes in many long-lived trees may even have evolved as a way for plants to defend themselves against insect herbivores and pathogens that can rapidly evolve adaptations to an individual host tree over hundreds of years (Arnold et al. 2003; Rodriguez et al. 2009). These endophytes, as well as mycorrhizae, are known to have strong effects on interactions between plants and insect herbivores. Sometimes the endophytes have negative effects on generalist herbivores but either no effect or positive effects on specialist herbivores (Hartley and Gange 2009). The hyphae of endophytes, in turn, have bacteria within them, although the effects of these bacteria on fungal and plant fitness are mostly unknown (Hoffman and Arnold 2010). These symbioses have the potential to shape the trajectories of evolution in plant populations both directly and indirectly, placing selection on plants to control their symbionts.

Similarly, fungi have diversified to a great degree through the symbiotic associations they have made with other eukaryotes and with bacteria. Fungal symbioses run the gamut from species that form highly specific mutualistic associations to more complex multispecific networks. Lichens are generally an association between a fungal genotype and an algal genotype, whereas mycorrhizal species often form complex networks in their interactions with plants. There seems to be no end to the variety of ways in which fungi have evolved to manipulate other species. Some fungi that use decomposing wood for growth live in the same habitats as termites, and some of these fungi have evolved forms called termite balls that mimic termite eggs in shape and have a chemical compound that the termites use to recognize their eggs (Matsuura et al. 2009). The termites tend the mimetic forms of these parasitic fungi as if the structures were termite eggs.

Mycorrhizal fungi, though, are the preeminent fungal symbionts (Allen et al. 2003; Bruns and Shefferson 2004: Hoeksema 2010). They formed associations with plants over 400 million years ago. The complex process of genetic handshaking used by these fungi and plants in establishing these symbiotic associations became the basis for the development of subsequent plant endosymbioses, such as those with rhizobial bacteria (Parniske 2008). Mycorrhizal fungi depend obligately on plant roots, and the majority of plants participate in these symbioses. That great diversity of arbuscular mycorrhizal relationships with plants is generally thought to have diversified from a monophyletic group of fungi (phylum Glomeromycota), although some studies suggest that another group of fungi may have interacted with the earliest plants (Bidartondo et al. 2011).

Other mycorrhizal relationships, such as ectomycorrhizae, could have arisen independently as many as 66 times (Tedersoo et al. 2010). The complex cell structure of these fungi includes hundreds of nuclei within the same cytoplasm as well as vertically inherited bacteria in an association that appears to be as old as the mycorrhizal lifestyle itself (Naumann et al. 2010). How the fungal–bacterial symbiosis affects the fungal–plant symbiosis is still not clear. In addition, the mycorrhizal symbiosis involves free-living soil bacteria, sometimes called helper bacteria, that are involved in the symbiosis (Frey-Klett et al. 2007). Mycorrhizal interactions, then, have accumulated multiple genomes over time, while fostering yet new interactions.

ORGANIZING CO-OPTED GENOMES

The process of co-opting genomes, and preventing or controlling the process of being co-opted by other species, creates a never-ending source of conflicting selection pressures on species. Coevolving simultaneously with multiple genomes requires ways of partitioning the expression of genes so that genes favored by one interaction do not negatively affect genes favored by another interaction. One solution is for natural selection to compartmentalize the expression of genes to particular tissues or developmental stages.

The evolution of complex life histories may have evolved to a large extent directly as an evolutionary solution to the problem of managing interactions with multiple other species. Organisms partition their interactions into discrete developmental stages (Thompson 1994). Larval and adult stages of animals, juvenile and reproductive stages of plants, and sporophyte and gametophyte stages of fungi are often adapted to different species. Partitioning interactions among life history stages allows a species to coevolve with multiple species by turning on and off different genes at different developmental stages. This partitioning minimizes the trade-offs in adaptation that can result from coevolving with multiple mutualists and antagonists throughout a lifetime.

The genomes of multicellular species are therefore often a collection of subgenomes that are temporally segregated. A caterpillar that must confront the defenses of the plant on which it feeds and thwart attacks by parasitoids and predators needs genes different from those required by an adult butterfly that sucks nectar from the flowers of yet other plants while fending off attack by a different set of predators. We have yet to come to grips with how to assess the sum total of the many evolutionary rates that occur in species across these kinds of complex life histories.

It is often not even clear from which perspective we should evaluate these rates if we want to understand the overall pace and dynamics of evolution, now that we realize how often species co-opt genomes as a mechanism to co-opt yet other genomes. When piercing the body of a caterpillar to lay their eggs, some parasitic wasps inject polydnaviruses at the same time, which thwart the caterpillar's immune system (Moreau et al. 2009). The wasps use the viruses to attack caterpillars, but the viruses use the wasps and caterpillars to make more viruses. We now know that pathogens and parasites are capable of altering the behaviors of the species they infect in ways that improve their transmission to new hosts (Poulin 2010). That view is captured in scientific papers with titles such as "Invasion of the Body Snatchers" (Lefevre et al. 2009) and "The Life of a Dead Ant" (Andersen et al. 2009). Even so, some of these interactions have persisted for millions of years, which implies that the top parasite is not always in control.

This problem is becoming more important in evolution as we come to realize that many supposedly pairwise interactions between species are coevolved multispecific interactions. That realization has come about through a growing group of studies asking a simple-sounding question: How many other species does it take for a species to survive and reproduce in the wild? For example, the interaction between pea aphids (*Acyrthosiphon pisum*) and parasitoid wasps (*Aphidius ervi*) involves least six species: plant, aphid, two bacterial symbionts, bacteriophage, and parasitoid wasp. The aphids feed on the plants with the aid of an obligate bacterial symbiont (*Buchnera*); the wasps lay their eggs into the bodies of the aphids; and the aphids are able to kill the developing wasp larvae through a bacterial symbiont (*Hamiltonella defensa*) by producing a toxin that is encoded by a bacteriophage (Moran et al. 2005; Oliver et al. 2009; Peccoud et al. 2009). This six-species interaction leaves out the rhizobial and other soil microbial genomes that it takes to build the plant on which the aphid feeds.

Similar webs of interaction occur between wood-boring beetles and microbial species as the beetles attack trees such as pines and spruces. It takes multiple life history stages of southern pine beetles (*Dendroctonus frontalis*) and two mutualists to successfully attack a host tree (Scott et al. 2008). Each pair of adult beetles constructs a tunnel in the inner bark of a pine tree, into which the female lays her eggs. The developing larvae feed on the inner bark. The larvae, though, need to co-opt the genetic machinery of at least two other species to complete development. When a female lays an egg, she takes bits of a beneficial fungus (*Entomocorticium* sp.) that she then carries in a specialized storage compartment on her body and places in the tunnel. After the eggs hatch and the larvae begin to chew along the inner bark, the fungi propagate and spread throughout the chamber. The developing larvae feed on this fungus.

This mutualism, though, can be disrupted by another fungus, which competes with the beneficial fungus and prevents development of the beetle larvae. That interaction has favored beetles that have co-opted yet another species. In the compartment where they carry the beneficial fungus, the beetles also carry a bacterial species that produces antibiotics capable of selectively inhibiting the spread of the antagonistic fungus. To complete development, then, a southern pine beetle larva requires the inner bark of a pine tree, coevolved beneficial fungi, and specialized bacteria (Scott et al. 2008).

Trees respond to pine beetle species by using monoterpenes induced by beetle attack. These chemical defenses kill the eggs and the fungi. Multiple pine beetle species use an effective counterdefense by emitting a pheromone to attract other adult beetles to individual trees (Pureswaran et al. 2006; Erbilgin et al. 2007; Boone et al. 2011). This mass attack swamps the defenses of the pines. In multiple ways, then, the seemingly pairwise interaction between pines and pine beetles is a collection of coevolved interactions distributed among species and life history stages. Moreover, even this description of the structure of the interaction underplays the genomic complexity, which continues to increase as research continues on these interactions (Six and Wingfield 2011).

The number of genomes needed for plant development is equally as great. Most plants require the use of at least six genomes spread over their life histories to survive and reproduce in natural populations. These include a nuclear genome, a mitochondrial genome, a chloroplast genome, one or more mycorrhizal fungal genomes, a pollinator genome to move gametes, and a seed disperser genome to disperse offspring. Add to this list the fungal endophytes in leaves, the ants that some plants use for defense by evolving rewards to attract them, and the rhizobia that legumes use for nitrogen fixation. The evolutionary rate of a plant population depends on the combined evolution of all these species and their interactions. The loss of even one of these symbioses would have huge effects on the diversity of life. Some plants could not even begin to grow. The 25,000 orchid species in the world—two and half times the number of bird species—cannot germinate without their mycorrhizal fungi (Dearnaley 2007), and most of these orchids also rely on complicated interactions with pollinators, as Darwin documented so carefully in the book he wrote on orchids immediately following *The Origin of Species* (Darwin 1862).

This same exercise in listing the multiple genomes needed by organisms can be done for any multicellular species and for some unicellular species as well. As molecular and ecological studies probe deeper into the factors shaping the evolution of species, the web of genomic connections needed for survival and reproduction continues to increase in almost any species that is being studied in depth. That is true even in species and interactions with a long history of study. The well-known pairwise interactions between leafcutter ants and the fungus gardens they cultivate for food have turned out to involve interactions among the ants, their cultivated fungal, specialized *Escovopsis* fungi that consume the cultivated fungus, *Pseudonocardia* bacteria that produce antibiotics used by the ants to combat the *Escovopsis* fungi, and black yeasts that grow at the specialized location on the ants' cuticle where the mutualistic bacteria are maintained (Currie et al. 2006; Little and Currie 2007). There are conflicting views about the full range of potential roles of the bacteria (Mueller et al. 2008; Cafaro et al. 2011; Six and Wingfield 2011), suggesting that the interaction is even more complex.

Even the interactions between the ants and their fungal associates are turning out to be more genomically and ecologically complex than previously suspected. Some microfungi and other microbes occur in these fungus gardens, and the *Escovopsis* parasites do not occur in some populations near the northern range limit of attine ants and their gardens (Rodrigues et al. 2008, 2011). Moreover, during the evolutionary history of attines, some species have changed the fungi that they cultivate (De Fine Licht et al. 2010). As emphasized in a review of microbial evolutionary ecology (Little et al. 2008), our understanding of the "rules of engagement" is changing rapidly as we come to appreciate how often species use microbial species to battle other microbial species.

COEVOLVED GENOMES IN ECOSYSTEMS

As coevolved symbioses have proliferated, they have become the base of all major ecosystems. Those ecological relationships have solidified the importance of coevolved genomes in the organization and continuing evolution of life. This is not, however, the general impression one gets from reading studies devoted to the roles of competition or predation among free-living species in community assembly. Until recent years, these interactions formed not only the cornerstones, but also much of the overall structure, of the field of community ecology. Yet, the functional structure of major ecosystems cannot be decomposed below the fundamental symbiotic, often mutualistic, interactions that convert inorganic nutrients into organic nutrients. Although populations are the basic unit of evolution, the interactions of disparate genomes, not the component species, are the basic ecological and genetic unit of the organization of biological communities. These species at the base of ecosystems cannot function without their interactions with other species. In a deeply ecological sense, the dynamics of evolution within all major biological communities begin with a set of coevolved interactions.

From the ocean surface to its floor, coevolved interactions form the base of oceanic food webs. Over two-thirds of the surface of the earth is covered by oceans. Microscopic assemblages of photosynthetic plankton form the base of the food web in the open ocean, and cyanobacteria such as *Prochlorococcus* and *Synechococcus* are among the major photosynthetic organisms in many regions. Different ecotypes or species of *Prochlorococcus* occur in different oceanic zones, and some estimates suggest that collectively these species are the most abundant photosynthetic organisms on earth (Johnson et al. 2006). They are, however, highly dependent on other genomes to maintain their position within these ecosystems. That dependence is reflected to some degree in their tiny genomes, which are the smallest of any photoautotroph (Scanlan et al. 2009). Such a small genome is unusual among free-living species (Partensky and Garczarek 2010).

Photosynthesis in *Prochlorococcus* and *Synechococcus* involves complex coevolved interactions with viruses. Genes associated with photosynthesis appear in the cyanobacterial genome and the genomes of its viruses, resulting in a complex distribution of genes for photosynthesis among these coevolving genomes (Lindell et al. 2004, 2007; Sharon et al. 2009). How this interaction affects carbon fixation during photosynthesis, and how selection acts on this interaction, is still being sorted out (Thompson et al. 2011).

The interactions between these abundant cyanobacteria and other species are just as puzzling (Morris et al. 2011; Sher et al. 2011). *Prochlorococcus* grown in pure culture in water chemistry similar to ocean water fare poorly, but these populations fare better when grown in the presence of some members of the heterotroph community. One possible interpretation is that the heterotrophs mediate the local environment for *Prochlorococcus*, which live in zones in the ocean in which they are subject to damage by hydrogen peroxide produced from photooxidation of dissolved organic carbon (Morris et al. 2011). They lack, however, the catalases and other protective mechanisms needed to protect themselves from this compound. In normal ocean environments the surrounding heterotroph community reduces the levels of hydrogen peroxide, thereby reducing oxidative stress in *Prochlorococcus*. The most abundant photosynthetic organisms on earth therefore cannot form viable populations without the surrounding genomes on which they rely.

Similar codependence of species, some highly coevolved and some more indirect, occurs at the base of the other major oceanic environments. Fixation of atmospheric nitrogen into a biologically useable form is a prokaryotic process, which eukaryotes have exploited repeatedly by establishing symbioses (Fiore et al. 2010). Much of the diversity of oceanic life is clustered within and near reef-building corals, which rely on symbiotic dinoflagellates, and the resulting coral reefs harbor thousands of other species. The importance of these symbioses is evident in the environmental problem of coral bleaching now found in many reef corals, which results from the loss of these dinoflagellates and the subsequent death of the corals (Weis 2008).

Farther down in the depths of the ocean where sunlight never reaches, yet other communities are built directly on different coevolved interactions. As deep-sea vents and cold seeps exude sulfur and methane, these chemicals are converted to usable energy by specialized microbial species that live in symbiotic association with mussels, clams, shrimps, gastropods, polychaetes, and tubeworms (Duperron et al. 2007; Nakagawa and Takai 2008). These taxa form the base of the web of life surrounding these vents and seeps.

In terrestrial communities, much of the base of community assembly occurs belowground and just on the surface through coevolved mutualistic symbioses. Lichens, which are coevolved interactions between fungi and algae, form an important part of primary succession in many communities. Half of the more than 13,000 species of ascomycete fungi form lichen associations that began up to 600 million years ago (Yuan et al. 2005; Usher et al. 2007), and by one estimate lichens are the dominant vegetation over about 8 percent of terrestrial environments (Brodo et al. 2001). Mycorrhizae, which are ancient coevolved interactions between fungi and plants, form on the roots of most plants and affect plant nutrition and growth in all major terrestrial environments. Rhizobia, which are coevolved interactions between bacteria and plants, form on the roots of legumes and some other plants worldwide and fix nitrogen that then becomes available for plant growth. Although all these interactions are often mutualistic, they can also be antagonistic under some environmental conditions (Piculell et al. 2008), resulting in complex coevolutionary dynamics in these species at the base of the food web. Without symbioses between these highly disparate genomes, there would be no complex terrestrial communities of eukaryotes.

At the decomposing end of food webs, symbioses are again crucial to maintaining the web of life. In oceanic environments, cellulose-digesting shipworms rely upon symbiotic microbes with cellulases (Luyten et al. 2006), as do termites in terrestrial communities (Tokuda and Watanabe 2007). Without symbiotic interactions between highly different genomes, there would be no eukaryotic food webs, and even webs of interaction among prokaryotes (bacteria and archaea) would be much reduced. We know all these things, but we are only now starting to internalize the deep importance of these interactions for the dynamics of communities and the dynamics of evolutionary change.

THE CHALLENGES AHEAD

The web of coevolving genomes is not a Gaia-like world functioning as a self-sustaining unit. Rather, it is a wildly dynamic world of constantly coevolving mutualistic and antagonistic interactions. The history of adaptive evolution is, to a greater extent than we realized even a decade ago, the story of co-opting and co-opted genomes. The pace of growth of our knowledge about the interdependencies of genomes is fast, but we are only now getting beyond scratching the surface. We still know little about the multiple evolutionary routes by which species have evolved into obligate, vertically transmitted mutualistic or parasitic symbionts (Antonovics et al. 2011a), and we know little about the genetic mechanisms by which species have evolved to control their symbionts.

We have made great progress in recent decades in understanding symbioses and other interactions between species, but we are still in the early stages of understanding the many ways in which they favor evolutionary change. We know only a tiny fraction of the earth's microbial diversity, and we know even less about how these microbial species interact and impose selection on each other within natural environments (Little et al. 2008). The sheer speed at which microbial species can evolve guarantees that there are surprises ahead as we continue to try to understand the ecological basis of evolution across the major domains of life.

For now, what we can say is that the history of evolution is, at its deepest level, about the ways in which species have taken different routes to assemble genomes. Some rely mostly on genes in the core genome inherited from parents, and others rely heavily on co-opting the genomes of other species. As we learn more about the molecular and ecological structure of species, we are realizing that interactions are an inherent part of the genetic composition of most species. We are long past the point at which we can consider symbioses, whether mutualistic or parasitic, as something added on to the evolution of species. We are also long past the point at which we can consider coevolution among species as something that has surely happened but perhaps not too often. The pendulum has swung the other way. It now seems that coevolution is one of the major drivers of relentless evolutionary change. It results in the evolution of mutualistic interactions and antagonistic interaction, and it can create genomic conflicts at every level in the hierarchy of life. It is to those conflicts that we now turn.

6

Conflicting Genomes

As species evolve to co-opt and manipulate other genomes, conflicts are inevitable. In recent years, we have become increasingly aware of just how fundamental is the flux of selection between conflict and cooperation at every level in the formation of genomes. The ramifications of this evolutionary tension extend deeply into the history of life, and those ramifications have continued to proliferate over time. The web of life has not evolved toward more conflict or more cooperation; it has evolved toward ever more complex interplays between conflict and cooperation.

That interplay begins at the intragenomic level in interactions among genes, extends to the intergenomic conflicts among sexes, and becomes further amplified in the intergenomic conflicts among species (Trivers 1974; Dawkins 1982; Foster and Kokko 2006; Wade and Goodnight 2006; Rice et al. 2009; Alonzo and Pizzari 2010). The tension is evident among selfish and cooperative bacterial cells that form biofilms, and among the interactions that shape every aspect of the continuum between antagonism and mutualism throughout the web of life. Each level in the genomic hierarchy of life mitigates some aspects of conflicts among lower genomic levels and magnifies others.

This chapter explores the implications of ongoing intergenomic conflict for the evolution of fundamental aspects of species ranging from sex ratios in populations to the sizes of genomes.

BACKGROUND: CONFLICT AND MANIPULATION

Interactions between plants and pollinators provide some of the clearest examples of evolution along a continuum from pure conflict to mutualism. Although many of these interactions are mutualistic, some have evolved strictly through deception. Some plants, for example, exploit the sexual systems of their pollinators either by producing flowers that visually mimic the mates of their pollinators or by producing chemicals used by these insects in their interactions with each other (Raguso 2008; Peakall et al. 2010; Schiestl 2010; Armbruster et al. 2011).

This is pure manipulation of another genome. If the deceptions are common and lower the fitness of the deceived pollinators (e.g., through lost time spent searching for mates), they could favor the evolution of avoidance responses. But if the deceiving population is uncommon and the deceptions impose no or low fitness cost on the pollinator, then a species could continue to manipulate another species without a coevolutionary response. These deceptions can be such an important part of the biology of a species that they could even be the basis for speciation. Ecological and molecular studies of sexually deceptive orchids and their pollinators have shown that most of these plant species attract and exploit a single pollinator species and that speciation in these orchids is associated with a change in chemical scent and a switch to a new pollinator (Peakall et al. 2010; Griffiths et al. 2011).

Intergenomic conflicts occur even between interacting genomes that would seem, superficially, to be under the same, rather than conflicting, selection pressures. Symbionts transmitted through the eggs of their hosts (i.e., vertically transmitted in the jargon of parasitology) have the same fate as their hosts. If a host does not reproduce, neither does the symbiont. There is, though, still plenty of room for conflict within that process, as is evident from the fact that vertically transmitted symbionts in nature range from mutualistic to antagonistic in their interactions with hosts (Hosokawa et al. 2010; Saikkonen et al. 2010).

WOLBACHIA AND OTHER SYMBIONTS

In recent years, the range of potential conflicts and outcomes in interactions between symbionts and hosts has been studied especially well in the interactions between invertebrates and *Wolbachia* bacteria. These bacteria are found in a high proportion of nematodes and arthropods, and they also occur in some other invertebrates—which is another way of saying that they occur in a high proportion of animal species worldwide (Hilgenboecker et al. 2008). *Wolbachia* bacteria are mostly mutualistic with nematodes but are generally antagonistic with most insects and other arthropods. Exceptions to this general pattern are appearing as the evolutionary ecology of more of these interactions is studied in detail (Weeks et al. 2007; Hosokawa et al. 2010).

Antagonistic selection on interactions between *Wolbachia* and arthropods has taken multiple forms (fig. 6.1), favoring symbionts that kill male gametes, turn male hosts into females, or create parthenogenetic host popula-

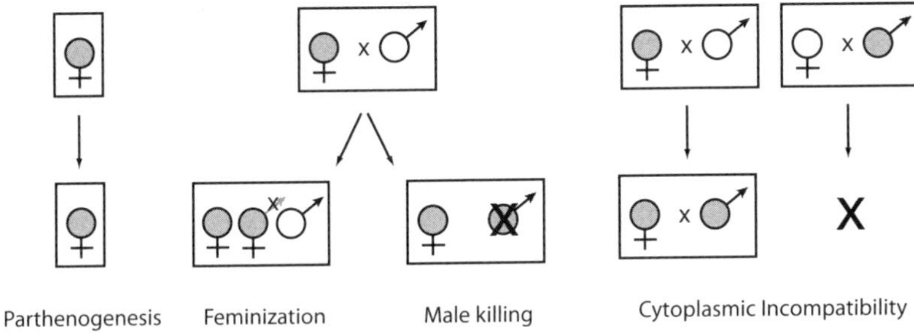

Fig. 6.1 Four ways in which *Wolbachia* bacteria manipulate reproduction in their hosts. Filled circles indicate infected individuals; unfilled circles indicate uninfected individuals. After Werren et al. (2008).

tions (Werren et al. 2008). Selection often acts on *Wolbachia* to drive host populations to produce females rather than males, because these symbionts are transmitted only through the eggs produced by females. In host populations with *Wolbachia*, infected males can successfully mate only with infected females.

Wolbachia and other symbionts have now been shown to be the cause of many instances of skewed sex ratios in insects and other arthropods (Werren et al. 2008; Engelstädter and Hurst 2009; Russell and Stouthamer 2011; Tabata et al. 2011; Vanthournout et al. 2011). These interactions may also lead to multiple other changes in host reproduction, resulting in host populations that differ in their patterns of reproduction (Stouthamer et al. 2010). When *Wolbachia* colonize a species, they can spread quickly within and among populations. Because *Wolbachia* infection can be determined in museum specimens, it is possible to get a sense of the short-term evolutionary dynamics across host and *Wolbachia* populations over time. For example, museum and current populations of common eggfly butterflies (*Hypolimnas bolina*) in eastern Asia and the Pacific islands show a pattern of correlated change in *Wolbachia* infection and butterfly sex ratio over the past century (Charlat et al. 2007; Hornett et al. 2009). Studies using molecular markers suggest that this *Wolbachia* infection spread across the South Pacific Islands at most 3,000 years ago and probably more recently, creating different current patterns of reproduction in different host populations (Duplouy et al. 2010).

As infection continues, there are multiple potential coevolutionary responses by hosts to the intergenomic conflict created by the *Wolbachia*. One is selection on hosts to suppress male-killing and restore a normal one-to-one sex ratio. Crosses in common eggfly butterflies (Hornett et al. 2006) and also in *Cheilomenes sexmaculata* ladybird beetles (Majerus and Majerus 2010) suggest that these species have evolved a dominant male restorer gene or multiple genes. Models of the evolution of these restorer genes suggest that the genes could spread quickly through a host population over just a few decades or a few hundred years, depending on how strongly the bacteria skew the sex ratio of their host population (Hornett et al. 2006).

The evolutionary process can become more complicated as yet other genomes become involved in these interactions. Some common eggfly populations have acquired two strains of sex-distorting bacteria (Charlat et al. 2006), and competition among these genomes can increase the intergenomic conflict. Alternatively, crosses among hosts with different *Wolbachia* strains can convert one type of sex distortion into another type (e.g., from cytoplasmic incompatibility to male-killing), thereby changing the structure of selection on the host–symbiont interaction, as has been found in experiments with *Drosophila* (Jaenike 2007).

Viruses add yet another dimension to the conflict. In some species *Wolbachia* contribute to defenses against a host's other parasites such as viruses (Hedges et al. 2008; Teixeira et al. 2008; Moreira et al. 2009). Of course, *Wolbachia* have their own viruses, whose roles in these interactions are still being unraveled (Bordenstein et al. 2006). As more becomes known about the relationships between *Wolbachia* and invertebrates, it is becoming clear that these are common, highly dynamic, evolving interactions with profound ecological effects on the many thousands of host species in which they occur.

NUCLEAR–MITOCHONDRIAL CONFLICTS

Conflicts with symbiotic genomes have also shaped the evolution of reproduction in plants. Most flowering plants are hermaphroditic, but almost every possible alternative distribution of gender within and among plants has been found (Charlesworth 2006b; Obbard et al. 2006; Ming et al. 2011). Although sex chromosomes occur in plant species, they have been found in less than fifty species (Ming et al. 2011). When plants have evolved differences in the distribution of genders, other mechanisms are generally involved. Among the most intriguing, from the perspective of intergenomic conflict, are those in which plant populations are composed of a mix of female individuals and hermaphroditic individuals, a condition called gynodioecy (Case and Ashman 2007; Delph et al. 2007; McCauley and Olson 2008; Bernasconi et al. 2009).

In some gynodioecious species the distribution of gender is shaped by intergenomic conflict between nuclear and mitochondrial genes (McCauley and Olson 2008; Bernasconi et al. 2009). These genomic incompatibilities are interesting because they suggest that conflict between mitochondrial and host genomes is continuing even now after many millions of years of obligate association. The conflict, commonly called cytoplasmic incompatibility, seems often to be driven by cytoplasmic male-sterilizing (CMS) genes in the mitochondria. These genes prevent pollen production, causing females to produce female offspring. This effect has been countered in plant populations, which have evolved nuclear restorer genes that restore pollen production, thereby making individuals hermaphrodites rather than females.

The frequency of female plants varies among gynodioecious species and even among populations. These differences may be due, in part, to costs associated with different kinds of restorer genes (Delph et al. 2007). The differences may also reflect how selection, gene flow from pollen and seeds, and genetic drift differ among populations in their combined effects (Dufay and Pannell 2010). In addition, the expression of gender is developmentally plastic in some species and the frequency of female plants varies along environmental gradients, suggesting that ecological context is important in shaping the occurrence of gynodioecy within species (Delph 2003; Delph and Wolf 2005).

As nuclear–mitochondrial interactions continue to evolve, they can escalate in the number of genes involved (Bailey and Delph 2007). Populations of the perennial herb *Silene nutans* in France have at least two cytoplasmic male sterility genes in the mitochondria and up to four restorer genes in the nuclear genome (Garraud et al. 2011). The current expectation from theoretical and empirical studies is that gynodioecious species will have multiple cytoplasmic male sterility genes and multiple nuclear restorer genes. The evolving balance in populations between female and hermaphroditic plants will depend mostly on the costs of restorer genes (Bailey and Delph 2007).

There are, though, some additional twists. Although mitochondria are inherited solely through females in most species, there are exceptions. Inheritance of some mitochondria through fathers, sometimes called paternal leakage, has been found in the well-studied gynodioecious species *Silene vulgaris* (Bentley et al. 2010). The frequency of paternal leakage in that species varies among populations from 10 percent to zero (fig. 6.2). The degree of conflict between mitochondrial and nuclear genomes is therefore likely to vary among populations of this species, but the conflict would remain because most mitochondria would still be inherited through females. Moderate paternal leakage would tend to maintain mitochondria that cause

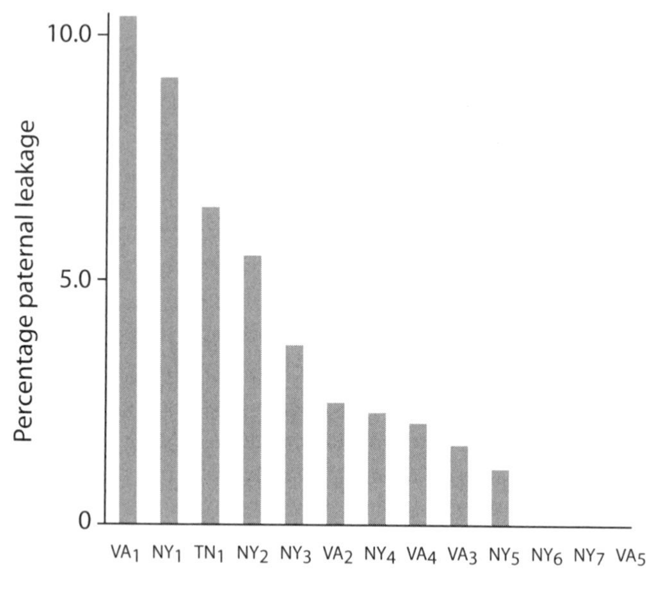

Fig. 6.2 The proportion of offspring in *Silene vulgaris* plants showing evidence of paternally inherited mitochondria in controlled experimental crosses of plants from thirteen populations collected in New York, Tennessee, and Virginia. After Bentley et al. (2010).

male sterility but also those that do not, thereby enhancing the stability of gynodioecy in a population (Wade and McCauley 2005).

Mitochondria of gynodioecious *Silene* species have relatively high levels of variation in comparison with hermaphroditic species (Städler and Delph 2002; Touzet and Delph 2009). Moreover, populations often have a combination of females and hermaphrodites. It therefore seems likely that selection is continuing to act on the conflict between these mitochondrial and nuclear genomes.

There are multiple ways beyond nuclear–mitochondrial interactions or *Wolbachia* by which intergenomic conflicts restrict reproduction in other species. Some fungal endophytes sterilize their plant hosts, resulting in plants that propagate vegetatively and thereby provide the endophytes with an unchanging host genotype for a longer time (Clay 2009). Some ants that live in domatia on tropical plants and protect the plants from herbivores and vines, clip the inflorescences of their host plants. The sterilized plants grow faster and larger, leading to larger ant colonies on those plants (Frederickson 2009). As we learn more about genomic interactions, we are likely to continue to find new ways by which genomic conflicts have evolved and continue to evolve.

THE CONSEQUENCES FOR GENOME EVOLUTION

The tension between conflict and cooperation ultimately affects the sizes of genomes. Hosts build up arsenals of genes that aid in protection against parasites. Parasites acquire genes, and sometimes whole genomes, that counter host defenses, while shedding genes that are redundant with functions they can acquire from their hosts. Mutualistic symbionts often take the shedding of genes even further, shrinking in size over millions of years as they transfer some of their genes to their hosts. Partly as a result of this ecologically driven molecular evolution, genome sizes vary greatly among species.

The smallest genomes are generally found among species that rely most heavily on the genomes of other species, including parasitic and mutualistic symbionts living within other species (fig. 6.3). The coefficient of variation in genome size is also lower in obligate symbionts than in facultative symbionts, reinforcing the view that obligate symbionts in general tend to evolve toward small genomes. These patterns, though, should be viewed as tantalizing but tentative, because relatively few of the hundreds of thousands, or millions, of symbiont species have so far been sampled.

Bacterial symbionts with the smallest genomes have shrunk their genome sizes partially by losing genes. Selection has favored symbionts that use more of their host's genes and fewer of their own genes for even the most basic aspects of survival, growth, and reproduction (Moran et al. 2008). This can be a slow process distributed over millions or tens of millions of years. Some of these changes in symbionts occur through selection favoring loss of biosynthetic pathways that can be obtained from the host. Some other changes may result from relaxed selection or from the genetic bottlenecks to which vertically transmitted symbionts often are subject in each generation. An insect can harbor ten million cells of a particular intracellular bacteria symbiont, but only a tiny fraction of those bacterial cells will make it into the eggs that a host produces (Douglas 2010). Unless countered by strong selection, some genes eventually will be lost from the symbiont population.

This pattern of gene loss has been studied in detail in *Buchnera* bacteria, which are obligate symbionts of some aphids and aid in aphid nutrition. The symbiosis formed about 150 million years ago, and the bacteria evolved

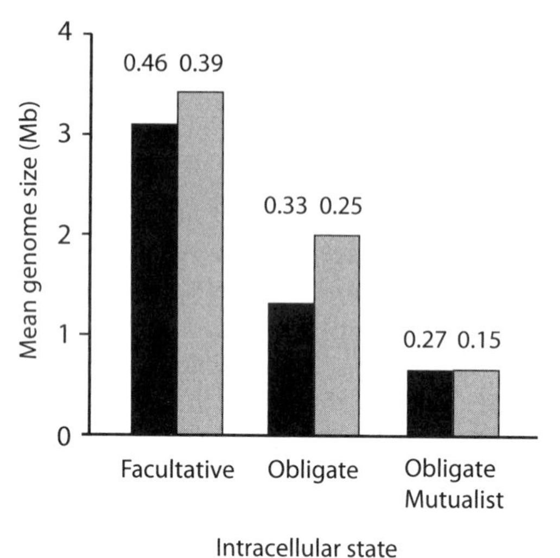

Fig. 6.3 Mean genome sizes (in megabases) of bacteria that live as facultative or obligate intracellular symbionts in other species. The obligate symbionts are divided into those that are obligate mutualists with their hosts and those that are not. Black bars indicate all bacteria; gray bars indicate γ-Proteobacteria. The numbers above the bars are coefficients of variation in genome sizes for each group. After Toft and Andersson (2010).

over time to a small genome size (Tamas et al. 2002). During the past 50 million years *Buchnera* populations have neither acquired new genes nor undergone chromosomal rearrangements (Tamas et al. 2002). Comparison among *Buchnera* species suggests that these genomes are still shrinking (Gil et al. 2002).

Amid this loss, there has been continued evolution in the remaining genes. The *Buchnera* genome in pea aphids has evolved quickly, as have the pea aphids, since the Pleistocene, with different genotypes of *Buchnera* associated with pea aphids that attack different plant species (Peccoud et al. 2009). Pea aphids have diverged in their adaptations to different hosts, but the pattern of divergence among these *Buchnera* lineages is consistent mostly with genetic bottlenecks, although other evolutionary processes may also be involved (Funk et al. 2001).

Over long periods of time, genomic loss could become so severe that a mutualism between a symbiont and host would decay completely. There are at least four potential pathways, though, by which the process can be slowed or alleviated. Strong selection could act to maintain crucial biochemical pathways involved in the maintenance of the symbiosis. Alternatively, selection on the host or the symbiont populations could favor alleles of other genes that individually or collectively compensate for any loss of gene function (Douglas 2010). Or, the symbiont could acquire new genes through horizontal gene transfer from other symbionts, if it is living in a host that has other symbionts (Bordenstein and Reznikoff 2005; Chafee et al. 2010).

The fourth alternative is to acquire yet more symbiont species (Douglas 2010), thereby continuing the evolutionary process of co-opting genomes. The result can be strong codependency not only between the host and the symbiont, but also between symbionts. Sharpshooter leafhoppers harbor two intracellular bacterial symbionts, of which one, *Sulcia muelleri*, produces eight of the ten essential amino acids and the other, *Baumannia cicadellinicola*, produces the remaining two as well as some vitamins (McCutcheon and Moran 2007). Cicadas also harbor *Sulcia* bacteria, but not *Baumannia*, and instead have different bacteria, *Hodgkinia cicadicola*, which are strikingly similar to *Baumannia* in their amino acid biosynthesis but different in vitamin production (McCutcheon et al. 2009). *Baumannia* and *Hodgkinia* are only distantly related and differ in their genome size and architecture. Their remarkable biosynthetic similarity has therefore resulted from convergence, rather than from divergence from an ancestor with similar characteristics.

Selection for complementarity in coexisting symbiont genomes, however, is not inevitable. The acquisition of additional symbionts can bring additional conflicts if the symbionts are not completely codependent. Even though vertically inherited symbionts share the same fate, they can still compete among themselves to increase their relative representation in the next generation of gametes (Vautrin and Vavre 2009). Competition therefore could counteract selection for genome reduction, if selection favors genes that increase competitive ability. A host population may be limited in the number of vertically inherited mutualistic symbionts that can be inherited through gametes.

Although competition generally favors divergence of competitive species rather than increased competitive ability, studies in experimental evolution of species such as *Drosophila* have shown that selection can favor the evolution of increased competitive ability if the species are not given the option of diverging (Joshi and Thompson 1996). Vertically inherited symbionts may have some options for divergence, but the options available during the critical stage of passage through gametes sometimes may be very limited. There may be, then, occasional opportunities during the history of a species when selection could favor complementary symbionts.

GENOME EVOLUTION IN PARASITES

Genome evolution in parasites is even more complicated than in mutualistic symbionts, because selection can simultaneously favor gene expansion and gene reduction. Parasites are constantly under selection to develop novel ways to escape host detection during invasion, ward off host defenses during development, and find new hosts. That can lead to the accumulation of genes over time. If a parasite population has successfully breached host defenses, though, they can also shed some genes for basic functions already performed by their hosts. At any moment in evolutionary time, it can look as though a parasite genome has either a high or low proportion of its genes devoted directly to its interactions with hosts relative to closely related species. It all depends on the dynamics of gene gain and gene loss as parasites continue to coevolve with their hosts.

The evolution of an obligate parasitic lifestyle has therefore resulted both in large genomes and small genomes. Rust fungi, which are pathogens of a wide range of plants, including cereal crops, have some of the largest genomes found in fungi (Baxter et al. 2010), and some powdery mildews, which commonly attack angiosperms, have genomes that are four times as large as the median size of other ascomycete fungi (Spanu et al. 2010). In contrast, microsporidians, which are fungus-like intracellular parasites of many animal species, have some of the smallest genomes among eukaryotes (Texier et al. 2010). Even among the more than 1,200 species of microsporidians, genome size varies by an order of magnitude, although gene number varies less (Texier et al. 2010). Among parasitic bacteria, *Coxiella burnetii*, the causal agent of human Q fever, has a genome size that is twice the size of many other intracellular bacteria (Omsland and Heinzen 2011).

As genomic sequencing becomes routine in studies of interactions among species, it will become increasingly possible to ask questions about how evolutionary change in genome size reshapes coevolution between hosts and symbionts. One possibility is that coordinated genome reduction stabilizes assemblages of interacting parasitic and mutualistic symbionts as each species undergoes gene loss relative to free-living ancestors. There is some evidence for parallel genome reduction in such assemblages. For example, human body lice (*Pediculus humanus humanus*) have the smallest known genome size of any insect species. Their primary mutualistic endosymbiont, which aids the lice in living on nutrient-poor blood, and the pathogenic bacterial species that they transmit to human hosts also show evidence of reduced genome size relative to their free-living relatives (Kirkness et al. 2011). In this and similar cases, however, the degree of biosynthetic coordination in genome loss is mostly unknown.

As we learn more about the extent to which all life depends on other species, the question of what constitutes "minimum life" takes on a new meaning. Most of the smallest known genomes are found in species that live symbiotically, either as mutualists or as parasites, within species. The process of coevolution, played out over billions of years, makes it difficult to decide in many instances just what constitutes the minimum of what we should call a living organism. For many years biologists and philosophers debated what constitutes life, and the definition often involved the ability to self-replicate or have the genetic machinery for some other fundamental processes. Viruses were often excluded from that definition of life, partly because they use the genetic machinery of their hosts for replication and other essential functions. But most species have to use some aspects of the genetic machinery of other species at some point in their life cycle. In fact, some viruses are more complicated than some other forms of life, because they often have a complex life history that involves a parasitic stage and a free-living stage used to find new hosts.

There is, then, no clear place to draw the line. Viruses are in many ways the most successful lifestyle on earth. They have outsourced to other species almost everything needed to sustain them, and in doing so they have become the most numerous form of life. Some obligate bacterial parasites and mutualists have genomes so small and lifestyles so dependent on a host that they are at the interface between organisms and organelles (Douglas and Raven 2003). Whether viruses are living organisms or something else becomes less relevant when viewed in this way. These populations are simply the most accomplished at using the genetic machinery of other species for survival and reproduction.

MOBILE GENETIC ELEMENTS

Beyond viruses, there is a bewildering array of mobile genetic elements with potentially important effects on evolutionary dynamics. Appreciation of those effects has required an infusion of ecological views and Darwinian approaches into molecular biology (Werren 2011). Richard Dawkins's influential book *The Selfish Gene* (1976) forced a focus on the inherent conflict in interactions at all levels in the hierarchy of life. The discovery of highly repetitive DNA opened the possibility that some genetic elements may be selfish or parasitic, replicating at the potential expense of the large genomes within which they reside (Doolittle and Sapienza 1980; Orgel and Crick 1980). Some DNA elements—transposons—move about in genomes and replicate in different places; importantly, these are not rare events. Transposons occupy a large percentage of the genomes of some species, making up two-thirds of genomes of maize and about 45 percent of the genome of humans (fig. 6.4). In bacteria, mobile genetic elements compose as much as 20 percent of some genomes (Newton and Bordenstein 2011). In turn,

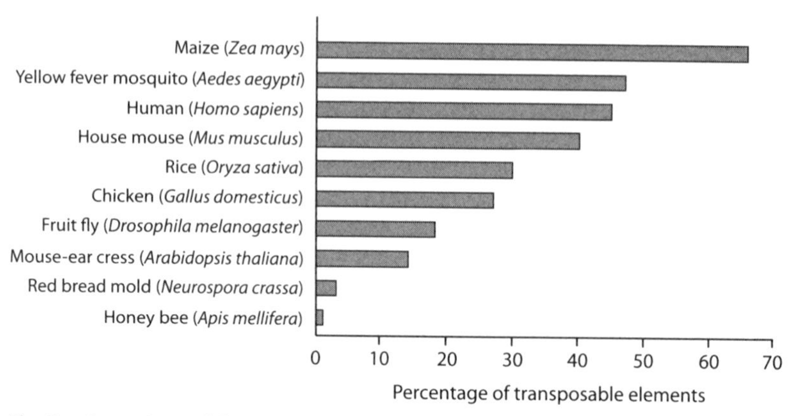

Fig. 6.4 Percentage of the genomes of eukaryotic organisms composed of transposable elements. After Biémont (2010).

organisms harbor a variety of mechanisms that can act to control mobile genetic elements (Johnson 2007).

Genomes are therefore not as tightly integrated as once thought by many biologists. It has taken decades for that view to shift. As John Werren and colleagues (Werren et al. 1988; Werren 2011) have noted in reviews of the study of mobile genetic elements, even as recently as the late 1980s the idea that aspects of genomes could be selfish or parasitic remained foreign to many molecular biologists. At the turn of this new century, biologists were still trying to sort out how to apply the concepts of evolutionary ecology and coevolutionary biology to many aspects of molecular biology, as suggested by the title of John Avise's (2001) article in *Science*, "Evolving Genomic Metaphors: A New Look at the Language of DNA." Since then, it is becoming clearer each year that mobile genetic elements of various sorts are involved in genomic interactions at multiple levels, including pathogenicity in parasites and resistance in hosts. The integration of molecular biology and evolutionary ecology is proceeding at an increasing pace as the study of adaptation continues to expand through application of Darwinian approaches in molecular biology, the incorporation of genomic approaches in ecology, and the development of conceptual frameworks that span these long-separate research traditions (Dirzo and Loreau 2005; Peay et al. 2008; Johnson et al. 2009; Venner et al. 2009).

Intergenomic conflict at the molecular level in eukaryotes is, in effect, the coevolutionary process mediated by transmission. Components of the genome that differ in transmission patterns are under different selection pressures (Werren 2011). Hence, nuclear genes, mitochondrial genes, chloroplast genes, *Wolbachia*, and mobile genetic elements are all in potential conflict as selection acts on each to maximize representation in the gametes of the next generation of a eukaryotic population. Bacterial populations have equally complex intergenomic relationships involving mobile elements that affect their evolution and their ecological relationships with other species (Newton and Bordenstein 2011). Appreciation of the role of molecular intergenomic conflicts in evolutionary change is likely to increase in the coming years as approaches from molecular biology and evolutionary ecology continue to unravel the complex relationships between these conflicts and the Darwinian fitness of the organisms within which these conflicts take place.

THE DEEPER UNKNOWNS OF MOLECULAR AND MICROBIAL ADAPTATION

We are probably still underestimating the extent to which the adaptation and diversification of life have been driven by coevolving and sometimes conflicting genomes. Some interactions extend so far back to the earliest eons of life that they can barely be identified now as interspecific interactions. There are even indications that some bacteria may themselves be the result of an ancient coevolved interaction. Bacteria are prokaryotes, which means they lack the nucleus found in more complex eukaryotic organisms. Some of the most common and important bacteria, including the ones that are often themselves symbionts within more complex organisms, have double membranes rather than single membranes.

James Lake (2009) has proposed that the double membrane is the result of an ancient coevolved symbiosis between two single-membrane prokaryotes, one from the group called Clostridia and the other from the group called Actinobacteria. He has found evidence for this hypothesis in the combination of proteins found in these species in comparison with single-membrane prokaryotes. The double-membrane species have a mix of protein families consistent with the view that they originated by symbiosis between species from two different prokaryotic groups. If these double-membrane prokaryotes originated from symbiosis, they did so more than two billion years ago (Lake 2009). They are the species responsible for the origination of the earth's oxygen atmosphere, which started to rise more than 2.7 billion years ago. With more than two billion years of coevolution, it is no wonder that is so difficult now to tease apart whether these organisms that we view as one species were once separate species.

There are vast areas of coevolution of interactions between microbial species and other species that are only now being explored—some conflicting, some mutualistic. As one major example, macroalgae such as kelps form oceanic forests along continental coasts and are the base of some of the richest food webs in oceans. These species harbor a diverse array of bacteria, most of which are unstudied (Goecke et al. 2010). There are even greater gaps in our understanding of the archaea, which have now been recognized as fundamentally different from bacteria and eukaryotes for over a third of a century. Research on these species is growing quickly, but much remains unknown about the ecological interactions of these species, which are involved in multiple ecosystem processes and are common components of some communities (Orcutt et al. 2011; Steele et al. 2011). Although initially thought to occur mostly in extreme environments such as terrestrial hot springs, hydrothermal vents, and salt brines, they have now been found in a wide range of terrestrial and marine habitats (Jarrell et al. 2011). There is, though, a curious lack of archaea that are known pathogens of eukaryotes (Cavicchioli et al. 2003).

Erin Gill and Fiona Brinkman (2011) have argued that lack of archaeal pathogens is not simply due to inadequate sampling. They have estimated that about 0.36 percent of bacterial species are known to be pathogenic in humans. If a similar percentage of archaea were pathogenic, then there should be about sixteen archaeal species known to be pathogenic. They have hypothesized that the lack of pathogenic archaea is due to differences in the phages that attack archaeal and bacterial species. Bacteria are effective as pathogens often because their virulence factors are carried by phages or mobile elements. From the phage perspective, bacteria are simply vehicles that allow phages to infect eukaryotes. The cell surfaces of bacteria and archaea differ, however, which would make it difficult for bacteriophages to attach to archaea and transfer virulence factors to these species. Hence, phages have the potential to transfer virulence factors among bacterial species but not between bacterial species and archaeal species.

That begs the question of why phages have not coevolved with archaea to infect eukaryotes, and there is no current answer to that question. It is certainly not that archaea lack viruses. They harbor a complex array of viruses (Snyder and Young 2011), and some of these viruses seem to have genes of bacterial origin (Prangishvili et al. 2006). The possibility that phages are, in some way, involved in maintaining the divide among archaea, bacteria, and eukaryotes—at least with respect to pathogenesis—is an intriguing view worth pursuing.

Given the great diversity of species and what we know about the importance of coevolution in the diversification of life, it would be surprising if it turns out that archaea are not involved in coevolved interactions with many other species. As the search continues, they are being found in more environments in large numbers. By one estimate, up to 20 percent of the microbial cells in the ocean may be archaea (DeLong and Pace 2001). Archaeal and bacterial species can act in complementary ways in ecosystem processes such as nitrification (Raghoebarsing et al. 2006), suggesting that they have the potential to form mutualistic interactions. In an experimental microcosm study that imposed mutualistic selection on an archaeal species and a bacterial species, the cultures with the two species rapidly evolved to higher levels of productivity under some experimental conditions (Hillesland and Stahl 2010). Moreover, archaea are inhabitants of the bodies of eukaryotes, including the human digestive systems, but how they interact with the other species as part of the microbiota of eukaryotes is not yet known (Dridi et al. 2011). As genomic tools continue to improve, archaea are likely to be implicated in a much wider range of ecological interactions.

One of the other deep unknowns along the continuum of conflict to cooperation in microbial evolution is the breadth of ways in which social interactions among microbial species shape evolution and intergenomic conflicts. It is now known that microbial species can engage in social interactions mediated by chemical signals (West et al. 2007; Queller and Strassmann 2009; Bourke 2011; Xavier 2011). The discovery of quorum sensing in bacteria showed that interactions among bacteria can be controlled by gene expression that depends on the density, usually, or some other related properties of the surrounding population or the composition of the surrounding community (Williams et al. 2007). Quorum sensing involves the production of small hormone-like molecules, called autoinducers, which diffuse into the surrounding environments and are perceived by neighboring cells. Once concentrations of these molecules have exceeded a threshold, the bacterial population undergoes a coordinated response through changes in gene expression, allowing better access to nutrients, defense against enemies, or some other qualitatively different response to environmental challenges.

Quorum sensing was first described in *Vibrio fischeri* bacteria that inhabit the light organs of some marine species, and this sensing is a central component of some marine symbioses that produce bioluminescence (Nealson et al. 1970; McFall-Ngai 2008). Once the bacterial population reaches a threshold density each day, the bacteria alter their gene expression and begin to bioluminesce. More generally, quorum sensing has become an increasingly important topic in microbial research because it can affect so many aspects of microbial biology and evolution. It is involved in the production of mucous to form the microbial mats called biofilms, and it may affect the complex microbial interactions that occur in the mucous layer of corals (Goldberg et al. 2011). The formation of biofilms by *Vibrio cholerae* reduces grazing by protozoa and therefore may function in part as an antipredator defense (Erken et al. 2011). Quorum sensing also has the potential to affect patterns of kin selection within microbial populations (Strassmann et al. 2011), and it allows bacteria to act in some ways like multicellular organisms (Waters and Bassler 2005).

Even at the seemingly simplest levels of life, then, there is the potential for sophisticated communication that can affect how genomes interact and how those interactions may shape interactions with other genomes. There is a sociobiology of molecular systems, as Kevin Foster (2011) has called it, that pervades every aspect of the process of the evolutionary ecology of coopting genomes.

THE CHALLENGE AHEAD

At every level in the hierarchy of life, intergenomic conflict appears to be one of the major drivers of continuing evolutionary change. All three domains of life have been influenced by interactions with potentially conflicting disparate genomes, whether those other genomes are members of the same domain, species within other domains, or viruses. This tension between conflict and cooperation between genomes is woven deeply into the history of all lineages. It affects important aspects of evolution ranging from sexual reproduction to the formation of microbial biofilms and the evolution of genome size.

Although there are many current challenges in molecular adaptive evolution, among the most important is the need for a better understanding of when intergenomic conflicts are amplified or mitigated as species continue to co-opt additional genomes and defend themselves against continual attacks by yet other genomes. These evolutionary dynamics are made possible by the tremendous reservoirs of genetic variation found within species. The variation is itself a result of continual selection, and we turn now to the ecological processes that maintain that variation.

Part 3

Variable Selection and Adaptation

7

Adaptive Variation

Even after the past century of studies in population genetics and evolutionary ecology, it remains important to stress that evolution and directional evolution are not synonyms. There remains a lingering, often unstated, view that real evolution is about directional change in populations and that fluctuating evolution, which has been found in almost all species that have been studied in detail, somehow does not truly count in any deeply important way in the grand scheme of things. These evolutionary meanderings are implicitly viewed as clouding the larger picture of change. Consequently, discussions about evolutionary rates are often restricted to net change in traits as measured by darwins or haldanes (see chap. 1), and conclusions about what drives evolutionary rates often remain focused on what factors are most likely to drive net evolutionary change in traits or lineages. Other common evolutionary outcomes produced by local adaptation and selection mosaics seem unimportant from that perspective.

There is nothing inherently wrong with a focus on directional evolutionary change for some major questions in evolutionary biology, as long as it remains clear that net evolutionary change captures only one aspect of the evolutionary dynamics of species. There is no question that directional selection is common and has been a driving force in the diversification of life. It is also clear that populations may sometimes undergo directional change even when selection fluctuates from year to year or generation to generation (Kingsolver and Diamond 2011; Morrissey and Hadfield 2012). Taken to the extreme, though, a focus on directional evolutionary change hides much of the ecology, and much of the dynamics, of evolutionary change. If evolution were almost solely about directional change, then species would not be as genetically variable as we often find them in nature.

In this chapter, I consider the ecological mechanisms that maintain adaptive variation in populations and what these mechanisms tell us about the major drivers of evolution.

BACKGROUND

Evolution continues to occur in populations partly because there is considerable standing genetic variation on which selection can act. Directional evolutionary change erodes some of that variation over time, unless it is replenished by mutation, gene flow from other populations, fluctuating selection, or various forms of balancing selection that act in ways that maintain genetic variation. One of the major advances in evolutionary biology in the past two decades has been the demonstration, first through phenotypic studies and more recently through molecular studies, that natural selection often acts in ways that maintain adaptive genetic variation even within local populations. The major mechanisms have long been known, including frequency-dependent selection, density-dependent selection, heterozygote advantage, recombination and sexual reproduction in many species, and horizontal gene transfer in some species.

Evolutionary biologists implicitly understand that these evolutionary changes are a crucial part of the evolutionary process, yet all this is lost on nonbiologists. They often take literally statements indicating that horseshoe crabs have evolved little in the past several million years. They interpret phrases such as "living fossils" as descriptive of species that have not undergone evolutionary change. If we could observe the finer-scale evolutionary history of horseshoe crabs, it would surely show considerable evolution over time in multiple traits, although much of the evolution would not have been sustained directional change.

At this point in the 200 million years of evolution in horseshoe crabs, there are four species, three Asian and one North American (Avise et al. 1994). The North American species, *Limulus polyphemus*, shows considerable variation at microsatellite loci (King et al. 2005). There is evidence of much gene flow among the populations, yet the populations cluster into geographic groups that are distinct both genetically and morphologically (Riska 1981; King et al. 2005). Less is known about genetic variation in the Asian species, but there is evidence for genetic differences among at least some populations (Yang et al. 2007). At least some of the genetic and morphological variation in these species has likely been maintained over time through fluctuating selection on these remarkable animals. The eggs of horseshoe crabs are the major food of some migratory shorebirds (Walls et al. 2002) and are therefore likely subject to strong selection by predators. It seems unlikely that crabs have escaped selection and evolution throughout their long history.

That is a safe guess to make, given what we now know about selection on populations. The accumulating terabytes of genomic data suggest that populations are often more genetically variable than we thought, yet molecular analyses have identified much evidence of selection acting on the genomes of species. If most of that selection was directional selection, genetic variation within populations would be much more restricted than we observe. Forms of selection other than directional must be important in the ongoing dynamics of evolutionary change.

The high levels of genetic variation found in many species continue to surprise us all as the genomic data continue to accumulate. The data are, of course, best for humans. Early genome-wide analyses of positive selection (i.e., directional selection acting on advantageous mutations) in humans identified hundreds of candidate genes, including more than twenty that were outside the range of sequences likely to be changing through neutral evolution (Sabeti et al. 2007). At that time, one estimate suggested that 10–13 percent of the amino acid substitutions between humans and chimpanzees may be adaptive (Gojobori et al. 2007). Since then, more reports of positive selection and potentially rapid evolution in humans have accumulated as more individuals have been sequenced and genome-wide analyses of selection have become more powerful (Hawks et al. 2007; Auton et al. 2009; Crisci et al. 2011).

A progress report by the 1000 Genomes Project Consortium found, within 179 individuals from four populations, approximately 15 million single nucleotide polymorphisms (SNPs)—that is, differences among individuals in a nucleotide found at a particular DNA location (Altshuler et al. 2010). There were also one million short insertions and deletions, and 20,000 structural variants. Most of this variation was previously unknown. Moreover, their analyses suggest that much local adaptation has occurred in human populations and that much of the selection on humans has been on standing genetic variation rather than on new mutations. As more human genomes are studied with the new generation of sequencing methods, estimates of standing genetic variation continue to increase (Coventry et al. 2010). Although the human genome is the most thoroughly studied of all genomes, there is no reason to suspect that humans are special in the number of genes under selection. Large amounts of variation are maintained within species, and a surprisingly large number of genes show the signatures of selection.

The possibility that some genes are under varying, rather than long-term directional, selection comes from studies showing that some genetic polymorphisms are ancient. In *Arabidopsis* plants, some plants are susceptible to particular pathogens and others are not, and some populations and species harbor a combination of susceptible and resistant individuals. These genetic polymorphisms have been maintained within *Arabidopsis* for 0.35–5.1 million years (Bakker et al. 2006). These resistance genes (R-genes) are among the most variable genes found in *Arabidopsis* (Borevitz et al. 2007; Clark et al. 2007). Studies of the signatures of natural selection in two *Arabidopsis* species have suggested that selection on these R-genes has been common, variable over time, and variable among the genes (Bergelson et al. 2001; Bakker et al. 2006; J. Wang et al. 2011). Moreover, experimental studies on plants manipulated for presence or absence of R-genes have demonstrated that selection mediated by pathogens can act strongly on these genes and that, in the absence of pathogen attack, there is a cost to plants of maintaining these alleles that confer resistance (Tian et al. 2003; Roux et al. 2010). These genes are therefore likely to have experienced considerable fluctuating selection during this long history.

The increasing number of studies showing even more genetic variation than previously suspected, together with studies showing the signatures of variable selection on genes and the ancient age of some genetic polymorphisms, suggests that an important part of ongoing evolution is about variable selection rather than long-term directional selection. That is an important result for our understanding of the interplay between evolutionary and ecological processes because it highlights two points. Some of the ecological dynamics found within biological communities are likely to result from variable selection that constantly changes the levels and patterns of genetic variation within and among species. And the disconnect between the rapid evolution often found in populations by evolutionary ecologists and the slower pace of evolution found in the fossil record by paleobiologists probably reflects true differences in the kinds of evolutionary change detectable at these different timescales, as long suspected by evolutionary biologists.

THE COMPLEXITY OF FLUCTUATING EVOLUTION

In the simplest form of fluctuating selection, the intensity, direction, or target of natural selection varies among generations. Evolutionary theoreticians have found in mathematical models almost every possible outcome of fluctuating selection during the evolution of species. The models show that fluctuating selection can lead to the loss of genetic variation, or it can act as a major mechanism by which genetic variation is maintained within populations. In some cases, it can serve as a major mechanism by which populations persist in variable environments, and in others it can be inconsequential to long-term evolutionary trends (Haldane and Jayakar 1963; Kondrashov and Yampolsky 1996; Burger 1999; Johnson and Barton 2005; Bell 2010).

The reason for such a wide range of possible outcomes is that the consequences of fluctuating selection depend on the assumptions used in the mathematical models and the ecological conditions being modeled. The pace of change in genetic variation under fluctuating selection often differs among single gene models, oligogenic models, and quantitative genetic models. The models differ in outcome depending on mutation rates, gene flow, the strength of natural selection, and the rate at which the modes and directions of selection change over time. Most models have been of single populations or of single populations with some additional movement of genes into them from other populations. Some models suggest that fluctuating selection may be especially likely to maintain genetic variation in traits controlled by only a few genes within natural populations in constantly changing environments (Bell 2010). Generally, it seems that fluctuating selection can result in almost any adaptive outcome, but averaged across many populations within a species, it can maintain genetic variation within a species and sometimes prevent the divergence of populations into separate species.

Part of the problem in understanding fluctuating selection is that it has often been difficult to design experiments to evaluate how the evolution of phenotypic variation in traits relates to the evolution of genetic variation. Some models, for example, assume that selection acts on traits controlled by many genes and that a population occurs in an environment that is otherwise unchanging except for selection on those traits. Under those conditions, fluctuating selection can lead to decreased phenotypic variation in a population, but it is uncertain whether that necessarily results in much loss of genetic variation.

The problem can be illustrated in carefully controlled laboratory studies (Pélabon et al. 2010). The experiment is straightforward, although time-consuming. Take *Drosophila melanogaster* flies from a genetically variable population and measure a set of traits known to be controlled by many genes—in this particular experiment a set of characters that define the shape of the wing. Set up replicated cultures of the fly population and measure 100 flies of each gender in each subsequent generation. After measuring the flies, choose 25 from each gender as parents for the next generation and repeat that process for 20 generations. For fluctuating selection, alternate the direction of selection in each generation by choosing parents for the next generation that are either above or below the mean. For the control populations, choose individuals haphazardly from each generation to be parents of the next generation. To put the effects of fluctuating selection in context of other forms of selection, test for the effects of stabilizing selection by choosing as parents for the next generation the 25 individuals of each gender whose trait values are closest to the mean at the start of the experiment. For disruptive selection, choose as parents those individuals with the most extreme values at both ends of the distribution of values.

After 20 generations of fluctuating and stabilizing selection, the fly populations generally show slightly decreased phenotypic variation in the wing traits relative to the control populations, whereas disruptive selection increases variation in most traits. Exactly why phenotypic variation decreases in this experiment is not certain, but Pélabon et al. (2010) suggest that it may be due to selection against genes with large phenotypic effects. That is, the experiment favors genes and gene combinations that do not create phenotypes far from the mean. Despite the changes in phenotypic variation, the overall effect of selection on genetic variation could be small if the most extreme phenotypes were caused by a few alleles with large phenotypic effects. Under these stringent conditions, then, fluctuating selection is essentially a form of time-delayed stabilizing selection. Both forms of selection could maintain and even continue to accumulate much genetic variation in a population, but the retained genes would be those that individually, in combination with other genes, have small effects on shifting phenotypic traits away from the mean.

When populations are studied in nature, researchers often find evidence that the traits favored by selection differ among years, seasons, habitats, or local abundance or frequencies of other species (Grant and Grant 2008b; Ehrlén and Münzbergova 2009; Siepielski et al. 2009). These studies have analyzed many different kinds of traits, including the seasonal timing of diapause in copepods in response to probability of predation (Hairston and Dillon 1990; Ellner et al. 1999), the major histocompatibility genes that may confer resistance to parasites in feral sheep (Charbonnel and Pemberton 2005), the wing pattern of *Callimorpha (Panaxia) dominula* moths (O'Hara 2005), the color and banding pattern of *Cepaea* snails (Cook and Pettitt 2008), and the enzyme polymorphisms that affect dispersal ability in butterflies (Hanski 2011), among others (see Bell 2010 for review). Fluctuating selection is therefore likely to contribute to the ongoing evolution of many traits in many species.

Fluctuating selection, however, often occurs in concert with other evolutionary processes, making it difficult to determine how to partition the importance of these various processes. In one of the most famous cases, involving the differences among individuals in wing color in brightly colored tiger moths (*Callimorpha* spp.) in Europe, the frequency of these differing wing morphs varies within and among populations. Interpretations of this fluctuating polymorphism in wing color have included fluctuating selection (Fisher and Ford 1947), genetic drift acting in concert with other evolutionary processes (Wright 1948; O'Hara 2005), and relaxed selection in some populations (Brakefield and Liebert 1985). It is possible that all these evolutionary processes contribute to the maintenance of this color polymorphism within each species and that polymorphism is sometimes transient within local populations.

Many ecologically important polymorphisms vary geographically and appear to be maintained in species at least partly through fluctuating selection that favors different morph frequencies in different habitats. The classic example is that of *Cepaea nemoralis* snails, which have been studied at multiple sites throughout Europe and show complex patterns of polymorphic variation in color and banding patterns on their shells (Cain and Sheppard 1950; Cook and Pettitt 2008; Ozgo 2011; Silvertown et al. 2011). Although there is a great deal of variation in morph frequencies among habitats, an analysis of multiple past studies has shown that yellow-banded patterns are found more often in open habitats and nonyellow unbanded patterns are found more often in wooded habitats (Cook 2008). In addition, there are continent-wide patterns in the distribution of morphs. Some distributions appear to have changed during the past half century (Silvertown et al. 2011). These changes do not seem to be directly related to climate but may instead be related more to changes in predation pressures and habitats. Interestingly, these continent-wide patterns have been identified through citizen-based science organized by the Evolution Megalab, which has involved participants in fifteen European countries (Silvertown et al. 2011).

In some cases, the distribution of morphs among populations can change at fast rates, mirroring a patchwork of rapid environmental change across ecosystems. In Sweden, pygmy grasshoppers (*Tetrix subulata*) vary genetically in color from light gray to black, and the frequency of morphs varies among the shore meadows and cleared forests where these insects live (Forsman et al. 2011). The black morph is favored in environments that have recently burned. That morph increases on average to about 50 percent in the year following a fire and then rapidly decreases to about 30 percent after four years (fig. 7.1).

These changes suggest that polymorphisms are highly dynamic within and among populations and that some, but not always all, of these changes

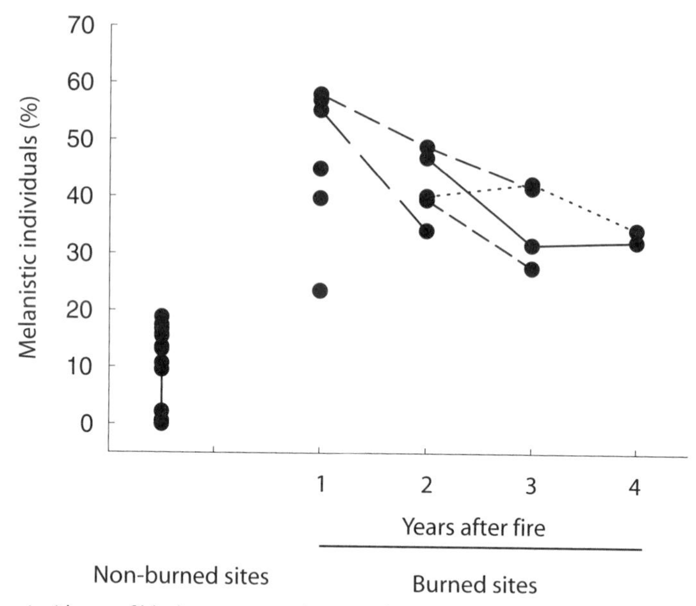

Fig. 7.1 Incidence of black pygmy grasshoppers (*Tetrix subulata*) in Sweden in non-burned sites and in burned sites during the first four years after a fire. Values for burned sites connected by lines are for sites sampled in consecutive years. After Forsman et al. (2011).

are due to continuing selection. Equally important, the rate of change in the frequency of morphs is often quick, sometimes apparent over just a few years. These fast rates suggest that fluctuating polymorphisms could be both cause and consequence of some of the ecological dynamics found within species as the frequencies of morphs change among years.

Studies like this one also highlight the limits of what we infer from molecular surveys of the overall strength and forms of selection within genomes. The black morph may result from a few genes that are under strong selection only during the several years following each fire. Although a genome-wide analysis may conclude that there is little overall evidence of fluctuating selection in the genome of this species, the occasional rapid evolutionary increase in the frequency of black morphs following fires could be what allows these populations to persist in nature.

LESSONS FROM THE MOST FAMOUS CASE STUDY: DARWIN'S FINCHES

The most famous examples of rapidly fluctuating evolution are those involving Darwin's finches in the Galápagos Islands, and these studies are continuing to provide new insights into the process of adaptation. Building on the work of David Lack (1947), Peter and Rosemary Grant began studying this adaptive radiation in detail in the early 1970s. Since then, they have evaluated the pattern of adaptive divergence of populations among the islands and exactly how and why populations on one small island, Daphne Major, continue to evolve from year to year and decade to decade.

The results that they and their colleagues have carefully accumulated over multiple decades constitute the most important studies of adaptation and divergence in natural populations ever undertaken. Their extraordinary research program is so well known—and so beautifully summarized in Grant and Grant (2008b) and in books by others (Weiner 1995; Donohue 2011)—that these rapidly evolving birds are one of the major reasons why views on relentless evolution have begun to change. The results of these studies have become the touchstone for any discussion about the drivers, rates, and mechanisms of evolutionary change in adaptive traits.

Between two and three million years ago a group of finch-like tanagers arrived on the Galápagos Islands and began diversifying into the fourteen currently recognized species. The birds now range in size and shape from the small, thin-beaked warbler finch to the larger, thick-beaked large ground finch. They have diversified as the number of islands has increased and climates have fluctuated, especially over the past one million years (Grant and Grant 1996b). Even among the currently recognized species, there are large differences among populations on different islands. Sharp-beaked ground finches on Santiago have an average size that is more than twice that of these birds on Genovesa (Grant and Grant 2008b). Superficially, the diversification of Darwin's finches looks like a story of slow directional selection of populations under different environmental conditions. The Grants have shown the real situation to be much more evolutionarily dynamic.

Their studies suggest that long-term diversification has resulted from repeated cycles of speciation, with plenty of additional evolutionary dynamics along the way (Grant and Grant 2008b). The islands differ in the range of available food resources, which favors birds with different beak morphologies on different islands. Songs also diverged among populations on different islands, partly as a consequence of the evolution of different beak sizes and shapes, partly through errors made by young birds in copying the songs they heard, and partly through sexual selection (Grant and Grant 1996a, 2010a). As populations from different islands came into contact, songs provided only partial barriers to mating, resulting in some hybridization between populations during these early stages of speciation. Competition for resources, however, disfavored birds with intermediate morphologies, thereby favoring individuals that mated only with other, similar individuals. Selection during this stage also may have contributed to further divergence among sympatric populations.

Even over the short term, all the different parts of this process have been observed in detailed studies of medium ground finches, *Geospiza fortis*, and cactus finches, *Geospiza scandens*, on the small island of Daphne Major (fig. 7.2). These studies have shown that the directions of adaptive evolution can fluctuate repeatedly within just a few decades. Since the middle 1970s, the finch populations have undergone multiple evolutionary changes through a combination of natural selection on morphology, hybridization, and also changes in song through cultural transmission (Grant and Grant 2002, 2008a, 2008b, 2008c, 2010b). During a major drought in 1977, 85 percent of medium ground finches died, and those that survived were larger individuals that were able to crack the seeds of *Tribulus cistoides*. Because beak size is a highly heritable trait in these birds, evolution toward birds with large beak size occurred quickly.

During the wet year of 1983, selection and evolution occurred in the opposite direction. *Tribulus* plants were smothered by the growth of other plants, especially the vine *Merremia aegyptia*. These plants had small seeds, providing a selective advantage to *G. fortis* individuals with small, pointed beaks. During these years, the cactus finch population, which feeds on different plants, evolved in the opposite direction. Over the next two decades *G. fortis* birds remained smaller and had more pointed beaks, but beak size slowly decreased back to the range found in 1973. Meanwhile, cactus finches remained large and retained blunter beaks than in 1983, but overall beak size continued to decrease (Grant and Grant 2008b).

Evolution driven directly by available food resources explains only some of the observed evolutionary changes in the birds on Daphne Major. In 1982, two females and three males of large ground finches, *G. magnirostris*, became established on the island. Like *G. fortis*, large ground finches commonly feed on *Tribulus* seeds, but they do so much more efficiently than *G. fortis*. They also eat more, because they are almost twice the size of *G. fortis*. As the *G. magnirostris* population grew through breeding and further immigration, these birds increasingly competed for food with *G. fortis*. During the drought of 2003 and 2004, *G. magnirostris* outcompeted *G. fortis*, so that among *G. fortis* individuals, those with small bills were favored because they were less likely to compete for the same foods eaten by *G. magnirostris*. Hence, although the *G. fortis* population evolved to larger beaks during the 1977 drought due to selection imposed directly by availability of particular food resources, it evolved toward smaller beaks in 2004 due to competition

Fig. 7.2 Three Darwin's finches from Daphne Major in the Galápagos Islands. *From top to bottom: Geospiza fortis* (medium ground finch), *G. scandens* (cactus finch), and *G. magnirostris* (large ground finch). Photographs are not to scale. *Geospiza fortis* and *G. scandens* weigh on average about 20–21 grams, but *G. magnirostris* weighs about 34 grams (Grant and Grant 2008b). Photographs courtesy of Rosemary Grant and Peter Grant.

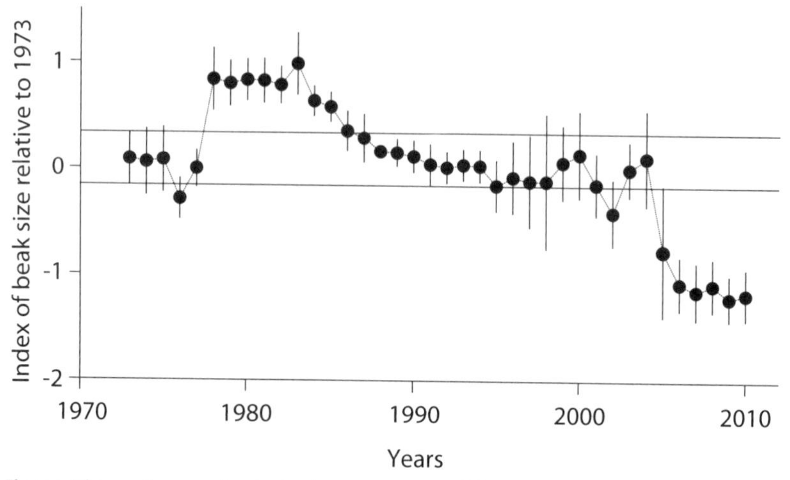

Fig. 7.3 Change in mean beak size (±95 percent confidence intervals) over time in medium ground finches (*Geospiza fortis*) on the island of Daphne Major in the Galápagos. Values are deviations in the first principal components of beak size. The horizontal lines indicate the upper and lower confidence limits of mean beak size in the 1973 sample. After Grant and Grant (2010a).

with another species that altered food availability in the opposite direction (fig. 7.3).

As population sizes have fluctuated on Daphne Major, *G. fortis* and *G. scandens* individuals have sometimes mated with each other. In the 1970s none of these hybrids survived to breed, but some did in the 1980s, contributing to the subsequent evolution of morphology in these species. By 2003 about 30 percent of *G. scandens* individuals contained some *G. fortis* alleles, and the two species had also become more similar morphologically (Grant et al. 2004). *Geospiza fortis* also sometimes breeds with *G. fuliginosa*, called the small ground finch, which occurs on the island as an immigrant in small numbers. These hybridization events contributed importantly to the standing genetic variation of *G. fortis* and *G. scandens* on Daphne Major (Grant and Grant 2010b). Culturally inherited songs have played an important role in these hybridization events, as some individuals have mis-imprinted on the song of the other species. Over a remarkably short length of time, the songs of these birds have changed on the island (Grant and Grant 2010a).

In less than four decades, then, *G. fortis* and *G. scandens* have undergone evolution through selection driven by changing food resources associated with droughts and El Niño events, competition from the immigrant *G. magnirostris* population, and interspecific hybridization as population numbers fluctuated and some individuals mis-imprinted on the wrong species. In every decade, *G. fortis*, *G. scandens*, or both species have evolved significantly in body size, beak size, or beak shape.

BALANCING SELECTION, HETEROZYGOTE ADVANTAGE, AND SELECTION AGAINST INBREEDING

Although fluctuating directional selection can maintain polymorphisms at the species level, it is likely often insufficient to maintain variation over long periods of time at the population level. Some form of balancing selection must often be involved. The two major hypotheses on how selection acts directly to maintain genetic variation through balancing selection are heterozygote advantage and frequency-dependent selection. With heterozygote advantage, individuals heterozygous at loci under selection are favored over individuals homozygous at those loci. With frequency-dependent selection, individuals with rare forms of alleles at loci under selection are favored over individuals with common forms of alleles.

These two mechanisms of balancing selection have been notoriously difficult to distinguish in nature solely from studies of the current pattern of genetic variation found within populations. By definition, selection driven by heterozygous advantage favors heterozygosity within a population, but natural selection favoring relatively rare alleles through frequency-dependent selection will also result in greater heterozygosity than occurs in a randomly breeding population. Both forms of balancing selection can therefore result in intermediate frequencies of alleles at a locus.

Some studies showing evidence for balancing selection therefore conservatively conclude that one or both processes may be involved (Huchard et al. 2010). Molecular and statistical methods that allow evaluation of the expected levels of genetic variation at each locus relative to other loci in the genome are providing new ways of searching for the signatures of balancing selection (Charlesworth 2006a; Buzbas et al. 2011). Through these studies, the number of examples of traits thought to be under balancing selection continues to grow (Makinen et al. 2008; Andrés et al. 2009; Hanski 2011; Linnenbrink et al. 2011).

Heterozygote advantage is the simplest version of balancing selection. It results in consistent differences among individuals in their relative fitness in a particular environment. Individuals with two different allelic forms of a gene (heterozygote individuals) have higher Darwinian fitness in that environment than do individuals with only one form (homozygotes). Most well-studied examples involve selection driven by interactions among species rather than by selection imposed by variation in the physical environment, and the clearest example is human resistance to malaria. Multiple human genes have been shown, or are suspected, to be maintained in a polymorphic state by heterozygote advantage within populations under risk of malaria (Wellems et al. 2009). The first example was reported in 1954, when Anthony Allison found that individuals whose red blood corpuscles have a heterozygous trait called sickle cell are less susceptible to one form of malaria in Kenya than homozygous individuals (Allison 1954). The sickle hemoglobin gene is distributed throughout Africa and parts of Asia, and its distribution matches fairly closely the distribution of transmission intensity of malaria in Africa. Similar examples of polymorphisms are known from other species, but many of these may be maintained by forms of balancing selection other than heterozygote advantage (Hedrick 2007).

One of the other most convincing cases of heterozygous advantage is the maintenance of different forms of the phosphoglucose isomerase (*Pgi*) in butterflies. This enzyme has been shown to be important in multiple aspects of the ecology of butterflies. In *Colias* butterflies, heterozygotes for *Pgi* have higher male mating success and are able to fly at lower temperatures (Watt and Cassin 1983; Watt et al. 2003). There is also striking similarity of *Pgi* allelic frequencies over large geographic areas, although frequencies vary somewhat among local populations (Watt et al. 2003). In Glanville fritillary butterflies (*Melitaea cinxia*), heterozygotes have a higher dispersal rate, a higher body temperature when flying at low ambient temperatures, a larger clutch size than homozygotes, and allelic frequencies that depart from expectations of neutral evolution (Saastamoinen and Hanski 2008; Niitepıld et al. 2009; Wheat et al. 2010; Hanski 2011). Both in *Colias* and *Melitaea*, individuals homozygous for one or more of the alleles have relatively low survival (Watt 1977; Orsini et al. 2009).

Pgi in butterflies and sickle cell in humans both seem to be under strong selection that directly favors heterozygotes over homozygotes. Heterozygosity, however, can also be favored in inbred populations through selection against inbreeding depression. Although heterozygote advantage is driven by the fitness advantage of heterozygotes, inbreeding depression is driven by the fitness disadvantage of homozygotes. The higher the level of inbreeding, the greater the chance that deleterious recessive mutants will occur in the homozygous state.

Consequently, selection favors individuals that avoid breeding with close relatives. The great variety of mechanisms that have evolved to avoid inbreeding suggests that selection against inbreeding has been a strong selective force in some species (Charlesworth and Charlesworth 1987; Sherborne et al. 2007; Teixeira et al. 2009; Barrett 2010). At the other extreme, however, outbreeding depression arising from mating of populations adapted to different environments puts a limit on selection against inbreeding.

Through selection against inbreeding, deleterious recessive mutations remain in the population at frequencies determined by the mutation rate and the strength of selection against homozygous individuals that harbor those deleterious mutations. Heterozygosity maintained by this form of mutation-selection balance will generally maintain deleterious alleles at low frequencies, whereas heterozygosity maintained by heterozygote advantage will maintain alleles at intermediate frequency (Charlesworth and Willis 2009). Selection driven by either heterozygote advantage or inbreeding depression therefore can act as powerful means of maintaining genetic variation in populations.

FREQUENCY-DEPENDENT SELECTION AND PARASITES

Frequency-dependent selection has long been considered as effective as heterozygote advantage in maintaining genetic variation within populations. Many of the inferred examples are of genes that contribute to resistance against pathogens or parasites. The theory of parasite-mediated frequency-dependent selection has been explored for decades, and many mathematical models show that the process can produce cycles of change in allelic frequencies on both sides of an interaction between two species. This adaptive process can therefore result in continual evolutionary change in populations. The pace of change depends on the number of alleles involved, the ability of the parasite to adapt to multiple host alleles, the relative strength of selection on the host and parasite populations, and other factors (Van Doorn and Dieckmann 2006; Matessi and Schneider 2009; Trotter and Spencer 2009).

During negative frequency-dependent selection between hosts and their parasites, parasites become adapted to the most frequent genetic forms of a local host population. This adaptation favors rare mutant forms of the host to which the local parasite population is not adapted. Over multiple generations, one genetic form of a host can cycle from rare to common to rare repeatedly (fig. 7.4). More generally, though, this cyclical process is likely to lead to the evolution and maintenance of multiple host genotypes within a population as selection constantly changes, favoring first one host allele and then another and then another. Meanwhile, the parasite population will also evolve multiple forms that are cycling up and down, as the selection on the parasites tracks changes in allele frequencies in the host population.

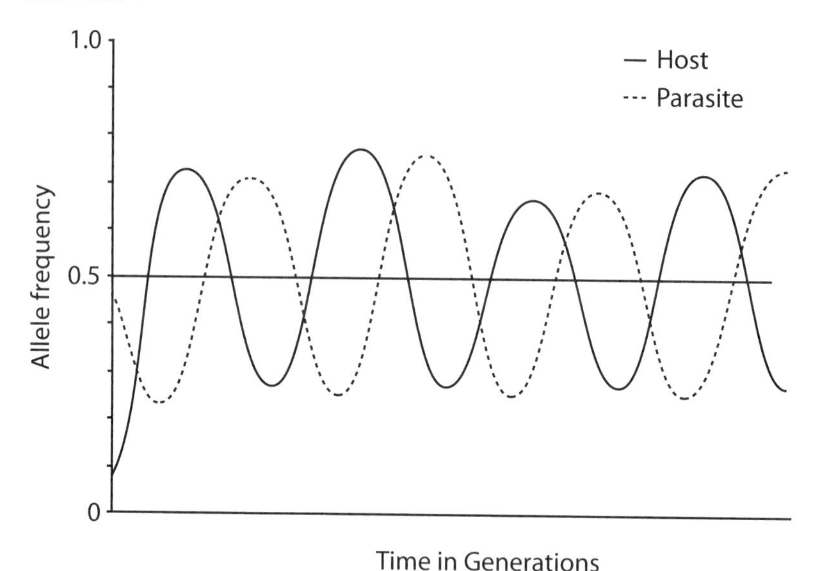

Fig. 7.4 Cyclical change in host and parasite alleles due to frequency-dependent selection. One allele at one locus is shown for each species. In natural populations, selection would commonly be distributed across multiple alleles, creating more complex swings in allele frequencies.

Through this coevolutionary process, natural selection can maintain much genetic variation in populations, although we still do not know how much of the variation is due to this particular form of balancing selection.

This form of evolution is therefore different from the popular view of evolution as a process that shifts populations slowly and continually toward greater defenses or larger sizes or more complex forms. Evolution can be cyclical rather than escalatory, and it can directly favor further change (Tellier and Brown 2007; Brown and Tellier 2011). The further a host population moves in one evolutionary direction, the greater the strength of natural selection to move the population in another direction. As a result, populations retain a mix of defenses and counterdefenses that are favored or disfavored at different points in time.

Studies of the ecological drivers of fluctuating selection point to natural selection driven by parasites and pathogens as one of the major reasons why species continue to harbor so much genetic variation. A large worldwide analysis of variation in humans at the level of individual nucleotide sites (single nucleotide polymorphisms) has indicated that the strongest evidence of selection on variation on the genome is associated with sites involved in immunity (McEvoy et al. 2009). Although some of the variation is likely maintained within populations or at the species level, substantial variation is probably maintained by selective processes that operate at the genomic level with little effect at the level of individuals. Together, these processes provide the raw material for yet further evolution.

COPING WITH RAPIDLY CHANGING GENETIC VARIATION

Many species have regions of their genome that evolve particularly fast, and those regions are often those involved in coevolution with pathogens and parasites. That seems to be the case, for example, with the fungus-like *Phytophthora infestans*, which caused the Irish potato famine in the 1880s and continues to ravage potato crops worldwide. The entire genome of this pathogen has been sequenced and, compared with other *Phytophthora* species, it has regions that include many repeated sequences of genes and gene families involved in attacking hosts and evading host defenses (Haas et al. 2009). These fast-evolving genes produce disease effector proteins that alter the physiology of their hosts, making it easier for these pathogens to colonize potatoes. Rapid genetic changes in these genes continue to add new variation to these pathogen populations, providing raw material for natural selection to favor forms of the pathogen able to overcome each new resistant form of potato introduced by plant breeders.

The evolutionary dynamics of continually renewed genetic variation become more complicated when the relative generation times of hosts and parasites differ greatly. In species like humans, a parasite may go through tens or even thousands of generations within a host, with natural selection acting during each generation as a small parasite population adapts to that particular individual. The parasites in that host individual are part of a larger population associated with a local population of hosts, which are in turn connected to parasite populations over larger geographic scales. The result can be a great deal of genetic variation that shows a clear geographic structure.

For example, HIV types in humans have diversified quickly into geographically different types over just the past several decades. HIV-1, which may have transferred from chimpanzees to humans, has three major genetic forms and at least thirteen subtypes or sub-subtypes (Cohen et al. 2008; Lynch et al. 2009). The major subtypes differ in their DNA sequences by more than 15 percent, and they differ in the geographic regions in which they are most common. The genetic diversity of HIV has continued to increase through the current epidemics, but much of that diversity appears to result from natural selection that happens within individual humans during the course of an infection (Pybus and Rambaut 2009). Following its infection of an individual, HIV undergoes rapid evolutionary changes in response to selection imposed by the human immune system. The result is continual turnover of HIV genetic lineages within humans infected with HIV. This is rapid evolutionary change in its most dramatic form.

Human influenza virus has a different global pattern of genetic variation that arises from a combination of localized sources of new genetic forms, global air travel, rapid development of immune response in human populations, and further rapid evolution of new forms in the virus (Pybus and Rambaut 2009). In any year there are few genetic strains of the virus within and among human populations. New strains tend to arise in Southeast Asia and spread to temperate regions as we fly in airplanes and spread the strains to destinations around the world. Our immune system quickly develops antibodies against each new strain as it spreads through populations during the following winter, leading to the demise of that form. This course favors the spread of a new form of the virus to which we do not yet have antibodies. The process repeats itself as the virus continues to mutate, which it does quickly (Suzuki and Nei 2002). Unlike HIV, then, human influenza tends to have low genetic variation globally and within individuals, and it has fast genetic turnover worldwide.

Human influenza, though, is just one version of a genetically diverse group of type A influenza viruses that infect humans, birds, and some other mammals. These viruses harbor a great deal of genetic diversity that is constantly changing as new mutations appear, the different types of the virus reassort genetic segments, and selection acts on these viruses in different ways in different hosts. The H1N1 (swine flu) epidemic, for instance, appears to have come from pigs through an earlier triple reassortment of avian, human, and pig influenza viruses into a novel genetic form (Chen et al. 2009).

It is therefore not surprising that the battle against the evolution of diseases must continually be fought but can never truly be won. New genetic variation continually enters pathogen populations, and nowadays that variation can come from far-distant populations. Meanwhile, genetic variation is maintained within local populations through mutation, recombination of genes within populations, reassortment of genes among strains, and sometimes frequency-dependent selection.

We and other long-lived organisms have, in turn, evolved to cope with genetically diverse and rapidly evolving parasites through genetic mechanisms that maintain genetic diversity within our own populations. Among these mechanisms, the vertebrate immune system is the most impressive. The major histocompatibility complex (MHC) is the part of our immune system that generates customizable adaptive immunity against specific infection. When a new infection occurs, the class I molecules from the MHC system bind to peptides on the foreign invader, causing the infected cells to die. Class II molecules bind to extracellular foreign peptides and stimulate antibody and macrophage production. Through this process the body produces customizable antibodies and defenses against foreign invaders (Meyer and Thomson 2001).

There are so many foreign invaders and so much variation in selection on MHC genes that this part of our genome harbors an astonishing amount of genetic diversity. It is the part of the human genome that shows the strongest signal of selection in some recent genome-wide analyses of signatures of recent selection (Albrechtsen et al. 2010). In humans, the MHC system is located in a region on chromosome 6 called the human leukocyte antigen (HLA) system, which contains more than 220 genes. In the late 1990s the international ImMunoGeneTics project's IMGT/HLA database, which is the repository for reports on genetic variation for this region of the human genome, included 964 known allelic variants within this region. A decade later, the number had risen to over 3,300 alleles, with more than 450 new alleles being reported each year (Robinson et al. 2009). The number continues to expand. The maintenance of so much variation in such a small region of the genome may be due to multiple sources of natural selection, but there is now enough data to know that selection imposed by parasites is one of the major drivers.

Fish, amphibians, reptiles, birds, and other mammals all have versions of the MHC system, and there is growing evidence that selection imposed by parasites maintains at least some of the variation found in these species as well. MHC alleles have been shown to affect susceptibility to particular pathogens or parasites in wild Atlantic salmon (Dionne et al. 2009). The pattern of MHC polymorphism found in other species such as horses is also consistent with selection driven by pathogens or parasites (Janova et al. 2009). Some of these studies indicate that selection favors relatively rare forms of these alleles, thereby maintaining genetic variation in these populations as selection favors first one allele and then another.

High levels of genetic variation also seem to be maintained through selection caused by parasites in other parts of our genome and probably that of other vertebrates. It has long been suspected that the diversity found in our ABO blood group genes is maintained to some degree by selection favoring individuals with blood types to which local parasites are not adapted or, alternatively, by highly variable selection driven by other, constantly changing environmental factors. Some studies have shown that individuals with particular blood types are more resistant to some diseases than others, and there is some evidence that blood types are maintained by negative frequency-dependent selection in populations (Fumagalli et al. 2009).

Genetic variation maintained by interactions with parasites also appears to extend to variation in the number of copies of genes found within individuals. Copy number variants (CNV) are repeated DNA sequences of about 1,000 base pairs or more. They are common in some species, including mammals and flies. Somewhere between 1,500 and 2,900 human genes involve copy number polymorphisms (CNPs) (Nguyen et al. 2008). Common fruit flies (*Drosophila melanogaster*) have over 2,600 independent CNVs, and a little over half of those overlap with known genes (Emerson et al. 2008). Some of these are known toxin-resistance genes.

It is unknown how much of that variation is simply neutral with respect to natural selection, but we should expect that some CNVs are maintained by selection. Individuals with higher copy numbers of some genes have higher levels of the proteins produced by those genes. Copy number can also affect gene structure and the regulation of genes. If some of those genes are associated with defenses against parasites, which we now know they are, then individuals with different copy numbers will differ in their susceptibility of parasites. Similarly, copy number may affect the ability of parasites to overcome host defenses.

The rapid development of genomic techniques has made it easier to identify these polymorphisms and then evaluate whether they are maintained through natural selection. This is now an active area of research that links molecular, evolutionary, and ecological approaches, and examples of CNPs associated with disease are increasing quickly. About 5 percent of the human genome has duplicated gene copies associated with immune responses to pathogens (Bailey et al. 2002). The human gene *CCL3L1* has a copy number polymorphism that affects the susceptibility of individuals to HIV infection (Gonzalez et al. 2005). A gene crucial in folate synthesis in the malarial parasite *Plasmodium* has high levels of copy number polymorphism and shows evidence of having been under strong selection imposed by antifolate drugs used to combat malaria (Nair et al. 2008).

It is unclear, though, whether these examples are the exceptions or the rule for how copy number polymorphisms are maintained within species. Selection driven by ecological factors other than parasites, such as sexual selection or climate-mediated selection, could be equally important. In some cases it is puzzling why particular variants persist, because they are associated with increased susceptibility to diseases such as asthma, allergic responses, or psoriasis (Ionita-Laza et al. 2009; Wain et al. 2009). It seems probable, though, that some of these variants are interacting with the rest of the genome in complex ways. In addition, relaxation of selection could account for why some populations currently have high levels of polymorphisms (Nguyen et al. 2008). Natural selection has yet to eliminate variants that may have been favored in the past. The real answer to what maintains this variation is likely "all of the above." The current problem to solve is which of these answers is the most important in maintaining so many copy number polymorphisms within species.

MOLECULAR MIMICRY AND GENETIC VARIATION

The problem of controlling diverse assemblages of enemies is increased by the evolution of viral pathogens that have evolved specifically to mimic normal parts of the physiological processes of their hosts, thereby escaping detection (Elde and Malik 2009; Domingo-Gil et al. 2011). Even interactions that we might guess would favor escalating coevolutionary changes are often driven instead by the evolution of novel tactics in the pathogen to escape detection—such as mimicry or camouflage—and by novel tactics in the host to detect these pathogens. These forms of selection provide an evolutionary alternative to accumulation of increasingly expensive arsenals of defenses and counterdefense. There is always the potential, though, for some escalation, because molecular mimicry could sometimes demand the use of relatively expensive metabolic pathways by pathogens in order to remain mimetic, and it may sometimes demand the evolution of expensive detection mechanisms in hosts.

All cells are bombarded with parasites that have evolved to co-opt various cell functions in their hosts. These parasites are constantly evolving new ways of attacking hosts while remaining undetected. Every major component of the eukaryotic cell cycle is subjected to viruses and bacteria that mimic the normal molecular cues used by host cells. These pathogens mimic molecules that control cell death, the cell cycle, immunity, movement through the cytoskeleton, and movement across membranes (Elde and Malik 2009; Davey et al. 2011; Ludin et al. 2011). In doing so, they manipulate host functions to their own advantage. Some viruses prevent the process of cell death that organisms commonly use to prevent the spread of infections. Other viruses mimic cyclin activity in the cell cycle, thereby keeping cells at stages most conducive to viral propagation. Yet other viruses mimic molecules that control passage of any particles through the cell wall into the cytoplasm or across interior membrane walls. This mimicry allows the pathogen to enter a cell. The result is a never-ending set of molecular coevolutionary battles involving every part of host cells.

Selection can favor pathogens that are perfect mimics of host molecules and simply co-opt the host functions to their own advantage, or selection can favor pathogens that are imperfect mimics and trick the host into performing new functions (Elde and Malik 2009). Although perfect mimicry would superficially seem to be the route most likely favored by selection, it can restrict evolution of the pathogen. It makes the pathogen subject to all the controls of the host's cellular regulation mechanisms. A greater range of options is available to imperfect mimics, because they can change the cellular processes to their own ends. Imperfect mimicry, though, is not a guaranteed route to parasite success. It opens the way for hosts to evolve mechanisms of discriminating their own normal molecules from the pathogen's imperfectly mimetic molecules.

One of the most impressive examples of molecular coevolution with viral mimics is that between primates and poxviruses (Elde et al. 2009; Elde and Malik 2009). These poxviruses have evolved to subvert immune response by mimicking eIF2α, which is the substrate of protein kinase R (PKR). This protein blocks protein translation in infected cells by phosphorylating eIF2α. Poxviruses, however, produce an imperfect mimic called K3L that competes with eIF2α, thereby restoring translation and viral replication. PKR must therefore discriminate between eIF2α and the viral K3L. These three-way interactions create multiple possible coevolutionary scenarios, but eIF2α is a highly conserved molecule. As a result the coevolutionary dynamics occur between PKR and the viral K3L. PKR must discriminate between the essentially unchanging substrate eIF2α and the rapidly evolving K3L. Amid coevolution with K3L, though, it must retain the ability to interact with its normal substrate.

The coevolutionary process has favored the evolution of PKR at multiple places on its surface to discriminate eIF2α from K3L. That process has also favored the evolution of K3L at the sites being used by PKR to make that discrimination. By using multiple points along the molecular surface for discrimination, PKR makes it more difficult for the viruses to escape detection. It has resulted in adaptive diversification of the sites of mimicry and discrimination during the phylogenetic divergence of primates. Comparison of the molecular substitutions found in the PKR of Old World monkeys and their poxviruses suggests strong episodic positive selection on both sides of this interaction (Elde et al. 2009). Remarkably, the rapid evolution of PKR has been so selective that divergent PKRs respond most strongly to the viral K3L to which they are normally subject, yet they can recognize eIF2α from multiple primate species (Elde et al. 2009).

The number of examples of molecular mimicry continues to expand, as does the range of molecular mechanisms by which coevolution between eukaryotes and prokaryotes generates adaptive variation and diversification. These studies suggest a molecular landscape that is under constant and intense selection as hosts evolve to detect and defend against pathogens, and pathogens evolve to escape detection and circumvent defenses if detected. The forms of molecular defense and counterdefense vary among taxa (Rolff and Reynolds 2009; Schneider and Collmer 2010; Scanlan et al. 2011), but genomic analyses are showing that significant parts of genomes are devoted to these evolutionary battles.

THE CHALLENGES AHEAD

Selection favoring adaptive genetic variation is now evident at every level in the hierarchy of life. Most of our current understanding is from studies of selection favoring variation at a genetic locus, a phenotypic trait, or a set of genetically correlated traits. With so much variation maintained by selection in so many ways, one of the current challenges is to understand how the many different forms of selection, acting together, shape the evolution of variation in populations. There are undoubtedly many ripple effects throughout genomes as selection favors greater or less variation in particularly crucial parts of genomes.

The important message of ecological, genetic, and genomic studies of selection is that variation is not just something that selection acts upon. Selection often acts directly in ways that maintain variation. The maintenance of adaptive genetic variation is so important that selection has favored the evolution of processes by which genomes acquire new genetic variation. It is to those processes that we now turn.

8

Recombination and Reproduction

Most species have evolved mechanisms for exchanging genes among individuals. Once the ability to exchange genes in a systematic way had evolved, it resulted in three of the most important outcomes in the history of life: the evolution of sexual reproduction, the evolution of mating types, and differential selection on those mating types. These three aspects of populations are central to any understanding of why evolution is so relentless: they enhance genetic variation on which natural selection can act, and they often lead to frequency-dependent selection that maintains variation in populations. The variation can be expressed in complex ways. For example, a population may have two sexes that differ in the types of gametes they make and also have multiple mating types within a sex. Or a population may show little evidence of separate sexes and still have multiple mating types that are incapable of mating with other individuals of the same type.

This chapter explores the ecological underpinnings of the evolution of genetic recombination and sexual reproduction and their roles as drivers of evolution.

BACKGROUND: SEXUAL REPRODUCTION, MATING TYPES, AND FREQUENCY DEPENDENCE

The evolution of sexual reproduction often favors the evolution of separate sexes or mating types, and that generates multiple mechanisms by which selection continually varies within populations. The presence of males and females in a population results in selection favoring the minority sex, because those individuals often will have easier access to mates and thereby contribute disproportionately to the next generation. All else being equal, this frequency-dependent selection should result in an equal sex ratio in an outbreeding population, as selection seesaws between favoring either males or females (Fisher 1930).

The expected sex ratio, however, differs in highly inbred populations, because selection often favors a higher proportion of females. The expected sex ratio is also skewed in species in which genetic relatedness among females differs from that among males, or in environments in which it is energetically more costly to produce highly fit offspring of one sex than the other. The presence of sex-determination genes in some species prevents wild swings in sex ratios, but some degree of variation in sex ratios is found even among these species. The theory of the evolution of sex ratios, based in part on frequency-dependent selection, has become a well-developed and highly successful branch of evolutionary biology. It has been successful at predicting the dynamics of evolutionary changes in males and females in multiple species. It has also been successful at explaining many of the observed patterns of skewed sex ratios found in populations, when those ratios are subject to selection on nuclear genomes (Hamilton 1967; Charnov 1982; West 2009) rather than interactions caused by symbionts (see chap. 6).

The situation becomes more complicated when there are multiple mating types. Although Ronald Fisher wrote in 1930 that there were always two sexes in sexually reproducing species, we now know that the situation can be more complicated. Some species include multiple mating types rather than identifiable sexes. Individuals cannot mate with other individuals of their own mating type, but they can mate with any individual that is not of the same mating type. The ciliate *Tetrahymena thermophila*, for example, has seven self-incompatible mating types (Paixão et al. 2011). In this species, genes affecting mating type determine only the probability that an individual will be of a certain type (Arslanyolu and Doerder 2000). Although the occurrence within a species of so many self-incompatible mating times is rare in nature, it has been found in multiple other ciliates. *Tetrahymena* includes species with as few as 2 mating types and others with as many as 8 or 9, and *Euplotes* includes a species with 12 mating types (Phadke and Zufall 2009). Species with multiple mating types show a complicated pattern of biased sex ratios in which both selection and genetic drift may be involved, but frequency-dependent selection may help maintain multiple forms within a population (Paixão et al. 2011).

Even more complicated is the situation in many fungi, in which different lineages have evolved different mechanisms of sex determination and different numbers of mating types (Lee et al. 2010; Billiard et al. 2011). At the one extreme *Saccharomyces cerevisiae* yeasts have two mating types, and reproduction occurs when the two mating types meet. They can, however, also undergo self-fertile sexual reproduction. At the other extreme, basidiomycete mushrooms often have many mating types in which cells fuse after mating but nuclear fusion does not occur. Among these species, there is no mechanism for asexual production of spores. In between these extremes is a bewildering array of mechanisms of sexual reproduction that involve many different complex patterns of haploid, diploid, and tetraploid life history stages.

By some estimates, basidiomycetes can have up to thousands of mating types (Brown 2001; Fraser et al. 2007). The mating-type genes of mushrooms often involve many alleles, and individuals with different allelic combinations constitute different mating types. After mating, the fused cells generally form a mycelium, called a dikaryon, which includes the paired, but unfused, haploid nuclei. The genetics and molecular biology of these complex mating systems have been studied in detail (Brown 2001), but it is still unclear what combination of selection pressures could have favored this complicated form of sexual reproduction.

Whatever its origin, the result is a mechanism producing high levels of genetic variation in many basidiomycete species. Rare genotypes would have a high probability of encountering individuals of different mating types and may therefore be favored by selection. The co-occurrence of so many mating types in some taxa makes sense only if frequency-dependent selection is acting on the populations (May et al. 1999; Engh et al. 2010).

SEXUAL REPRODUCTION AND FREQUENCY-DEPENDENT SELECTION

Frequency-dependent selection on mating types is just one of multiple ways in which frequency-dependent sexual reproduction shapes adaptive variation and the evolution of populations. The even more basic question is why species reproduce sexually rather than asexually, which has been a difficult problem to solve in evolutionary biology (Williams 1975; Maynard Smith 1978; Bell 1982; Otto 2009; Lively 2010b; Misevic et al. 2010). All else being equal, natural selection should favor females that reproduce asexually because they leave more genetic copies of themselves than a sexual female. If a sexual female produces, on average, an equal number of female and male offspring, she has a twofold disadvantage over an asexual female that produces only female offspring. Over time, asexual females should increase in frequency in a population over sexual females. Adding to the direct genetic costs of sexual reproduction are the ecological costs that can also affect fitness, such as time spent in searching for mates and the risk of sexually transmitted diseases.

Despite these costs, sexual reproduction remains the most common, and often the only, form of reproduction in many species. Through sexual reproduction, genetic variation within each population is rearranged every generation as males and females mate and their genes recombine in novel combinations in their offspring. In species that can reproduce both sexually and asexually, the frequency of sexual reproduction often varies among populations, suggesting that sexual reproduction remains under selection in some species (L. Barrett et al. 2008; King and Lively 2009).

There are two main views of how sexual reproduction is maintained by selection within species, and both are based on the fitness advantage resulting from the recombination of genes during sexual reproduction. As first noted by Ronald Fisher (1930) and H. J. Muller (1932), recombination allows combinations of advantageous genes to appear more rapidly than would occur in species lacking recombination. In the decades since then, two complementary versions of that view have come to dominate subsequent research, one focusing on the selective disadvantage of asexual reproduction and the other on the selective advantage of sexual reproduction. In 1964 Muller noted that asexual clones tend to accumulate deleterious mutations over time, resulting in a ratchet-like decrease in fitness (Muller 1964; Felsenstein 1974). This process, known as Muller's ratchet, occurs because there is no mechanism of DNA repair to excise deleterious mutations from the genome. Muller argued that recombination during sexual reproduction prevents mutational load from building up in populations by providing an effective mechanism of DNA repair.

By this view, sexual reproduction is favored because the alternative is generally worse, at least over the long term. The effects of Muller's ratchet could decrease through long-term selection favoring clones with lower mutation rates (Soderberg and Berg 2011). Nevertheless, the fact that most species are sexual, or have some alternative ways to allow recombination, suggests that asexual species are generally at a fitness disadvantage.

The major alternative view is that sexual reproduction is common because it produces individuals with rare gene combinations that are more likely to escape from constantly evolving virulent parasites and pathogens. By this view, sexual reproduction is a result of frequency-dependent antagonistic coevolution (Lively 2010a). It favors parasites that become adapted to the most common host genotypes, and it favors rare host genotypes to which the parasites are not adapted. There is no escalating arms race (Neiman and Koskella 2009). Sexual reproduction is about being different rather than being better defended, which is an expanding theme in evolutionary biology.

This view of the evolution of sexual reproduction relies on the spatial structure of populations, because it is driven by the adaptation of parasite populations to their local hosts. As parasites become adapted quickly to their local hosts, each generation of a host population becomes more susceptible to its local parasites than the previous generation, unless the host population continues to coevolve with its parasites. Local adaptation of parasites puts asexual host females at a disadvantage, because each female produces offspring genetically identical to herself, except for the occasional mutation. If an asexual female has a common genotype, her offspring will also have that genotype and be at risk of attack by the same parasites. If, instead, a female mates with a male, her offspring will have a novel mix of genes from both parents. One or more of those offspring may be a rare genetic form resistant to attack by the currently most common forms of the parasite.

Some mathematical models of the evolution of sexual reproduction suggest that both coevolution with enemies and Muller's ratchet could contribute to its long-term maintenance (Howard and Lively 2002). The beneficial effects, moreover, are spread among generations, with some having immediate effects in the next generation and others having more delayed effects (Peters and Lively 2007). Other models suggest that the importance of coevolution in favoring sexual reproduction depends on the virulence of a parasite and the genetics of the coevolving traits (Otto and Nuismer 2004; Peters and Lively 2007). Overall, it seems there are multiple conditions under which selection can directly favor sexual reproduction and the continuing evolution that it fosters.

TESTING FOR FREQUENCY-DEPENDENT SELECTION FAVORING SEX

Analyzing how parasites favor sexual reproduction in nature has been difficult, because most species are strictly sexual. That quickly narrows the potential choices for experimental studies. The ideal candidates are populations with both sexual and asexual individuals, and New Zealand mudsnails (*Potamopyrgus antipodarum*) have become one of the premier models for these studies. The snails have been the subject of long-term careful observations and experiments by Curtis Lively (1987, 2010a) and colleagues. For more than two decades, they have taken advantage of the fact that sexual and asexual forms of the snails occur together in multiple alpine lakes in New Zealand. The frequency of sexual snails varies among lakes. These snails are heavily attacked in some lakes by parasites, especially by a trematode species (*Microphallus* sp.) that prevents reproduction in attacked individuals.

It has taken multiple kinds of field and laboratory analysis to piece together the evidence that attack by these trematodes favors natural selection for sexual forms of the snail (Koskella and Lively 2007; Dybdahl et al. 2008;

Lively 2010a). Sexual snails are more common in lakes where the trematodes are abundant than in lakes where trematodes are rare, suggesting that the trematodes favor the sexual forms of the snails. There are multiple genetic forms of the snail within the sexual and asexual populations, and the lakes differ in any year in which genetic forms are the most common. Yearly surveys of the lakes have shown that rare forms of the snail tend to increase until they become common and then decrease over time. During the past twenty years, there have been multiple evolutionary changes in the relative proportions of different genetic forms of the hosts in the lakes (Jokela et al. 2009).

Field studies have shown that the trematodes in each lake are adapted to the most common genetic forms of the snail in that lake (Lively and Dybdahl 2000). Moreover, studies of experimental coevolution in laboratory mesocosms have shown that the most common snail genotype becomes more susceptible to the parasites over time and then decreases in frequency, because the parasites become adapted to it (Koskella and Lively 2007, 2009). In these controlled experiments, the most common host genotype at the beginning of the experiment decreases in frequency over subsequent generations in mesocosms with trematodes but not in mesocosms lacking trematodes. In addition, in mesocosms with parasites the initially most common genotype becomes more susceptible to its coevolving parasites during the multiple generations of the experiment. There is a time lag in adaptation of the parasites to the changing genotypic frequencies of the hosts, as expected if parasites are evolutionarily tracking host frequencies.

Taken together, these studies suggest that attack by highly virulent parasites favors sexual reproduction. The parasites favor rare forms of the snail less susceptible to attack by the locally adapted parasites. Additional evidence for the role of parasites in maintaining sexual reproduction comes from studies of snails in shallow water as compared with deep water (King et al. 2009). A trematode must pass through two hosts during its lifetime. The snails serve as the intermediate host, and dabbling ducks and wading birds serve as the final host. The snails occur both in shallow water and in deep water, but the birds occur only in shallow water. Snails living in shallow water suffer high rates of infection by the trematodes, because the parasites can cycle between the snails and the ducks. Coevolution between the snails and parasites is strong in these shallow-water habitats, and selection favors sexual snails. The dabbling ducks and waders, however, do not forage in deep-water habitats, which breaks the parasite life cycle and results in chronically low levels of infection in snails living in deep water. Snails in the deep-water habitats are more often asexual (fig. 8.1).

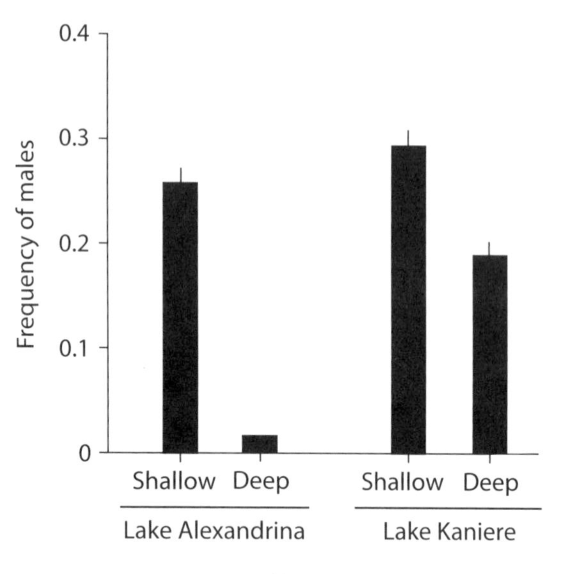

Fig. 8.1 Frequency of males among New Zealand mudsnails (*Potamopyrgus antipodarum*) in two alpine lakes. Values are means ±1 SE. After King et al. (2009).

Through a set of experiments that challenged deep- and shallow-water snails with parasites within and among lakes, King et al. (2009) showed that the parasites are not only adapted to snails within their local lake but, even more finely, to subpopulations of snails within their local lakes. Local adaptation driven by coevolution is restricted to shallow-water snails and their parasites. The trematodes are better adapted to their local shallow-water snails in their home lake than to deep-water snails from that lake or to shallow- or deep-water snails from another lake (fig. 8.2).

One additional experiment was needed to complete the argument: expose the parasites from both lakes to snails from yet another lake. That experiment showed that snails from shallow and deep water did not show any inherent differences in susceptibility to parasite infection in the absence of local parasite adaptation. Taken together, this and the other experiments eliminated the non-coevolutionary alternative hypothesis that sexual snails or shallow-water snails are simply more susceptible to attack by the trematodes.

These experiments also show that the parasites were no more infective to sympatric, deep-water snails than to allopatric snails from either habitat. Hence, unlike in shallow water, coevolution is not occurring in deep water. Follow-up experiments have shown that these patterns have remained stable across three lakes over at least two years and are robust to the parasite dose used in the trials (King et al. 2011). The long-term observational studies and carefully designed experiments on these snails and trematodes have produced a set of results that argue convincingly for the role of frequency-dependent selection driven by parasites in the maintenance of sexual reproduction in at least this species.

Additional evidence for the role of parasites and coevolution in the evolution of sexual reproduction comes from a multigenerational experiment evaluating the persistence of populations and the rates of outcrossing. Using *Caenorhabditis elegans* nematodes and virulent *Serratia marcescens* bacteria, Levi Morran and colleagues established three nematode populations in the laboratory: obligately selfing, obligately outcrossing, and wild type (Morran et al. 2011). This last population allowed individuals the freedom either to self-cross or outcross, and the baseline outcrossing rate was about 20 to 30 percent. They introduced novel genetic variation into the populations and then subjected each population to three treatments in five replicate cultures. The two extreme treatments had either nematodes with no bacteria or nematodes with coevolving bacteria. A third treatment had nematodes with bacteria, but the bacteria used in each generation were from a fixed, non-evolving strain. After thirty generations, all the obligately selfing populations in the coevolution treatment were extinct, but all the outcrossing and wild-type populations still persisted in all three treatments. Also, the

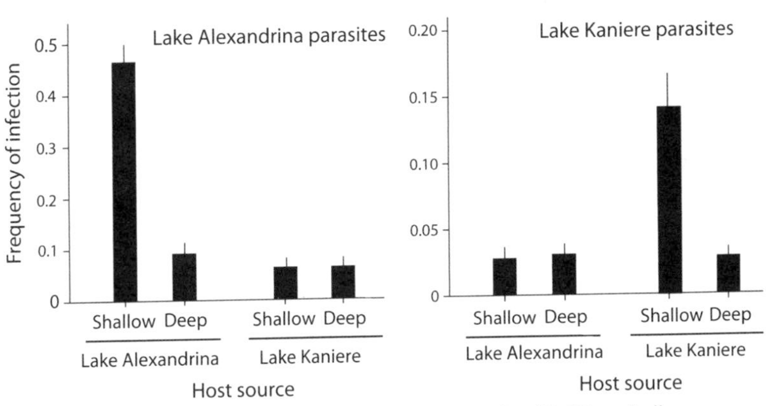

Fig. 8.2 Experimental infection of New Zealand mudsnails with *Microphallus* trematodes. Shallow-water and deep-water snails in both lakes were challenged with trematodes collected from shallow water in these lakes. Parasites show local adaptation to the shallow-water snails in their home lake. Values are means ±1 SE. After King et al. (2009).

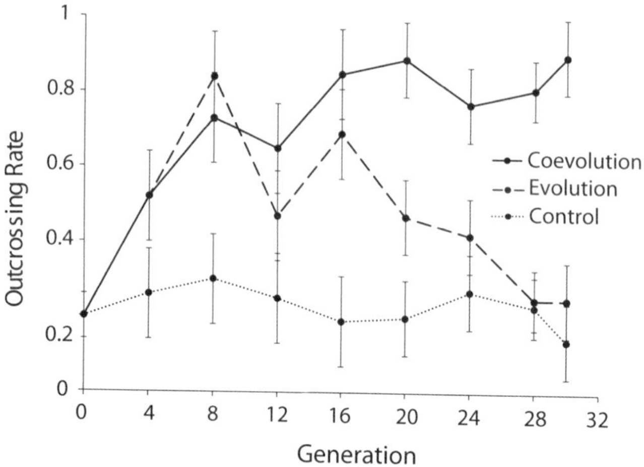

Fig. 8.3 Change over time in the rate of outcrossing in wild-type populations of *Caenorhabditis elegans* nematodes coevolving with highly virulent *Serratia marcescens* bacteria in laboratory cultures. The changes are compared with control populations that did not have bacteria and with "evolution" populations that were subjected each generation to a fixed, non-evolving strain of the bacteria. Values are the means ±2 SE of five replicates for each treatment. After Morran et al. (2011).

rate of outcrossing increased over time in the coevolution treatment of the wild-type populations (fig. 8.3).

Together, the field and laboratory studies suggest that coevolution with parasites favors genetic variation in host populations through multiple mechanisms, including increased sexual reproduction, high levels of outcrossing, and frequency-dependent selection. Whether coevolution with parasites is the major cause of the maintenance of sexual reproduction, rather than one of several important causes, is not known. These studies, though, show convincingly that selection mediated by parasites is a potentially powerful selective force in maintaining genetic variation in populations.

SEX AND SEXUALLY TRANSMITTED PARASITES

Parasites affect the evolution of sexual reproduction in yet other ways that favor ongoing evolution. Sexually transmitted pathogens and parasites are common in nature. Humans alone are subject to more than a dozen species transmitted during sexual contact, including viruses, bacteria, yeast, protozoa, and insects (Nahmias and Danielsson 2011). Sexually transmitted diseases, however, are not universal in sexual species. They generally require hosts that come into physical contact during mating, remain alive through more than one mating season, and mate with more than one partner (Antonovics et al. 2011a). Sexual organisms with these characteristics tend to accumulate sexually transmitted parasites, thereby expanding the complexity of selection pressures on sexual species.

In some cases, the sexual contact can be indirect, as in plants in which fungal pathogens are transmitted among flowers by pollinating insects. In such cases, selection can be pulled in conflicting directions as it acts to favor traits that attract mutualistic pollinators and thereby maximize the benefits of sexual reproduction, while simultaneously favoring traits that minimize disease transmission (Bernasconi et al. 2009; Sasu et al. 2010). Even without sexually transmitted diseases, selection on plants during sexual reproduction can be driven by a balance between optimizing traits to attract pollinators while optimizing traits that help minimize attack by herbivores that specialize on flowers or other tissues (Adler and Bronstein 2004; Brody et al. 2008; Irwin 2009; Galen et al. 2011). The result can be selection on reproductive traits that varies across space and time from positive to negative frequency-dependence (Toräng et al. 2008).

As species coevolve with parasites that favor sexual reproduction and other parasites that exploit sexual reproduction, they often must be subject to shifting selection pressures associated with reproductive traits. How species balance those potentially conflicting selection pressures is still poorly known. Theoretical and experimental work on the evolution of interactions between parasites and hosts is one of the fastest growing topics in evolutionary biology. Much of it, however, is devoted to other questions: the evolution of parasites in the absence of host evolution, the evolution of hosts in the absence of parasite evolution, the evolution of competing parasites, coevolution between a host species and a parasite species, or mechanisms of dealing with broad classes of parasites (e.g., MHC resistance). We therefore still have little understanding of how coevolution proceeds when a population is attacked simultaneously by sexually transmitted and non–sexually transmitted parasites.

This is a problem with no simple solution, but at least some of the components involved in the process are becoming more evident. What we know, though, can sound complicated, because it is complicated and we know only some parts of the answers. In general, the evolution of a parasite depends on whether its rate of spread increases with host population density or is independent of population density. If it is independent of population density, then parasites could drive host populations to extinction (Boots and Sasaki 2003). That result, though, depends on whether the ecological and evolutionary dynamics are distributed among populations or restricted to a local population (Best et al. 2011). In addition, the dynamics of coevolution depend on how selection varies among populations and how much gene flow occurs among host and parasite populations (Thompson 1994; Gomulkiewicz et al. 2000; Nuismer et al. 2000; Best et al. 2012).

Consequently, whether parasites drive their host populations to extinction locally or globally, or continue to coevolve with them in some or all populations, depends on the interaction of evolutionary and ecological processes. The current mathematical models suggest that the crucial ones include the transmission rate and virulence of the parasites, the rates of evolution of resistance in the hosts, the genetics of the coevolving traits, the sizes and spatial configuration of the populations, the relative rates of gene flow among parasite and host populations, the structure of selection within populations, the occurrence of selection mosaics among populations, and the distribution of coevolutionary hotspots and coldspots (Nuismer et al. 2007; Gandon and Day 2009; Gandon and Nuismer 2009; Boots 2011). With so many factors shaping interactions between parasites and hosts, it is likely that selection within and among sexual populations will continue to change over time through the combination of selection pressures imposed by sexually and non–sexually transmitted parasites.

FREQUENCY DEPENDENCE AND SEXUAL SELECTION

The evolution of sexual reproduction generates yet another class of opportunities for frequency-dependent selection on populations driven by parasites. Sexual selection conjures images of the evolution of highly exaggerated traits such as the tails of peacocks, the antlers of deer, and the lekking behaviors of male grouse and manakins. It can, however, be involved in the maintenance of genetic variation through frequency-dependent selection mediated by parasites.

One of the hypotheses for the maintenance of high levels of variation at MHC loci in vertebrates is that individuals choose mates with rare combinations of MHC alleles. Those individuals produce offspring more likely to escape serious attack by the local parasites, because the parasites are adapted to more common MHC alleles. The alternative sexual selection hypothesis is that individuals choose mates with higher levels of MHC heterozygosity. Some, but not all, species that have been studied show evidence for balancing selection at these loci driven by mate choice. The dozens of studies that have tried to separate the roles of frequency-dependent selection and heterozygous advantage have differed in their conclusions (reviewed in Huchard et al. 2010).

Sexual selection can also be frequency dependent even without parasite-mediated selection. Frequency-dependent sexual selection could act on multiple mating morphs and mating tactics within each sex (Sinervo and Lively 1996; Shuster and Wade 2003). In these cases, frequency-dependent selection occurs because different mating strategies or tactics are favored in different social environments. There has been a long-standing debate over whether relative fitnesses can be equal among morphs or must be different among morphs for the process to persist (Shuster 2011), but it seems generally that frequency-dependent selection can maintain two or more mating morphs of males or females in some populations.

Selection can be so strong on mating tactics that allele frequencies in some populations have been observed to undergo significant evolution in some species over just a few generations. Among the longest-term field studies showing these rapid evolutionary changes are those on the three male morphs of western side-blotched lizards, *Uta stansburiana*. Morph frequencies in these lizards tend to oscillate over a four- or five-year period through frequency-dependent selection that acts on three morphs in a rock-paper-scissors fashion (Sinervo and Lively 1996; Bleay et al. 2007). Over larger geographic scales, some populations have lost a morph, but three-morph populations remain common in this species (Corl et al. 2010a, 2010b). Within these populations, females are sexually dimorphic and differ in sexual strategies, which increases the complexity of selection within and between the sexes (Sinervo et al. 2000).

Maintenance of two or three mating morphs through frequency-dependent selection has also been found in other species. Female color morphs of common lizards (*Lacerta vivipara*) in Eurasia differ in aggressive behaviors (Vercken et al. 2010). The advantage of each type of behavior decreases as that morph becomes more common in the population. Female color morphs of *Ischnura elegans* damselflies in Sweden experience frequency-dependent selection (E. Svensson et al. 2005). When a morph becomes common, it suffers a loss in fitness through harassment by males that constantly pursue females of that year's most common morph. This harassment lowers her feeding rate. The damselfly morphs may also differ in other traits such as larval survival, and hence selection may act at multiple stages of the life history of these insects.

In some other cases, though, multiple morphs may persist because selection does not disfavor any of the current morphs. Male morphs of *Paracerceis sculpta* isopods have evolved multiple mating strategies to gain access to females (Shuster and Wade 1991). Some males defend harems within sponges; some mimic female behavior and morphology; and yet others invade harems. All the morphs seem to have equal fitness, at least under the ecological conditions in which they have been studied.

ALTERNATIVES TO SEXUAL REPRODUCTION

Abandoning sexual reproduction completely is an alternative evolutionary solution to the complex selection pressures associated with sexual reproduction and sexual traits. Although asexual lineages occur in many taxa, the number of confirmed obligately asexual lineages continues to shrink. In recent years, multiple taxa thought to be asexual have turned out to have cryptic or novel modes of sexual reproduction or at least the genetic machinery suggesting that they may sometimes reproduce sexually. These include multiple species of fungal and protozoan pathogens of humans (e.g., Poxleitner et al. 2008; O'Gorman et al. 2009; Heitman 2010) and a fungus-growing ant species that is now known to consist of sexual and asexual populations (Rabeling et al. 2011). Molecular methods to infer sexual reproduction continue to improve, and these methods are showing evidence of a history of sexual reproduction in multiple taxa whose life cycles have been difficult to study directly (Schurko et al. 2009; Grimsley et al. 2010).

The lineages of asexual parasites often appear to be shorter lived than those of their sexual relatives (Bengtsson 2009; Hörvandl 2009). A few lineages, however, seem to have been obligately asexual for anywhere from a million years to over a hundred million years. They have sometimes been called ancient asexual scandals, because they seem to defy the argument that asexual lineages are doomed to early failure relative to sexual lineages (Judson and Normack 1996). The evidence for obligate asexuality in these taxa comes most often from the failure to detect males. In some cases, though, so many individuals have been sampled that it seems likely that males do not occur or they occur rarely (Birky 2010). Among animals, the potentially ancient asexual lineages include bdelloid rotifers with about 460 species, root-knot nematodes in the genus *Meloidogyne* with about 80 species, darwinulid ostracods with about 35 species, one or more stick insects in the genus *Timena*, and some orabatid mites, which is a lineage of about 10,000 species that include sexual species and some asexual species, some of which are thought to be ancient asexuals (Danchin et al. 2011; Schwander et al. 2011). These taxa are distributed throughout the animal web of life.

In a world of more than a million animal species, that is not very many. Moreover, what constitutes a species in these asexual lineages is problematic. Calculating the number of described species within an asexual lineage involves interpretation based on gaps among populations in genetics, morphology, and ecology (Birky and Barraclough 2009). Nevertheless, the fact that these lineages occur at all suggests that there are alternatives to sexual reproduction. With so few taxa, any explanation for how a lineage has managed to remain asexual is bound to be a description of that special case. But these special cases are important, because they help us understand why sexual reproduction is so common.

One route to long-term asexuality seems to be to adopt an extreme run-and-hide strategy. If an individual can escape locally coadapted parasites by moving elsewhere, then the advantage of sexual reproduction would be reduced (Ladle et al. 1993). That appears to be the evolutionary solution adopted by bdelloid rotifers, which are among the few completely asexual lineages of animals. This puzzling group of freshwater invertebrates is thought to have been asexual for tens of millions of years (Welch et al. 2009; Wilson and Sherman 2010). They have three unusual characteristics. They can survive in almost any moist habitat, including ephemeral moist patches of vegetation and rainwater. They are easily dispersed by wind and occur on every continent. And they can survive complete desiccation for years at any life history stage. Their only major enemies are fungal parasites, most of which are a group of obligate, lethal endoparasites within a single genus, *Rotiferophthora*, which the rotifers acquire by eating spores. In microcosm experiments that allow for desiccation and dispersal, the rotifers are able to escape their fungal parasites (Wilson and Sherman 2010). If it takes such extreme tactics to maintain an asexual lifestyle through tens of millions of years, then perhaps this example is telling us that parasites truly are pervasive in nature and sexual reproduction is generally the only viable lifestyle for most species.

Studies of the mechanisms by which most other asexual lineages persist are inconclusive (Bengtsson 2009; Schön et al. 2009). Some ancient and recent asexual populations are polyploid (Stenberg and Saura 2009), and some are restricted to high elevations, high latitudes, islands, or disturbed habitats (Vrijenhoek and Parker 2009). In some mathematical models, asexuals are most likely to persist over the long term if their population sizes are large, their intrinsic reproductive potentials are high, and they do not compete with closely related sexual populations (Rice and Friberg 2009).

Much more commonly, many species have the potential to be facultatively sexual, either occasionally undergoing sexual generations or cycling between asexual and sexual generations. Many aphids disperse to new sites as sexual adults, undergo multiple asexual generations while living on host plants, and then form sexual adults to start the cycle again. Within some plant lineages, species differ in how they incorporate sexual and asexual reproduction in their life histories. In the saxifragaceous plant genus *Lithophragma*, all nine species are capable of producing a combination of sexual flowers on the floral stalk and underground asexual bulbils on the roots. Two of the species also sometimes produce asexual bulbils rather than flowers at some nodes on the floral stalk, and one species (*L. glabrum*) sometimes abandons sexual reproduction completely and produces only asexual bulbils both underground and on the floral stalk. Over a short distance along a hillside, populations can vary from individuals producing only asexual aerial bulbils to those producing both flowers and bulbils (Thompson, pers. obs.). The number of ways in which sexual and asexual reproduction have diversified within and among lineages seems so vast that the fine-tuning of sexual reproduction, and plasticity in the ability to be sexual or asexual, must be under continual selection in many populations.

BACTERIA, ARCHAEA, AND HORIZONTAL GENE TRANSFER

The mechanisms by which prokaryotes acquire novel genetic diversity differ greatly from those of eukaryotes. Bacteria and archaea do not undergo recombination of whole sets of chromosomes through meiosis. Instead, they acquire snippets of DNA from each other through multiple routes, including direct uptake from the surrounding environment, transfer by phage, or transfer of plasmids between bacteria in contact with each other. These mechanisms are called horizontal gene transfer (also called lateral gene transfer) because they involve transfer of genetic material from one individual to another within a generation rather than through the reproductive process.

Horizontal gene transfer occurs throughout genomes and is responsible for a wide range of adaptive traits (Haegeman et al. 2011; Wiedenbeck and Cohan 2011). It has been implicated, for example, in the disease severity and host specificity of *Pseudomonas* bacteria (Sibly et al. 2011), shifts of *Staphylococcus* among mammalian hosts (Guinane et al. 2011), and the occurrence of phosphate acquisition genes in the major genera of oceanic bacteria within and between oceans (Coleman and Chisholm 2010). In some cases, horizontal gene transfer can drive evolution on short timescales. Multiple cases of antibiotic resistance are due to the transfer of plasmids, phages, or other mobile genetic elements (Andam et al. 2011; Svara and Rankin 2011). Despite this influx of new genes, the genomes of prokaryotes remain relatively small, which suggests that misfit genes must be continually eliminated by selection even as favorable genes are retained (Kuo and Ochman 2009).

Marine and terrestrial environments are awash in viruses and mobile genetic elements, providing a supermarket of opportunities for prokaryotes to incorporate new genes through these horizontal mechanisms. The result, however, is not a single supergenomic species, because gene transfer among individuals is biased. Just how biased, and what the bias means for our ability to reconstruct the history of microbial life, is currently a hotly debated question.

By some views, horizontal gene transfer has been so widespread that it undermines any realistic reconstruction of the tree of life. By these views, species form a genetically complex net of life that masks shared ancestry at least at the deeper levels of phylogenetic history (Doolittle 2010; Zhaxybayeva and Doolittle 2011). By other views the exchange of genes has been sufficiently restricted to more closely related lineages that it does not undermine the overall patterns found in analyses of shared ancestry of genes and could, instead, reinforce that ancestry (Keeling and Palmer 2008; Andam and Gogarten 2011).

Horizontal gene transfer is so pervasive in prokaryotes that some biologists have argued that bacteria and archaea cannot even be viewed as individuals with their own characteristics (Goldenfield and Woese 2007). By continually exchanging pieces of DNA, populations and communities of bacteria are in constant genetic flux. These different views of prokaryotic life all highlight the importance of gene exchange in the dynamics of evolution. One of the major goals of current genomic analyses of the history of life is therefore to understand how the balance between vertical and horizontal gene exchange has shaped the organization of life and the pace of evolutionary change.

Rates of gene exchange in bacteria vary widely, but the majority of current estimates are lower than or near the rate of mutation per capita per gene, and less than 20 percent are greater than ten times the mutation rate (Vos and Didelot 2009). These rates suggest that gene exchange has the potential to provide raw material for natural selection similar to that provided by mutation, but the exchange is not generally high enough to swamp selection and adaptation (Wiedenbeck and Cohan 2011). Under experimental conditions, however, rates of gene transfer can be high (McDaniel et al. 2010), which suggests that high rates could occur under some particularly favorable ecological conditions in nature.

There are some suggested patterns and general conclusions in the accumulating data. Horizontal gene transfer appears to be more common among closely related prokaryotes than among distantly related taxa

(Andam and Gogarten 2011). Even so, horizontal gene transfer has been suggested between almost every domain of life, including bacteria to archaea, archaea to bacteria, bacteria to eukaryotes, archaea to eukaryotes, and within eukaryotes (Keeling and Palmer 2008; Andersson 2009; Bock 2010; Boto 2010). Highways of gene sharing occur between some major lineages, such as between thermophilic archaea and hyperthermophilic bacteria that often live in similar environments (Aravind et al. 1998; Beiko et al. 2005; Zhaxybayeva et al. 2009).

Within eukaryotes, horizontal gene transfer has been documented most often between unicellular species, and most prokaryotic genes acquired by unicellular eukaryotes encode proteins that affect metabolism (Andersson 2009). All eukaryotes, however, have had some horizontal gene transfer from prokaryotes in their evolutionary history (Keeling and Palmer 2008). Nematodes, which are among the most common parasites of plants, might have acquired their parasitic ability through horizontal gene transfer from prokaryotes (Haegeman et al. 2011). Pea aphids (*Acyrthosiphon pisum*) have carotenoids that were acquired through horizontal transfer from fungi (Moran and Jarvik 2010). Fungi themselves show evidence of several hundred events of horizontal gene transfer from prokaryotes, as suggested by genomic analyses (Richards et al. 2011). Horizontal gene transfer could also be part of the explanation for the persistence of some ancient asexual eukaryotes. Some eukaryotic lineages that include ancient asexual species, including bdelloid rotifers and *Meloidogyne* nematodes, show evidence of gene acquisition from prokaryotes by horizontal gene transfer (Scholl et al. 2003; Gladyshev et al. 2008).

How horizontal gene transfer has shaped coevolution among species within and across the domains remains largely unknown. Its role, however, could be large and, if so, would affect our views of the coevolutionary process. If, for example, nematodes have acquired their ability to parasitize plants through horizontal gene transfer from prokaryotes, then the genetics of coevolution between plants and nematodes derives from an interaction between fungi and prokaryotes.

More generally, horizontal gene transfer offers to some eukaryotic species an alternative to developing a long-term coevolving symbiosis with a prokaryotic species. When possible, a eukaryote could acquire instead the crucial metabolic genes from prokaryotes. It is, though, probably a rare event. Coevolution with prokaryotic symbionts is pervasive in eukaryotes, with each species retaining its own genome, but horizontal gene transfer between prokaryotes and eukaryotes appears to be relatively rare. It is unknown when and how horizontal gene transfer becomes a possible alternative to long-term coevolution with a symbiont, but genomic studies are opening up new avenues in the search for these evolutionary events.

Coevolution may be linked to horizontal gene transfer in other ways. When long-term coevolution occurs between eukaryotes and prokaryotes, horizontal gene transfer could delay genomic decay, which sometimes accompanies evolution of an obligate symbiotic lifestyle in intracellular bacteria, whether mutualistic or parasitic. It has long been thought that obligate intracellular bacteria are generally free of bacteriophages and other genetic parasites. That view, though, is changing. When these bacteria have access to novel gene pools, they can acquire new genes (Bordenstein and Reznikoff 2005; Kent et al. 2011). The genomes of *Wolbachia* bacteria, which are widespread symbionts of arthropods and nematodes, have more genomic diversity than expected (Ishmael et al. 2009). Therefore, horizontal gene transfer may have played a major role in the maintenance of genetic diversity in the interactions between *Wolbachia* and at least some insects (Kent and Bordenstein 2010; Chafee et al. 2011). The genomes of *Rickettsia*, which are intracellular bacteria commonly transmitted among vertebrates by arthropods, also include multiple mechanisms by which genes can potentially move about within and among genomes, including transposons, plasmids, and phages (Merhej and Raoult 2011).

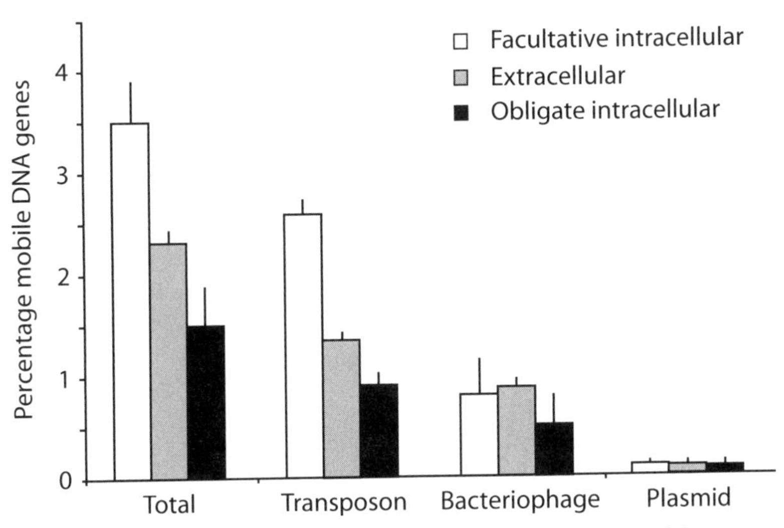

Fig. 8.4 The percentage of the genomes of symbiotic bacteria that are mobile genetic elements of any of three types: transposons, bacteriophages, or plasmids. The values are the normalized means ±1 SE for three lifestyles: facultative intracellular, extracellular, and obligate intracellular. After Newton and Bordenstein (2011).

There is, though, a pattern in the tendency of bacteria with different lifestyles to acquire mobile genetic elements. Bacteria that live as symbionts have anywhere from zero to 21 percent of their genomes as mobile elements, but facultative intracellular symbionts generally have more genetic elements than extracellular symbionts or obligately intracellular symbionts (fig. 8.4). The simplest, and possibly most likely, interpretation is that opportunities for acquiring mobile genetic elements are greater in populations living in environments in which they are exposed to a greater diversity of mobile elements.

Sorting out how selection favors horizontal gene transfer is not straightforward. Many cases of gene transfer are mediated by plasmids or phages, and the transfer may often be a coincidental by-product of the infection process and the host's gene repair process (Levin and Cornejo 2009). Hence, selection need not directly favor high rates of horizontal gene transfer any more than it need favor directly higher rates of mutation, although both are possible. Rather, selection may act primarily on the plasmids or phages at the infection stage, and on the bacteria that have acquired the new genes. In contrast, selection could act directly on the rate at which bacteria acquire exogenous DNA from the surrounding environment, but the mechanisms by which this would occur are unresolved (Levin and Cornejo 2009).

THE CHALLENGES AHEAD

Sexual reproduction and horizontal gene transfer favor further evolutionary change. These processes are themselves the outcomes of selection on populations. Both processes are tied to other processes that add to the dynamics of evolution: frequency-dependent selection, sexual selection in sexual species, coevolution between eukaryotes and parasites, and coevolution between bacteria and phages. As genomic analyses continue to reveal the complex chimerical structure of genomes, and as ecological analyses reveal the complex patterns of selection on interacting genomes, the study of the relationships among recombination, selection, and evolution is entering a new phase that should provide novel insights into our understanding of the dynamics of evolutionary change.

The tremendous variation generated and maintained by these processes creates the potential for selection to mold evolution in different ways in different populations, because genes differ among environments in how they are expressed and affect the fitness of individuals. We now turn to what we are learning about the ways in which interactions between genes and environments contribute to ongoing evolution.

9

Divergence and Selection across Environments

Selection on local populations is the fundamental building block of the relentlessness of adaptive evolution. Gene combinations are expressed in different ways in different environments, which generates the chance for selection to vary among environments in the genotypes it favors. In addition, species interact with other species in different ways in different environments, altering the patterns of selection within the web of life in ecosystem after ecosystem. Each species, and each interaction between species, is therefore a collection of semi-independent evolutionary experiments taking place in different environments.

This chapter explores our growing understanding of variation in selection among environments and why that variation is important for our understanding of relentless evolutionary change.

BACKGROUND: A WORLD OF DIVERGENT POPULATIONS

One of the most important results of studies in molecular population genetics has been the demonstration that almost all species are groups of genetically distinct populations. Biologists since Darwin have appreciated that species often differ among geographic regions in their traits, and taxonomists commonly acknowledged those differences by naming subspecies within species. Beginning in the 1960s, however, studies of protein variation among populations showed that populations were more genetically variable than previously suspected (Lewontin and Hubby 1966), and subsequent studies showed even greater variation among populations. Since then, the continued development of molecular tools and studies of the evolutionary ecology of populations in nature have shown that selection can favor differences among populations over surprisingly small scales (Laine 2006; King et al. 2009; Koskella et al. 2011).

A species, then, is not a single evolutionary experiment or even a small number of evolutionary experiments. It is often a large collection of semi-independent evolutionary experiments distributed across many environments. Selection constantly acts not only on tens of millions of species but also on hundreds of millions of populations that differ in how they express their inherited traits and how they interact with other species.

Some years ago, Jennifer Hughes, Gretchen Daily, and Paul Ehrlich used the results of molecular studies available at the time to calculate how many genetically distinct populations occur in the world (Hughes et al. 1997). Their estimate was 1.1 billion to 6.6 billion populations among eukaryotic species. No one since then has been brave enough to try an updated estimate. Regardless of whether the actual number is closer to 1 billion or 10 billion, ongoing evolution is distributed across a staggering number of genetically distinct populations.

There are some species, such as mussels in the eastern Pacific Ocean and a species of Antarctic krill, that show little or no evidence of genetic differences over large spatial scales (Addison et al. 2008; Bortolotto et al. 2011), but these are becoming the exceptions as studies accumulate for many species. Species showing little differentiation among populations tend to be those that broadcast their gametes or young in great numbers into the air or the ocean currents, or they are large as adults and move over great distances. Yet even some large and wide-ranging species, such as grizzly bears in North America and cod in the North Atlantic Ocean, show geographic differences among populations at multiple spatial scales (Pogson et al. 2001; Proctor et al. 2005). As researchers probe deeper for genetic differences among populations, they often tend to find them. An intercontinental survey of *Arabidopsis thaliana* plant populations found evidence of divergence among European populations at every spatial scale included in the analysis and also divergence among North America populations at local scales (Platt et al. 2010). This result is typical for widespread species.

ADAPTATION IN METAPOPULATIONS

As species diverge into genetically distinct populations, groups of populations remain connected as metapopulations. These metapopulations are linked over time by varying degrees of gene flow, extinction, and recolonization (Levins 1969; Hanski et al. 2011). These clusters of populations create the potential for adaptation at the metapopulation level rather than simply at the local population level. That adds another level to the dynamics of adaptive evolution, and these dynamics have now been studied from several perspectives in mathematical models.

The simplest metapopulation is one in which two neighboring populations living in different environments are subject to different forms of natural selection but are connected by gene flow (e.g., Ronce and Kirkpatrick 2001). Alternatively, this simple two-environment world can be modeled as a miniature metacommunity in which two species coevolve in different ways in different environments (Nuismer et al. 1999; Gomulkiewicz et al. 2000). As populations are allowed to evolve in these models, they often produce patterns and dynamics of adaptation that differ from those found in single, isolated populations. Evolutionary change is often much more dynamic. It depends on the relative strength and direction of natural selection in each population, the rate of mutation, and the level of gene flow between the populations. Selection sometimes favors traits that are compromises in adaptation to the different environments.

In metapopulations of three or more local populations, the potential evolutionary dynamics and patterns expand even more. There are two extremes in how local populations can be connected with many possible intermediates. At the simpler end of the continuum, a stable local population acts as a source of immigrants to surrounding populations that cannot maintain themselves. The sink populations repeatedly receive colonists from the source population, persist for some time, and then become extinct whenever insufficient new immigrants arrive from the source population (Harrison et al. 1988; Manier and Arnold 2005; Kawecki 2008). Some populations at edges of the geographic ranges of species are sink populations that rely on immigrants from habitats more suitable for that species, but sink populations can occur anywhere.

These sink populations are often evolutionary sinks as well as demographic sinks. They often have reduced genetic variation that is a subset of the genetic variation found in the source population. Although natural selection may act strongly within these sink populations, any evolution is short lived. Local adaptation is lost every time the population goes extinct and the habitat is recolonized by the source population.

At the other end of the continuum are metapopulations with continually shifting source and sink populations. Extinction of local populations and recolonization of habitats are distributed throughout the metapopulation. No local population consistently acts as the source of immigrants to the other populations. This is the traditional view of a metapopulation envisioned by Richard Levins (1969), who developed the first mathematical models of these kinds of population dynamics. These models provided an initial formal framework for capturing important ecological aspects of metapopulation dynamics.

The original metapopulation models did not allow for any genetic differences among populations. Studies in the 1960s and 1970s, though, were making it clear that it was important to incorporate genetic structure if we were to understand the dynamics of metapopulations. The most detailed studies pointing in that direction were those of Paul Ehrlich and colleagues working on Bay checkerspot butterflies, *Euphydryas editha*, at and near the Jasper Ridge Biological Reserve of Stanford University. Those studies showed that, even at this local scale, these butterflies were not a single randomly breeding population. They were a collection of semi-independent populations both ecologically and genetically (Ehrlich 1965; Ehrlich and Mason 1966; Ehrlich et al. 1975; Ehrlich and Hanski 2004).

These kinds of results turned out to be important to our understanding of evolutionary dynamics. Although natural selection can act strongly at the level of a local population, the adaptations are also filtered by selection at the metapopulation level. Adaptation begins at the transient level of the local population level, is reshaped at the metapopulation level, and forms larger-scale patterns at regional scales that connect metapopulations (Thrall et al. 2002; Laine 2005; Toju 2008). The genotypes best able to survive and reproduce within the dynamic structure of metapopulations are the ones that will proliferate over time. The overall trajectory of evolution becomes the combined total of each of the trajectories across all the local populations as extinction and colonization continues.

If selection acts differently among local populations within a metapopulation, then adaptation in metapopulations has four possible outcomes (Hanski et al. 2011). Natural selection can favor convergence of populations such that individuals are, on average, specialized for one of the environments, or it can favor convergence of populations on individuals that can fare well across the multiple environments within range of the metapopulation. Alternatively it can favor more complex outcomes in which adaptations vary among populations within the metapopulation. Selection can result in a metapopulation in which each population is adapted to a different environment, or it can favor a metapopulation composed of populations whose adaptations reflect primarily the patterns of local gene flow rather than the underlying differences in selection on neighboring populations among habitats. In this case, the pattern of local adaptation would be at least slightly mismatched to the underlying template of environmentally variable selection.

This complicated fourth outcome would arise if the spatial scale of colonization and gene flow were short: similar adaptations would occur more often among neighboring populations than among more distant populations. The result would be groups of similarly adapted populations distributed as a patchwork over the metapopulation. These models assume that local populations do not experience random genetic drift or coevolve with other species. Genetic drift and coevolution, however, can also strongly alter patterns of adaptation across a metapopulation (Gandon and Nuismer 2009).

PHYLOGEOGRAPHIC STRUCTURE

At even larger spatial scales, populations are connected through their complex past histories of range expansion and contraction, producing regional patterns of traits and adaptation that extend beyond metapopulations. Molecular analyses have made it possible to determine how populations are genetically connected across time as well as space, forming a phylogeographic hypothesis for each species (Avise 2000; Edwards and Bensch 2009). These analyses have also made it possible to determine which populations within regions show evidence of population expansion, contraction, or stability over timescales ranging from decades to thousands of years or even longer (e.g., Biek et al. 2006; Murray et al. 2010; Thompson and Rich 2011). Phylogeographic analyses have now been performed on hundreds of species, making it a standard part of any detailed analysis of the historical and current genetic structure of species. These studies show that species are evolutionarily dynamic beyond the scales of local populations or metapopulations. They reinforce the long-held view that populations at the edges of geographic ranges often diverge from central populations partly because peripheral populations often have tenuous genetic connections to more central populations (Mayr 1992; Grant and Grant 2008b; Price 2008; Thompson and Rich 2011).

Most phylogeographic studies assume that the genes used in the analyses are not subject to strong natural selection. The goal is often to consider the spatial and temporal history of a species as viewed primarily through the processes of mutation, random genetic drift, and gene exchange with other populations. These studies generate hypotheses about the history of diversification of a species into populations that do not interbreed freely with each other. They also provide an estimate of the number of separately evolving units within a species.

Phylogeographic studies capture some of the continual flux of populations as they are diverging, combining, losing contact with each other in some places, and regaining contact in other places. The pattern of genetic connectedness among populations usually reflects the idiosyncratic history of each species. Evolution is so dynamic that even closely related species rarely have exactly the same phylogeographic history across a region. They

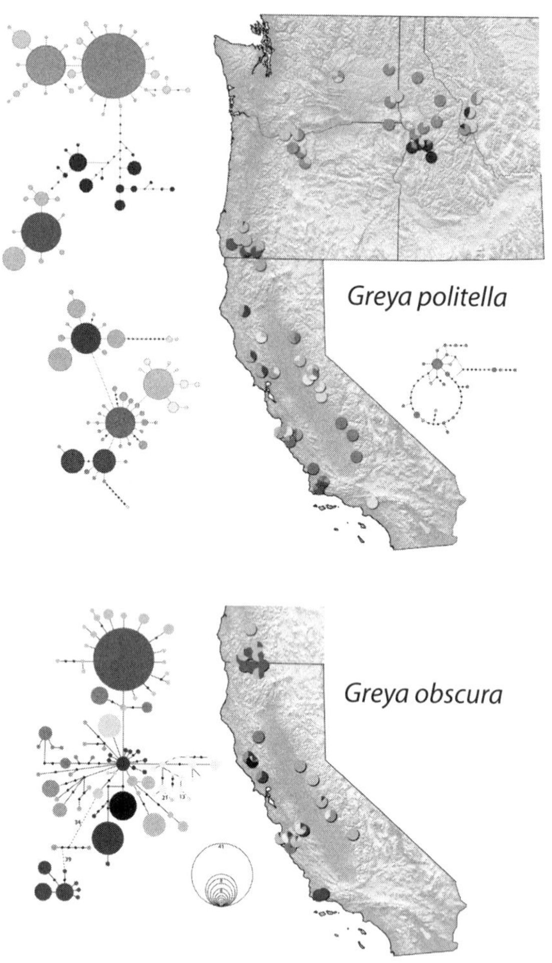

Fig 9.1 The pattern of geographic divergence in two closely related moth species, *Greya politella* and *Greya obscura*, in western North America. Both species range from California north to Oregon, and *G. politella* extends into Washington State, Idaho, and Montana. A few additional populations of *G. politella* have been found farther east in the Rocky Mountains (not shown). Divergence in both species has been estimated using DNA sequencing of the cytochrome oxidase gene (shown here) and amplified fragment length polymorphisms. In the region where the two species overlap, they use the same habitats and host plants, yet they differ in some respects in their pattern of genetic connectedness among populations. The network diagrams indicate the number of nucleotide base pairs separating populations. Each line connecting two black dots or larger circles indicates a difference of one base pair. The shades of gray in the circles indicate groups of genetically similar populations within the network diagram and on the map. The size of the circles represents the number of sampled individuals with the same DNA sequence (haplotype). After Rich et al. (2008) and Thompson and Rich (2011).

may show some broad similarities in how their populations are connected, but they usually show many small dissimilarities as well.

Similarity in broad pattern and dissimilarity in the details have been shown in some *Greya* moth species in North America (fig. 9.1). These species are widely distributed in western North America, and a few species extend across the Bering Sea in eastern Asia. *Greya politella*, which is the most widespread species, has evolved over its several-million-year history into multiple genetically distinct populations, which cluster into three regional groups: the Pacific Northwest populations, the southern Sierra Nevada populations, and the populations throughout the rest of California (Brown et al. 1997; Rich et al. 2008). Within and among these geographic regions, the moth populations differ not only in their molecular signatures of divergence but also in the traits they use in their interactions with host plants and the effects they have on fitness of their hosts (Thompson and Cunningham 2002; Thompson and Merg 2008).

The California and southwestern Oregon populations of *G. politella* overlap with *Greya obscura*, which is a geographically more restricted species that uses the same habitats and host plants. The phylogeographic histories of these two species are similar in that they show evidence of past range expansion within the central part of this region (Thompson and Rich 2011). Both species also show strong divergence of populations at one or more edges of their geographic ranges. These edge populations differ by as much as 3 to 5 percent in DNA sequences from the central populations. *Greya politella* and *G. obscura*, though, differ in many of the finer-scale patterns of genetic divergence within the central populations. *G. obscura* shows more genetic divergence and less regional clustering of populations than *G. politella*.

Phylogeographic patterns of divergence are often likely to underestimate the dynamics and divergence of adaptive evolution among populations. Populations can be virtually identical according to neutral gene markers but differ markedly in one or more adaptive traits. This difference in genetic signal occurs for at least two reasons. Traits under natural selection often evolve faster than selectively neutral traits, and a genetic sample of a small portion of a genome is likely to miss the traits evolving quickly by natural selection. Regions in which seemingly adaptive traits differ among populations that appear identical by phylogeographic methods are important clues that natural selection may be responsible for the differences among traits.

SELECTION MOSAICS

The geographic structure of a species becomes magnified as it coevolves with other species. Populations adapt to their local environments through what are commonly called genotype-by-environment interactions ($G \times E$), and interactions between species add another level: genotype-by-genotype-by-environment interactions ($G \times G \times E$). These $G \times G \times E$ interactions are known as a geographic selection mosaic, and the E can be anything from temperature to the surrounding web of other interacting species (Thompson 1994, 2005).

Selection mosaics are at the core of why coevolution among species drives so much evolutionary change in nature. Coevolution between (or among) species depends not only on the genes of those species, but also on how those genes are expressed in different environments. An interaction between, say, a parasite and host population may depend on the average environmental temperature at which the interaction takes place. A particular parasite genotype could be highly virulent at some temperatures but relatively benign in other temperatures. The difference in virulence could be due to differences among environments in the expression of parasite genes or differences in the expression of host defenses.

Differences among environments could be so great that an interaction usually antagonistic in one environment could be commensalistic or even mutualistic in another environment. The ecological differences could occur even if the interacting species initially have the same distributions of genotypes in both environments. Because genes are expressed in different ways in different environments, and each population is subjected to a different surrounding web of life, selection can act immediately on those differences and drive populations along different evolutionary trajectories. Hence, even without new mutations, an environmental change could result in large alterations in how different populations of a species coevolve with other species.

Selection mosaics may often be the major driver of divergence in interactions among species. In the past, gene flow has been considered the major factor shaping adaptation of populations across environments, but it now seems that selection mosaics control the effects of gene flow on adaptation between interacting species (Gandon and Nuismer 2009). It is the combined effects of selection mosaics and gene flow that set the stage for ongoing coevolutionary dynamics in interactions.

The simplest way to visualize the potential importance of selection mosaics in driving evolutionary change is through an experiment involving just a few interacting genotypes in different environments. Bishop pines (*Pinus muricata*) and mycorrhizal fungi (*Rhizopogon occidentalis*) interact along the coast of California (Kennedy et al. 2007; Peay et al. 2010). The fungi interact with several coniferous species and show some evidence of local adaptation (Grubisha et al. 2007; Hoeksema and Thompson 2007). When the fungi and the plants grow together, they form mycorrhizae on the plant roots. This interaction can result in higher plant growth because the fungi are able to increase nutrition to the plant. In return, the plants provide sugars that can increase fungal growth. The result is a mutualistic interaction that increases the fitness of the plants and the fungi. That does not mean, however, that the interaction between every bishop pine genotype and every *Rhizopogon* fungus genotype is mutualistic within a population. Nor does it mean that a set of interacting genotypes will be mutualistic in all environments.

The potential variation in ecological outcomes can be shown by constructing a simple experiment using just two genetically distinct fungal individuals (sporocarps) and seeds from two plant individuals. Although each seed may not be genetically identical to every other seed on that plant, these seeds will all have very similar genotypes. The seeds and fungi are planted in four different environments: sterilized field and lab soil, and unsterilized lab and field soil that contains other nonmycorrhizal microbes. The experiment therefore involves two fungal genotypes, seeds from two trees, two physical environments (different soils), and two biotic environments (soils with or without nonmycorrhizal microbes).

Root length can be used as an index of how this interaction affects the plants, and there are three possible outcomes. The mycorrhizal relationship can increase plant growth, decrease it, or have no effect. The experiments show that the outcome for the plants depends on the particular genotypes of the plants and on the fungi and the environments in which they have been grown (fig. 9.2). All these ecological outcomes occur in just this restricted set of genotypes and environments. Fungal success also depends on the genotypes and the environments (Piculell et al. 2008; Hoeksema 2010). In a real population, there would be many more fungal and plant genotypes, creating a distribution of ecological outcomes in any population that could range broadly from antagonistic to mutualistic (fig. 9.3).

Similar experimental results, but not controlled so precisely for genotype, suggest that selection mosaics occur in interactions between plants and rhizobial bacteria, fungal pathogens, ants, and birds (Rudgers and Strauss 2004; Laine 2007; Benkman 2010; Heath et al. 2010), between bacteria or fungi and their viruses (Vogwill et al. 2009; Bryner and Rigling 2011), and between animal species (Nash et al. 2008; Bauer et al. 2009). In natural populations, we can never study geographic selection mosaics with the same degree of genetic precision and completeness that we can study them in laboratory experiments or in field plots where a few clones are grown in

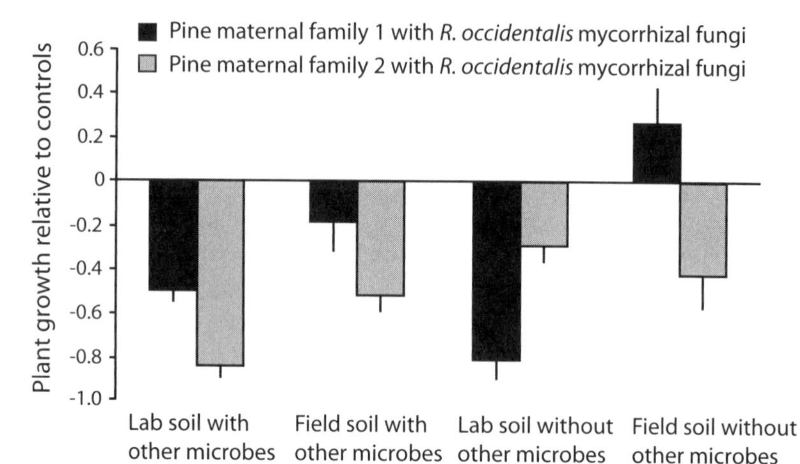

Fig. 9.2 The effect on plant growth of the mycorrhizal fungus *Rhizopogon occidentalis* on the growth of bishop pine (*Pinus muricata*) seedlings in laboratory soil or field soil, and in the presence of nonmycorrhizal soil microbes or in sterilized soil. Results are scaled to show the effects of mycorrhizal fungi on plant growth as compared with control plants lacking these fungi. Results are shown for seeds collected from two trees (i.e., two maternal families) at the same locality where the fungal sporocarps were collected. After Piculell et al. (2008).

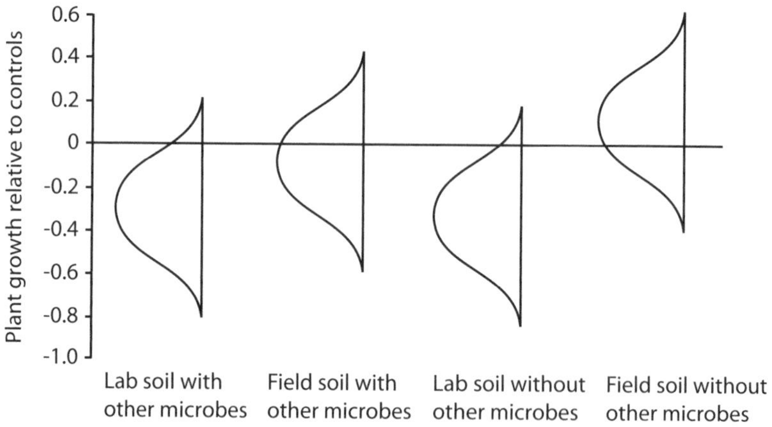

Fig. 9.3 Hypothetical distribution of the ecological outcomes of two interacting species in four environments, generalized from fig. 9.2. The figure assumes that the interacting species have the same distributions of genotypes in all environments and that the outcome for fitness (in this case, plant growth) depends on the physical and biotic environments in which the interaction takes place. Each point in each distribution is the ecological outcome of the interaction between a plant genotype and a mycorrhizal genotype, and the distribution of outcomes in each environment reflects the relative frequencies of outcomes among all plant individuals in a population.

different environments. It is simply impossible to evaluate the effects of every gene or genotype of one species on every genotype of another species in two or more environments. Even a sample of dozens of genotypes may give only a partial answer. Some genotypes in one species will probably have similar effects on many genotypes in the other species in any particular environment, requiring the testing of many genotypes to get a robust answer. Consequently, it will generally not be possible to determine the exact distribution of a genotype-by-genotype-by-environment interaction for any coevolving species. At best we can explore the distribution in models and, to a limited extent, in small, fast-growing species (Gomulkiewicz et al. 2007; Laine 2007; Hoeksema 2010; Nuismer et al. 2010).

That level of detail, though, is not necessary for most studies of how selection mosaics may contribute to the dynamics of evolution. What we often care about is whether the structure of selection on interacting species varies among environments. The importance of geographic selection mosaics is that two or more species are likely to evolve in different ways in different environments, even when the genetic composition of the populations is initially similar in all environments. Used in that ecologically relevant sense, selection mosaics are common in coevolving species in nature (Laine 2009; Thompson 2009a). Some interactions involve variable selection between pairs of species, and others involve complex variable selection among multiple species (Gómez et al. 2009; Muola et al. 2010; Ruano et al. 2011; J. Soler et al. 2011). The more we learn about species interactions, the more it seems that selection mosaics are almost inevitable: at least some genotypes will differ among environments in the expression of their genes, and those differences will affect interactions with other species. Natural selection will act on those differences.

SELECTION MOSAICS AS INTERACTIONS DIVERSIFY IN SPECIES: AN EXAMPLE

Selection mosaics often appear to be caused not so much by differences among physical environments as by the presence of yet other species that alter the interaction. The *Greya* moths discussed earlier in this chapter provide useful insights into this process. The coevolved mutualism between *Greya* moths and woodland star (*Lithophragma* spp.) plants differs geographically depending on the presence of some other insects that can also act as pollinators. Where some species of solitary bees and bee-flies (Bombyliidae) are common, they can overwhelm the mutualism between *Greya* and woodland stars (Thompson and Cunningham 2002; Thompson and Fernandez 2006). Their pollination is sufficient to mask any positive effects of the moths on

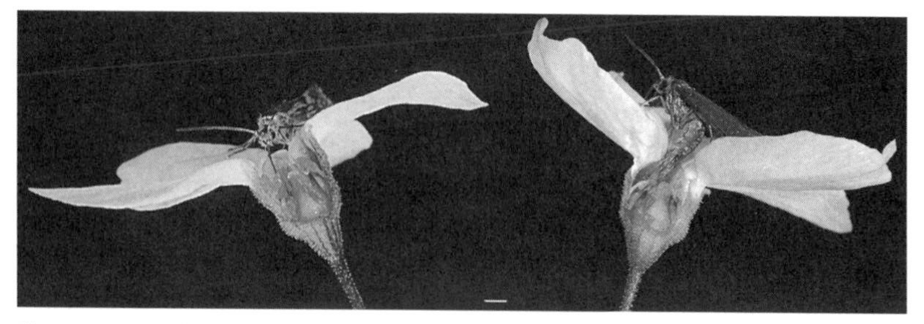

Fig. 9.4 *Greya obscura* taking nectar (*left*) and *G. politella* preparing to lay eggs (*right*) into *Lithophragma cymbalaria*. The flowers have been cut longitudinally from the top of the flower to the nectary disk. The scale at the bottom middle is 1 mm.

plant fitness. Field studies of multiple populations have shown that the interaction between the moths and the plants is mutualistic where effective co-pollinators are rare, but the interaction can be mutualistic, commensalistic, or even antagonistic in environments where the co-pollinators are common.

The mutualism between *Greya* moths and woodland stars involves two different pollination mechanisms. *Greya obscura* simply pollinates while nectaring, like most pollinators. *Greya politella*, however, pollinates through an unusual mechanism (fig. 9.4). A female first drinks nectar from the flower and then turns around and lays her eggs by inserting her abdomen through the corolla. As she extends her abdomen into the corolla, pollen adhering to the lower abdomen and membrane attached to the ovipositor rubs off onto the floral stigma, thereby pollinating the flower. The moth larvae develop in the floral ovary and eat a small percentage of the hundreds of developing seeds. Those eaten seeds impose a direct fitness cost to the plant. That cost is small when there are no other co-pollinators, because the flower may still produce a hundred or more seeds after the larvae have finished eating some seeds. The loss, however, can be an important cost in environments where plants have the option of pollination by insects that do not lay eggs in the flowers.

Greya politella is the most common pollinator of woodland stars through the geographic range of these plant species. Initially, it seemed that *G. obscura* was simply a cheater in this mutualism. Its geographic range is embedded within the larger range of *G. politella* (fig. 9.1), and it is an inefficient pollinator in comparison with *G. politella* (Thompson et al. 2010). As it turns out, however, things are more complicated than that. *Greya obscura* females lay eggs in the ovary wall or stem rather than in the ovary, and the larvae often eat ovary wall or stem tissue rather than seeds. They therefore impose a lower direct fitness cost on the plants. These moths generally co-occur with *G. politella* in a region where few co-pollinating bees and bee-flies visit woodland stars. The plants therefore rely on visits by *Greya* moths. Where the two moths co-occur, *G. obscura* often has higher population densities than *G. politella*, and it fluctuates much less in numbers among years. That difference in population dynamics turns out to be important in understanding this mutualism.

In years when *G. politella* is rare, *G. obscura* remains common and much of the pollination is due to this inefficiently pollinating species (Thompson et al. 2010). The moth makes up in sheer numbers what it lacks in efficiency as a pollinator. Current studies on this interaction suggest that selection differs yet again in places where there are two, rather than one, woodland star species, and these plant species co-occur with either *G. politella* alone or with both *Greya* species. These studies are aimed at understanding how the moth and plant traits are pulled in different directions depending on which combination of *Greya* moths and woodland star plants co-occurs locally.

There is, then, a selection mosaic in this interaction that clusters into major geographic regions: pairwise coevolution between *G. politella* and either one or two woodland star species; interactions between *G. politella* and one woodland star species potentially swamped by co-pollinating bees and bee-flies; and interactions involving both *Greya* species and one or two woodland stars species (fig. 9.5). At the northern edge of the geographic range of these interactions, the interaction has diversified even more to include another closely related plant genus. Over the decades that this interaction has been studied, almost every possible ecological outcome has been observed in the pairwise interactions between *G. politella* and woodland stars, depending on the local presence of other species (e.g., Thompson and Pellmyr 1992; Janz and Thompson 2002; Thompson and Fernandez 2006; Thompson and Merg 2008; Cuautle and Thompson 2010; Thompson et al. 2010). Similar multispecific mosaics have been observed in other mutualisms involving pollinating floral parasitic insects, but these often include a pollinating species and one or more nonpollinating "cheater" species. These include interactions between yucca moths, which are close relatives of *Greya* moths, and yuccas (Althoff and Thompson 1999; Segraves et al. 2005); between fig wasps and figs (Herre and Jandér 2008; Jandér and Herre 2010); and between globeflower flies and globeflowers (*Trollius*) (Després and Jaeger 1999).

The selection mosaic in the interaction between *Greya* moths and woodland star plants is not random. It results from a combination of differences among ecosystems in the physical environment and the web of life. Some

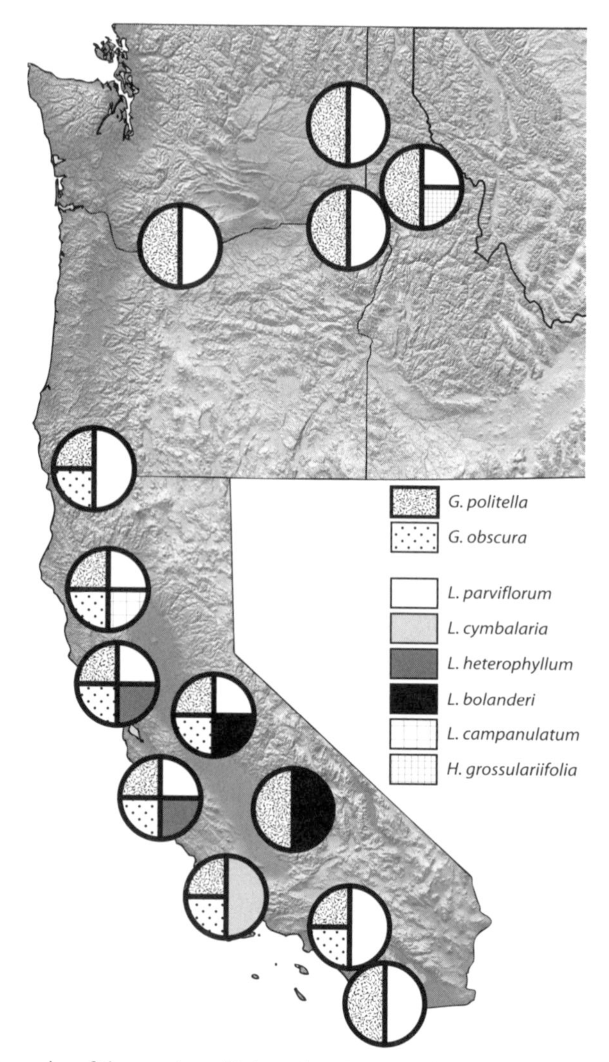

Fig. 9.5 Examples of the number of interacting *Greya* moth species and woodland star (*Lithophragma*) species in local ecosystems in western United States. Also shown are interactions with the closely related plant genus *Heuchera* along part of the northern edge of the interaction. Several cryptic species may be embedded within *G. politella* and *G. obscura*, adding to the complex geographic structure of these interactions.

physical environments are particularly good as nesting sites for ground-nesting solitary bees. Bee-flies lay their eggs in the nests of solitary bees and are parasites on the bee larvae. The physical environments that produce an abundance of solitary bees and bee-flies in these environments are the ones in which the mutualism is swamped between *Greya* moths on woodland stars. It is less clear what limits the distribution of *G. obscura* to only a subset of the regions inhabited by *G. politella*, but it is likely a result of Pleistocene events. The northern limit of *G. obscura* coincides with the northern limit of

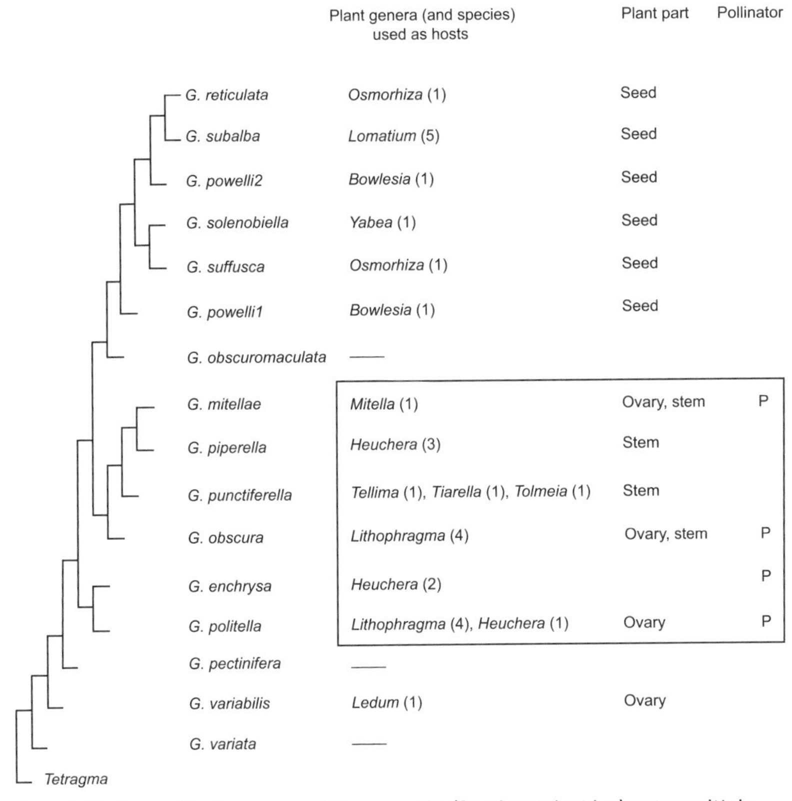

Fig. 9.6 Phylogenetic divergence of *Greya* moths (family Prodoxidae) onto multiple plant families. The branches of the phylogeny are collapsed here to equal length, but the estimated lengths are given in Brown et al. (1994) and Pellmyr et al. (1998). The solid box encloses the moths that feed on plants in the Saxifragaceae. Plant genera above the box are in the Apiaceae. A dash indicates that the host of that *Greya* species is not yet known. The third column shows the plant part eaten by the larvae, and the fourth column indicates whether the moth also pollinates its host plant. Additional complexities of these interactions are explained in Thompson (2010).

multiple other species (Calsbeek et al. 2003), and northern populations of *G. politella* above this region are genetically distinct from those south of it.

These small interaction webs—in which groups of closely related species interact with other groups of closely related species—are among the most common features of species-rich biological communities. They show how local adaptation, the process of speciation, the attraction of yet other species to an interaction, and the changing geographic ranges of species all contribute to the formation of selection mosaics. These small webs are likely to be highly dynamic over thousands or millions of years.

Phylogenetic studies make it possible to track some of these changes. Speciation in *Greya* moths is usually associated with colonization of different host plant species. As the moth populations have diversified in species over millions of years, they have colonized three plant families (fig. 9.6). One of the oldest species in the genus feeds on Labrador tea (*Ledum*) in the Ericaceae, another group of species feeds on saxifrages (Saxifragaceae), and yet another group feeds on plants in the carrot family (Apiaceae). These have been the only known major shifts among host plants that these moths have made throughout their history. These are large shifts, because they occur among plant families with different chemical compounds and different morphologies to which the insects must adapt.

Amid these major shifts, however, *Greya* moths have repeatedly made multiple smaller shifts among closely related plant species. For example, they have colonized the same genera of a small section of the Saxifragaceae more than once (i.e., the group of plant genera within the rectangle in fig. 9.6). Hence, there is a multispecific selection mosaic at levels beyond *G. politella* and woodland stars: *Greya* species and their hosts have combined and recombined in different ways in different ecosystems and at different times during their evolutionary history.

ADDING TILES TO SELECTION MOSAICS

Multispecific selection mosaics can include even fundamentally different forms of interaction, rather than simple variations on a theme. Each type of interaction imposes a different form of selection on a population. The evolution of a population then becomes the weighted average of all these different forms of selection, and that weighted average will rarely be exactly the same in any two environments.

Studies of the assemblages of species associated with goldenrod (*Solidago* spp.) plants have been among the most detailed in showing how the structure of selection on species varies in these kinds of complex multispecific interactions (Abrahamson and Weis 1997; Craig et al. 2007; Ando and

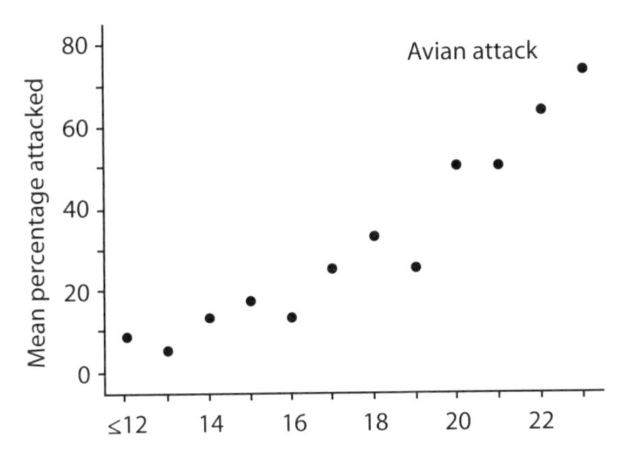

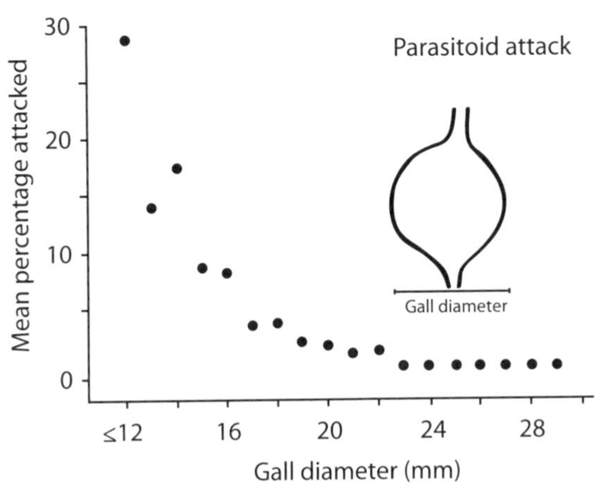

Fig. 9.7 Pattern of attack on goldrenrod (*Solidago altissima*) galls of different size (i.e., diameter) by birds and by *Eurytoma gigantea* parasitoid wasps. *Mordellistena convicta* beetles have a pattern of attack similar to that of the parasitoids. Only the pattern of attack in forests is shown. After Craig et al. (2007).

Ohgushi 2008; Ohgushi 2008; Tooker et al. 2008; Utsumi et al. 2011). Selection on goldenrod-feeding insects depends on the local plant genotypes and on the other herbivores, predators, parasites, and ants that interact directly and indirectly with these species. In Minnesota and North Dakota, one species of goldenrod (*Solidago altissima*) is attacked by *Eurosta solidaginis* flies that induce spherical galls in the stems (fig. 9.7). The galls attract multiple enemies: birds that prey on the fly larvae, parasitic wasps (*Eurytoma gigantea*) that lay their eggs in the fly larvae by piercing the wall of the gall, and *Mordellistena convicta* beetles that lay their eggs on the gall wall, chew into the gall as larvae, and often kill the fly larva. Fly larvae in large galls are more likely to be killed by birds, but larvae in small galls are more likely to be killed by the parasitoids or the beetles (Craig et al. 2007).

Parasitoids and beetles are common on goldenrod plants in the forests and prairies, and they exert strong selection in both environments for flies that induce large galls (fig. 9.7). Bird predation, however, is common only in the forests, and it favors flies that induce small galls. Consequently, selection in prairies favors insects that induce large galls, because that selection pressure is not countered by avian predation; selection in forests favors galls of intermediate size. These patterns of selection are consistent among years and among multiple forests and prairies in the north central United States. The mean sizes of galls in these environments are consistent with these contrasting patterns of selection (Craig et al. 2007)

Similar complex selection mosaics are being reported for an increasingly wide range of interactions (Siepielski and Benkman 2004; Gómez et al. 2009; Muola et al. 2010; Martén-Rodríguez et al. 2011). Some of these studies evaluate selection mosaics in a broad sense by analyzing how selection differs among populations of the same species. Other studies evaluate selection mosaics in the more formal and narrow sense, by experimentally evaluating how selection acts on a genotype or how selection on a genotype differs among environments or how selection on the web of species interacting with that genotype differs among environments. In these experiments, clones or offspring of the same maternal family are grown in different environments, and the resulting interactions are then monitored over time.

Among the clearest of these experiments are those on narrowleaf cottonwood (*Populus angustifolia*). For these studies, multiple individuals of six plant clones were collected from two different sites in Utah between 1982 and 1990 and planted as saplings in several common gardens along an elevational gradient (Smith et al. 2011). After the plants were fully grown and sexually mature, they were studied in 2005 for the pattern of attack on the plants by galling aphids and attack on the galls by birds. Aphid abundance and the intensity of predation by birds differed among genotypes at each site. Aphid fecundity on each clone also differed among the sites, and gall size and gall abundance varied among the environments. Natural selection on gall size differed among the sites from directional to stabilizing. This is one of a remarkable set of studies on the genotypic and environmental effects on the web of life associated with cottonwoods that have continued now over multiple decades. These studies have all shown strong effects of genotype, environment, genotype-by-environment interactions, and, to varying direct and indirect degrees, genotype-by-genotype-by-environment interactions (Whitham 1989; Floate et al. 1997; Whitham et al. 2006; Keith et al. 2010).

CLIMATE ENVELOPES AND SELECTION MOSAICS

If selection mosaics are common and the intensity of selection on species in evolving interactions is strong, then the geographic distributions of species are probably often determined by the combined effects of their local adaptations. Neither biotic interactions nor physical environmental effects are likely to predominate in shaping the geographic ranges of species, except at the extremes of temperature, humidity, salinity, and nutrients. This is why predictions about the long-term effects of climate change on individual species are difficult. Geographic ranges are determined by the ability of populations to sustain their numbers in some environments but not in others. Local populations go extinct wherever adaptive evolution cannot keep pace with the increased mortality and reduced fecundity that often accompany changing environments.

Despite this complexity of adaptation within species, climate envelope models are often highly successful at delineating the current ranges of many species (Duncan et al. 2009; Hodkinson et al. 2011). Using a small number of abiotic factors, these models often are able to capture with remarkable detail where species occur within a specified region. The models then predict how the distributions of those species will become altered with climate change. If the continuing evolution of interactions among species is important in local populations, then why are climate envelope models so successful? The simplest non-evolutionary interpretation of these models is that populations of interacting species just shift their position across environments as they collectively track changes in their physical environments. When it gets too hot to dance inside, they just move into the streets and continue to dance. Some shifts in the geographic ranges of species are therefore likely due to that purely ecological process at least for modest changes over short periods of time (Bartomeus et al. 2011). Most shifts, though, probably involve evolutionary change.

Climate envelope models are probably successful because they capture the current limits on phenotypic plasticity and the limits of environments in which species can persist amid their interactions with other species. They describe the outcome of a huge natural experiment in which populations have been placed in different physical and biotic environments and then forced to survive or go extinct. The climatic variables used in the models are therefore a proxy for all the G × E and G × G × E interactions found within a species and the limits of short-term evolution possible in species in different environments. The success of these models therefore does not suggest that climate itself directly determines the geographic ranges of species.

One of the most detailed international efforts to evaluate how climate change may alter the geographic ranges of particular species is the work on the tiny herb *Arabidopsis thaliana*. This species is native to Eurasia, where it occurs from the southern Mediterranean to Fennoscandia, but it is now a cosmopolitan weed. It shows a wide range of life histories, with individuals in some populations germinating in the fall and individuals in other populations germinating in the spring. Even within a small part of its geographic range, such as the gradient from coastal Spain to the Pyrenees, populations differ in life history traits (Montesinos et al. 2009). In common garden experiments, populations differ in germination season, growth rate, and flowering time. Some of these differences are controlled by genes such as *FRIGIDA*, which affects the ability of plants to respond to environmental signals that trigger different life history stages (Wilczek et al. 2009, 2010).

Using these results, researchers have been able to determine in which kinds of physical environment this species might be successful and also how life histories might evolve amid climate change. It was already known that selection on life history traits in *Arabidopsis* is strong in some populations but also highly variable among populations (Donohue et al. 2005). These newer studies suggest that the evolution of germination timing could influence the rate at which this species expands its geographic range once established in a region.

Even if physical environments control selection on some life history events, the timing of those events must simultaneously pass through the filter of biotic interactions. Slight shifts in life histories can affect interactions with potential mates, competitors, parasites, predators, and mutualists (Boggs and Ehrlich 2008; Bartomeus et al. 2011). A plant that flowers when no pollinators are available simply does not reproduce. Hence, one of the growing concerns in conservation biology is how climate change might change the geographic ranges of species indirectly by altering selection mosaics among interacting species. The more complex the local web of specialized interactions on which a population depends, the greater the number of potential constraints placed on that population as it responds to climate change.

That complexity has been highlighted in the efforts to conserve endangered large blue (*Maculinea arion*) butterflies in the United Kingdom. Jeremy Thomas and colleagues have worked for decades to understand the ecological conditions that allow the successful persistence of populations of this species (Thomas et al. 2009). These butterflies are restricted to habitats that have two other taxa: thyme plants on which the larvae feed early in development and *Myrmica sabuleti* ants. The ants are as crucial as the plants, because later in development the butterfly caterpillars live in the nests of these ants as social parasites, feeding on ant brood. This butterfly species, then, can maintain viable populations only where *M. sabuleti* ants co-occur with thyme plants. The ants, though, are limited to places where soil temperature remains above a certain threshold. In this part of the geographic range of the butterflies, that means habitats in which herbivores trim the vegetation enough to allow sunlight to reach the soil surface. There is, then, a complex set of relationships between abiotic and biotic conditions that allow this butterfly species to persist in this part of its geographic range. It could be captured to some degree by a map of the physical conditions, but the actual direct limiting factors to the butterfly populations are the other associated species.

Similar complex relationships between abiotic and biotic conditions determine the distribution of this butterfly species in other parts of its geographic range in Europe. The butterfly larvae parasitize different *Myrmica* ant species in different regions (Sielezniew et al. 2010; Casacci et al. 2011). Each population uses specialized tactile, morphological, and chemical signals that allow it to mimic ants and escape detection as a predator while in an ant nest (Elmes et al. 2002; Barbero et al. 2009). Other *Maculinea* species show similar complexity in their life history requirements and interactions, and they show evidence of local adaptation to their ant populations (Nash et al. 2008). For multiple *Maculinea* species, then, their geographic ranges depend on their local adaptation to interactions with other species.

THE CHALLENGES AHEAD

Many interactions between species have persisted for millions of years, which indicates that species and even assemblages of species often track each other amid changes in the physical environments in which they live (Vrba 2005; Eldredge 2008). The problem we still need to solve is how these species and their interactions persist for so long amid so much environmental change. Increasingly, it seems likely that the process involves a combination of selection that continually fine-tunes populations to local environments, phenotypic plasticity (G × E) that may itself be a result of selection, and the great range of mini-evolutionary and mini-coevolutionary experiments generated by selection mosaics (G × G × E). Species experience their environments as much through interactions with other species as through the direct effects of physical environments in which they live. Selection mosaics amplify the local evolution of populations and diversify the variety of ways in which pairs or groups of species interact with each other. Along the way, some populations go extinct and others proliferate through selection and adaptation.

Selection mosaics generate the potential for selection to favor not just local adaptation in species but sometimes adaptation at remarkably small scales in some populations. Those studies show us that selection is continually fine-tuning the adaptation of populations. It is to the new generation of studies of very local adaptation that we now turn.

10

Local Adaptation

The major premise of *The Origin of Species* was that populations adapt to their local environments through natural selection. Speciation occurs because populations diverge as they become adapted to different environments. A century and half later, we are still trying to understand the geographic scales at which natural selection shapes local adaptation and how all those local adaptations collectively drive the further evolution and diversification of species. Getting the answers is important, because the spatial scale of selection and local adaptation has the potential to affect the overall pace and trajectories of evolutionary change within species, the ecological dynamics of populations, and the process of speciation.

This chapter discusses what we have been learning in recent years about the spatial scale and structure of local adaptation and its implications for evolutionary dynamics. The examples are biased toward organisms such as microbes, plants, insects, small freshwater fish, and some marine species in order to consider at just how small a scale local adaptation can occur in nature.

BACKGROUND: THE PROLIFERATION OF STUDIES OF LOCAL ADAPTATION

The rise of ecological genetics and evolutionary ecology from the 1930s to the 1960s produced studies showing that species could adapt to local environments over surprisingly small spatial scales and short periods of time. It took a long time, however, for the effects of those studies to have major impacts on the fields of ecology and evolutionary biology. In the early stages of the reincorporation of evolution into ecology and ecology into evolutionary biology, Allee and colleagues argued in their 1949 textbook *Principles of Animal Ecology* that "modern ecologists have been somewhat reticent in developing evolutionary principles." That reticence, though, was already being pushed aside through the work of a new generation of field-oriented evolutionary biologists.

Some of these researchers were trying to understand the ecological factors maintaining genetic polymorphisms in nature, including the occurrence of dark-winged and light-winged morphs of butterflies and moths, color patterns of moths, and the banding patterns of *Cepaea* snails (Ford 1945; Cain and Sheppard 1950; Kettlewell 1959). Others were trying to understand the spatial scales at which selection favored different traits. Among the most detailed of these studies were those that analyzed the traits of plant species growing in different environments of California, ranging from the Pacific coast to the mountain tops of the Sierra Nevada. The initial studies showed that differences among populations in some morphological traits were retained when the plants were grown in common gardens (Clausen et al. 1941). Subsequent studies showed that these geographic differences were genetically based (Clausen et al. 1947). By 1966, Subodh Jain and Anthony Bradshaw, writing in a joint paper from their personal perspectives in North America and Europe, could summarize with confidence the current state of knowledge by saying that "population studies . . . in many plant species have revealed the existence of localized races" (Jain and Bradshaw 1966).

Some of these studies were undertaken explicitly to ask if natural selection or random genetic drift was the cause of differences among populations (Epling and Dobzhansky 1942; Cain and Sheppard 1950). Sewall Wright (1931) showed that in small populations, mutant genes can be either lost or fixed by chance alone unless they are favored strongly by natural selection. Moreover, genetic drift can result in populations differing genetically even over very small spatial scales. These chance effects can be captured in simple and elegant laboratory experiments in which bacterial populations that differ only in a selectively neutral fluorescent-marker color are allowed to expand from the center to the edge of a Petri dish where individuals of both colors occur (fig. 10.1). The contrasting colors within the dish (shown here as contrasting shades) develop because, as the population expands from its origin in the center, the red cells become fixed by chance in some sections of the expanding circle and the green cells become fixed in other sections. Once the cells become fixed for one color within a section of the dish, all subsequent expansion outward from that region has only one or the other color.

As studies of the relative roles of selection and drift continued, some population differences thought initially to be due to genetic drift were demonstrated decades later to be driven by divergent natural selection over surprisingly small spatial scales (Schemske and Bierzychudek 2001). Gradually, the view of many population geneticists that selection in nature

Fig. 10.1 Random genetic drift in an expanding colony of *Pseudomonas aeruginosa* bacteria obtained by mixing two strains that differ only in their fluorescent markers. Here they are shown as dark and light regions but are red and green in the actual experiments. These markers do not affect selection on the populations. After Xavier (2011), following from experiments by Hallatschek et al. (2007).

is often weak was challenged by field biologists who found evidence that selection was often strong and sometimes divergent even among neighboring populations that seemed to be in similar habitats. The results of this midcentury research were incorporated, directly and indirectly, into the developing fields of ecological genetics (Ford 1964) and evolutionary ecology (Lack 1947; Brown and Wilson 1956; Hutchinson 1959; Orians 1962; Janzen 1966; Harper 1967).

At the same time, more ecological approaches were entering into the study of speciation. It became clear that the central issues in the speciation process were about the roles of natural selection, gene flow, and geographic isolation in the divergence of populations (Mayr 1947; Stebbins 1950). The interplay among these ecological and evolutionary disciplines continued to broaden in scope, generating highly influential studies such as Paul Ehrlich and Peter Raven's (1964) attempt to suggest how local novel mutations affecting ecological interactions could eventually result in large-scale adaptive radiation of species. By late in the twentieth century the rapidly accumulating ecological and molecular evidence had indicated that populations of most species are constantly under one or more forms of selection, and many populations undergo at least some degree of evolutionary change generation after generation.

There are now more ways of studying local adaptation than ever before. The most common ways remain common garden experiments, in which individuals from multiple populations are grown in the same environment and evaluated for differences in the expression of traits thought to affect fitness (e.g., growth rate, size at maturity, number of offspring). Also common, and more informative, are reciprocal transplant experiments, in which individuals from two or more populations are grown in their own environment and in the environment of the other populations. This latter type of experiment is more effective at determining whether there is evidence of local adaptation, because it allows a clearer separation of genetic and environmental effects on traits (Ebert 2004; Nuismer and Gandon 2008).

These studies have helped to determine how much of the difference among populations in traits is due to consistent differences in genes and gene expression rather than to phenotypic plasticity. The application of these experiments, though, is limited. Experiments on local adaptation are easiest on small, fast-growing species (Hoeksema and Forde 2008). Moreover, reciprocal transplants cannot (or should not) be performed on some species, because the potential for introducing foreign genes into other populations is high—thereby potentially destroying the local adaptation that is being studied.

New molecular and statistical methods are providing additional ways of assessing how at least the intensity of selection varies not only among individuals but also among genes within genomes, and how the current intensity of selection compares with selection in the past (Mitchell-Olds and Schmitt 2006). No single method can capture the complexity of selection on local adaptation. Collectively, though, the expanding range of techniques is showing more convincingly than ever before that local adaptation is common although not universal, often involves multiple traits correlated with each other to varying degrees, and occurs at spatial scales from large to small (Leimu and Fischer 2008; Burdon and Thrall 2009; Hereford 2009; Sanford and Worth 2010).

THE SPEED OF LOCAL ADAPTATION

There are now hundreds of studies showing that individuals often survive, grow, or reproduce better in their normal environment than in other environments, thereby at least suggesting local adaptation driven by natural selection. The number of studies of local adaptation has continued to grow so quickly that papers now appear each year reviewing only studies relating to particular habitats, life histories, taxa, or the range of factors driving or limiting the process of particular forms of adaptation (e.g., Greischar and Koskella 2007; Hoeksema and Forde 2008; Prentis et al. 2008; Hereford 2009; Futuyma 2010). The study of local adaptation has become a growth industry, not because these studies are easy to carry out well (they are not) but because local adaptation of populations is the central process in evolution. It has the potential to drive the ecological dynamics and persistence of populations and the diversification of species, and we therefore need to know the speeds and spatial scales at which it occurs.

Particularly notable in recent years is the expansion of studies of local adaptation in marine species. The study of local adaptation has long been the purview of terrestrial and freshwater biologists. Studies of salmonid fish that live both in streams and in oceans are the major examples among partially marine species. A decade ago, only a few studies of local adaptation had been carried out in fully marine species, partly because there had been a long-standing view among many marine biologists that local adaptation was unlikely to occur in many marine taxa. That view was based on the observation that many marine species have planktonic larvae, and those larvae have the potential to range widely in ocean currents. Gene flow would therefore swamp any selection for local adaptation. That view has turned out to be wrong.

Local adaptation of populations has now been shown in a wide range of marine species, including species as different as giant kelp, marine midges, and predatory snails (Alberto et al. 2010; Sanford and Worth 2010; Kaiser and Heckel 2012). There are enough examples of local adaptation in marine fishes (Conover et al. 2006) and invertebrates (Sanford and Kelley 2011) to merit full reviews of the published studies of these species. The studies of marine invertebrates suggest that local adaptation in these species is most evident at scales of hundreds of kilometers (fig 10.2), which is larger than that observed in many terrestrial studies. Nevertheless, some marine

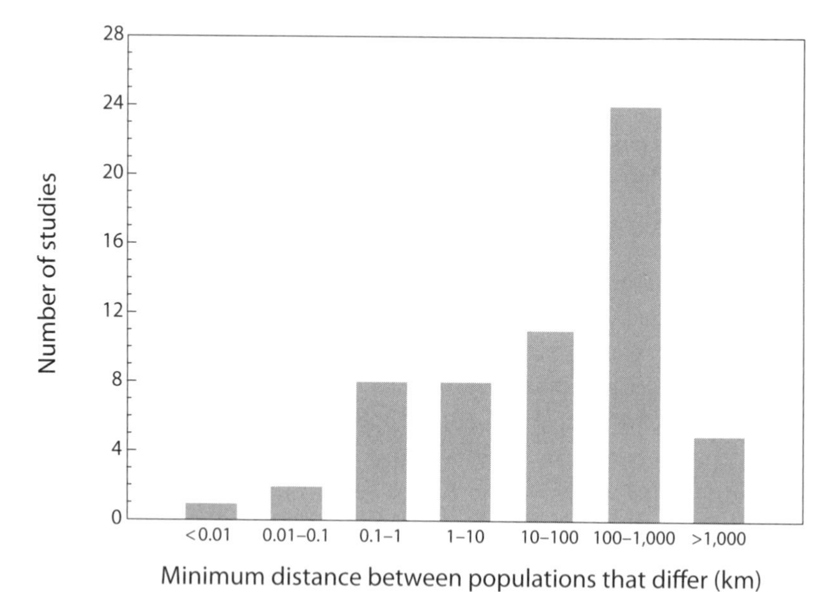

Fig. 10.2 Distribution of the geographic scale at which adaptive differences have been found in 59 studies of marine invertebrates. Most studies have found differences at the scale of 100–1,000 km, but some species show local adaptation at very small spatial scales. After Sanford and Kelly (2011).

species show evidence of local adaptation at spatial scales of less than a square kilometer (e.g., Sherman and Ayre 2008), as has also been observed in some terrestrial species (Laine 2005; Koskella et al. 2011).

As studies of local adaptation continue to accumulate, they are showing many complex geographic patterns. Some show mosaics of adaptation across landscapes in multiple traits. Others show smooth gradients—that is, clines—in those traits. Yet others show smooth transitions in adaptive traits across large regions with occasional abrupt changes among neighboring populations—that is, step clines. Still others show indications of a high ability to respond to selection in some populations but reduced ability to respond to selection in other populations, such as those near the leading edges of geographically expanding species where genetic variation may be reduced (Pujol and Pannell 2008). Collectively, these studies suggest that selection continually reshapes species in ways sometimes so subtle that they can be detected only by careful experiments that involve reciprocally transplanting individuals among environments followed by careful monitoring of survival, growth rates, and reproductive rates.

Many of the traits that show smooth transitions among neighboring populations mirror similar smooth transitions in the physical environments, as the populations adapt to gradients of temperature, salinity, soil nutrients, or light. Some clines reflect rapid sweeps of populations across large geographic regions, with local adaptation evolving almost immediately. One of the most remarkable examples is that of *Drosophila subobscura*, which shows clines in morphological and physiological traits across its native environments in Europe. In the late 1970s these flies were accidentally introduced into North and South America and spread rapidly along the west coasts of both continents (Prevosti et al. 1988). Molecular studies of their colonization history suggest that the flies were first introduced from the Mediterranean region of Europe into South America and from there into North America (Pascual et al. 2007). Since then, these populations have evolved rapidly and developed clines in morphology and physiology along environmental gradients similar to those found in their native ranges in Europe.

On all three continents, wing size increases gradually with latitude. These populations have also developed clines in chromosomal rearrangements that are thought to be linked to adaptive traits (Prevosti et al. 1990; Balanyá et al. 2009). Not all traits, however, have evolved in the same way. The fly populations on all three continents have differences in their tolerance to desiccation (Gilchrist et al. 2008). Local adaptation in these flies is therefore a complex mix of evolving traits. Moreover, the traits have evolved at different rates. The clines in chromosomal inversions formed only a few years after colonization, but the clines in wing size formed over about two decades.

Environmental temperature appears to be a major driver of these clines. Ambient temperatures affect the body temperature of flies and therefore their physiological performance, which in turn favors selection on traits that regulate body temperature (Huey and Pascual 2009). Superimposed over these continent-level clinal patterns are also more global patterns of evolutionary change in these flies. Over the past quarter century, the populations have been evolving in response to rising global temperatures. In 22 of 26 monitored populations on three continents, local ambient temperatures have increased in recent decades. In all but one of the populations, the frequency of chromosomal inversions characteristic of warmer climates has also increased (Balanyá et al. 2006).

Some of the evolved changes occurred during the first decade after the flies were found in the wild in the Americas. The rate of directional evolution in Chile in the wing size of females, as measured using haldanes, was among the highest ever recorded for a trait controlled by many genes (Gilchrist et al. 2004). Equally interesting, although wing size also varies clinally along environmental gradients in Europe and in North America, different sections of the wing have evolved to produce the cline on the three continents (Gilchrist et al. 2004). Hence, natural selection has favored geographic differences in wing size, but different genes have been under selection on each continent to produce that effect.

THE ECOLOGICAL TRAJECTORIES OF LOCAL ADAPTATION

Across the same environments in western North America, another cline involving *Drosophila* has developed over the past several decades, but this one involves the evolution of ecological outcomes in interactions with symbionts. Maternally inherited *Wolbachia* are widespread among *Drosophila* species, and they commonly lead to female-biased sex ratios by selectively favoring production of female offspring (see chap. 6). These bacteria can spread through populations even if they reduce the fecundity of the host females because, once the infection starts to spread, infected females have an advantage over uninfected females. Any uninfected female that mates with an infected male or with a male carrying a different strain of the bacteria generally produces embryos that die (Turelli and Hoffmann 1995; Werren and Jaenike 1995; Bourtzis et al. 1996). Eventually, the host population is made up almost exclusively of infected individuals. Although these symbionts have evolved relationships ranging from parasitism to mutualism in various taxa, they are commonly parasitic in insects (Moran et al. 2008; Serbus et al. 2008).

Drosophila simulans flies in southern California were first discovered to be infected with *Wolbachia* in the late 1980s (Hoffman et al. 1986). These *Wolbachia* caused cytoplasmic incompatibility, reducing the fecundity of infected females by 15 to 20 percent in laboratory tests. Over the next decades, the bacteria spread north by more than 700 kilometers (Turelli and Hoffmann 1991). The evolutionary theory of virulence predicts that maternally inherited parasites should evolve toward lower levels of virulence, because natural selection will favor less virulent strains that allow hosts to persist until the symbiont is passed on to the host's offspring (Bull 1994). This is in contrast to symbionts that are passed on through infectious spread in which selection often favors higher levels of virulence. Since the 1980s, the *Wolbachia* infecting the southern populations have evolved not only reduced virulence but also a mutualistic relationship with their *Drosophila* hosts. *Wolbachia*-infected females now have 10 percent higher fecundity than uninfected females, at least in laboratory tests (Weeks et al. 2007). More recent analyses are consistent with these results, although the extent to which these effects are found consistently in nature is still under investigation (Carrington et

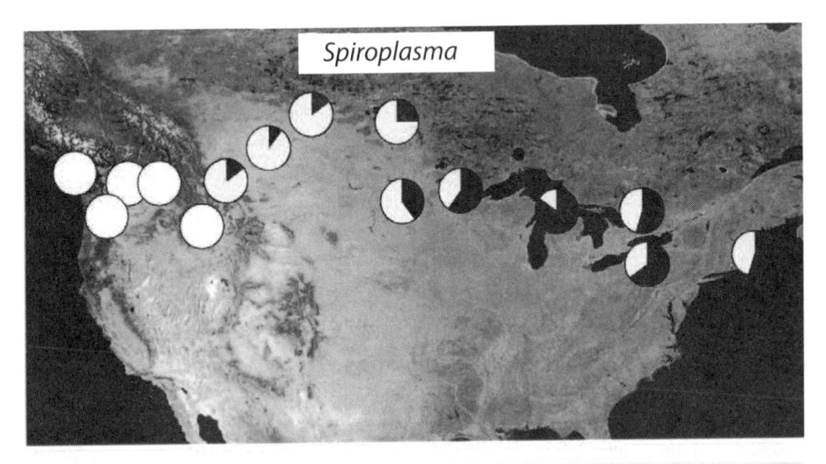

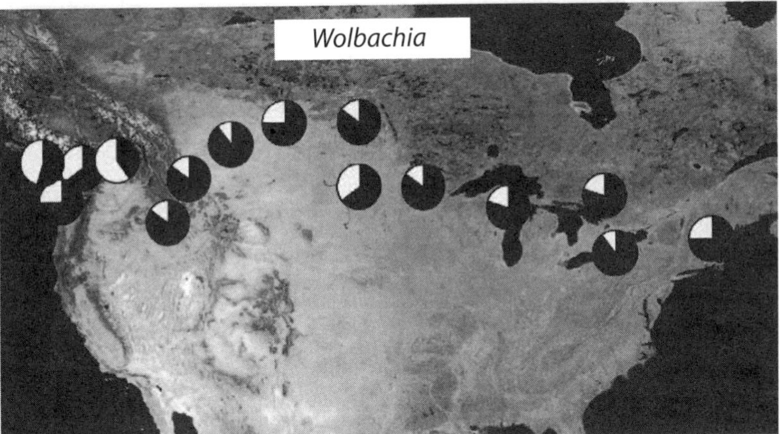

Fig. 10.3 Frequency of *Spiroplasma* and *Wolbachia* symbionts in *Drosophila neotestacea* populations across southern Canada and northern United States. Black sections of each pie diagram indicate the proportion of flies infected in each population. After Jaenike et al. (2010b)

al. 2011). It will be interesting to see if, in the coming years, the northern populations also evolve mutualism or if natural selection on the ecological outcome varies among the environments of California.

Recent studies of *Drosophila neotestacea* have suggested that spreading waves of adaptation between *Drosophila* and their symbionts can continue through multiple routes, sometimes involving interactions between symbiont species. Such waves of change in species interactions have the potential to rapidly alter local, regional, and even continent-wide patterns of adaptation. *Drosophila neotestacea*, a mushroom-feeding fly species, occurs across

Canada and northern United States (Jaenike et al. 2010b). It is infected by *Wolbachia* across this latitudinal band. It is also infected by another maternally inherited symbiont, a species of *Spiroplasma*, which is spreading westward (fig. 10.3). These two symbionts co-occur in eastern North America at a high frequency, approaching 75 percent in some populations. Almost all doubly infected female flies pass on both symbionts to their offspring. That creates the potential for a mutualism to evolve between these symbionts because specific genetic lineages will remain and evolve together for generation after generation (Jaenike et al. 2010a). Alternatively, it could create the opportunity for conflicts between symbionts (Jones et al. 2010).

In this particular case, the presence of the *Spiroplasma* reduces infection by a parasitic nematode that can occur at high frequencies in natural populations of this *Drosophila* species. This positive effect on its host may be responsible for the recent westward spread of *Spiroplasma* among *Drosophila* populations (Jaenike et al. 2010b). These studies may have caught this interaction early in its formation, offering a glimpse into how new mutualisms develop in nature as local adaptations and then spread to other populations.

FLEXIBILITY OF LOCAL ADAPTATION

The accumulating studies of local adaptation show that species can adapt quickly in more than one direction. One of the clearest examples is the rapid evolution of mouthpart ("beak") lengths in soapberry bugs, *Jadera haematoloma*, over the past half century as they have evolved to exploit new plant species introduced into the habitats in which these bugs live. Soapberry bugs feed on seeds of soapberry (Sapindaceae) plants by piercing the fruit wall to get to the developing seed (fig. 10.4). The space between the outer fruit wall and the seed is hollow, and the distance varies among plant species. A bug's mouthparts must be long enough to span the distance from the fruit wall to the seed to feed successfully. Populations of the bugs differ in the average length of their beaks, and those differences correspond to differences among local host plant populations in the fruit-to-seed distance (Carroll and Boyd 1992).

When nonnative soapberry species were introduced into parts of North America during the twentieth century, these plants became potential new hosts for the bugs. In Florida, the introduced plants have shorter fruit-to-seed distances than do the native plants, whereas in Louisiana they have longer distances than the native plants. Within fifty years (about one hundred generations), some populations of the bugs in Florida evolved shorter beaks, allowing them to use this new source of food, and some populations in Louisiana evolved longer beaks (Carroll and Boyd 1992). Other traits are

Fig. 10.4 Balloon vine (Sapindaceae) in Guanacaste Province, Costa Rica.

also involved, some of which are associated with feeding and others with flight. The result is a complex pattern of evolution among multiple genetically correlated traits in these insects (Carroll et al. 1998, 2001; Dingle et al. 2009).

That part of the story was known a decade ago, but work since then has demonstrated that the evolution of mouthpart length and other correlated traits truly is possible in a short number of generations. Laboratory selection experiments have shown that beak length is a highly heritable trait that can evolve rapidly under strong selection (fig. 10.5). It requires only six generations of selection to re-create the evolution of longer or shorter beaks (Dingle et al. 2009). Reverse evolution back to lengths found before the introduction of new hosts can also be produced experimentally in the same length of time.

In nature, these rates could differ if the evolution of these traits is positively or negatively correlated with the evolution of other traits also under selection. For example, in this species, mouthpart length and wing length are highly correlated, and these traits are in turn correlated with egg mass and early egg production. Hence, the evolution of beak length brings with it a set of additional traits that affect other ecologically important aspects of the life history of these insects and could affect the actual rates of evolution in nature (Dingle et al. 2009).

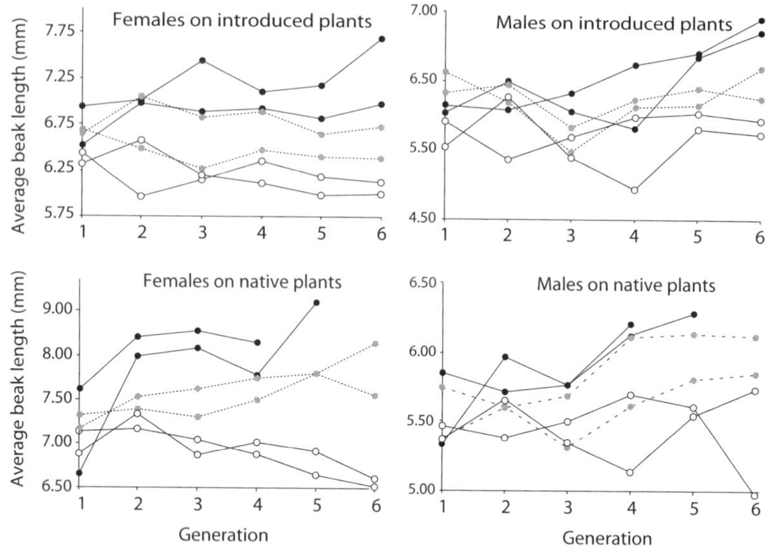

Fig. 10.5 Evolution of beak length in soapberry bugs, *Jadera haematoloma*, from two geographic regions during six generations of artificial selection for longer beaks (*solid circles*) or shorter beaks (*open circles*). Control populations (*gray circles and dashed lines*) were maintained under the same laboratory conditions but with no selection on beak length. The top two panels show evolution of populations that are now adapted to an introduced host and therefore represent the potential for further evolution in a population recently adapted to a new host. The bottom two panels show evolution of populations adapted to the native host and therefore represent a reenactment of the rapid evolution that has happened in recent decades. After Dingle et al. (2009).

Rapid evolution onto newly introduced hosts also appears to be occurring in different soapberry bugs in Australia (Carroll et al. 2005), and similar examples of shifts among related hosts have been demonstrated in multiple other insect species (e.g., Feder and Forbes 2008; Singer et al. 2008). These studies suggest that adaptations of insects to local plant species and populations have been highly dynamic over evolutionary time. These are subtle evolutionary shifts. Given the dominance of insects and plants in the web of terrestrial life, however, these shifts are probably among the most important in driving changes in those webs.

LOCAL ADAPTATION AT VERY SMALL SCALES

Perhaps the most remarkable discovery about local adaptation is that it can sometimes occur over surprisingly small spatial scales in nature. In some species, populations differ in adaptation over just a few kilometers (Toju 2009; Laine et al. 2011; Sanford and Kelley 2011). A few studies have found indications of local adaptation at even finer scales. In 1978 George Edmunds and Don Alstad used the results of transplant experiments to argue that black pineleaf scales, *Nuculaspis californica*, become adapted to individual ponderosa pine trees as the scales produce generation after generation on individual trees. That particular example turned out to be much more complicated, because subsequent work showed that multiple factors other than local adaptation to pines are likely involved in the fine-scale genetic differences found among black pineleaf scale populations (Alstad 1998). Nevertheless, those initial results and related studies by others in subsequent years motivated similar analyses and a broadening interest in local adaptation in insects (e.g., Fox and Morrow 1981). By the late 1990s, there was general consensus that very local adaptation may occur in some insect species, but the spatial scale of local adaptation varies widely among species (Mopper and Strauss 1998; Van Zandt and Mopper 1998).

Some studies have provided evidence that adaptation in some plant-feeding insect populations is so local that it may occur at the level of individual trees (e.g., Egan and Ott 2007; Tack and Roslin 2010). In principle, there is no reason why relatively sedentary populations of insect, microbial, and or other small species should not become adapted to individual long-lived trees, small patches of trees, or any other long-lived host. If a group of breeding individuals is subject to consistent natural selection on or within one host, and selection is strong enough to swamp the diluting effects of immigrants coming from other host individuals, then very local adaptation should occur. The spatial scale of local adaptation therefore depends on the spatial configuration of the hosts, the distances among hosts, and the distances over which insects or microbes tend to move (Tack and Roslin 2010).

Because insect species differ in their tendencies to disperse, even the same spatial configuration of hosts can result in different levels of local adaptation to trees in particular locations. In a detailed study of six insect species that mine or gall the leaves of pedunculate oak (*Quercus robur*) in southern Finland, each insect species showed a different pattern of local adaptation across an experimental woodland that was monitored for multiple years (fig. 10.6). In five out of six species, local adaptation was higher in trees receiving fewer immigrants from surrounding populations. In the sixth species, no signs of local adaptation were detected. For this species, previous studies suggest that local selection pressures may be too weak or inconsistent to allow local adaptation amid gene flow from neighboring populations (Roslin et al. 2006; Gripenberg et al. 2007).

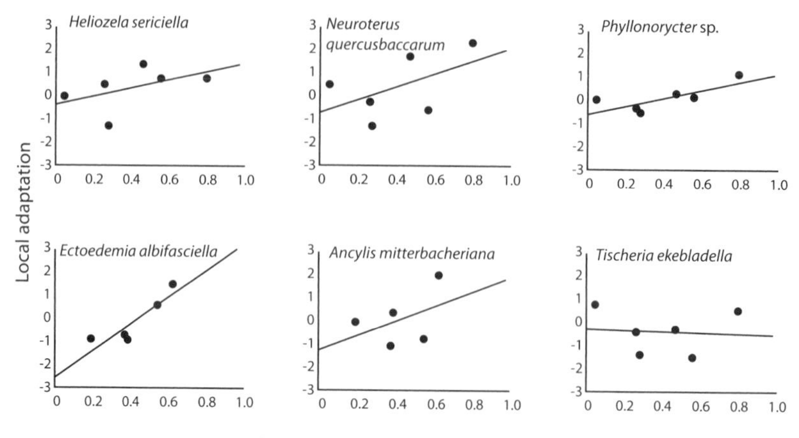

Fig. 10.6 Local adaptation in six insect species that produce mines or galls on pedunculate oak in southern Finland. The x-axis indicates the estimate from related studies of the proportion of the local breeding population that came from that location. The y-axis indicates an estimate of local adaptation obtained by estimating the difference (in logarithms) between the abundance of insects with the local genotypes as compared with the abundance of insects with foreign genotypes. Zero on the y-axis indicates lack of evidence for local adaptation; positive values indicate evidence for local adaptation; negative values indicate that insect individuals originating from other trees do better than local individuals. After Tack and Roslin (2010).

Some microbial species also show evidence of local adaptation at the level of individual trees. Horse chestnut (*Aesculus hippocastanum*) trees in Europe harbor multiple bacterial species on the surfaces of their tissues and within their tissues. These bacteria are, in turn, attacked by bacteriophages. When the phages are challenged in experiments with leaf-inhabiting bacteria from their tree or from nearby trees, the phages are better able to infect bacteria from their normal host tree (Koskella et al. 2011). Local adaptation occurs in phage populations on the leaf surface and in the leaf interior, but it is stronger in phage populations in the interior (fig. 10.7).

One interpretation of this result is that the leaf interior is a relatively stable environment during a growing season, providing time for multiple generations of the phage to evolve adaptation to their hosts. In contrast, local phage populations on the leaf surface are continually diluted throughout the season by a continual rain of phages from the surrounding landscape. Additional studies are needed to understand the mix of selective forces acting on these interactions, but it appears that the spatial scale of local adap-

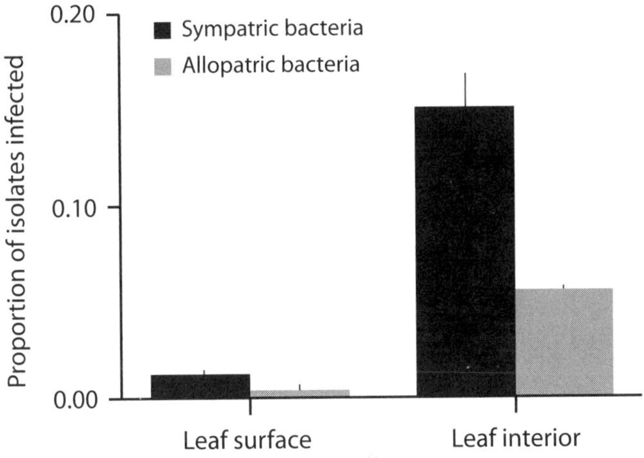

Fig. 10.7 Local adaptation of phage populations attacking bacteria on the leaf surface and leaf interior of the horse chestnut (*Aesculus hippocastanum*) trees from which they were collected in comparison to phage populations collected from nearby trees. Values are means ±1 SE. After Koskella et al. (2011).

tation in these phage populations differs between microbial population in the interior and those on the surface of the same plant.

Recent work on other species has indicated local adaptation of pathogens even on small host plants, although at the level of patches of plants rather than individual plants. The genetic structure of resistance and virulence in the interactions between Australian wild flax (*Linum marginale*) and flax rust (*Melampsora lini*) can differ among populations just hundreds of meters apart (Thrall and Burdon 2003), suggesting differential local adaptation although other evolutionary forces also appear to generate some of the divergence (Burdon and Thompson 1995). At an even smaller spatial scale, resistance of narrowleaf plantain plants (*Plantago lanceolata*) to powdery mildew fungi (*Podosphaera plantaginis*) can vary across tens of meters within a population, with the highest resistance occurring within micropatches where the disease tends to be more prevalent year after year (Laine 2006). Because temperature is known to have strong effects on the performance of the fungi in this interaction (Laine 2008), some of this variation in local adaptation may be due to variation in environmental temperature over small scales.

Australian wild flax and narrowleaf plantain both show considerable resistance diversity against their major pathogens. That diversity accumulates as sampling scales up from populations to metapopulations to even

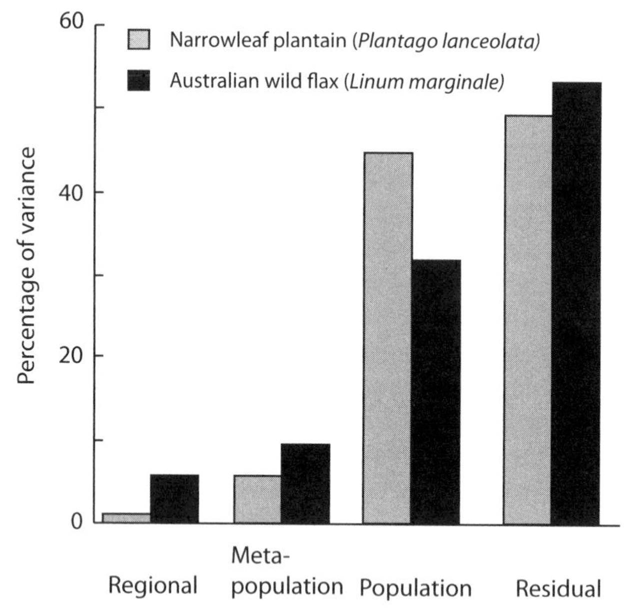

Fig. 10.8 Percentage of variance in resistance diversity found in narrowleaf plantain plants (*Plantago lanceolata*) in Finland and wild flax (*Linum marginale*) in Australia at three spatial scales: regional (among metapopulations), within metapopulations, and within populations. Also shown is the unexplained variance. After Laine et al. (2011).

larger scales among metapopulations. The greatest diversity, though, occurs at the within-population scale in both species (fig. 10.8). Since local adaptation relies on genetic variation, this distribution of variation suggests that the potential for local adaptation is greatest at this scale if selection is strong. Some mathematical models suggest that the effectiveness of selection on adaptation at higher spatial scales depends on the demographic costs associated with maladaptation as individuals move among populations (Hanski et al. 2011).

LOCAL ADAPTATION SHAPED BY THE WEB OF LIFE

Studies of adaptation at small scales highlight one of the emerging major points about local adaptation: although differences among physical environments favor the local adaptation of populations (through genotype-by-environment interactions), interactions with other species amplify those differences (through genotype-by-genotype-by-environment interactions). As a result, the geographic scale of local adaptation in many species is probably much smaller than we could predict based on adaptation to physical environments alone. In some cases, the major effect of the physical environment on the geographic scale of adaptation is primarily to act as a selective sieve on the web of life rather than as a direct driver of local adaptation.

Among the clearest examples is the evolution of divergent life histories among populations of guppies, *Poecilia reticulata*, in the streams of Trinidad. These are the guppies commonly found in pet stores, and they are excellent for experimental studies because they are easy to grow and breed. Their native range extends from northeastern South America to some Caribbean islands.

In Trinidad, guppies occur in mountain streams, where waterfalls block the upstream movement of some species but not others. At lower elevations below the waterfalls, guppy populations are subjected to predation by a wide range of predators, but above the waterfalls killifish (*Rivulus hartii*) are the only predators. Killifish are able to get above the waterfalls by leaving the streams on rainy nights and hopping on land to sites upstream (Reznick 2011). During the 1970s John Endler showed that predatory fish can act as strong selective agents on the guppy populations by selecting against males that use conspicuous color patterns to attract female mates (Endler 1980). He found that where predation was high in natural populations, males were inconspicuous relative to the background color and pattern of their stream environments; where predation was low, however, males had patterns that made them more conspicuous. Predation and sexual selection acted as conflicting selection pressures on these populations. In experimental populations, he found that guppy populations evolved changed color patterns after only about two years. Almost three decades later populations introduced into river sections without predators had evolved in their conspicuous color patterns but in different ways, suggesting the idiosyncratic directions of evolution among populations in sexually selected characters (Kemp et al. 2009). These studies have shown that guppy populations adapt rapidly to the presence of high or low levels of predation and the visual cues used by females in selecting among males.

Also in the 1970s David Reznick began using these Trinidadian populations to ask if differences in predation along the streams could also affect other aspects of the life history of guppies. Each of the streams is a replicated natural experiment. Reznick and colleagues found that guppies from high-predation regions below the waterfalls had evolved to mature early and devote a high proportion of their resources to reproduction. Guppies from low-predation regions above the waterfalls mature later and invest less in reproduction. Reznick and colleagues followed up on those observations by rearing guppies from low-predation sites and high-predation sites and introducing them into the opposite kind of site. For controls, they also reared

Table 10.1 *The number of years and generations required for guppy populations to evolve detectably after introduction from high-predation sections of streams into low-predation sections*

Trait	*Years*	*Generations*
Male coloration	2.5	4.4
Male age and size at maturity	4.0	7.0
Female age and size at maturity	7.5	13.0
Offspring number and size	11.0	19.1
Reproductive effort	11.0	19.1
Predator escape behavior	20.0	35.0
Schooling and predator inspection	34.0	59.2

Sources: After Reznick et al. (2008), based on results by Endler (1980), Reznick et al. (1990, 1997), Magurran et al. (1992), and O'Steen et al. (2002).

guppies and introduced them into the same kind of site. They then monitored the evolution of these experimental populations.

Over the next decade the populations evolved toward life histories consistent with selection acting either in naturally occurring low-predation or high-predation environments (Reznick et al. 1990). In subsequent years other traits have evolved, such as the ability to escape predators (table 10.1). The color patterns, life histories, sizes, and behaviors of these guppies differ along these streams not because there is an important physical gradient along them but because a few physical barriers create sections of river that differ in the intensity of predation. These studies show not only the speed at which multiple traits can become locally adapted in nature, but also the interpretative power of studies of experimental evolution in nature (Reznick and Ghalambor 2005). More generally, these studies show that it is possible for populations to rapidly adapt in multiple ways over short periods of time, even when in natural environments that impose multiple, sometimes conflicting, selection pressures.

GENE EXPRESSION AND PHENOTYPIC PLASTICITY

A half century ago Jens Clausen and William Hiesey (1958) mused that "basic to an understanding of . . . the processes of natural selection, is knowledge concerning the phenotypic expression of genes and of gene combinations over a range of environments . . . [but this] aspect of genetics has been explored but little." We have come a long way since then, but we are still only beginning to explore the full implications of how variation among environments in the expression of genes shapes the structure, dynamics, and spatial scales of local adaptation. We are still working toward understanding when variation in the expression of genes among environments is favored or disfavored by selection and what kinds of variation in expression are adaptive (Ghalambor et al. 2007).

The simplest demonstrations of environmentally determined gene expression are in experiments that place genetically identical or near-identical individuals from a population into different environments and evaluate whether the traits differ in expression in those different environments. There is now a long history of these and similar studies (Snell-Rood et al. 2010) showing differences among environments in average growth rate, height, weight, defenses against parasites, and in many other morphological, physiological, behavioral, and life history traits. These genotype-by-environment (G × E) effects—sometimes called reaction norms or, more generally, phenotypic plasticity—are sometimes due not to any one environmental effect. Rather, they are sometimes due to multiple environmental influences that affect the expression of genes at different stages in the development of individuals.

Environmental effects on the expression of traits early in development can be magnified in traits expressed later in development. For example, environmental conditions experienced by some insects during larval development affect later pupal and adult stages. In some cases, environmental effects on larval physiology determine whether an individual develops quickly into an adult after a brief pupal period or, instead, remains as a pupa throughout a long winter (Friberg and Wiklund 2010; Friberg et al. 2011). The phenotypic plasticity found in some species is so extreme that individuals growing in different environments can easily be mistaken for different species. *Bicyclus* butterflies in Africa, for example, eclose with large eyespots during the warm wet season but emerge with highly cryptic wings in the cool dry season (Oostra et al. 2011). This may reflect selection favoring phenotypic plasticity in expression of eyespots under different environmental conditions, but that is not yet certain.

The usual way to study the genetics of complex traits is to evaluate how traits covary phenotypically (called the P-matrix in population genetics) or genetically (i.e., the G-matrix—see box 2.1). If those patterns vary among environments, then understanding how selection acts on that covariation gets complicated. New mathematical tools in recent years (e.g., structural equation models; techniques from network theory) have made it possible to begin to evaluate the genetic and developmental pathways by which individuals develop different life histories or traits. These models can track how expression of multiple traits at each developmental stage affects expression at each later developmental stage. They can evaluate how plasticity in gene expression affects gene interactions at each developmental stage, leading to different adult phenotypes (e.g., Tonsor and Scheiner 2007).

Even so, the selective forces shaping flexible developmental pathways are only partially understood. At the simplest level, it is easy to understand why selection should favor some flexibility in the expression of traits in unpredictable environments. There are, though, multiple views and results on whether phenotypic plasticity is often directly favored by natural selection or, alternatively, is a consequence of relaxed selection on populations. It may be both. How and why plasticity evolves may depend on the costs and benefits of alternative traits, when the alternatives begin to be expressed during development, and the adaptive landscape over which the traits are expressed (Price et al. 2003; West-Eberhard 2005; Snell-Rood et al. 2010; Hunt et al. 2011).

There are three aspects of the phenotypic plasticity of genes that are immediately important for understanding evolution and are the focus of much current research. One is that strong environmental effects on gene expression could slow natural selection. In a varying environment, the same gene may be expressed in a different way in each generation. That variation in expression makes it difficult for natural selection to favor any particular gene or gene combination over another because, over multiple generations, many gene combinations may produce the same phenotypes. If, however, genetically based traits are expressed in predictably different ways in different environments, then natural selection can act on those differences (Pfennig et al. 2010).

The second point is that the degree of plasticity found in a trait may itself be shaped by natural selection. Laboratory experiments have shown that the phenotypic plasticity of some genetically controlled traits can evolve quickly over the course of less than twenty generations (fig. 10.9). In one set of experiments, *Drosophila melanogaster* flies were subjected to artificial selection for either increased or decreased phenotypic plasticity for thorax length when grown at different temperatures (Scheiner 2002). These flies usually mature at smaller sizes at higher temperatures than at lower temperatures. Selection against phenotypic plasticity in thorax length, used as an index of body size, eliminated the temperature-dependent variation after only eleven

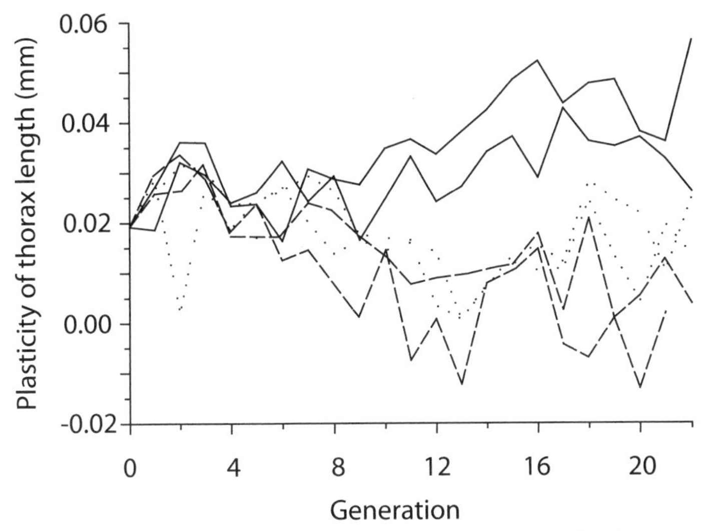

Fig. 10.9 Response to artificial selection on *Drosophila melanogaster* flies for increased (solid lines) or decreased (dashed lines) phenotypic plasticity of thorax length for populations raised under temperatures ranging from 19°C to 25°C. Control populations (dotted lines) were maintained in the same way but without artificial selection. At the start of the experiment, populations showed a linear negative response to temperature in thorax length. Hence, the value of the index for phenotypic plasticity at the start of the experiment is greater than zero. After Scheiner (2002), based on results in Scheiner and Lyman (1991).

generations. At that point, individuals from populations raised at temperatures ranging from 19°C to 25°C all had the same average thorax length. In contrast, individuals from populations that had been selected for greater phenotypic plasticity had increased the degree to which their thorax lengths differed across the temperature range (Scheiner and Lyman 1991; Scheiner 2002). These experiments also indicated that plasticity for this trait is controlled by complex interactions across multiple loci.

The third important point about phenotypic plasticity is that natural selection can favor the full expression of a trait in some environments and no expression or reduced expression of that trait in other environments. Some water fleas (*Daphnia*) develop a helmet or spines or a "crown-of-thorns" only in environments in which the developing larvae perceive the chemical scents of predators in the water (Petrusek et al. 2009). Tadpoles of some spadefoot toad populations develop as a large carnivorous form when raised in the presence of fairy shrimp but develop as small-headed omnivorous forms when raised without fairy shrimp in their environment (Pfennig et al. 2007). Many plants produce particular chemical defenses only after attack by pathogens or herbivores (Wu and Baldwin 2010; Karban 2011), and the immune systems of most animals have highly inducible components (Litman et al. 2007). Even some sexually selected characters, such as the exaggerated horns on some dung beetles, show extreme plasticity, with individuals developing into distinct morphological forms (disproportionately large or small horns), depending on their genes and the nutritional environments in which they developed (Rowland and Emlen 2009). In some cases, even different generations of the same population developing at different times of year express different morphological or physiological forms, including aphids with or without wings, or butterflies with different color patterns or life histories (e.g., Brisson 2010; Larsdotter Mellstrom et al. 2010; Oostra et al. 2010).

Through these means, different populations with the same distributions of genotypes sometimes differ in fundamental ways in their morphologies and physiologies and even their life histories, generating different potential trajectories for further evolution. Phenotypic plasticity, then, is not just variation, and it is probably not always an alternative to the ongoing evolution of populations amid environmental change. It may be just as often a form of adaptive variation molded by natural selection in different ways, in different local environments, or at different life history stages.

THE CHALLENGES AHEAD

Getting the spatial scale of local adaptation correct is crucial for our understanding of the structure and dynamics of evolution. Studying the dynamics at too large a scale or too small a scale can result in puzzling and erroneous interpretations. If populations are grouped at too large a scale, then a study will lump multiple populations with different local adaptations. It will appear that there are genetic polymorphisms within a large regional population rather than separate populations with their own local adaptations. If populations are divided in studies at too small a scale, then nearby populations will appear similar in their adaptations, suggesting that local adaptation is not important. There is no simple way of determining the right spatial scale for a study of local adaptation. Improved populational, molecular, and statistical tools, however, are helping in these assessments by focusing on evaluation of adaptation under different ecologically relevant conditions such as in growing, stable, or declining populations (North et al. 2011) and in metapopulations with different patterns of gene flow (Hanski et al. 2011).

Equally important is our need to gain a deeper understanding of the ecological causes of local adaptation and the kinds of traits, including phenotypic plasticity in traits, favored under different selection pressures. Most studies are designed to evaluate the pattern and scale of local adaptation. The underlying causes and traits are often inferred only from the most obvious differences among the habitats and populations. Differences in physical environments sometimes provide the template for differences in selection on populations, but the actual direct source of differential selection favoring local adaptation may be due to differences in interactions among species.

Sorting out the direct causes of local adaptation is becoming increasingly important as debate continues on how changing physical environments and webs of interaction among species are altering selection on species. In the coming decades, we will likely find that we are still underestimating the commonness of local adaptation, because it can take so many forms. We have, though, at least reached the point at which we expect to find it when we take the time to evaluate it carefully. We turn in the next chapter to what we are learning about the ways in which selection mosaics and local adaptation fuel ongoing evolution, and sometimes coevolution, in interactions among species.

Part 4

The Dynamics of Coadaptation

11

Coevolutionary Dynamics

Selection mosaics and local adaptation set the stage for geographic mosaics of evolution and coevolution in interacting species. These mosaics are built on a template of ever-shifting physical environments. Species vary in size, physiology, life history traits, and behaviors across gradients of temperature, rainfall, seasonality, salinity in marine environments, oxygen levels in soils and water, and multiple other characteristics of the physical world. Selection on evolving interactions can either amplify the selection imposed by physical environments or counteract it. All interactions between species are therefore likely to be in constant evolutionary flux.

The challenge is not just to partition the effects of physical environments from the effects of selection imposed by other species in driving evolutionary change. It is also to understand how selection imposed by physical environments and selection imposed by species interactions act together to continually reshape the traits of species and the web of life. Similar combinations of physical and biotic environments appear repeatedly, creating at least the potential for repeated geographic patterns in how species evolve in their interactions with each other.

This chapter considers what we currently know about the processes that drive the ongoing evolution of interactions between species and the mosaics of adaptation and coadaptation that develop across environments.

BACKGROUND: GEOGRAPHIC MOSAIC THEORY

Evolutionary biologists and ecologists have long known that selection on interacting species differs among environments. Nevertheless, for decades the mathematical theory of coevolution was based on interactions between a single population of one species and a single population of another species. Those results were extrapolated to expectations about how species should evolve and coevolve. It was a species-level, rather than population-level, view of the coevolutionary process. By that view, once an evolving interaction reached some kind of equilibrium, further evolution would likely be erratic. The implicit assumption was that, despite minor differences, any pair or group of species would interact, and impose selection on each other, in much the same way everywhere as they evolved toward an equilibrium. As field studies accumulated showing tremendous geographic variation in some interactions between species, and as population geneticists showed the genetic distinctness of populations within species, it became clear that evolving interactions are much more dynamic. We needed a better way to formally think about how ecological variation in interactions continually shapes the evolution, and sometimes coevolution, of species.

The geographic mosaic theory of coevolution was developed as a framework for envisioning how coevolution proceeds among locally adapted populations living in contrasting environments (Thompson 1994, 2005). The theory is an attempt to incorporate the minimum components of population biology needed for an ecologically and evolutionarily realistic theory of coevolution, and evolving interactions in general. Toward that end, geographic mosaic theory is based on three observations common to most interactions between species: almost all species are groups of genetically distinct populations; interacting species differ in their geographic ranges; and interactions among species differ among environments in their ecological outcomes.

The evolutionary hypothesis that arises from those observations is that coevolution draws from three sources of variation on which natural selection acts. The structure of selection on interacting species varies among populations and environments, and we call that variation a geographic selection mosaic. The strength of reciprocal selection also varies among environments. Interactions will commonly have coevolutionary hotspots where selection is reciprocal on the interacting species. These hotspots will be interspersed among coevolutionary coldspots, where selection acts either on one or neither species in an interaction. In addition, traits will vary among environments as they are continually remixed among populations due to gene flow, random genetic drift, and metapopulation dynamics. Each of these three components of coevolutionary variation contributes to the coevolutionary process (fig. 11.1).

Geographic mosaic theory therefore provides a formal way of describing all the components of a genotype-by-genotype-by-environment (G × G × E) interaction between species. We can consider a G × G × E interaction in its most formal way at the genetic level, or we can use it more generally at the phenotypic level to consider how the ecological and evolutionary processes shape interactions across environments. Species may coevolve

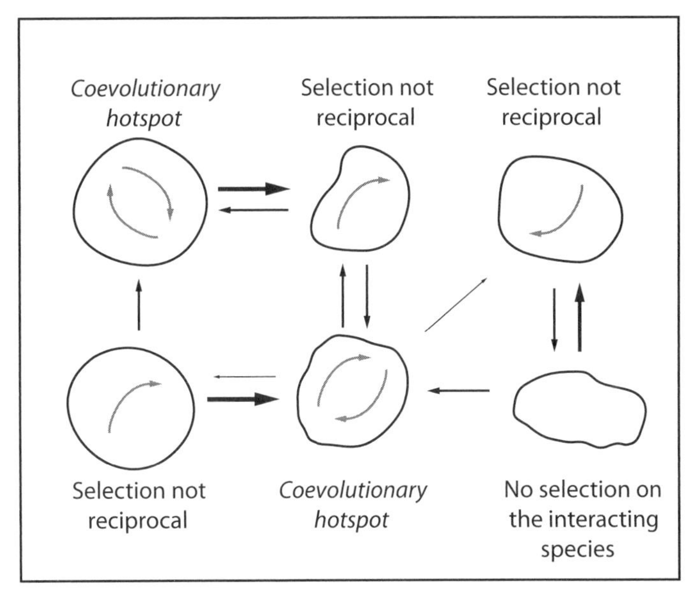

Fig. 11.1 Diagrammatic representation of the geographic mosaic of coevolution between two species. Selection mosaics are shown as gray arrows drawn at different angles. Coevolutionary hotspots are shown as double gray arrows indicating reciprocal selection within a site. Coevolutionary coldspots are shown as single gray arrows (selection on only one of the species) or no gray arrows (selection on neither species) within a site. Trait remixing is captured in this diagram only as different widths of black arrows indicating differences in gene flow between sites, but it also includes the other evolutionary processes that affect the geographic distribution of traits, including random genetic drift and metapopulation dynamics. The diagram shows only a pair of species, but many interactions involve geographic mosaics in groups of coevolving species.

with each other (G × G), but each genotype may vary among environments in its expression (G × E), and the outcome of the interaction between each pair of interacting genotypes may differ among environments (G × G × E). A population is made up of multiple genotypes, and each environment is a collection of multiple mini-environments. Every locality where a set of species co-occurs therefore has a unique distribution of G × G × E interactions on which natural selection can act. That distribution changes over time as selection acts on local populations, and traits become remixed among populations over time across landscapes through gene flow, occasional local loss through random genetic drift, and metapopulation dynamics.

Geographic mosaic theory argues that the evolution of interactions among species is an ecological process that depends on how fitness is distributed in interactions between species within and among environments. By this view, evolving interactions are relentless partly because the structure of selection, the intensity of reciprocal selection, and the distribution of traits within and among populations of interacting species are always changing, along with their changing environments. Even without new mutations, interactions will continue to coevolve in genetically variable species because every change in the local frequency of genotypes in one of the interacting species and every change in the local physical environment or surrounding web of life ripples through the distribution of G × G × E interactions at that locality.

COMMONNESS AND EXPECTATIONS OF GEOGRAPHIC MOSAICS

Scores of studies have now analyzed various components of the geographic mosaic of coevolution in multiple environmental settings. For example, in reviewing twenty-nine studies involving plants, Anna-Liisa Laine (2009) found that these studies analyzed on average thirteen plant populations and an equal or smaller number of parasite, pathogen, or mutualist populations. Most studies have found evidence for spatial variation in the strength of selection, and some have found evidence for variation in the structure of reciprocal selection, including variation in the direction of selection acting on the species. Most studies have been on pairs of interacting species living either in different abiotic environments or in environments that differ in the surrounding web of other species. The studies have ranged in scale from a few meters to continent-wide. Collectively, these studies suggest that geographic mosaics are common in coevolving interactions in nature.

Mathematical models of the geographic mosaic of coevolution suggest that interactions spatially structured in this way are much more dynamic than those in isolated settings. Gene flow between coevolutionary hotspots and coldspots can favor further evolutionary change (Gomulkiewicz et al. 2000). Gene flow between sites in which interactions differ in ecological outcome—for example, antagonism versus mutualism—can produce ongoing evolutionary change (Nuismer et al. 1999, 2003). Selection mosaics, coevolutionary hotspots, and trait remixing together can result in local maladaptation in one or the other species at some localities, which drives further evolution in an interaction (Thompson et al. 2002). Local patterns of adaptation and coadaptation also change as the degree of symmetry in gene flow between populations changes and as the sizes of populations change (Gandon and Nuismer 2009). The geographic mosaic of coevolution almost guarantees that species undergo continual evolutionary change.

Experimental and mathematical studies show that species interacting in a geographic mosaic sometimes coevolve faster than species coevolving in isolation from other populations of the same species (Brockhurst et al. 2003; Lopez-Pascua and Buckling 2008; Best et al. 2011; Paterson et al. 2010). Species coevolving as geographic mosaics also sometimes coevolve toward equilibrial states that differ from those found in cases in which coevolution is only local. In addition, coevolutionary mosaics often maintain polymorphisms for longer lengths of time (Burdon and Thrall 2000; Kniskern and Rausher 2007). If the coevolving populations are distributed along an environmental gradient, such as a nutrient gradient, the populations evolve at different rates and in different ways than when coevolving in isolation in a single environment. These populations form geographic gradients in traits such as host resistance, parasite infectivity, and ecological outcomes (Hochberg and van Baalen 1998; Hochberg et al. 2000; Forde et al. 2007; Lopez-Pascua and Buckling 2008).

Geographic mosaics can also affect how population dynamics and coevolution affect each other. In models of local coevolution, population dynamics and coevolutionary dynamics can have strong reciprocal effects (Jones et al. 2009; Holland and DeAngelis 2010; Mougi and Iwasa 2011; Poisot et al. 2011). In geographic mosaic models, those effects depend on whether populations experience hard selection—essentially, extra deaths in a population resulting from selection—or soft selection—selective deaths occurring instead of nonselective deaths (Gomulkiewicz et al. 2000). The relationships between coevolution and population dynamics have only recently been the focus of detailed studies in nature, but they are increasing as an important area of coevolutionary research (Urban and Skelly 2006; Palkovacs et al. 2009; Morris 2011; Schoener 2011).

MOSAICS ACROSS CONTRASTING PHYSICAL ENVIRONMENTS

All geographically widespread interactions between species are subject to physical gradients that can produce clines in adaptation and coadaptation. Those settings provide opportunities to analyze how coevolutionary selection occurs amid selection imposed by physical environments. Studies of camellias (*Camellia japonica*) and camellia weevils (*Curculio camelliae*) in Japan have been among the most insightful at teasing apart how geographic differences in physical environments and coevolution shape geographic mosaics of traits and ecological outcomes (fig. 11.2). These plants and insects have undergone remarkable geographic divergence in their coevolving traits through selection mediated by the interaction itself, clinal differences in physical environments, and historical separation of populations.

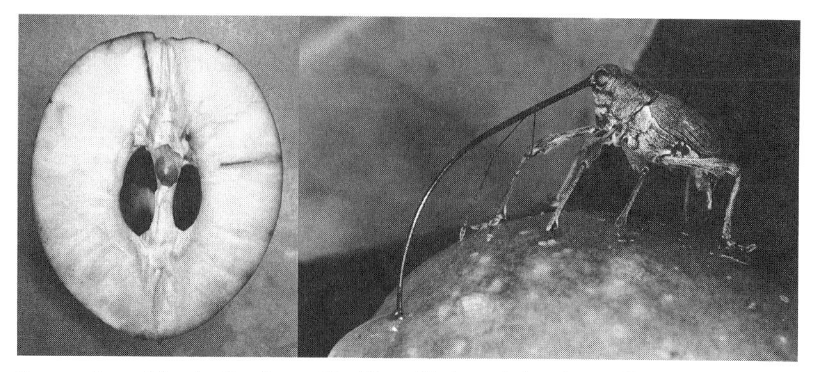

Fig. 11.2 Coevolving traits of camellias (*Camellia japonica*) and camellia weevils (*Curculio camelliae*) in Japan. *Left*, a cross section of a camellia fruit showing the thick pericarp through which a camellia weevil chews with her rostrum before laying an egg on a seed in the center of the fruit. Two attempts that were too shallow to reach the seed are visible in the photograph. *Right*, a female camellia weevil beginning excavation into the fruit. Note that the fruit on the left is shown at a much smaller scale than the fruit and weevil on the right. Photographs courtesy of Hirokazu Toju.

Camellia weevils have the longest rostrum of any species in the genus *Curculio*, which includes more than 300 species distributed over multiple continents (Hughes and Vogler 2004). The weevil's rostrum varies from 9 to 21 mm among populations, and the thickness of camellia fruits shows a correspondingly wide range of sizes among populations (Toju and Sota 2006a, 2009). A camellia weevil lays her eggs on developing seeds by chewing through the fruit (pericarp) and then placing an egg into the hole. She must be able to chew all the way to the seed to be successful.

The length of weevil rostrums and the thickness of camellia fruits vary geographically in ways that result in greater success for the weevils at higher latitudes than at lower latitudes (fig. 11.3). Rostrum lengths tend to be more exaggerated relative to fruit thickness at higher latitudes than at lower latitudes, making it possible for more high-latitude weevils to penetrate a fruit fully and reach a seed (Toju and Sota 2006a). This latitudinal pattern could be due to any of several causes: historical differences in how the interaction has evolved in different regions; a physical gradient that imposes selection on one or the other species independent of the interaction; or an interaction with yet another species that imposes selection on the plants or the weevils. In this case, it is the physical environment that seems to establish the basic template. Analysis of variation in fruit and weevil traits relative to environmental temperatures has suggested that fruit thickness is greater in regions with warmer temperatures (Toju and Sota 2006b). Hence, there seems to an underlying selection pressure favoring larger fruits in warm climates.

That analysis explains only part of the pattern. Phylogeographic studies of weevils indicate that this species has historically been subdivided into two groups, each of which went through a population bottleneck and subsequent population expansion (Toju and Sota 2006b). Consequently, the underlying latitudinal pattern is a composite of two semi-independent geographic patterns. Even so, these two variables together cannot explain the overall geographic pattern. The most powerful predictor of the local trait values of each species is, instead, the trait value of the other species. Coevolution against a backdrop of a physical gradient and the historical separation of populations therefore appears to explain the observed geographic patterns.

The same combined influences of variation in the physical environment, separation of populations, and coevolutionary selection occur at more local spatial scales (Toju 2008, 2009; Toju et al. 2011). On the island of Yakushima, which is only about thirty kilometers in diameter, fruits tend to have thinner pericarps in cool, highland environments than in lowland environments. Nevertheless, weevils with short rostrums and plants with thin fruits are found both at low- and high-elevation localities. Overall, geographic variation throughout the island in rostrum length and fruit thickness is explained better by the pattern of variation in the traits of the other species than by the pattern of variation in the physical environment, although both factors are involved (Toju 2008).

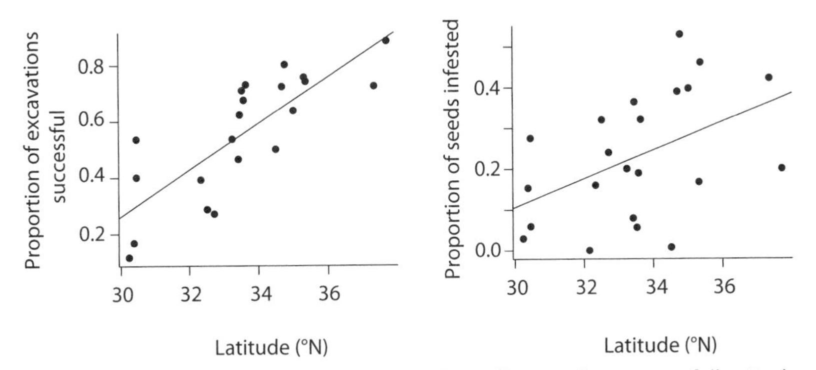

Fig. 11.3 Geographic differences in the ability of camellia weevils to successfully attack camellia fruits across eight degrees of latitude in Japan. The graph on the left shows the proportion of excavations that reach the seed; the graph on the right shows the proportion of seeds infested by larvae. After Toju (2009).

As an alternative to coevolution, the large-scale and small-scale patterns of trait matching between these two species could have resulted simply from evolution in the weevils as they track evolution in the plants. As fruits evolve toward larger or smaller sizes due to many causes other than the weevils, selection might simply have favored weevils that matched the local fruit sizes. That explanation, however, seems unlikely for several reasons. Studies of thirteen sites across Japan have shown that weevils differ geographically in the fruit sizes they prefer (Toju 2007). Weevil populations are under strong selection to choose fruits that they can successfully excavate. That selectivity places strong selection on plants to diverge from fruit sizes that the weevils can attack successfully. Pericarp thickness is a highly heritable trait in these camellias (Toju et al. 2011). The strong selection imposed by the weevils on this highly heritable trait suggests that weevils are driving at least some of the geographic differences found in fruit thickness. Consistent with this view is the observation that camellias generally have thicker pericarps on islands where beetles are present than on islands where weevils are absent (Toju et al. 2011). Hence, the selectivity of weevil attack, which imposes strong selection on this highly heritable plant trait, coupled with differences in pericarp thickness on islands with weevils as compared with islands without weevils, suggests that coevolution, rather than simply evolutionary tracking by the weevils, explains the geographic differences in pericarp thickness.

Coevolution between camellias and camellia weevils has been undergoing geographic diversification for about 6 million years, based on molecular clock estimates (Toju and Sota 2009). This is about a quarter of the 24 million years that extant *Curculio* species in Japan have been diverging from each other. For about 17 million years there was little net evolution of rostrum length. Then, over a few million years, coevolution between camellias and camellia weevils went through a period of directional change that resulted in the exaggerated traits seen today in some of these populations. The exaggerated rostrums arose through only moderate evolutionary rates, as estimated using haldanes (Toju and Sota 2009). Hence, the current range of exaggerated traits appears to have arisen through sustained coevolution over thousands of generations. Major differences between the two monophyletic clades occurred even more recently, showing major divergence only 14,000 years ago. These phylogeographic breaks suggest the importance of Pleistocene events in establishing some of the current geographic patterns (Toju and Sota 2009).

Similar examples of selection imposed both by physical environments and other species are being discovered as more studies analyze the same interactions in multiple places. The most difficult to analyze are those in which directional selection on traits imposed by coevolution follows the same geographic pattern as that imposed by the physical environment. For example, natural selection has commonly favored spines in plants living in arid environments, and the evolution of these structures has been attributed to multiple selection pressures, including selection to reduce the effects of high- and low-temperature extremes, to capture mist and direct water to roots, or to reduce herbivory (Loik 2008). Once spines are favored by selection for any reason, they can be acted on and reshaped in response to new selection pressures.

The classic example is the spines of acacia plants, which have undergone modification in different species in response to various selection pressures. Some species have hollow, inflated thorns evolutionarily modified as homes for the mutualistic ants that protect these plants (Janzen 1966). Some other species have spines that regrow to longer length following herbivory by grazing mammals but produce fewer thorns without continued herbivory (Young and Stanton 2003; Huntzinger et al. 2004).

In Chile, spines are at the center of a complex geographic mosaic that involves columnar cacti (*Echinopsis* and *Eulychnia*), the parasitic mistletoes that live on these plants (*Tristerix aphyllus*), and the mockingbirds (*Mimus thenca*) that eat the mistletoe seeds (Martinez del Rio et al. 1996; Medel et al. 2010). These interactions occur along a long aridity gradient that forms the underlying template for the mosaic. The cacti produce spines that can be remarkably long in some populations (fig. 11.4). Mockingbirds eat the mistletoe fruits and then defecate the seeds onto the spines of the surrounding cactus plants, where they commonly perch. When the mistletoe seeds germinate, the radicle must grow long enough to reach the cactus stem and develop roots.

The length of cactus spines and the ability of mistletoes to develop long radicles vary geographically in ways suggesting selection imposed by a combination of aridity and coevolution. Spine length tends to be greater in more arid regions than in less arid regions, but there is more variation in spine length than can be accounted for by physical gradients in aridity alone. Cacti within the range of the interaction have longer spines than those living outside the range of the interaction. The pattern is consistent across the two cactus genera in which these interactions have been studied (Medel et al. 2010).

Long spines decrease the chance of mistletoe infection, because birds are less likely to perch and defecate on long-spined plants (Medel 2000). Even if birds defecate on long-spined plants, only mistletoe individuals with

Fig. 11.4 The interaction in Chile involving (*a*) the fruits of parasitic mistletoes (*Tristerix aphyllus*) that grow on columnar cacti (*Echinopsis* and *Eulychnia*), (*b*) the fruit-eating specialist mockingbird species (*Mimus thenca*) that eats the seeds and defecates on the cacti, and (*c*) the mistletoe seeds with long radicles that must reach beyond the spines to the cactus stem. Photographs courtesy of Rodrigo Medel.

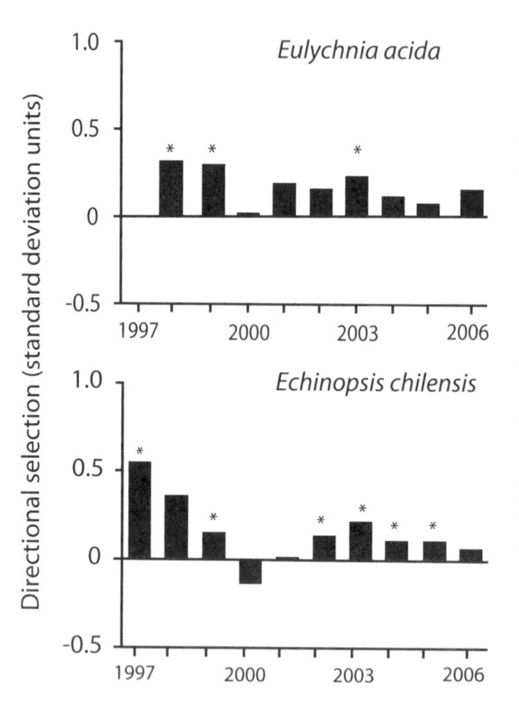

Fig. 11.5 Variation over ten years in selection on spine length in two cactus species at Las Chinchillas National Reserve, Chile, driven by mistletoe infection. Values indicate the trend of directional selection measured in units of standard deviation (standardized selection gradients). Positive values indicate selection favoring cacti with longer spines; negative values indicate selection favoring cacti with shorter spines. After Medel et al. (2010).

long radicles, as are often found in these populations, are able to reach the cactus stem. If mistletoes are able to establish on a plant, they can greatly reduce, and sometimes completely suppress, fruit and seed production (Medel 2000). The interaction therefore imposes strong selection favoring plants with long spines, and a decadelong study has indicated that the selection favors plants with long spines in most years (fig. 11.5).

The combined physical and biotic selection pressures produce a geographic pattern in which the traits of the plants and the mistletoes are well matched in some areas and mismatched to varying degrees in other areas (Medel et al. 2010). The cacti and mistletoe populations tend to be better matched in the northern regions of the interactions and less so in some southern regions. Aridity, mockingbird density, and mistletoe density all vary geographically, turning selection imposed by physical environmental gradients into more complex mosaics.

REPEATED TILES WITHIN GEOGRAPHIC MOSAICS

Another approach to understanding the combined effects of selection imposed by physical environments and coevolved interactions is to ask if similar interactions coevolve in similar or different ways in contrasting physical environments. One way of doing this is to study how closely related species coevolve with other species in different geographic regions. Closely related taxa have many of the same traits, and it is therefore not surprising that natural selection should often shape traits in similar ways among similar groups of interacting species. Demonstrating this important aspect of coevolution, however, has been difficult, because it demands studying multiple similar interactions across multiple environments.

Among the most geographically far-reaching studies are those of the diversification of interactions among crossbills, squirrels, and pines throughout North America and Eurasia. The work began with studies in northwestern North America, and those results generated predictions on trait evolution in these species that now have been tested for the same or similar interactions in other parts of North America and in Europe. Crossbills (*Loxia* spp.) and many squirrel species, such as pine squirrels (*Tamiasciurus*), harvest seeds from conifer cones across Eurasia and North America. Crossbills extract seeds by using their crossed bill to pry apart the scales that surround the seeds within each cone (fig. 11.6). Crossbill populations have bills with sizes and shapes specialized to different conifer species (Benkman 1993). In contrast to crossbills, squirrels bite off scales to reach the seeds, and some squirrel populations and species differ in jaw structure and feeding habits, depending on the traits of the conifers on which they feed

Fig. 11.6 Male (*left*) and female (*right*) red crossbills (*Loxia curvirostra*) extracting seeds from a closed lodgepole pine (*Pinus contorta*) cone. The bill crossing enables the birds to bite and create gaps between hard woody cone scales, as done by the female in the photograph. The birds then spread apart the bill, further widening the gap and exposing the underlying seeds. Photograph courtesy of Craig W. Benkman.

(Smith 1970; Benkman and Balda 1984). The plants, in turn, show multiple adaptations to reduce successful attack by these seed predators, and their adaptations against crossbills differ from their adaptations against squirrels (Benkman 1995, 2010; Coffey et al. 1999).

The presence and relative abundance of squirrels and crossbills vary geographically. As a result, there are multiple possibilities for complex selection mosaics and coevolutionary hotspots and coldspots in these interactions at continent-wide scales. In some regions of North America, red squirrels are the major predators of conifer seeds, cutting the cones from trees and placing them in large middens for future use. Where they are abundant, squirrels impose stronger selection on cones than do crossbills. Where squirrels are rare, conifers lack these defenses and often coevolve with crossbills (Benkman et al. 2008).

Crossbills and red squirrels differ in the selection pressures they impose on the shapes of conifer cones. When red squirrels extract seeds from cones, they start at the base and chew off scales until they reach the seeds, which are concentrated at the other end of the cone. Squirrels concentrate their foraging on trees with cones that have many seeds relative to the mass of the cone, because it minimizes the work needed to get to the seeds. Predation by squirrels therefore favors trees with cones of relatively large mass relative to the number of seeds.

Where squirrels are absent, conifers tend to produce more seeds relative to the mass of a cone (Benkman 1999; Parchman and Benkman 2002). Crossbills impose selection primarily on the thickness of individual cone scales, rather than on cone mass and seed number. When feeding on some conifers, such as lodgepole pine, these birds begin feeding on the tip of the cone rather than its base, and they avoid trees with cones that have relatively thick scales. Selection imposed by crossbills therefore favors trees whose cones have relatively thick scales over the entire cone.

Coevolution with squirrels or crossbills favors similar traits in conifers, whether they are pines in the Rocky Mountains of western North America or black spruces in Newfoundland of eastern Canada (Benkman 1999, 2010; Parchman and Benkman 2002). Local adaptation in these interactions can sometimes occur at surprisingly small scales. Within the Rocky Mountains, some isolated mountain ranges have lodgepole pines but no squirrels, and these populations show reduced defenses against squirrels (Edelaar and Benkman 2006). In other pine species, the geographic mosaic of selection occurs at large scales and involves yet other squirrels. Selection on ponderosa pines (*Pinus ponderosa*) involves not only pine squirrels but also western gray squirrels (*Sciurus griseus*) in the western part their range and Abert's squirrels (*Sciurus aberti*) in the eastern part of their range. Western gray squirrels eat seeds, but Abert's squirrels feed on phloem tissue of the inner bark, thereby indirectly affecting cone production (Snyder and Linhart 1998; Murphy and Linhart 1999). These species differ in the degrees to which they affect selection imposed by crossbills on ponderosa pine populations, but, within local populations, the presence of squirrels generally decreases the evolution of traits adapted specifically to crossbills (Parchman and Benkman 2008).

Similarly, complex geographic mosaics in the interactions among crossbills, squirrels, and conifers have now been found in the Iberian Peninsula and on the islands of the Mediterranean Sea (Mezquida and Benkman 2005). The situation in the Caribbean, however, is different. Coevolution of pines and crossbills involves traits similar to those in North America, but presence or absence of squirrels does not confound coevolution with crossbills in the Caribbean (Parchman et al. 2007).

Fire places an additional major selection pressure on the evolution of cones and the geographic mosaic of coevolution among these species. In fire-prone regions, selection has often favored cones that remain tightly closed until exposed to the kind of high heat associated with fire (serotiny), whereas selection on pines in other regions has favored pines with cones that open more readily. There is no simple relationship between selection imposed by the physical environment and that imposed by seed predators on the evolution of serotiny, but these agents of selection are sometimes conflicting. In regions where squirrels are absent, fire-prone populations of lodgepole pines (*Pinus contorta*) have a high frequency of serotinous cones, but in fine-prone regions with squirrels, the frequency of serotinous cones is generally much less than 50 percent (Benkman and Siepielski 2004).

The geographic mosaic of coevolution between crossbill and pine species has resulted in so much differentiation among crossbill populations that some of these divergent populations may now function as separate species. In North America, the crossbills in the South Hills of Idaho have now been described as a separate species (Benkman et al. 2009). Multiple other populations, each specialized to a different conifer species, appear to be at least in the early stages of speciation (Parchman et al. 2006). In Europe, the common crossbill (*Loxia curvirostra*) may also include multiple species (Summers et al. 2007; Edelaar 2008), with up to three species overlapping in some geographic regions (Summers et al. 2010).

Yet a different version of geographically varying selection among birds, squirrels, and conifers occurs in the interaction between Clark's nutcracker (*Nucifraga columbiana*), pine squirrels (*Tamiasciurus* spp.), and some pines with large seeds in western North America, including limber pine (*Pinus flexilis*) and whitebark pine (*P. albicaulis*). In these interactions, the pines and the birds are mutualistic. The pines produce large seeds, which the nutcrackers collect and distribute widely in small caches in the ground. By one estimate, a single nutcracker can store 32,000 whitebark pine seeds in a year, with each cache holding between 3 and 7 seeds (Tomback 1982). The next generation of pines comes from seeds in the subset of caches never revisited by the birds. In contrast to the birds, the squirrels harvest large quantities of cones, which they place in large middens in the center of their territories. Few, if any, seeds within these large middens are likely to survive to germination.

Where nutcrackers are abundant and squirrels are rare, natural selection favors plants with traits that allow the birds to extract seeds efficiently from cones (Siepielski and Benkman 2007). Where squirrels are more abundant than the birds, selection favors plants whose cones have defensive traits. Where the birds and squirrels are both common, they impose conflicting

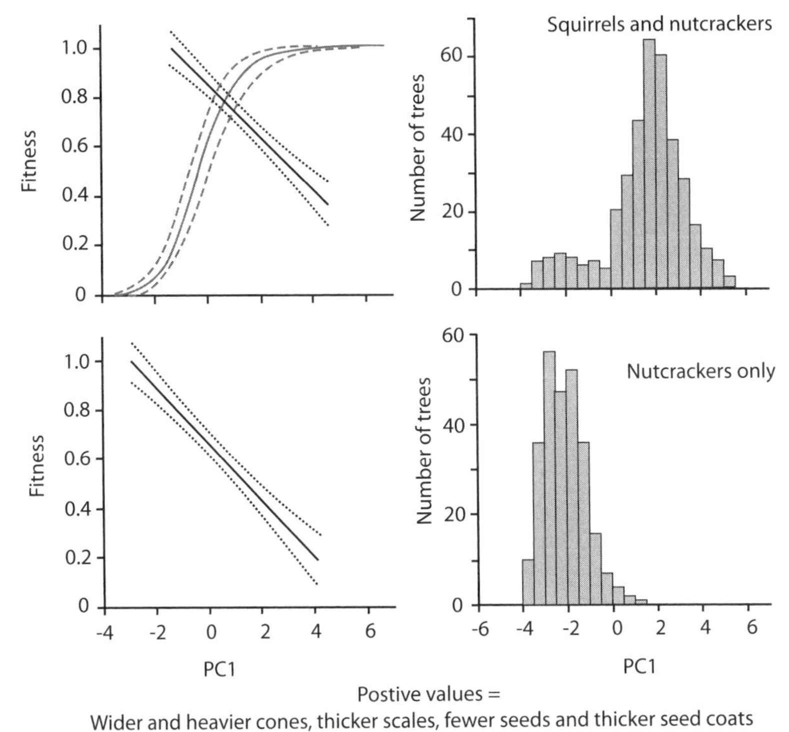

Fig. 11.7 Variation in the range of traits found among cones of limber pine (*Pinus flexilis*) in regions in which only nutcrackers are present compared with regions in which both nutcrackers and squirrels are present. Seed traits are represented here as a composite of multiple traits, called PC1, obtained using principal components analysis. Positive values of PC1 have the combination of traits indicated on the graph; negative values have the opposite combination. Fitness is estimated as the proportion of seeds removed from trees having different values along the principal components axis. Curved solid line is fitness moderated by squirrels; straight solid lines are fitness mediated by nutcrackers. After Siepielski and Benkman (2010).

selection on seed morphology (fig. 11.7). In these regions, selection on seed morphology is therefore much more variable. The outcomes of this variable selection can be seen in the distribution of traits within populations. In limber pine, seed traits are up to two times more variable in regions with squirrels and nutcrackers than in regions where only the birds are common (Siepielski and Benkman 2010). The geographic mosaic of coevolution in this interaction therefore results not only in regional differences in the mean values of traits but also in differences in the variation of those traits within populations (fig. 11.7).

Other mammals that hoard cones and seeds for future use can increase the geographic variation in selection pressures on conifers. Small seed caches placed in the soil by some mammalian species such as pine chipmunks (*Tamias*) or deer mice (*Peromyscus*) can contribute to successful seed dispersal and germination, because the animals do not always retrieve all the seeds they cache (Vander Wall 2002). In the dry forests of the eastern Sierra Nevada in North America, up to four pine species with winged seeds co-occur. Pre-dispersal seed predators, mostly mammals but also some insects, kill a high percentage of these seeds, ranging in one study from 7 percent to 66 percent, depending on the pine species. Once the seeds are dispersed, anywhere from a third to almost all seeds are removed from the ground by animals. Studies of radioactively labeled seeds have shown that many seeds of some of these pines end up in small seed caches (Vander Wall 2008). In Jeffrey pine (*Pinus jeffreyi*) and sugar pine (*Pinus lambertiana*) almost all seedlings come from these caches, but in ponderosa pine (*Pinus ponderosa*) only some seedlings come from caches, and in lodgepole pines (*P. contorta*) no seedlings arose from caches. The selection pressures imposed on conifers by caching seed predators are therefore distributed across several stages of seed development and dispersal, and those various selection pressures are bound to vary geographically.

REPLICATED CHARACTER DISPLACEMENT

One of the potentially most common forms of replicated coevolution across environments is among species that have radiated from a common ancestor and then have come back together in various combinations in different ecosystems as potential competitors. This situation occurs in most species-rich biological communities worldwide. Rather than one species from a genus within a community, there are commonly two or more species of some genera, and sometimes a single family can dominate a community. Some Australian landscapes are dominated by species of *Eucalyptus*, *Acacia*, or *Protea* plants, and some forests in Southeast Asia are dominated by plants in the Dipterocarpaceae.

In the forest dynamics plot on Barro Colorado Island, Panama, most (58 percent) of the 328 tree and shrub species are represented by genera that have two or more species within the community (Kembel and Hubbell 2006). Within the one plot are 16 *Inga* species, 13 *Psychotria* species, and 12 fig species. DNA barcoding of most of the tree species within the plot has confirmed that the assemblage includes species spanning a wide range of phylogenetic relatedness, including some genera with more than 10 species (Kress et al. 2009). Similar patterns occur in most plant and animal assemblages worldwide, with many genera and sometimes families represented by one or a few species but others represented by multiple closely related species within communities.

This fundamental aspect of community organization is receiving increased attention from two perspectives. One is a deeper evaluation of the long-standing problem of how species competing for limiting resources diverge in their traits or use of habitats. A standard tenet of evolutionary biology is that where closely related species come into contact, they will compete for limiting resources and their traits will evolve in ways that reduce competition (i.e., competitive character displacement). The other is a broader question about whether species-rich assemblages show evidence of a strong phylogenetic signal in their organization, with the probability of coexistence increasing or decreasing among species with degree of relatedness (Kraft et al. 2008; Verdú and Valiente-Banuet 2008; Kress et al. 2009). The latter question is explored in chapter 17. The more restricted question for this chapter is whether coevolution leads to similar or different patterns of trait evolution wherever the same species come into contact. That is, do competing species coevolve in the same ways regardless of the physical and biotic environments in which they co-occur?

At least for *Anolis* lizards, the answer seems to be that divergence in trait evolution among the species is often the same, but each species may diverge in different ways in different places. *Anolis* lizards have diversified into about 400 species in northern South America and the Caribbean islands, and their diversification has become one of the major models in evolutionary biology (Losos 2009). Throughout the Caribbean, anoles colonized the large islands through rare instances of dispersal over water from the mainland and then, in some cases, from larger islands such as Cuba to smaller islands (Glor et al. 2005). The large islands of the Greater Antilles were initially colonized from the mainland by a single anole species (Nicholson et al. 2005). Diversification began on Cuba and Hispaniola and spread to Puerto Rica and Jamaica (Mahler et al. 2010). On each island, the lizards have diversified into groups of coexisting species that differ in their morphological traits and their use of habitats. Each island is therefore a semi-independent experiment, making it possible to ask whether the same traits that minimize competition evolve repeatedly across ecosystems.

In these anoles, the number of evolutionary options appears to be restricted. The lizards have partitioned the microhabitats within each island by diverging into four to six morphological forms (fig. 11.8). These ecomorphs include forms called trunk-crown, trunk-ground, grass-bush, trunk, twig, and crown-giant (Moermond 1981; Losos 1990). The morphs differ from

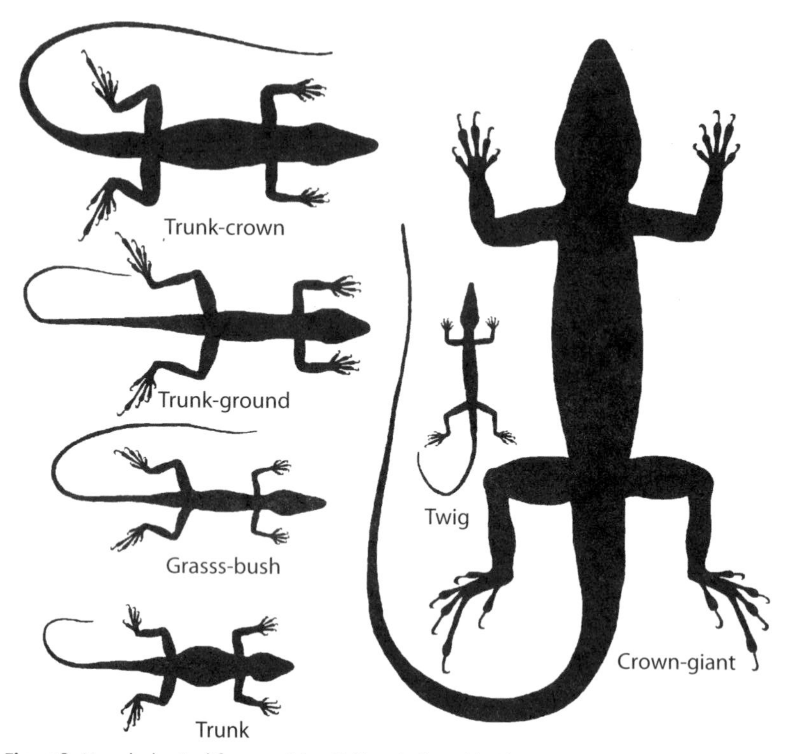

Fig. 11.8 Morphological forms of *Anolis* lizards found in the Greater Antilles islands, highlighting the differences in size and shape found among these species. The silhouettes are images drawn from museum specimens collected from the island of Hispaniola. After Losos (2009).

each other in where individuals perch and how they move—jump, crawl, or walk—in addition to how they look. Grass-bush morphs, for example, are small anoles with long hind legs and tails that perch in grasses or bushes and jump after prey. Trunk morphs have relatively long forelimbs and small, short tails. They perch on tree trunks and run after prey.

There are two ways in which these islands could have accumulated similar sets of ecomorphs. Either each morphological form could have evolved once and then dispersed to the other islands, or each form could have evolved anew on each island. Molecular phylogenetic studies suggest that the forms have evolved repeatedly and independently on the islands (Losos et al. 1998; Glor et al. 2001). Lizards on each island are generally more closely related to each other than to lizards on other islands. The exact sequence by which the morphs originated on each island is not known. It is therefore not yet possible to determine the fraction of morphs that originated through coevolutionary change rather than through unidirectional evolutionary change in which later colonizers simply fit within the range of morphs already there. It is clear, though, that the divergence of these species has been driven by selection that has repeatedly favored the displacement of traits among co-existing populations on these different islands.

The lizards have partitioned the habitats by diverging from each other in body size and limb dimension. The similar patterns of morphological divergence among islands are best explained by similar selection pressures on all islands rather than by other historical causes (Langerhans et al. 2006). The behaviors of these species have also diverged in similar ways, matching the particular habitats in which they occur (Johnson et al. 2010). Hence, the same combinations of morphology, behavior, and microhabitat use have all evolved repeatedly among these islands.

Morphological diversity among the coexisting lizards has increased over time on all the islands, but it has increased faster and been greater in Cuba and Hispaniola than in Puerto Rico or Jamaica (Mahler et al. 2010). The populations on Cuba and Hispaniola began to diversify when the number of competing lineages was low. In contrast, the populations on Puerto Rico and Jamaica diversified more slowly at a time when there were more competing lineages on these islands. Overall, however, the rates of morphological diversification found among these islands fall within the range of rates found among anoles in mainland communities (Pinto et al. 2008). There is nothing particularly fast about the rates on these islands.

Morphological diversification, though, has proceeded differently on the islands than on the mainland, suggesting that the patterns are not always completely repeatable. Island lizards have shorter limbs and better-developed toe pads than do lizards on the mainland. The differences in evolution between the mainland and island species suggest that the restricted morphological diversification found on the islands results from a restricted range of ecological opportunities on the islands rather than from any inherent constraints on the potential directions of evolution. Mainland anoles have been able to diversify in yet other directions, when presented with a greater range of physical or biotic environments.

The causes of these differences are not yet clear, but the possibilities include differences between the mainland and island communities in the predators or other competitors that impose selection on these species (Pinto et al. 2008). At least for the island populations, competition appears to be a major force imposing strong selection. Experimental studies on small islands in the Bahamas with a single anole species suggest that intraspecific competition acts as a stronger selection force than predation, at least under those conditions (Calsbeek and Cox 2010). Moreover, the pattern of intraspecific variation in morphology on single lizard islands in the Bahamas is similar to the pattern of interspecific divergence found on islands in the Greater Antilles (Calsbeek et al. 2007). These observational and experimental results suggest that competition, both intraspecific and interspecific, is likely a major driving force in diversification on islands.

The patterns also differ among island groups in the Caribbean, suggesting that there are limits to the repeatability of ecomorphs even on islands. In contrast to those on the Greater Antilles, *Anolis* lizards on the Lesser Antilles have undergone much less diversification. The cause cannot be due just to lack of diverse habitats on these islands or their small size, because multiple morphs coexist on some of the smaller islands of the Greater Antilles. Rather, the histories of these two island chains may have differed sufficiently to allow more speciation in the Greater Antilles than in the Lesser Antilles. The Greater Antilles islands are fragments of old continental crust, and colonization of these islands has been infrequent. In contrast, the Lesser Antilles islands are a more recent volcanic chain, and there is evidence of more frequent colonization of these islands by lizards and other taxa such as birds (Ricklefs and Bermingham 2008; Losos and Ricklefs 2009).

The faunas of the Greater and Lesser Antilles are therefore highly dynamic but in different ways. There seem to be several repeatable patterns in the evolutionary divergence of anoles. Each pattern is restricted to a particular region and, presumably, to a particular set of environmental variables. These studies of repeated evolution of similar traits are helping to clarify when traits are likely to evolve in similar ways, rather than different ways, across environments.

THE CHALLENGES AHEAD

We now have well-studied examples of the geographic mosaic of coevolution in nature and solid mathematical theory on how spatially structured coevolution differs from local coevolution. We need, though, an even greater range of empirical and theoretical studies if we are to more fully understand how mosaics of adaptation and coadaptation are built on the templates of physical and biotic environments. Each of these studies is a major undertaking. All the major empirical studies of the geographic mosaic of coevolution in nature have required decades of work. It is hard enough to understand the patterns of selection acting on a population or an interaction in one place. Expanding that work across multiple ecosystems or even whole continents has required a detailed understanding of natural history, first to identify traits that appear to be under selection, then to make the ecologically relevant observations and experiments to interpret how selection acts on the traits, and then to choose the most informative other populations to repeat those studies. Knowing how these traits have diversified among other closely related species deepens our understanding of the range of likely options for coevolving traits. There is no way around this time-consuming process of building a hierarchical understanding of how coevolving traits have diversified through selection acting on populations in different environmental settings.

The study of the environmental templates of geographic mosaics of coevolution has been mostly the work of terrestrial biologists. We also need studies of these patterns and processes in the oceans. Careful analyses of the evolutionary ecology of interactions among oceanic species have expanded in recent years, but most have remained focused on the evolution of traits on one side of an interaction. Some studies, however, are showing strong geographic differences in evolving interactions and are moving toward full analyses of the structure and dynamics of geographic mosaics of potentially coevolving species (e.g., Sotka 2005; Sanford and Worth 2010; Sanford and Kelley 2011).

The next generation of genomic studies has the potential to greatly expand our ability to evaluate how a combination of selective forces shapes geographic mosaics. Population genomics has been heralded for more than a decade as one of the most promising new tools in ecology and evolutionary biology (e.g., Black et al. 2001; DeLong 2002; Stinchcombe and Hoekstra 2007; Butlin 2010; Siol et al. 2010; Elmer and Meyer 2011). But its potential to help us understand the dynamics of selection and adaptation in nature is only now being realized. These studies should sharpen our understanding of how coevolution can reshape traits in different ways in different environments. Beginning with statistical outlier genes that deviate in the pattern of variation from most of the rest of the genome, these studies can start to evaluate which outliers are under selection within a population and how the traits under selection vary in distribution among populations. For studies of the geographic mosaic of coevolution, population genomics provides a potential way to delve into how selection proceeds simultaneously in adapting species to their physical environments and to other species. Genomic approaches therefore offer a new way to help identify geographic selection mosaics and coevolutionary hotspots and coldspots (Vermeer et al. 2010).

These studies are coming at a time when we are realizing just how complicated can be the relationships among traits and environments, when multiple traits are evolving simultaneously and doing so in different ways in different environments. It is to the often puzzling geography of traits and ecological outcomes that we now turn.

12

The Geography of Traits and Outcomes

As species evolve and coevolve in different ways in different environments, they develop an ever-changing kaleidoscope of traits that appear, to varying degrees, matched or mismatched to their local environments. The geographic mosaic of coevolution generates some of this kaleidoscope, but so do the other ecological and evolutionary forces that shape traits. The mismatches are important because they are part of the variation within and among populations that drives further evolutionary change. The problem we face is that organisms are a bundle of traits, which together determine the fitness of individuals. What we call mismatches between the traits of individuals and their environments can be real mismatches or only perceived mismatches. Selection may act on different suites of traits in different populations, even if the major driver of selection on that species—attack by a parasite population, for example, or environmental temperature—is the same in all environments.

This chapter considers what we are learning about how the traits of species evolve to fit the surrounding web of life, and why variation in traits—and ecological outcomes associated with those traits—sometimes appears to fit so loosely to local environments, even in populations under strong selection. The first part of the chapter considers coevolved traits, and the latter part considers traits in general as species adapt to novel or changing environments.

BACKGORUND: TRAIT MATCHING AND MISMATCHING

Our expectations about the evolution of trait matching and mismatching vary with the kind of interaction we are considering. As parasites coevolve with hosts, selection favors parasites that match the traits of their local host, and it favors hosts that confound (i.e., mismatch) the locally adapted traits of their parasites. The same holds for predators and prey. Local populations should therefore often cycle between periods in which their traits are well matched and poorly matched as populations continue to coevolve. When traits are well matched between parasites and hosts, selection is especially strong on hosts; when traits are mismatched, selection is especially strong on parasites. The only way to separate true local maladaptation from the temporal dynamics of trait matching and mismatching is by monitoring populations over time.

In contrast, competition generally favors divergence of traits rather than matching of traits. If potentially competing species (e.g., closely related species) are very similar in many of their traits, it suggests either that the species are not competing for resources or that those particular traits are not the focus of selection. For example, overall body size does not show evidence of competitive character divergence among some co-occurring species, but the specific traits associated with prey choice or use of other resources do show divergence (Dayan and Simberloff 2005).

In mutualisms, symbiotic interactions favor highly complementary traits between host and symbionts. Matching in mutualisms between free-living species, however, is more complicated because these mutualisms often evolve toward networks of interacting species. Trait matching in these interactions is often shaped by selection acting throughout the web of interacting species. To varying degrees, the same applies to all forms of interaction, but it seems especially to hold in mutualistic networks.

The geographic mosaic of coevolution adds to the complexity of interpretation of local traits. Trait matching in coevolved interactions reflects the dynamics of selection across a regional scale as well as the local scale. Demonstrating that the traits of interacting species are matched or mismatched, locally or regionally, is therefore not sufficient evidence that an interaction is, or is not, coevolving (Thompson 2005; Anderson et al. 2010; Nuismer et al. 2010). Some traits of interacting species could appear matched across multiple environments simply because both species are responding in similar ways to geographic differences in physical environments. Consequently, all the major studies of the geographic mosaic of coevolution have used experiments to probe the process of selection (e.g., studies of fitness effects or selection gradients) to assess whether geographic patterns in potentially coevolved traits are likely the result of coevolution.

That does not mean that every study needs to evaluate selection and fitness on both sides of any interaction in all populations in equal detail. In some cases reciprocal selection can be inferred to some degree, even if selection has been studied directly only on one side of the interaction. For example, experiments might focus on the evolutionary responses of a host to a parasite, if earlier studies show that the parasite attacks only that host. In that case, fitness in the parasite depends entirely on its ability to survive on that host, and so the important initial question is whether the interaction imposes selection on the host.

EVIDENCE OF TRAIT MATCHING

A wide range of detailed studies of interactions thought to be coevolving as a geographic mosaic have been examined on either one or both sides of the interaction. The suggested coevolving traits are as diverse as the interactions themselves (table 12.1).

The interactions between toxic newts and garter snakes in western North America have provided a detailed picture of trait matching and mismatching among populations of two coevolving species. These studies have also illustrated the complexities involved in interpretation of the geographic patterns. Rough-skinned newts (*Taricha granulosa*) contain a neurotoxin called tetrodotoxin, which is lethal to vertebrates even at small concentrations. Newt populations differ in the amount of tetrodotoxin contained within individuals. Common garter snakes (*Thamnophis sirtalis*), which prey on the newts, differ among populations in their degree of resistance to tetrodotoxin (Brodie et al. 2002; Hanifin et al. 2008). In experimental trials, the greater the level of tetrodotoxin in the newt, the greater the chance a garter snake will reject that newt (Williams et al. 2010).

Where snakes commonly prey on newts, selection favors newts with relatively high levels of tetrodotoxin. Selection on snakes favors individuals that either avoid highly toxic newts or have resistance to the toxin. The result has been a geographic mosaic in which toxin levels of the newts and resistance levels of snakes are low in some localities, high in other localities, and mismatched in yet other localities (Brodie et al. 2002; Hanifin et al. 2008). Intriguingly, all the known mismatches are in one direction: the snakes have a higher level of resistance than needed to overcome the level of tetrodotoxin found in the newts (fig. 12.1).

There are multiple possible explanations for these mismatches, and research is continuing to unravel the ecological causes of these patterns. Resistance in the snakes is under relatively simple genetic control, with about a quarter of the resistance attributable to allelic differences at a single gene (Geffeney et al. 2005; Feldman et al. 2010). Hence, fixation of a few extreme alleles could have resulted in extreme highly resistant phenotypes in some snake populations. Genes for fine-tuning resistance either have not evolved or have not yet had time to be favored by selection. The snake and newt populations also show strong genetic differences among populations (Janzen et al. 2002; Ridenhour et al. 2007), but gene flow could still swamp some of the local selection and contribute to the mismatches. In addition, the

Table 12.1 *Examples of interactions currently being analyzed to assess the structure and dynamics of local adaptation in interacting species in nature*

Taxa	*Traits under selection*	*References*
Australian wild flax–flax rust	Gene-for-gene interactions	Thrall and Burdon 2003; Barrett et al. 2009
Woodland stars–*Greya* moths	Plant responses–oviposition	Thompson and Fernandez 2006
Japanese camellias–camellia weevils	Fruit size–rostrum length	Toju and Sota 2009
Yellow rocket (*Barbarea*)–flea beetles	Plant resistance genes–insect genes	de Jong et al. 2009; Nielsen et al. 2010
Conifers–crossbills	Cone shape–bill shape	Parchman and Benkman 2008; Benkman 2010
Chilean cacti–mistletoes	Spine length–radicle length	Medel et al. 2010
Plantago–powdery mildew	Resistance–virulence	Laine 2005, 2006
Wild parsnip–parsnip webworm	Furanocoumarins–P450 gut enzymes	Zangerl and Berenbaum 2003; Berenbaum and Zangerl 2006
Zaluzianskya plants–long-tongued flies	Floral depth–proboscis length	Anderson and Johnson 2008
Irises–long-tongued flies	Floral depth–proboscis length	Pauw et al. 2008

Heliconia flowers–hummingbirds	Floral shape and color–bill size and shape	Temeles and Kress 2003; Temeles et al. 2009
Large blue butterflies–ants	Mimetic chemicals–recognition chemicals	Nash et al. 2008
Slave-making ants–other ants	Raiding and chemical traits–defense traits	Bauer et al. 2009; Foitzik et al. 2009a; Ruano et al. 2011
Polistes wasps–social parasitic wasps	Life history traits–life history traits	Lorenzi and Thompson 2011
Leafcutter ants–fungi	Ant specialization–fungal specialization	Mueller et al. 2011
Snails–trematodes	Genotype-by-genotype interactions	King et al. 2009; Koskella and Lively 2009; Lively 2010a
Newts–garter snakes	Tetrodotoxin production–resistance	Brodie et al. 2005; Hanifin et al. 2008
Bacteria–phages	Resistance–virulence	Koskella et al. 2011

Notes: In some cases, current selection on the traits has been demonstrated in both species, but in other cases it has been inferred in one of the species. In all these interactions, traits vary geographically in the degree of trait matching. Multiple traits are under selection in some of the interactions, but only one set of traits or variables is shown here for each interaction.

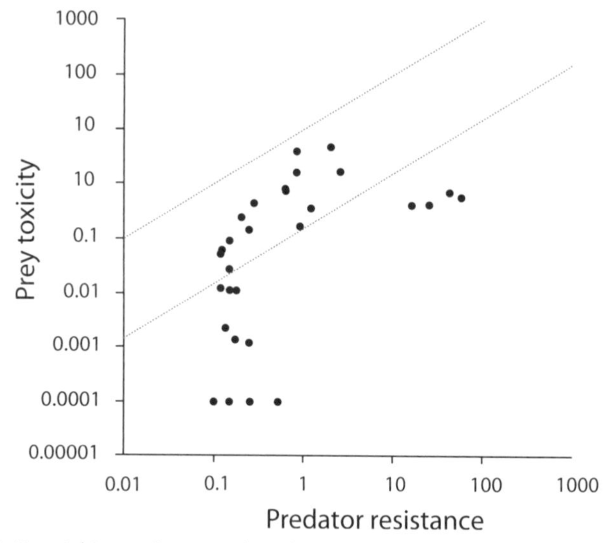

Fig. 12.1 Trait matching and mismatching between coevolving populations of rough-skinned newts (*Taricha granulosa*) and common garter snakes (*Thamnophis sirtalis*). Values are the means for interacting snake and newt populations at a site. Values above or below the dotted lines are considered to be trait mismatches. Values above the upper dotted line are levels of the neurotoxin tetrodotoxin so high relative to snake resistance that snakes would be incapacitated or die. Values below the lower dotted line are levels of tetrodotoxin sufficiently low relative to snake resistance that snakes would suffer no reduction in performance. All mismatches are in the direction of high levels of resistance in the snakes relative to levels of tetrodotoxin in the newts. Prey toxicity is calculated in milligrams on a logarithmic scale; predator resistance is calculated as the oral dose of tetrodotoxin, in milligrams on a logarithmic scale, needed to reduce the performance (speed) of a snake at that locality by 50 percent relative to the baseline performance for that population. After Hanifin et al. (2008).

newts have tetrodotoxin during their aquatic larval stages, and so defense against larval predators could drive some of the geographic patterns (Gall et al. 2011).

Adding to the geographic complexity, these toxic newts overlap in some regions with nontoxic *Ensatina* salamanders, which mimic the newts (Sinervo and Calsbeek 2006). This mimicry complex could create a three-way cycle of selection as some local populations of the newts, salamanders, and snakes oscillate in their patterns of matching and mismatching. On top of it all, there is another newt species (*Taricha rugosa*) and more garter snake species in some localities (Brodie et al. 2005), which could create more complex multispecific selection mosaic in these interactions. Given all the potential complexity of selection on these interactions, it is remarkable that there remains a general pattern of trait matching between rough-skinned newts and common garter snakes across the geographic range of this interaction.

PATTERNS IN TRAIT MISMATCHING

There are two major ways in which the traits of coevolving species can appear as mismatched. The traits on one side of the interaction may always be exaggerated in comparison to the traits on the other side of the interaction, as in the snakes and newts. Alternatively, the traits may be more exaggerated on one side of the interaction in some regions and more exaggerated on the other side in other regions. When traits are consistently mismatched more on one side of the interaction than on the other, it may indicate a real selective asymmetry. Alternatively, it may indicate that selection favoring increase in a trait in one of the species always demands a disproportionately large evolutionary change in the complementary traits of the other species to compensate for the change in fitness.

Decades ago Richard Dawkins and John Krebs (1979) proposed that coevolving interactions such as those between predators and prey are asymmetric with respect to selection, because failure for the prey means death and failure for the predator means the loss of dinner. By extrapolation, prey are more likely to evolve exaggerated defenses than predators are likely to evolve exaggerated counterdefenses. That argument has generated a wide range of qualifiers since it was first suggested (Brodie and Brodie 1999; Abrams 2000; Williams et al. 2010). It would not hold in species that depend obligately on each other. More broadly, whether it is applicable in less reciprocally obligate interactions depends on which species in the interaction is more obligately dependent on the other species. Hence, every possible direction and degree of asymmetry are possible in any form of interaction. Life–dinner asymmetry in coevolving interaction is only one of many possible forms of asymmetry.

There is, though, some indication from published studies that exaggeration of traits could be favored more on one side of an interaction than on the other in some interactions, including mutualisms. In an analysis comparing geographic and species differences in traits of multiple interacting pollinators and plants, Bruce Anderson and colleagues (2010) found that each increment in exaggeration of insect traits involved in pollination is matched by a disproportionate exaggeration in the corresponding floral traits. The species included in that study were mostly asymmetric in the degree to which they depend on each other, but that is common in these kinds of mutualism. That same study also suggests that some interactions between plants and insect herbivores follow the same pattern. These results are worth following up, because they suggest that some asymmetric mismatching of traits may be common in coevolving interactions.

Some mismatches, however, occur on both sides of an interaction. That could be due in some cases to relaxed selection among individuals with intermediate traits. In some pollination mutualisms, for example, coevolving traits may be more "forgiving" across a broad range of intermediate phenotypes, with selection concentrated mostly on extreme traits and extreme mismatches. Experiments on *Heliconia* plants and purple-throated carib hummingbirds in the Caribbean have shown that the rate at which the birds can extract nectar from flowers is compromised only at extreme mismatches between floral shape and bill shape (Temeles et al. 2009). If the effects on plant fitness are concentrated at the extremes of traits and

Fig. 12.2 A long-tongued *Prosoeca ganglbaueri* fly, a flower of *Zaluzianskya microsiphon* with which it has coevolved in southern Africa, and the rare mimetic orchid *Disa nivea*, which has evolved to exploit this interaction. The *Zaluzianskya* flower has a red corolla tube, and the mimetic flower has a white corolla tube. The fly is carrying pollinia from an orchid flower. Photograph courtesy of Bruce Anderson and Steven D. Johnson.

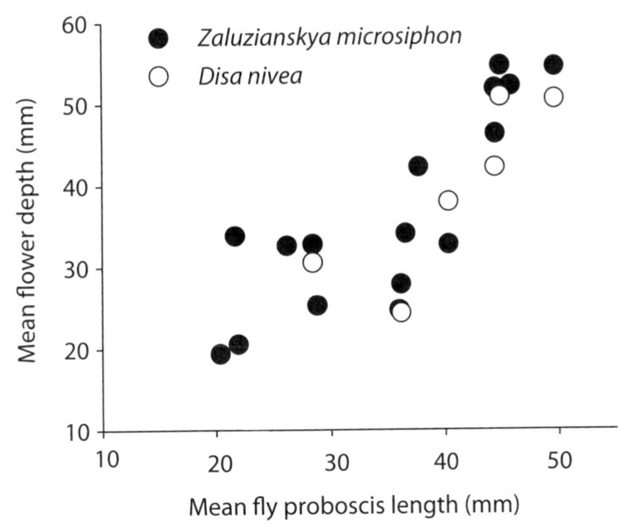

Fig. 12.3 Covariation among populations of coevolving long-tongued *Prosoeca ganglbaueri* flies and *Zaluzianskya microsiphon* plants, and additional covariation by the rare mimetic orchid *Disa nivea*. The proboscis and flower lengths are from a location in which individuals fall near the middle of the distribution of lengths found in these species. After Johnson and Anderson (2010); based on Anderson and Johnson (2008, 2009).

mismatches, then there may be little selection among plants and hummingbirds with intermediate phenotypes across a moderate range of mismatches.

The problem of the evolution of trait matching is made even more difficult by the fact that, as pairs of species coevolve, they often attract other species that exploit the interaction. These species can evolve to match the traits of one of the participants even if they do not impose significant selection on the coevolving species. This appears to have been the case in rare *Disa nivea* orchids, which offer no rewards to pollinators but instead mimic *Zaluzianskya microsiphon* plants that have coevolved with *Prosoeca ganglbauri* long-tongued flies in southern Africa (Anderson et al. 2005; Anderson and Johnson 2008, 2009). Coevolution in this interaction is driven by selection on corolla tube length in the plants and proboscis length in the flies (fig. 12.2). The *Z. microsiphon* plants and *P. ganglbaueri* flies covary geographically in their traits, but the mimetic orchid also covaries geographically in its traits with these other two species (fig. 12.3). These results show one way by which trait matching across environments has the potential to encompass multiple species within a web, even though coevolution may be occurring only between a single pair of species within the web.

Although local mismatching in some populations is an inevitable consequence of species coevolving as a geographic mosaic, the degree of local mismatching may depend to some extent on the genetics of the coevolving traits. Some models suggest that mismatches may be less common if the coevolving traits are controlled by multiple genes that each have a small phenotypic effect rather than by a few genes that each have a large effect (Ridenhour and Nuismer 2007). But it may not be quite that simple. Some non-coevolutionary models have indicated that selection in geographically structured populations connected by gene flow tends to become focused on fewer genes of large effect as the connected populations adapt to different environments (Yeaman and Whitlock 2011). These results are for populations evolving in response to fixed differences among environments rather than to coevolving species. How the genetics of adaptation and trait matching changes under different scenarios of geographic mosaics is currently unknown.

SOCIAL INTERACTIONS AND GEOGRAPHIC MOSAICS

In some cases, so many traits may be coevolving that they diversify among populations into a bewildering variety of combinations. At that point, it becomes biologically unmeaningful to match trait against trait among populations and environments. What matters is how fitness covaries, and that process may demand coevolution of multiple, seemingly unrelated, traits in interacting species.

Coevolution involving traits associated with social behavior seems especially prone to complex trait combinations. In fact, a characteristic of coevolution involving the social structure of species is that it often seems to involve multiple kinds of traits rather than a few traits. Studies in the past decade have begun to unravel some of this complexity, but it is likely that, for any interaction, we currently know only some of the traits that are coevolving and varying among populations.

Social traits are a particularly high-payoff target of selection in coevolving interactions. Once a species breaches the social structure of another species, it can follow multiple pathways of manipulating that species. When species evolve by exploiting the social structure of other species, it seems to open the floodgates for adaptation and counteradaptation through a tremendous diversity of morphological, chemical, physiological, behavioral, and life history adaptations that can vary among populations. As a result, the exploitation of social structures can take many routes: interspecific brood parasitism in which a species tricks another species into rearing its young; chemical mimicry in which predatory species are accepted as conspecifics within the colonies of social species; slave-making species that raid the colonies of other social species; and social parasites that invade the nests of other species and take over reproduction in the nest.

If we attempt to track a trait across environments and populations, we are sometimes left with puzzling results. It can become almost impossible to match the evolution of a specific trait against a specific countertrait among populations. Instead, the geographic pattern of trait matching is about the evolution of suites of traits on both sides of the interaction. The study of social trait matching in interacting species is therefore one of the areas of evolutionary biology in which the combined tools of behavioral ecology and evolutionary ecology are needed to give new insights on how to develop more integrated approaches (Gordon 2011).

The classic example of exploitation of the social system of other species is interspecific brood parasitism in birds (Davies 2000; Rothstein et al. 2002; Sorenson and Payne 2002; Kilner and Langmore 2011). Avian brood parasites have evolved to co-opt the parenting behavior of other birds. The parasites mimic the cues used by parents to trick hosts into rearing young that are not their own. Usually that mimicry involves a combination of traits that varies among species and populations. The traits include, for example, eggs that mimic host eggs, tossing of host eggs out of the nest, killing of nest mates after hatching, displays of colorful mouth interiors by nestlings in ways that elicit feeding by the foster parents, and begging behaviors. In turn, hosts have evolved tactics such as mobbing and egg rejection to prevent parasitism (Martín-Gálvez et al. 2007; Welbergen and Davies 2009). This kind of brood parasitism can also occur within species (Lyon and Eadle 2008; Shizuka and Lyon 2011; M. Soler et al. 2011), suggesting that the sophisticated traits associated with interspecific brood parasitism sometimes may evolve as an outgrowth of exploitation of other individuals within species. Regardless of the origin, the result is a geographic mosaic of multispecific coevolution in which different populations of some brood parasitic species use different traits to exploit hosts.

Some brood parasitic species have morphs with traits adapted to different host species (Antonov et al. 2010; Fossøy et al. 2011), suggesting that selection is strong on parasites to escape detection as parasite, and their adaptations sometimes need to be fine-tuned. In some populations the traits of brood parasites are therefore well matched. In other populations, however, some traits are not well matched (Davies and Brooke 1989; Avilés et al. 2011). For example, cuckoo species in Europe have eggs that match the color and spotting patterns on the eggs of some hosts but not others. Local mismatching could be due to recent shifts onto host populations that have not yet evolved defenses, or it could be due to time delays in the cyclical evolution of defenses and counterdefenses within host and parasite populations (Davies and Brooke 1989; Thompson 1994; Nuismer and Thompson 2006). Hence, even species like these that are clearly coevolving show considerable geographic variation in their degree of trait matching at any point in time.

Coevolution involving manipulation of social behaviors has extended even more deeply into the evolution of some taxa and shaped variation in an even greater range of traits. Coevolution may even have favored the evolution of highly social behaviors, which then became the target of coevolution with yet other species. For example, the complex social structures found in some eusocial insects are thought by some biologists to have been shaped in part as defenses against predators and parasites (Wilson 1971; Hamilton 1987; Boomsma and Franks 2006). Those social systems have then provided a multitude of ways for predators and parasites to exploit these highly structured societies. Ant, bee, wasp, and termite societies are all victim to assemblages of species with which they have coevolved, sometimes in different ways in different environments.

Some predatory species, for example, use chemical mimicry to invade the social organization of ants, generating geographic mosaics of chemical defense and counterdefense. Not only have large blue butterflies (*Maculinea* spp.) evolved to invade the social structure of *Myrmica* ants (see discussion in chap. 9), but they also have evolved suites of chemical compounds adapted to the local ant population they parasitize. *Maculinea* butterflies lay their eggs on host plants, where they hatch and begin to develop. Foraging *Myrmica* ants respond to the cuticular compounds on the caterpillars and carry them back to their nests (fig. 12.4). Most *Maculinea alcon* butterfly populations have evolved cuticular hydrocarbons that match those found on their local ant hosts, and these chemical profiles differ greatly among *Maculinea* populations in Denmark (fig. 12.5). The butterfly larvae impose strong selection on their ant hosts through predation on the ant brood, which significantly reduces colony size (Nash et al. 2008).

Ants have evolved in response to these co-opted cues by altering the chemical signals they use to identify conspecifics. *Myrmica rubra* ant populations parasitized by *Maculinea* are more chemically divergent from each other than ant populations free of parasitism by butterfly larvae (Nash et al. 2008; Fürst et al. 2011). Ant populations subject to *Maculinea* parasitism are also more aggressive toward intruders. The geographic mosaic of coevolution

Fig. 12.4 *Myrmica rubra* ant carrying a *Maculinea alcon* butterfly larva. Photograph copyright David R. Nash, used with permission.

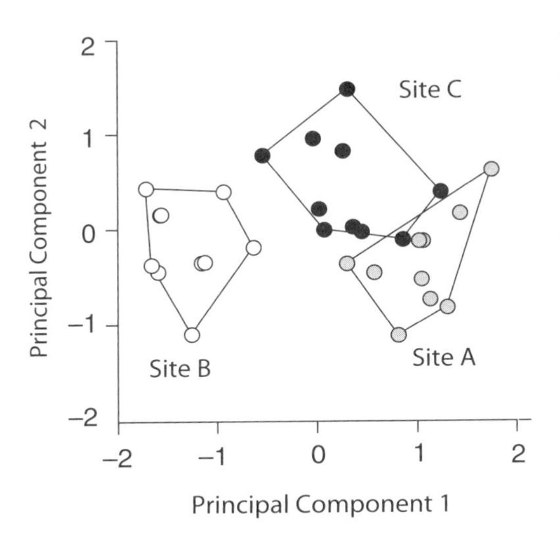

Fig. 12.5 The degree of difference in the chemical profiles among three populations of *M. alcon* in Denmark. The axes are principal components, which group suites of chemical compounds. Principal component 1 accounts for 60.8 percent of the variance; principal component 2 accounts for 18.4 percent. After Nash et al. (2008).

in this interaction therefore appears to be shaped by at least three important selective forces: selection on the butterflies to match their chemical profiles to their local ant host population, selection on the ants to mismatch the chemical profiles, and the presence of the butterflies in some habitats but not in others.

Ant species themselves have co-opted the chemical traits of other ant species to invade their social systems and, in this case, to enslave them. Slave-making ants are obligate social parasites that raid nests of other ants for brood and sometimes queens (Foitzik et al. 2009b). The slave-makers enlist the slaves to rear their own brood and forage for the nest (Hölldobler and Wilson 1990). The raids impose strong selection on the local populations of species whose nests they raid, because the life expectancy of a raided nest is often greatly reduced after an attack (Fischer-Blass et al. 2006; Johnson and Herbers 2006).

Slave-making in ants originated independently at least nine times, and new examples continue to be discovered (Hölldobler and Wilson 1990; Beibl et al. 2005). How slave-makers capture and enslave other ants varies among species and sometimes among populations. Some slave-makers use chemicals produced by the Dufour's gland to manipulate the behavior of their victims during raids. Ants commonly use these chemicals to lay trails for others to follow. Slave-makers directly use chemicals to disrupt the social organization of the colonies they attack. Some species spray the chemicals onto ants within the attacked nest, which induces those ants to attack each other and results in panic and disorganization within the nest (Bauer et al. 2009).

Other slave-makers are subtler in their use of chemicals. They have evolved cuticular hydrocarbon profiles similar to their local host species, which may reduce defensive responses in hosts (Ruano et al. 2011). Inevitably, the evolution of these chemical traits has resulted in coevolutionary responses, and those responses have involved different mechanisms in different species. Some species subject to slave-making raids have evolved defenses that help prevent successful raids (Herbers and Foitzik 2002); others have evolved the capacity to rebel and kill slave-maker pupae in the nest in which they are enslaved (Achenbach and Foitzik 2009).

Species of slave-making and host ants studied in detail in Europe and North America often have large populations that are genetically distinct from each other (Foitzik et al. 2009b; Pennings et al. 2011). These populations show local adaptation in their traits and ecological outcomes. In some cases, slave-makers are more successful in attacking local populations of their normal hosts than they are in attacking allopatric host populations

(Fischer and Foitzik 2004; Foitzik et al. 2009a). These studies also show that local host populations are more successful in resisting attack by local slave-makers than attack by allopatric slave-makers, suggesting local coadaptation. Local adaptation is mediated by multiple traits, including secretions in the Dufour's gland. The secretions, however, vary geographically in their effectiveness at disrupting the organization of raided colonies (Bauer et al. 2009). Hence, the combination of traits generating effective defenses and counterdefenses differs among populations and species, but not all combinations are equally effective. Different combinations of traits may therefore represent coordinated evolutionary solutions that may be difficult to abandon for other trait combinations.

SOCIAL PARASITISM, PREDATION, AND GEOGRAPHIC MOSAICS

Adding a third species to a coevolving social interaction can completely rearrange the focus of selection and pattern of trait matching found between a pair of coevolving species. This appears to have happened in the interactions between *Polistes* wasps and other socially parasitic *Polistes* wasps that co-opt the nests of founder females. In these species, the geographic mosaic of coevolving traits depends on the timing and intensity of parasitism and the presence in some environments of major nest predators (Lorenzi and Thompson 2011). The interaction coevolves through a geographically varying mix of divergent life histories, mechanisms of social exploitation, morphologies, and nest-building behaviors.

In the meadows of the Alps and Apennine Mountains of Italy at elevations above 1,000 m, each *Polistes biglumis* female mates in the fall, overwinters, and then builds in the following late spring a small paper nest on a south-facing rock near the ground. She lays eggs in some of the cells of the nest and cares for her developing young. By late July, the first adults emerge. In some populations, these new adults remain at the natal nest and behave as workers to help their mother raise more offspring. Offspring maturing late in the season become the sexual adults for the next generation.

The process can be interrupted by three kinds of enemy that lower the number of offspring a female produces or that destroy the nest entirely (fig. 12.6). Predators can attack the nest and kill all the immature brood. Another *P. biglumis* female can usurp the nest. Or a socially parasitic *P. atrimandibularis* female can colonize the nest and work alongside the founding *P. biglumis* female to develop the nest further and lay eggs of her own.

Polistes atrimandibularis females are obligate social parasites. A female parasite enslaves the founder female, and the two females continue to rear offspring. The fitness of the founder female is reduced by the presence of

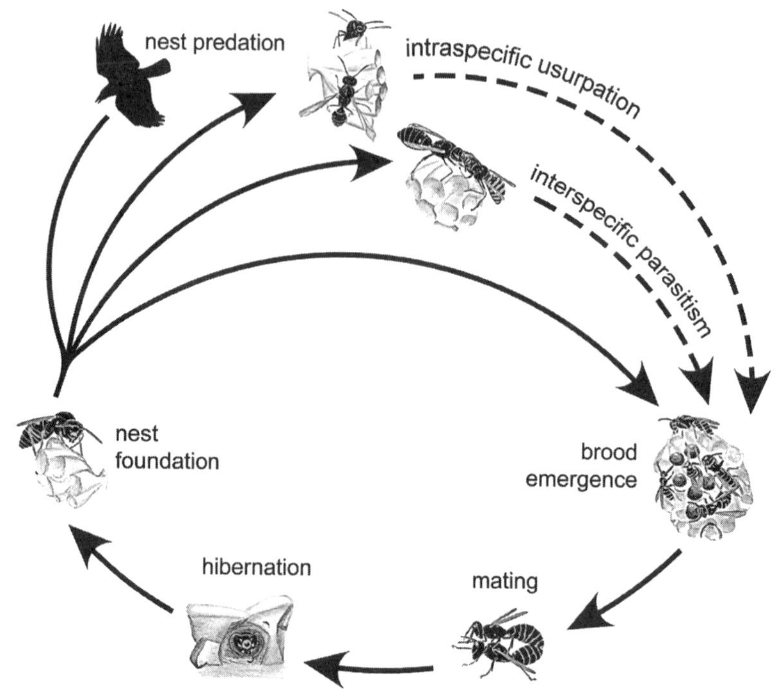

Fig. 12.6 The life cycle of the solitary wasp *Polistes biglumis* in the Alps and Apennine Mountains of Italy. Shown are the three kinds of enemy that can affect the evolution of traits in these insects. After Lorenzi and Thompson (2011).

the parasite as the two of them contribute to rearing the parasite's offspring (Lorenzi et al. 1992). The effect of the social parasite, however, is more ecologically complicated, because her presence can limit attack on the nest by predators.

Social parasites and predators therefore differ in the selection pressures they place on *P. biglumis* life histories and nest-building behaviors (Lorenzi and Thompson 2011). Consequently, *P. biglumis* and *P. atrimandibularis* populations have evolved in ways that reflect selection imposed by the parasites alone or by parasites and predators together. At Monte Mare, where predation and parasitism are low, *P. biglumis* females are small in size and build large nests with little protection. At Montgenèvre, where enemies are abundant, females are intermediate in size and build small, highly protected nests. Between these two extremes, female traits and nest traits vary with the relative importance of the enemies. Selection on the socially parasitic *P. atrimandibularis* also differs among sites. Selection on *P. atrimandibularis* females at Monte Mare favors multiple traits that sustain host *P. biglumis* reproduction, whereas selection on *P. atrimandibularis* females at Montgenèvre favors females that constrain host reproduction.

Studies of phenotypic selection show a combination of directional and nondirectional selection acting on foundress traits, parasite traits, and nest traits (Lorenzi and Thompson 2011). For example, although the presence of social parasites can result in lower predation on nests, it can also favor foundress females that make small nests and rear only sexual offspring rather than make large nests with workers. Although the production of workers would result in more sexual adults from a nest later in the season, the benefit would be lost if the nest were parasitized. The optimal suite of traits favored by selection in these populations is therefore a complex mix of life history, behavioral traits, and nest traits that vary geographically with selection imposed by social parasite and predator populations.

Overall, the relationships between coevolution and the social structure of species are among the most intriguing current questions on the geographic mosaic of coevolution. Because selection on social systems often varies among environments and among patterns of local kinship due to many causes, coevolution through manipulation and response to social systems may be a major mechanism by which geographic mosaics form in nature.

CHANGING THE TEMPLATES: EVOLUTION IN INVASIVE SPECIES

In trying to understand how selection favors different trait combinations in different environments, we often study species that have been evolving in those environments for thousands or millions of years. If those traits are the result of coevolution with other species, the problem is particularly difficult since coevolution tends to speed up the evolutionary process, rapidly masking clues to how the trait combinations initially assembled. One way of simplifying the search is to focus on how trait combinations evolve within a population in the generations soon after introduction into a new environment. In recent centuries we have created hundreds of inadvertent new evolutionary experiments that address that question, as we have introduced species into new continents and oceans far away from the physical and biotic environments in which they evolved.

Throughout the history of life, species have occasionally colonized very different environments from those of their ancestors. The extreme is the colonization of remote islands by plant, animal, fungal, and microbial species. We have, though, accelerated that process beyond the rates likely ever to have occurred previously in nature. We have moved over large distances species that would never have moved, in a single step, to a new continent, island, or ocean. It is one thing for a species to expand its geographic range by colonizing neighboring and often similar ecosystems, adapting along the way. It is quite another to suddenly be dropped into a completely different web of surrounding life.

These uncontrolled experiments have had devastating environmental effects on some ecosystems, as some introduced species have become invasive and outcompeted the previously dominant species in those communities. At the same time, they have also provided unparalleled opportunities for studying the pace and dynamics of the early stages of adaptive evolution in nature. Studies of rapid evolution in these species allow us to see whether changes in the environmental templates result in fundamentally different trait combinations in subsequent generations, or whether species tend to evolve repeatedly along the same trajectories under similar selection pressures in the early stages of adaptive change.

Interpreting the earliest stages of evolutionary change in introduced species is often difficult, though, because we often do not know the origin of the introduced populations (Colautti et al. 2009). We cannot compare the traits of the introduced populations to the traits of the local native populations from which the colonists were taken. Also, we rarely know how much genetic variation was initially present in the introduced populations. Founder effects, resulting from introduction of only a biased subset of genes found in the original population, and random genetic drift, resulting from the small sizes of founder populations in the early generations after introduction, can therefore be part of the early stages of evolution in an introduced population. What appears as potential rapid adaptive evolution could simply reflect these chance effects on gene frequencies. Nevertheless, as introduced populations expand they are also subject to natural selection, adapt to their new environments and, in some cases, become invasive.

There are two kinds of hypotheses of how some introduced species become so successful that they expand their population sizes and geographic ranges to become dominant, invasive members of communities in the new environments (Sakai et al. 2001; Lambrinos 2004). These hypotheses sometimes tend to focus on ecological outcomes or composite traits, such as dispersal ability or growth rate, rather than on the specific traits that produce those outcomes. At the one extreme are ecological hypotheses that argue that introduced species are free of their normal natural enemies and expand exponentially over time in their new environments. The initial stages of the expansion appear to be slow and the latter stages appear to be fast, because population growth is simply an exponential process. That hypothesis may apply to some invasions, but it is inadequate as a general explanation for two reasons. It cannot explain why some introduced species fail to become highly invasive species. Just as importantly, many introduced species have been shown to undergo rapid evolution after their introduction into a new geographic region, suggesting that invasions are commonly an evolutionary process (Sakai et al. 2001; Suarez and Tsutsui 2008; Whitney and Gabler 2008).

The alternative hypothesis that has received increasing support is that populations of introduced species have evolved to become locally adapted to their new physical and biotic environments. Implicit in this hypothesis is the view that introduced populations undergo continued evolutionary change as a species spreads across ecosystems in the new region. Selection on populations at the leading front of an invasion may differ from selection farther back from the front, where populations have been established for decades or hundreds of years. Moreover, populations in different parts of the invasive front could differ from each other in their adaptations. As selection continues, established populations could evolve different traits from those at the leading front of the invasion. By this view, invasive species become invasive through local adaptation as a species spreads among ecosystems.

Specific versions of how species evolve during invasions have been constructed for different taxa. For example, some hypotheses on animal invasions have focused on the evolution of dispersal ability in populations at the leading fronts of invasions (Alford et al. 2009; Phillips 2009), whereas hypotheses on plant invasions have often focused on the evolution of defenses against enemies (Orians and Ward 2010). Yet other hypotheses have focused on the evolution of traits associated with rapid growth or other life history traits that allow colonization of the increasingly disturbed habitats found worldwide, or the evolution of traits in physical environments that are either resource-limited or resource-rich.

These hypotheses differ in the suggested ecological drivers of selection. Hypotheses based on evolving defenses take two forms: evolution through increased competitive ability (EICA) or evolution through a shift in the balance between defenses against specialist and generalist enemies. According to EICA, species become invasive because selection favors individuals that redeploy for increased competitive ability the resources that in their native environments would be needed for defenses against herbivores and pathogens (Blossey and Nötzold 1995). Release from antagonistic interactions with higher trophic levels allows natural selection to focus on traits beneficial for interactions within trophic levels. Implicit in this hypothesis is the view that the evolution of competitive ability and the evolution of defenses against other enemies impose conflicting demands on natural selection. Also implicit in this hypothesis is the view that evolution of the ability to cope with antagonistic interactions is more important than evolution of the ability to cope with physical stresses or the ability to interact with mutualists. In contrast, the shifting defense hypothesis focuses more on defenses against herbivores and pathogens rather than competition. It argues that natural selection in the new environments favors individuals that devote more resources to defenses against generalist enemies rather than to specialist enemies, which the populations left behind in their native environment (Müller-Schärer et al. 2004; Joshi and Vrieling 2005; Doorduin and Vrieling 2011).

Studies of invasive species have shown almost every possible result, sometimes providing simultaneous support for one or more of the hypotheses. Some studies, for example, find support for the evolution of resistance against generalist enemies but also for the evolution of fast growth rates (Oduor et al. 2011). Other studies find complex relationships between growth rates, competitive interactions, novel defenses, alteration of surrounding resources or microbes, and resistance or tolerance to other enemies (Franks et al. 2008; Handley et al. 2008; Oduor et al. 2011; Thorpe and Callaway 2011). In spotted knapweed (*Centaurea maculosa*), which is a major invasive plant species in western North America, plants are larger in their introduced range than plants in their native range in Eurasia. North American plants also have stronger competitive effects on neighbors than do plants in the native range. The North American plants, however, also are better defended against generalist herbivores and slightly better defended against specialist herbivores (Ridenour et al. 2008).

Overall, these studies suggest that rapid evolution is likely to be important in many plant invasions, and there are multiple trait combinations that result in a population becoming a successful invader. This is not surprising, since there are many ways by which species can enter communities and suppress or otherwise fit in with native species (Bennett et al. 2011; Foxcroft et al. 2011). The importance of studies of rapid evolution in introduced species has been the demonstration that evolution is a normal part of the process of invasion and that some of the expected tradeoffs in evolution are not as simple as previously assumed. Multiple successful trait combinations are possible early in the evolution of adaptation to a new environment. Growth rates, resistance against enemies, tolerance of enemies, adaptation to physical stresses and disturbances, and adaptation to mutualists can become mixed and matched in novel and unexpected ways as populations adapt to completely novel environments. Some species diverge rapidly into a set of locally adapted populations within just a few centuries, and even a few decades, after their introduction into new regions.

COEVOLUTION WITH INVASIVE SPECIES

The evolutionary possibilities differ when species are introduced, either simultaneously or sequentially, with one or more other species from their native ranges. Part of the template of their past evolution has been introduced as well. These co-introductions provide opportunities for analyzing the repeatability of coevolution in novel environments. In one of the most thoroughly studied examples, two coevolving species from Eurasia have quickly coevolved as a geographic mosaic following introduction into North America. Parsnips (*Pastinaca sativa*) and parsnip webworms (*Depressaria pastinacella*) have become one of the models for studies of the ecological and evolutionary dynamics of invasive species. The plants were grown in North American gardens beginning in the 1600s and probably escaped from cultivation. By the nineteenth century, parsnips had become a major weed in the midwestern United States and had also spread to other parts of North America.

In the late 1800s, parsnip webworms, which are the major herbivores of parsnips in the native range, were found on parsnips in North America, first in Ontario and then in rapid succession in other populations in Canada and the United States. These highly specialized insects feed mostly on flowers and developing seeds and have major effects on the fitness of the plants they attack. Studies several decades ago showed a great deal of variation among plants in levels of attack and plant responses (Thompson 1978; Hendrix 1979). The basis of that variation, though, remained unknown until May Berenbaum and Arthur Zangerl demonstrated that the plants differ genetically in the combinations of defensive furanocoumarins they produce, and the parsnip webworms (*Depressaria pastinacella*) differ in the combinations of chemical counterdefenses they have evolved to combat these chemical defenses (Berenbaum et al. 1986; Zangerl and Berenbaum 2003).

Since that discovery, studies of many populations have shown that these plants and insects have coevolved as a geographic mosaic within North America. The populations differ in their degrees of local matching and mismatching of chemical traits. Mismatching depends in part on the local presence of a third species—a native plant species called cow parsnip (*Heracleum lanatum*)—that the webworms use as an alternative host in some populations (Berenbaum and Zangerl 1998; Zangerl and Berenbaum 2003). Moreover, comparison of the chemical profiles of parsnip plants in herbaria suggests that the chemical profiles of parsnip populations in North

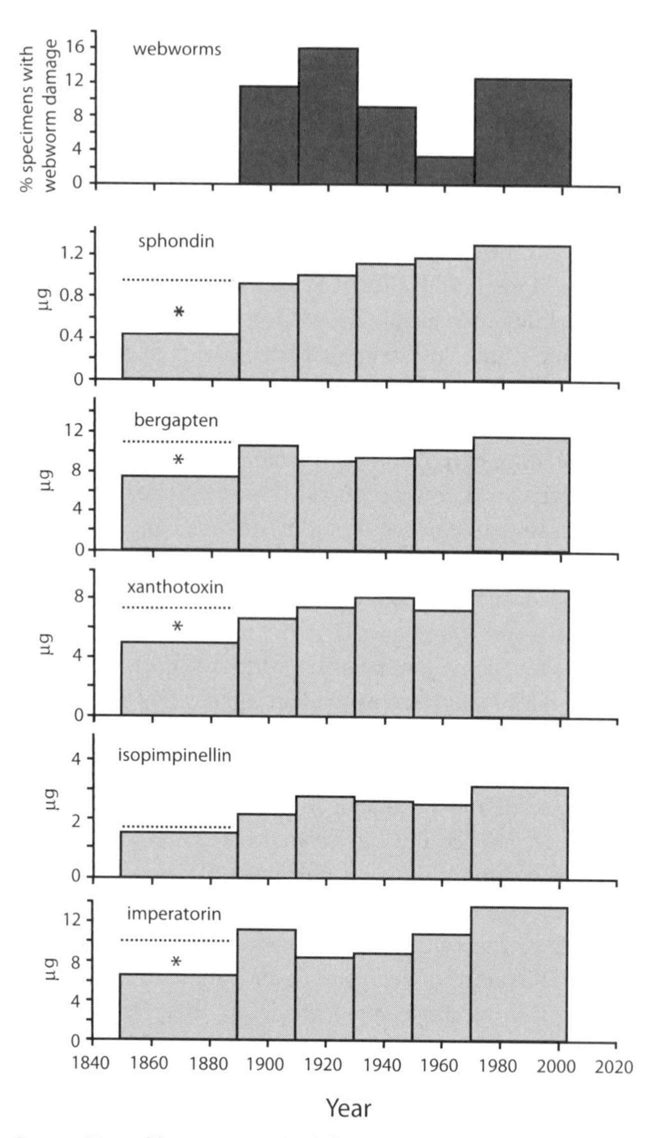

Fig. 12.7 Composition of furanocoumarin defensive compounds in wild parsnip (*Pastinaca sativa*) plants in North America before the introduction of parsnip webworms (*Depressaria pastinacella*) and after their introduction in the latter decades of the 1800s. Furanocoumarin levels increased significantly after the webworms were introduced. The dotted bars spanning the 1800s are the mean levels of the compound found in herbarium plants from Europe during that period. Asterisks indicate significant differences between American plants from the 1880s sampled from herbaria and European plants. After Zangerl and Berenbaum (2005).

America have evolved since the introduction of the webworms (fig. 12.7). In their native range these insects use both *Pastinaca* and *Heracleum* species, and *Heracleum sphondylium* is often the commonly attacked host.

It is intriguing that coevolution between these two co-introduced species seems to have been focused on a particular suite of chemical defenses and counterdefenses rather than on different traits in different populations. Chemical coevolution involving the same mechanisms also seem to dominate trait evolution in the native habitats in Europe as well (Berenbaum and Zangerl 2006). Hence, even when placed in completely new ecological settings, the overall focus of coevolutionary selection has remained the same. This could, though, be a short-term result. The coevolutionary process among populations could begin to diverge after more generations.

The same chemical coevolutionary process between parsnips and parsnip webworms seems to be repeating itself yet again in New Zealand. Parsnip webworms were first reported on weedy parsnips in New Zealand in 2004. Fortunately, studies of the interaction began soon after the invasion was discovered. The chemical profiles of the New Zealand plants differ from those in Europe and North America, but studies have already shown that the webworms are imposing strong selection on the parsnip populations for altered chemical profiles (Zangerl et al. 2008).

The studies of coevolution between parsnips and parsnip webworms have been important for our understanding of the dynamics of evolution for multiple reasons. They are the most thoroughly documented example of rapid formation of a geographic mosaic of interaction populations in nature. They are also the clearest example of rapid coevolution of chemical defenses and counterdefenses. In addition, although the effects of yet other species seem to matter in many interactions, these studies suggest that the overall process of selection on this pairwise interaction proceeds in similar ways even when an interaction is transported to a novel physical and biotic environment. In all environments, this interaction evolves, at least at these timescales, through selection on the relative frequency of a particular set of chemical defenses in the plants and a matching frequency of chemical counterdefenses in the insects. The specific frequencies differ among populations, but the chemical profiles remain the focus of coevolution even on different continents.

USING TRAITS TO CHANGE ENVIRONMENTS

We tend to consider rapid evolution and coevolution as processes that adapt populations to a particular place. There are, though, two alternatives. As populations disperse into surrounding environments, they could simply persist in some environments and go extinct in others, depending on their current traits. That is, species often just become sorted among environments based on their current traits, rather than evolve modified traits. Some approaches to the field of community ecology are based on species sorting in the absence of evolution. The other possibility is that traits evolve in ways that allow populations to move to new environments.

Rapid evolution of dispersal and migratory routes is an aspect of evolutionary biology that is becoming increasingly important, as species are being forced to adapt to climate change or human-mediated introduction onto new continents. Dispersal and migration have always been a part of the life histories of many species. Our activities, though, are imposing strong new selection pressures on populations either to disperse more or disperse less.

The many shifts in dispersal and migration we are currently observing in some species may be nothing more than the simple tracking of populations to environments. Rising global temperatures have caused populations of multiple species to shift northward in the Northern Hemisphere (Chen et al. 2011). Other range changes, however, seem to have resulted from rapid evolutionary change in the migratory behavior of individuals. One of most convincing examples has been the recent alteration of migratory routes by blackcap warblers (*Sylvia atricapilla*) in Europe (Able and Belthoff 1998; Bearhop et al. 2005; Rolshausen et al. 2009). Until recently blackcap populations in central Europe migrated in the fall to the southwest to reach their wintering grounds in the western Mediterranean. Blackcap populations farther east in Europe migrated to the southeast. Beginning in the 1960s, some individuals of the southwest-migrating blackcap populations in southern Germany and Austria began migrating northwest to the United Kingdom, 1,200 to 1,800 km north of their normal overwintering grounds on the Iberian Peninsula. Since then, those birds have undergone significant reproductive isolation from the southwest-migrating birds, and they cluster separately in genetic assignment tests (Rolshausen et al. 2009). The birds migrating along the shorter northwestern migration route also have rounder wings, narrower beaks, and some differences in plumage.

The proposed explanation for the change is not simply selection driven by climate change itself but rather selection driven by a combination of milder conditions and an increase in food provided by humans in winter. Experimental studies of blackcaps have documented that even the tendency to migrate at all can evolve quickly. In birds kept in aviaries, the behaviors associated with migratory activities have declined significantly over a little more than a decade. Moreover, birds subjected to just four generations of directional selection against migratory activity rapidly evolve significantly lower migratory activity (Pulido and Berthold 2010). These studies suggest that migratory behavior has a strong genetic basis that is highly responsive to natural selection.

Any major shifts in the migratory patterns of birds could drive rapid evolutionary change among interacting species throughout continents or across oceans. Scores of avian species have altered their migratory routes or the timing of migration during the past century, but it is unknown how many of these changes are due to the kinds of evolutionary change found in blackcap warblers (Gienapp et al. 2007a). Regardless of the causes, these alterations are likely to have broad effects on rapid evolution within the web of life, if they continue and even expand to include yet more species.

Consider just the complex geographic relationships between birds and fruits. The ripening of some plant species with bird-dispersed fruits appears to be timed to the peak of bird migrations in the autumn or the availability of migrants in overwintering regions, although multiple factors other than avian migration are involved as well (Thompson and Willson 1979; Willson and Whelan 1993; Guitian 1998; Poulin et al. 1999; Hampe and Bairlein 2000; Burns 2005). Plants that ripen fruits too early run the risk that their fruits will rot before birds find them. Plants that ripen fruits too late run the risk of having their fruits not dispersed, and plants that ripen at the same time as competing plants run the risk of fewer dispersed seeds. Altered avian migration patterns therefore have the potential to affect, at least to some degree, selection on the timing of fruit production for plant species along migratory routes and on wintering grounds. In this way, rapid evolution in avian species at one stage in their life cycle could have ripple effect on plant assemblages and plant traits stretching across thousands of kilometers.

It seems unlikely that we would ever detect such evolutionary changes over short timescales. Over even moderately long timescales of hundreds or thousands of years, though, it seems likely that rapid evolution imposed on a migratory species in one environment could have effects on evolution in far distant ecosystems on different continents or oceans. These potential large-scale effects are not limited to interactions involving birds. Large-scale migrations occur on all continents and in all oceans and include many plant and animal taxa. These migrations take a great range of forms, including multigenerational migrations in some insects (Dingle 2007; Bartel et al. 2011). Migrations by some animals impose huge effects on the ecosystems they use, such as wildebeests that migrate across the plains of Africa and whales that migrate across the northern and southern hemispheres of the Pacific Ocean. Many short-term changes in migration in these long-lived species undoubtedly result from plasticity in their behaviors rather than genetic changes, but some of the short-lived species with which they interact could undergo rapid evolutionary change.

THE CHALLENGES AHEAD

In studying the dynamics of evolution, we often focus on the temporal dynamics of traits within populations and the geographic differences found in traits among populations. Our assumption is that traits evolve through selection to match environments, and, in a broad sense, that is what we find. The process, though, is much more subtle, because selection can favor different trait combinations in different environments. Studies of any subset of these trait combinations can make it seem that the traits are mismatched to that particular local physical environment or the local population of a coevolving species. The possibility that evolution in one of the participants can favor disproportionate evolutionary responses in other species adds another component of complexity to interpreting trait matching and mismatching.

The challenge ahead is to understand how different trait combinations develop as populations evolve over time and diversify across environments. One way of understanding some of these complexities is to observe evolution under carefully controlled conditions over hundreds or thousands of generations. That is possible only in microcosms of small organisms. Those microcosms, however, have been providing new and important insights into the process of evolution. We turn now to what those microcosm experiments are telling us about the dynamics of evolutionary change.

13

Experimental Evolution

Our observations of the process of adaptive evolution in nature remain restricted to the changes that we have been able to observe over tens of generations. Experimental microcosms have extended our reach, allowing us to observe evolution over thousands or even tens of thousands of generations. These studies have been proliferating in recent years. They are providing insights into the ways in which populations adapt to new physical environments and to other species. Most of these experiments are simple to set up, but they require tremendous dedication to maintain them, if an experiment is to remain running for even moderately long periods of time. This chapter explores what these studies are telling us about ongoing adaptive evolution and coevolution.

BACKGROUND: MICROCOSMS

The following experiment is now done routinely in multiple research laboratories around the world each year. Take a group of bacterial cells that are all the same genetically and place them in a set of sterilized flasks or on Petri dishes with some nutrients. You now have a growing population of bacteria. Add some viruses (i.e., bacteriophages), again all genetically the same and capable of attacking that bacterial population, and then do something else for a week or two. Now test the bacteria for their ability to resist attack by the viruses. The bacterial population includes individuals that are still susceptible to the viruses, but it also has bacteria that are genetically resistant to the viruses.

The bacterial population has evolved. When you look closely at what happened, you find that new mutations have appeared in the bacterial population. Most of those mutations did not affect the ability of the bacteria to resist attack by phages, but at least one of those mutations conferred a new trait—often a slight change in the bacterial cell wall—that allows the individuals harboring that trait to resist attack by that particular phage type. During each of these generations the frequency of the resistant bacteria rapidly increases. Because bacteria produce multiple new generations within a single day, the bacterial population in a flask undergoes scores of generations within a single week. That is plenty of time for a resistant form to become the most common bacterial type in the flask, and multiple resistant forms may even have evolved.

But that is just the beginning. As you run the experiment, the phage population also evolves. Some new mutants appear that are able to overcome the new defenses in the bacteria. Natural selection favors those new mutant forms, which quickly become the most common form of virus in the flask. The bacterial and phage populations have now coevolved. They differ genetically from the populations in those flasks just a couple of weeks ago. You repeat the experiment as often as you wish, and the populations evolve and usually coevolve each time. The exact genetic mechanisms driving the evolutionary changes may differ among experiments, but each population will evolve in some way during that short period of time.

In a sense, laboratory microbial microcosms are living mathematical models. Unlike the simplified genetic architecture needed in models, though, these are real organisms with all the genes and attendant genetic complexity needed to make them truly viable. The environments are simplified in comparison to those found in nature, which provides a way of asking how evolution proceeds under different carefully specified ecological conditions. Experimental microcosms are therefore the important middle ground between mathematical models and field studies, making it possible to test hypotheses about evolution that cannot be tested directly in nature. These experiments tell us how evolution proceeds during the earliest stages of adaptive evolution to new physical and biotic environments, and how the evolutionary process may change over longer periods of time.

In addition, microbes have a key advantage as experimental evolutionary tools seldom found in other taxa. During an experiment, samples can be frozen every few generations, thereby preserving the genetic structure of the populations at that moment. These frozen samples are living fossil records of evolution. The ancestral population can be brought out of the freezer again months or years later, revived and allowed to grow in a flask or Petri dish with fresh nutrients, and then compared directly to descendent populations in their performance in that environment. That makes it possible to ask whether evolution has been adaptive during the experiment. Do the descendants do better in that environment, after multiple generations of selection, than their ancestors? The test is often done by competing the ancestral and descendent populations in the experimental environment.

Evolution in these experimental microcosms provides us with insights into the pace and genetic architecture of evolution and coevolution. Microcosm studies have therefore become an increasingly important tool in evolutionary research (Zimmer 2008). They are, though, more than just convenient tools. They are windows into the most common events of evolution in nature. The study of microbial evolution in microcosms is a way of glimpsing the evolutionary dynamics of the huge array of abundant and genetically diverse organisms that we cannot see.

THE INCREASING IMPORTANCE OF MICROBES AND MICROBIAL MICROCOSMS

Much of the relentless evolution in nature is microbial evolution and coevolution. The diversity and abundance of microbial species generate continuing evolution in every environment worldwide. It is therefore useful to recall a few observations that have transformed our views of genetic diversity and evolution in recent years through work on microbial populations. Although less than 8,000 prokaryotes (bacteria and archaea) and about 2,000 viruses have been formally described, some current estimates place the actual number of species closer to a million for each of these two groups (Chapman 2009). If the total number of species on earth is close to 11 million, as in some estimates, then perhaps 20 percent or more may be single-celled organisms (Chapman 2009).

That microbial diversity often occurs in natural microbial microcosms. Each of our bodies is a genetically diverse microbial microcosm connected by gene flow to other microcosms. Work on the human microbiome has suggested that a human gut may harbor about 100 trillion microbial cells distributed over a thousand microbial species (Bäckhed et al. 2005; Ley et al. 2006; Turnbaugh et al. 2007). Every human, and every other complex organism, is a microbial microcosm experiment. Within each of us, evolution in those microbial communities is proceeding month after month and year after year. In some taxa, such as hominids, these microbial assemblages are phylogenetically conserved, despite the new bacteria that are continually entering the gut (Ochman et al. 2010). There is therefore the opportunity for natural selection to shape these assemblages over long periods of time. It may do so in different ways in different kinds of organisms. Molecular studies suggest that organisms with different lifestyles, such as carnivores as compared with herbivores, or with different phylogenetic histories, such as vertebrates as compared with invertebrates, harbor different microbial assemblages (Ley et al. 2008). It is therefore not surprising that estimates of the number of microbial species continue to increase.

The tremendous genetic diversity found among these microbial species is continually reorganized through horizontal gene transfer, as genes move among individuals. These genetic processes add to the dynamics of evolution, creating microbial species that are genetic mosaics as they continue to evolve by acquiring genes from other microbial lineages (Rokyta et al. 2006; Hatfull 2008; Dobrindt et al. 2010). The microbial genes involved in adaptation to novel environments, including defenses and counterdefenses, are often acquired through horizontal gene transfer, whereas the genes involved in the core structure and replication of microbial species tend not to undergo such transfer. Estimates for some marine environments indicate that horizontal gene transfer is a frequent rather than rare event (McDaniel et al. 2010). Such studies have increasingly motivated the search for similar processes in more complex organisms (Keeling and Palmer 2008; Yoshida et al. 2010). As we might expect, most of these cases are thought to have been mediated by microbial species (see chap. 8).

The ever-widening studies of microbial species in nature have shown that microbes are not only genetically diverse but also astonishingly abundant (Bergh et al. 1989; Wommack and Colwell 2000). By some estimates there are 4×10^{30} viruses at any one time in the oceans (Suttle 2005), which is about ten times higher than some estimates for bacteria (Whitman et al. 1998). Studies of bacteriophage abundance in other environments such as in soil and in animal bodies have also indicated high abundances of viruses (Ashelford et al. 2003; Letarov and Kulikov 2009).

It is impossible for any of us to grasp a number with thirty zeros after it, but it is less difficult to understand the tremendous potential for novel mutation and natural selection in populations with those astonishing numbers. The numbers become even more astonishing when we link them to estimates suggesting that, in some environments, such as deep-sea ecosystems, viruses may be responsible for as much as 16–89 percent of bacteria deaths (Danovaro et al. 2008). Bacteria and viruses, then, are not only by far the most common organisms on earth; they are also pitted against each other in never-ending battles that are bound to favor ongoing coevolution. As studies accumulate, we should expect that, a few decades from now, the points of emphasis in evolution texts will differ considerably from current texts. Microbial evolution will get a much stronger spotlight. In the meantime, studies of microbial microcosms, together with the growing number of studies of microbial evolution in nature, are indicating that we have been underestimating the relentlessness of microbial evolution.

This ongoing microbial evolution occurs within genomes that are small compared with, say, plants or vertebrates. The most commonly used bacteria in studies of microcosm experiments are *Escherichia coli* and *Pseudomonas* species. The *E. coli* genome has more than 4,000 protein-coding genes, with the exact number varying with the strain (Blattner et al. 1997; Studier et al. 2009). *Pseudomonas* spp. have 5,000 to 6,000 coding genes, depending on the species and the strain within species (Silby et al. 2009; Mulet et al. 2010). By comparison, the human genome has about 20,500 protein-coding genes (Clamp et al. 2007). We therefore must be cautious in extrapolating the results of these experiments directly to the genetic architecture of evolution in more genetically complex organisms. The experiments, though, are important in themselves. Since much of evolution is microbial evolution, these experiments tell us a great deal about evolution in the microbial blanket that covers the earth.

Viral genomes are even smaller, and bacteriophages are generally only about 1 percent the size of the bacterial genomes they attack (Hatfull 2008). Yet natural populations of some phage species harbor high levels of genetic variation and can even recombine genes, as has been demonstrated in some wild populations of RNA viruses (Silander et al. 2005; O'Keefe et al. 2010). Many studies in experimental evolution have used DNA viruses, showing rapid evolution of the range of hosts they can attack or the site of attachment on the host cell wall (e.g., Turner et al. 2010; Meyer et al. 2012). Some of these viruses have two strands of DNA and others have only one strand. Some single-stranded DNA phages in the family Microviridae have fewer than 10 genes, but most have 11 genes and some have 5 more genes (Whitman et al. 1998; Brentlinger et al. 2002; Rokyta et al. 2006). In contrast, some double-stranded DNA phages have hundreds of genes, including the phage *Bacillus subtilis*, which has about 200 protein-coding genes (Stewart et al. 2009) and the phage T4, which has about 300 genes and has served as a workhorse in many studies in molecular biology and experimental evolution (Miller et al. 2003). In general, evolution and coevolution in bacteria and phages involve populations with tens to thousands of genes rather than tens of thousands of genes.

EVOLUTION IN ISOLATION

Evolutionary microcosm experiments are generally of five types: evolution of bacteria in novel physical environments; evolution of phages on bacterial populations that are not allowed to evolve; coevolution of bacteria and phages in a single physical environment; evolution of competing bacteria; and, more recently, coevolution of bacteria and phages across contrasting physical environments or among environments that vary in the presence of phages. The collective beauty of these studies is that we now have analyses of the evolution of bacterial populations in isolation from other species, and other studies of these same and other populations as they coevolve with phages in different environmental settings. We can use them to help us understand how evolution proceeds as populations adapt to new physical environments, how coevolution proceeds as species adapt to each other, and how coevolution proceeds across different physical environments.

The simplest place to start is with evolution of a single population to a new physical environment: evolution in isolation from other species. Many of these experiments have used *E. coli* bacteria, which generally lives commensalistically in the gut of many animals but has pathogenic forms that cause diseases. The species was first described by Escherich in 1885, and the best-studied strains, B and K12, were already in culture in laboratories in the early 1920s (Jeong et al. 2009). By the 1940s, these strains were being used extensively in laboratory studies, including those on the genetics of biochemical pathways, patterns of genetic recombination, and resistance to bacteriophages (Luria and Delbrück 1943; Tatum 1945; Lederberg and Tatum 1946). Studies since then have made this the most experimentally manipulated species for evolutionary microcosm and related studies (Zimmer 2008), with *Pseudomonas* bacteria and *Drosophila* fruit flies contributing an ever-growing number of results, and studies of yet other taxa also contributing important insights.

The genome of *E. coli* is sometimes characterized as having a core that encodes for essential cellular functions and a flexible, strain-specific part involved in molding adaptation to different environments (Dobrindt et al. 2010). Even the core, though, can be tweaked at least in laboratory environments and still maintain a functioning population. Because a number of biotechnological techniques use *E. coli*, engineering a "minimum" genome has been a goal. The genomes of some "workhorse" strains of *E. coli* used in bioengineering have been reduced in recent years by as much as 15 percent through selective deletions of parts of the genome (Posfai et al. 2006). When evolving naturally, however, *E. coli* often rapidly mutates, rearranges chromosomal sections, and acquires genes from the surrounding landscape. It is therefore an excellent model to use in studying the structure and dynamics of adaptation.

The most impressive of the experiments on the evolution of *E. coli* in microcosms are the long-term studies led by Richard Lenski at Michigan State University, which have continued for more than two decades (Lenski et al. 1991; Khan et al. 2011; Woods et al. 2011). On February 24, 1988, he set up twelve experimental populations of the B strain of *E. coli* (Lenski et al. 1991). Lenski wanted to ask some of the most fundamental questions in evolution:

Fig. 13.1 Experimental microcosms used in the long-term experiment with *E. coli* in Richard Lenski's laboratory. Photograph courtesy of Richard Lenski.

How fast can populations evolve, and how do the rates of evolution vary over time? Do the same evolutionary changes happen time and again in different populations? How does the rate of evolutionary change in the genome relate to the changes we observe in the morphology and physiology of individuals?

Each of the twelve experimental populations started from a single cell, grown in a covered flask containing a low level of nutrients, and transferred each day to a new flask (fig. 13.1). Six of the populations had a genetic marker that allowed the bacteria to be identified from the bacteria in the other six populations. Glucose was the nutrient that the bacteria relied on for growth in all the flasks, but the liquid medium also included citrate, which *E. coli* are generally unable to metabolize, and minimal salts. The level of glucose in the flasks was sufficiently low that, by the latter part of each day, the bacteria used up the available glucose. They were therefore living in an environment that favors those bacteria best able to compete for this limiting resource. After they exhausted the glucose, the bacteria remained in a stationary phase until they were transferred to a new environment with available glucose. Each population goes through about six or seven generations each day. Every 75 days, which is about 500 generations, samples of these cultures were placed in a freezer as a living fossil record of evolution in this new environment up to that moment.

In this highly competitive environment, all the populations evolved quickly as new mutations arose and natural selection acted on those mutations. They evolved to reproduce at a faster rate (Novak et al. 2006). They evolved larger average cell size (Stanek et al. 2009). Some populations evolved higher mutation rates (Sniegowski et al. 1997; Barrick and Lenski 2009).

This strain of *E. coli* does not exchange genes, and the bacteria in the experiment were initially genetically identical. Hence, evolution in these large populations resulted only from new mutations, followed by natural selection favoring some of those mutations over others. Mutations in *E. coli* happen at a low rate, but with so many genes and so many individuals in each flask, new mutations occur quickly.

The numbers are staggering. The *E. coli* genome in these experiments is made up of about 5 million pairs of nucleotides (Blattner et al. 1997). The chance that a mutation will occur in an individual bacterium at a particular nucleotide in any generation is about one in 10 billion (Lenski et al. 2003). If the populations were very small, new mutations would appear slowly within each population. Each flask, however, fluctuates in number of individuals each day from about 5 million to 500 million (Lenski 2004). The low point occurs during the daily transfer of bacteria to a new flask, and the high point occurs after the bacterial population has expanded later that day. Consequently, tens of thousands of new mutations are likely to occur each day within a flask (Richard Lenski, pers. comm.). The vast majority of these mutations have no effect on the Darwinian fitness of individuals, because the bacteria with these mutations are neither better nor worse than the other bacteria in the flask at surviving, competing, and reproducing. Most other mutations are detrimental. Over many generations, though, a few beneficial mutations have occurred, and these have spread rapidly through selection.

The experiments are now well beyond 55,000 generations. That number of generations is sufficient for mutations to have occurred repeatedly at every point in the genome. Lenski has estimated that during the first 20,000 generations of the experiment several hundred million point mutations occurred in each population (Lenski 2004). In addition to these mutations at individual nucleotide sites, other types of mutation are likely also to have occurred, as some particular sites are lost, added, or changed in their positions within the genome. These deletions, insertions, and inversions all have the potential to affect how genes are expressed and therefore how well individuals survive and reproduce.

With so much mutation occurring in these large populations, it may seem puzzling that the populations did not evolve even faster. One of the properties of rare events in all populations, however, is that even some potentially beneficial mutations will be lost due to chance alone during the early stages, when those mutations are present only in a tiny number of individuals within a population. Although these flasks contain up to 500 million individuals at the end of each day, only a few million of them are then transferred to the new flask. By chance alone no individuals harboring a new, but still very rare, beneficial mutation may be among those transferred to the new flask.

Because the ancestors are in the freezer, it has been possible over the years to check periodically just how well the current populations fare in this environment relative to how their ancestors would fare. This is a straightforward experiment: compete the ancestral and current population in the low-glucose environment and evaluate which does better and by how much. A sample of the ancestral population is taken from the freezer and placed with a sample of one of the current populations in a flask using the same nutrient conditions as in the long-term experiment. The relative fitness of the populations in that environment is measured as the relative growth rates of the current population in comparison to the ancestral population.

These experiments have shown that much of the initial evolution in these flasks was in the direction of improving the ability of these bacteria to survive and reproduce in this new environment. The descendants proliferate faster in this environment than their ancestors. The descendent populations in the flasks continued to undergo evolutionary change for thousands of generations, but their further evolution began to slow down after 20,000 generations (Cooper and Lenski 2000). At that point, the populations seemed to have exhausted many of the ways of exploiting this simple environment (fig. 13.2).

Evolution, though, continued to alter multiple traits, resulting in periods of slow change in traits followed by periods of rapid change (Elena et al. 1996; Blount et al. 2008; Barrick et al. 2009). The overall rate of directional adaptive change at the organismal level therefore has proceeded in a pulsed or step-like fashion. Multiple generations will go by within a population with little evidence of evolutionary change relative to the ancestral population, followed by a quick change as a new beneficial mutation takes hold and sweeps through the population by natural selection. One population, for example, showed three such pulsed changes during the first 2,000 generations of the experiment. The first occurred about generation 300, the second about generation 600, and the third about generation 1,300 (Lenski and Travisano 1994). Meanwhile, evolution at the molecular level remained nearly constant throughout 40,000 generations. Many of these molecular substitutions appeared to be beneficial, suggesting complicated relationships with patterns of molecular evolution and the patterns detectable in experiments at the organismal level (Barrick et al. 2009).

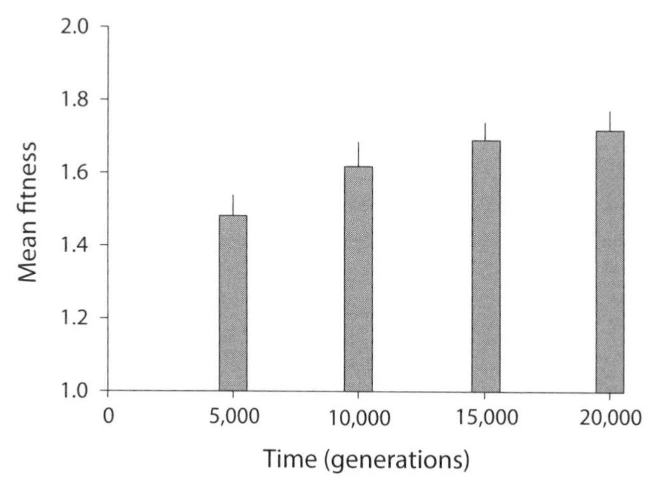

Fig. 13.2 Change in mean fitness in *E. coli* populations over 20,000 generations. Values are averages of twelve populations ±95 percent confidence intervals. Fitness is estimated as the ratio of growth of each current population relative to the ancestral population when they are put in direct competition in the selective environment. After Cooper and Lenski (2000).

All the observable phenotypic changes involved alteration of traits already known to occur in these bacterial populations. Then, a little after 30,000 generations, a major change occurred in one of the flasks (Blount et al. 2008). The medium in that flask would appear turbid by the end of the day, indicating a density of bacteria much higher than in the other flasks (fig. 13.1, flask in the center foreground). The first reaction was that there must have been some contamination, but experiments showed that the bacteria in that flask had evolved the ability to metabolize citrate. That was a remarkable result, because it showed that this particular population had evolved a completely new biochemical ability that allowed it access to a nutrient that had been in the experiment since day 1 but that no population until then had ever been able to exploit.

This important result demonstrated that, over a very short time in evolutionary history, a population could evolve a fundamentally new trait with a novel function. As Lenski said in an interview, the result was "outside what was normally considered the bounds of *E. coli* as a species" (Holmes 2008). This change did not simply involve evolution of a large average cell size in a population or some other slight modification. It involved a fundamental change in the biochemical pathways by which these populations metabolize nutrients found in the environment—a key innovation.

EVOLUTION OF SPECIALIZED FORMS WITHIN A FEW WEEKS

The long-term experiments with *E. coli* suggest that, in nature, there is probably no such thing as an actively growing bacterial species that has not undergone evolution every year of its million- or billion-year history. An alternative, though, could be that *E. coli* is unusually evolvable. That alternative now seems unlikely. Experiments with *Pseudomonas* bacteria have shown that rapid evolution also occurs in microcosms of other microbial species. *Pseudomonas* is a complex group of bacterial species that includes plant pathogens and animal pathogens. Some species can be grown easily in multiple kinds of laboratory cultures, and in some cases evolution in these cultures can be seen just by looking at the cultures after some days. One of the most visually striking cases of rapid evolution in *Pseudomonas* has become the basis for another of the most insightful experiments of microbial evolution in microcosms.

If a population of genetically identical *P. fluorescens* is placed in a small vial containing a nutrient-rich liquid, and the vial loosely capped and not shaken thereafter, natural selection favors morphologically distinct mutations in different parts of the vial (fig. 13.3). The smooth form resembles

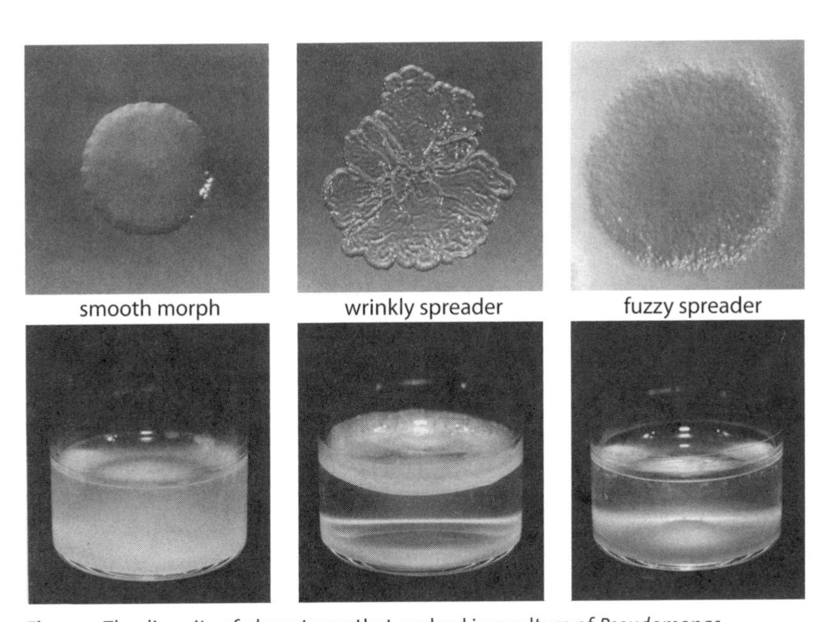

Fig. 13.3 The diversity of phenotypes that evolved in a culture of *Pseudomonas fluorescens* SBW25 placed in a novel environment (*top*) and the positions of those phenotypes as seen within a vial (*below*). The forms are called, from left to right, smooth morph, wrinkly spreader, and fuzzy spreader. After Rainey and Travisano (1998).

the initial population and dominates much of the liquid part of the vial. The wrinkly-spreader form produces a biofilm at the liquid-air interface. The fuzzy-spreader form is most visible as cells along the bottom of the vial, but their life history involves a complex use of the medium that is only now being unraveled (Paul Rainey, pers. comm.). These are truly newly evolved forms that are genetically determined and passed on to offspring (McDonald et al. 2009).

This experiment differs from the long-term experiment on *E. coli*, because the bacteria are not moved to a new vial each day. Selection therefore has the potential to favor bacteria that are adapted to different environments within a vial. Although evolution in a vial would not seem to offer many evolutionary options from a common ancestral form, these vials are not one environment. They are surprisingly complex ecological habitats, and the experiments with *P. fluorescens* show that selection can quickly favor the evolution of specialized forms with different ecological niches. These niches are not static, and current experiments are probing more deeply into the ways in which changes in these niches over time allow coexistence of the morphs (Paul Rainey, pers. comm.).

It is possible to manipulate how many of these forms evolve within a vial by changing the conditions. If the initial bacterial population is made up of one of the niche specialists, such as wrinkly spreaders, rather than the more generalist smooth form, then the other niche specialists do not evolve in that vial (Fukami et al. 2007). If niche specialists are introduced later in the experiment, they suppress diversification into novel forms. The evolution of niche specialists can be prevented simply by shaking the cultures during the experiment, which creates more homogeneous environments during the experiment (Rainey and Travisano 1998).

Experiments evaluating the fitness of wrinkly spreaders have shown that there are many mutations that can contribute to the fitness of these genotypes (McDonald et al. 2011). How those mutations are distributed among beneficial and deleterious effects depends on how the culture is maintained. In broth culture that is not shaken regularly, wrinkly spreaders can be favored at least for awhile because they can form a mat on top of the culture. In these cultures, beneficial mutants are normally distributed in their effects on fitness, with most having moderately strong beneficial effects. If those same mutants are placed in a broth culture that is regularly shaken, the wrinkly spreaders are generally disfavored by selection. The mutations show a normal distribution of effects on fitness, but most of the effects are deleterious.

These experiments demonstrate that even seemingly simple environments have enough spatial and temporal environmental complexity to favor rapid evolution and even the diversification of a population into multiple genetically distinct forms. The pattern of diversification depends on the history of colonization of the vials. If these extremely simple environments favor the rapid evolution and genetic diversification of populations, it is little wonder that life on earth has diversified into such a great variety of forms and that populations often retain so much genetic variation.

ADDING COEVOLUTION WITH OTHER SPECIES

The evolutionary changes observed in the experiments with *E. coli* and *Pseudomonas* spp. have occurred in microcosms in which the bacteria evolve alone. If bacteriophages are added to a microcosm, the bacterial population rapidly coevolves with them. These microcosm experiments on coevolution between bacteria and phages have provided a way of evaluating how adaptation to other species may differ from adaptation to physical environments.

We should expect that the pace, and perhaps even the genetics, of evolutionary change in coevolving populations will differ in some ways from that observed in populations evolving in response to physical environmental change. When a bacterial population evolves in response to a new physical environment, such as the level or kinds of nutrient within a flask, the resources simply get used up and the population enters a stationary state until more nutrients are added. But when a bacterial population is confronted with a phage population, not only does it evolve defenses against the phages, but the phage population also rapidly evolves within days or weeks. From the point of view of selection on the bacteria, the phage population is a constantly changing environment that changes specifically in response to each evolutionary change in the bacterial population. The adaptive landscape of the bacterial population continues to change as the phage evolves, and the same happens for the phage population. The result is a coevolutionary chase, with natural selection favoring bacteria with traits that make them resistant to attack by phages and, simultaneously, selection favoring phages able to overcome those defenses.

A successful phage must be able to attach onto a bacterium's cell wall, penetrate the cell wall, inject its DNA or RNA, and then take over the machinery of the bacterium's DNA to make new copies of itself. A bacterium's cell wall is therefore its major first line of defense against attack by phages. When an *E. coli* population is under attack, natural selection favors mutant forms with cell wall structures that make it difficult for phages to attach and penetrate. Exploration of the genetics of these interactions began more than half a century ago (e.g., Luria 1945), and experimental studies on the ecological and evolutionary dynamics of these interactions began soon thereafter (Campbell 1961; Chao et al. 1977; Levin et al. 1977; Lenski 1984). These studies have continued to expand, with some using the same B strain as those in Lenski's long-term experiments of *E. coli* evolving in isolation, and others have used the K-12 strain. Collectively, the experiments show that the bacteria are capable of evolving rapidly not only in response to changes in the physical environment but also in response to enemies, suggesting that multiple parts of the bacterial genome are capable of rapid evolution.

Rapid coevolution also occurs with *Pseudomonas* species. If *P. fluorescens* populations susceptible to attack by a particular phage type are set up in cultures that also include a phage population, it generally takes less than a few weeks for the bacterial populations to evolve forms resistant to the initial genetic form of the phage. The phage populations also evolve during these weeks and, as the experiment proceeds, coevolution between the bacteria and the phages follows different trajectories in different microcosms. Each bacterial population evolves adaptations to resist its own local phage population. If, later in the experiment, some bacteria are challenged with phages from their own microcosm and phages from another microcosm, they are better able to resist phages from their own microcosm (Buckling and Rainey 2002). As the experiment continues through hundreds of bacterial generations, the bacteria become resistant to a wide range of phage genotypes and the phage populations evolve the ability to infect a wider range of bacterial genotypes (Brockhurst et al. 2003).

Multiple forms of resistance can evolve quickly in the bacterial populations. *Escherichia coli* populations interacting with phages diverge into multiple resistant types within a few weeks (Forde et al. 2008b). During that time—about 150 generations—the bacterial populations and phage populations continually coevolve. Novel beneficial mutations that confer resistance accumulate in the bacteria population over time, and they begin to fluctuate in their relative frequency as selection continues to act on them. The dynamics of these mutants differ between high nutrient environments and low environments. Multiple resistant types appear quickly in bacterial populations in both environments, but by the end of the experiment the populations in low nutrient environments have more resistant types than those in high nutrient environments. That result, however, would not necessarily occur in all experiments with bacteria and phages. Mathematical models have indicated that the ways in which resource availability affects the pattern of change in resistance diversity over time depends on the genetic architecture of resistance in the bacteria and infectivity in the phages (Forde et al. 2008a).

ACCELERATED EVOLUTION THROUGH COEVOLUTION AND THE COST OF DEFENSE

One of the most crucial questions in these experiments is whether coevolution accelerates the overall rate of evolution. Reciprocal adaptation in the bacteria and the phages could favor an accelerated rate of evolution in both species, because each round of evolutionary change in the one species favors further evolution in the other. Alternatively, it could come at the expense of adaptation to the physical environment, thereby either reducing the overall rates of evolution or having no effect. Microcosm studies suggest that coevolution accelerates evolution at least under the experimental conditions studied so far.

One way of doing this experiment is to set up replicate populations of bacteria and phages and then allow the bacteria to coevolve with the phages in one set of replicates but prevent the bacteria from evolving in the other set. In the non-coevolutionary replicates, the phage populations are allowed to evolve, but the bacteria are replaced every two days with their ancestors. The phage populations in the non-coevolutionary replicates continue to adapt to the ancestral bacterial populations without any change throughout the experiment in the bacterial populations.

When this experimental design was applied to six replicate *Pseudomonas fluorescens* microcosms with coevolution and six without coevolution, the phage populations in the coevolutionary replicates diverged more, and they diverged faster, than those in the non-coevolutionary replicates (Paterson et al. 2010). The result is evident in two ways. Phages from coevolving populations are able to attack a wider range of hosts than phages from non-coevolved populations. Also, whole genome sequencing shows that phages from coevolved populations have a higher rate of molecular evolution than non-coevolved phages. The coevolved populations have a higher level of genetic diversity among replicates and greater molecular divergence among the replicates than the non-coevolved replicates. Although the non-coevolved populations also showed evolution, all the replicate populations evolved in much the same way as they adapted to the novel, but unchanging, physical environment.

The higher rates of evolution and genetic divergence in the phage populations coevolving with bacteria were concentrated on only four genes. In organisms with more complex genomes, the evolutionary rates could be even higher, because evolution could be distributed across yet more genes.

Alternatively, evolution could be the same or slower, because multiple evolving genes could interfere with each other, and that could be tested in future experiments. For now, these experiments show that coevolution can accelerate the rate of evolutionary change, and a great deal of rapid coevolution can occur even when only a few genes are under strong selection.

When resistance in bacteria or infectivity in phages evolves in these coevolutionary experiments, it can carry costs such as slower growth rates in those bacteria. If so, resistance in the bacteria would be favored only in the presence of phages. In the absence of phages, resistant individuals would be outcompeted by bacteria that lack the resistance genes. That is what occurs in *E. coli* microcosms, but the cost is sometimes only temporary. In one classic experiment, the cost of resistance decreased over 400 generations as natural selection favored yet other genes that decreased the cost (Lenski 1988).

Another way to explore the cost of resistance genes is to ask what happens to bacterial evolution if all the phages are removed. The *E. coli* from the long-term experiments in the Lenski laboratory are perhaps the best populations to use for these experiments, because they have been evolving without phages under carefully controlled and replicated conditions for tens of thousands of generations. That should be plenty of time to see any major effects on bacterial evolution. For this experiment, the six replicate lines of *E. coli* were challenged after 45,000 generations with three phage types (Meyer et al. 2010). The ancestral population of this *E. coli* strain was resistant to T6 phages, resistant to a lesser extent to a mutant form of T6 called T6*, and fully susceptible to attack by λ phages. If there is a cost to individuals that harbor genes conferring resistance to phages, such as a reduced growth rate, then natural selection should act against resistant individuals. The population should have evolved toward being completely susceptible to phage attack after many generations of selection.

The evolutionary result was much more nuanced. All six replicate populations remained just as resistant to T6 phages as their ancestor was, suggesting that resistance to T6 did not impose a cost on the bacteria even in the absence of phage, at least in the particular physical environment in which they had been growing for 45,000 generations. Resistance to T6 is conferred by a single point mutation in a gene called *tsx*. Meyer and colleagues sequenced that gene in all six populations and confirmed that the gene had not evolved over those many generations. In contrast, all six populations had evolved increased susceptibility to T6* phages, suggesting that selection had eliminated over time individuals resistant to that particular phage type. Most surprising of all, most of the replicate lines evolved resistance to λ phages despite their absence throughout the experiment. This intriguing result suggests that resistance had evolved as a by-product—a pleiotropic effect—of selection on some other trait that also happened to confer resistance to λ phages. Overall, these results suggest that bacterial populations are likely to show complex patterns of resistance and cross-resistance as they coevolve with phages across many different environments and phage types. If this same experiment had been done in a different physical environment, then the relative costs of resistance against these phages could have differed.

As bacteria and phages coevolve, we might expect that their defenses and counterdefenses would often initially show evidence of escalating arms races. But such arms races are costly and cannot continue indefinitely. In experiments with *P. fluorescens* and phages, coevolution initially proceeds both through arms races and through fluctuating selection. Over time, the arms races decelerate, leaving fluctuating selection on alternative phenotypes in the bacteria and the phages as the major form of selection (Hall et al. 2011). Molecular analyses of the phages show that the range of phage genotypes increases over time during coevolution in microcosms, and many molecular substitutions affect the tail fibers, which are involved in the crucial first step in attachment to the bacterial cell wall (Scanlan et al. 2011).

These results reinforce the developing view that antagonistic coevolution is often driven over the long term by forms of selection that maintain genetic variation in population rather than by sustained directional selection (see chap. 7). With such fast growth rates, the frequencies of bacterial genotypes have the potential to change quickly under strong selection. Under frequency-dependent selection, evolution inherently fosters further evolution as selection constantly acts to favor rare forms and disfavor common forms. In the literature of microbial evolution, this has in recent years sometimes been called kill the winner, which is simply a colloquial way of describing frequency-dependent selection.

COEVOLUTION IN COMPLEX MICROCOSMS

Coevolution in microcosms of *Pseudomonas* bacteria and phages speeds up even more if you simply shake the microcosm rather than leave it still, at least over the initial few hundred generations. For these experiments, some microcosms are placed in an incubator, and the bacteria and phages are allowed to coevolve for about a month. During that time they are periodically transferred to new microcosms with new growth media. Other microcosms are set up in the same way but are shaken by a mechanical shaker for one minute during every half hour throughout the day and night. At the end of the month, the microcosms that have been shaken have bacteria that are more resistant to phages, and the phages are more infective than those found in unshaken microcosms (Brockhurst et al. 2003). The exact reasons for this difference in the rate of coevolution are not yet known, but it may be that occasional shaking continually mixes different forms that are adapting to different parts of the microcosms. That introduces new genetic variation into each part of the microcosm, favoring further coevolution. Whatever the exact mechanism, these experiments show that the rate of coevolution can be further accelerated through manipulations of these seemingly simple environments.

The experiments can be made a little more ecologically complex by setting up separate microcosms and allowing some dispersal among them. For these experiments, some individuals are moved every few days between the microcosms. When these experiments have been done with *E. coli* and their phages, the result is the same as for mixing within microcosms: the rate of coevolution speeds up (Forde et al. 2004). The cause in this case seems to be that the individuals moved between the microcosms are too few to swamp local adaptation of bacteria and phages within each microcosm, but they are high enough that they contribute new genetic variation on which natural selection can act.

If more genetic variation is better, then why do bacterial populations not mutate even faster? Natural populations of bacteria sometimes harbor individuals called mutators, because they have high mutation rates. In coevolutionary experiments with *P. fluorescens* and phages, coevolution proceeds faster earlier in experiments with mutators than in those without mutators. During these early stages, phage evolution is not able to keep up with bacterial evolution. After about 85 bacterial generations, however, the levels of bacterial resistance do not differ between the trials (Morgan et al. 2010). There seems, then, to be an upper bound on the extent to which mutators will be favored in a bacterial population, at least in the relatively simple environments in which a bacterial population is coevolving with a single phage type.

The pace and dynamics of coevolution are governed not only by the level of genetic variation and the strength of selection, but also by how the genetic variation and the intensity of selection are distributed among interconnected microcosms. Both in *E. coli* and in *Pseudomonas*, setting up multiple microcosms with some movement of individuals between them results in

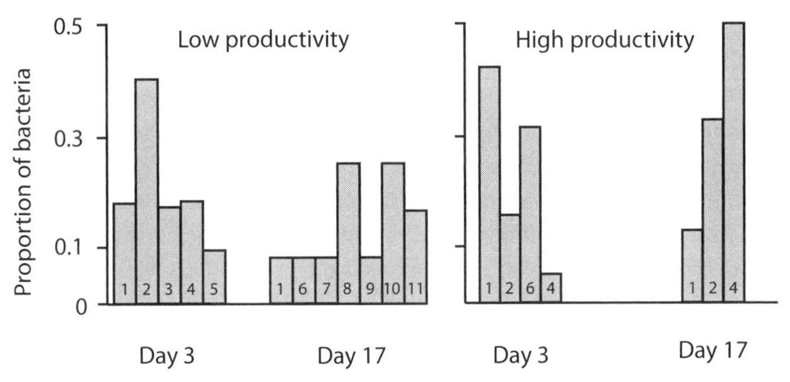

Fig. 13.4 Frequency of *E. coli* with different types of resistance against T3 phages after 3 days and 17 days in low-nutrient and high-nutrient microcosms. The observed resistance phenotypes are numbered 1 through 11 and indicated within the bars. After Forde et al. (2008b).

a dynamic geographic mosaic of coevolution (Forde et al. 2004, 2008b; A. Morgan et al. 2005). The microcosms coevolve in slightly different ways and at different rates depending on the new mutations that appear in each particular microcosm and the infusion of new genes as individuals disperse from other microcosms. On any given day, the bacteria are more resistant to the phages in some microcosms than in others, and the phages are more infective to their local bacteria in some microcosms than in others.

If microcosms of *E. coli* or *P. fluorescens* and phages also differ in the levels of nutrients within the cultures, then coevolution becomes even more dynamic. Soon after the microcosms are established, multiple resistant types evolve and continue to change in different ways in the different microcosms (fig. 13.4). In some experiments, coevolution occurs at faster rates in microcosms with high nutrient levels than in those with lower nutrient levels (Forde et al. 2004; Lopez-Pascua et al. 2010). If some individuals are allowed to disperse among the microcosms, then coevolution speeds up even more. This kind of experiment sets up a selection mosaic in which the structure of natural selection differs among environments, even when the experiments begin with the same bacterial and phage genotypes. Gene flow among the microcosms continues to mix the genotypes as the experiment proceeds.

In some experiments with connected microcosms that differ in nutrient levels, the bacteria coevolving in high nutrient environments have higher fitness than those in low nutrient environments (Lopez-Pascua et al. 2010). This difference increases the effect of coevolution in high productivity environments on the coevolution of populations in other environments, as genes move especially from populations in the high productivity environments to populations in the lower productivity environments. In some cases, the populations in the higher nutrient environment may have a greater effect on overall coevolution simply because those populations are larger and produce more mutations than populations in the lower nutrient environments. In all these experiments, the result is a constantly changing mix of adaptation and counteradaptation that is more dynamic than that observed in any one microcosm. These studies suggest that, in the heterogeneous environments in nature, evolution is likely to be even more dynamic than observed in these complex microcosms.

These linked microcosms can also be used to study how coevolutionary hotspots and coldspots affect the overall pace and direction of coevolution. In some experiments with *P. fluorescens* and SBW25φ2 phages, the cultures are allowed to coevolve strongly in some microcosms (hotspots) by periodically shaking the culture vessels. Shaking increases the rate of encounter between the species and the rate of coevolution (Brockhurst et al. 2003). Another set of culture tubes is started from the same bacterial and phage populations but not shaken during the experiment, resulting in lower encounter rates and low rates of coevolution (coldspots). If gene flow is allowed from the hotspot cultures to the coldspot cultures, coevolution is accelerated in the coldspots (Vogwill et al. 2009). If gene flow is allowed in the other direction, it slows the rate of coevolution. Coevolutionary hotspots embedded in a landscape therefore have the potential to act as "pacemakers" of the overall rate and direction of coevolutionary change in coevolving species (Vogwill et al. 2009).

EVOLUTION IN ORGANISMS WITH VERY FEW GENES

Escherichia coli, *Pseudomonas* spp., and their phages have relatively few genes, usually numbering in the hundreds or thousands. Some phages, though, have fewer than twenty genes. Experimental microcosm studies using these phages have provided additional insights into the process of adaptive evolution, because the mutational options available to these phages are even more restricted. Single-stranded DNA phages in the family Microviridae have become a major tool for these studies. They were first discovered in the 1940s and 1950s and shown soon thereafter to have a single-stranded genome (Sinsheimer 1959). The mechanisms by which the phages infect their bacterial hosts were already being analyzed in the 1960s (Burton and Sinsheimer 1963). By 1977 the DNA sequence of one of these phages, φX174, had been determined (Sanger et al. 1977). It attacks E. *coli* and some other gram-negative bacteria and has only eleven genes. Beginning in the 1990s, φX174 became increasingly used as a model in studies of experimental evolution (Bull et al. 1997; Wichman and Brown 2010).

Most of these experiments explore phage evolution in the absence of coevolution. Phage populations are established on a bacterial host population at the beginning of an experiment and allowed to evolve on that population. The bacterial population itself will likely also begin to evolve, but it is reset to its ancestral state about every two days by moving the evolving phage population to a new chemostat with ancestral bacteria. In this way the phage population experiences little evolution in its bacterial host, and whatever evolution occurs in the bacteria is quickly eliminated.

Under these conditions, parallel evolution is common among replicate phage populations. Because these phages have so few genes, it is possible to study parallel evolution at the level of individual molecular substitutions. In multiple experiments with φX174 phages, the same base substitutions have occurred more often than by chance among replicates within an experiment, and many of these substitutions appear to be adaptive (Wichman et al. 1999; Wichman and Brown 2010). There is also strong evidence of convergent evolution among populations starting from different common ancestors (Bull et al. 1997; Wichman et al. 2000). If the populations were simply adapting to a constant environment, then we might expect evolution to slow down after the populations had become adapted to that environment.

In one 13,000 generation experiment, however, the populations continued to evolve, and many of those later changes appeared to be adaptive (Wichman et al. 2005). Holly Wichman and colleagues have suggested that the most likely cause of this evolution is competition among the different genotypes that accumulate through evolution in the phage population over the course of the experiment. These different genotypes co-infect hosts and compete with each other. Hence, even in an apparently constant environment, evolution may continue as genetic diversity increases in the phage populations and the genotypes compete for hosts.

In experiments with microvirid phages, descendent lines in the new environment sometimes, but not always, show higher fitness than their ancestors, and the same result has occurred in experiments with other viruses (Bollback and Huelsenbeck 2009). Some of the differences among cultures in fitness may result from differences in the number and rate of appearance of beneficial mutations. The theoretical expectation is that the mutations will follow a simple exponential distribution in which most mutations are deleterious, a smaller number neutral, and a tiny number beneficial.

The results are much more varied than the expected distribution. In some experiments, mutations follow a more uniform distribution of beneficial effects, and beneficial mutations are more common in the early phases of adaptation than later (Rokyta et al. 2008). Descendent phage populations would then be more likely to show increases in fitness over their ancestors than generally expected by models that assume rare beneficial mutations. In other experiments with different phages, however, the distributions are more similar to an exponential distribution in which beneficial mutations are rare (Sanjuan et al. 2004; Orr 2010). There is, then, much still to be learned about what determines the distribution of potentially beneficial mutations in organisms with very small genomes.

In general, though, experiments with microbial species show that, regardless of genome size, rapid adaptive evolution is inevitable in any microcosm with sufficiently large population sizes. These experiments increasingly include not only bacteria and DNA viruses but also other microbes such as RNA viruses (Elena et al. 2008; Holmes 2009; Turner et al. 2010) and fungal pathogens (De Bruin et al. 2008). Moreover, some experiments involve rapid adaptation of microbial parasites to more genetically complex hosts such as invertebrates or vertebrates rather than to other microbial species (Little et al. 2006; Ebert 2008). Overall, these studies indicate that continuing microbial evolution is inevitable in environments ranging from simple homogenous environments to complex environments in living hosts.

BEYOND MICROBES

Experimental evolution is just as easy to show, although often even more time-consuming to undertake, in other organisms such as *Drosophila* fruit flies, *Tribolium* beetles, *Daphnia*, snails, nematodes, and laboratory mice, among others (e.g., Wade and Griesemer 1998; Mueller and Joshi 2000; T. Morgan et al. 2005; Ebert 2008; Zbinden et al. 2008; Garland and Rose 2009; Koskella and Lively 2009; Schulte et al. 2010). That fact is important, because these organisms are much more genetically complex than bacteria or viruses, have longer and more complex life histories, and, in most cases, reproduce sexually. Many studies have used *Drosophila*, because populations of multiple species have been domesticated over many decades for living and reproducing in laboratory cultures, and there is a great deal of genetic and genomic information on *Drosophila* and their traits (Burke and Rose 2009).

Studies of experimental evolution with *Drosophila* and other invertebrates, however, generally differ in the question posed from studies on microbes, because these studies often have different starting points. Many studies of microbial microcosms start with genetically identical individuals, and subsequent evolution relies on new mutations, which occur quickly because the bacterial and viral cultures are large. In contrast, studies of *Drosophila* evolution generally begin with a genetically variable population that is at least large enough so that natural selection rather than random genetic drift is the driving force in any subsequent evolution. Right from the start, then, evolution in *Drosophila* cultures can result from natural selection on standing genetic variation that is continually recombined during sexual reproduction, from selection on new mutations, or from some combination of these sources of genetic variation.

Drosophila also differ from microbial species in that ancestral populations cannot be frozen and then thawed later to test whether fitness in descendent populations has increased in a novel environment over that found in the ancestral population. Instead, the *Drosophila* populations that serve as the ancestral controls are maintained generation after generation in the laboratory in large populations in an environment as close as possible to the conditions at the start of the experiment. To serve as an effective control, the population chosen as the ancestral benchmark in the experiment needs to have become adapted to that control environment over many generations. Otherwise, it will still be adapting to the control environment at the beginning of the experiment.

Allowing the control population to adapt to its laboratory environment, before beginning experiments, is crucial. Evolutionary changes occur during the early stages of domestication in laboratories, and some are even predictable. For example, several times over the past twenty years, laboratory populations of *Drosophila subobscura* have been collected from pine woodlands in Portugal and established as new cultures. Each culture has been kept under the same laboratory conditions. The number of eggs laid by individual females has increased in all these cultures, as the populations have adapted to their new environments. Some other characteristics, such as the rate of development of juvenile flies, have not evolved in any predictable ways (Simões et al. 2006). No assumptions can therefore be made about how the control population will adapt to the new environment. Consequently, each control population must be allowed to adapt to its laboratory environment before the start of an experiment.

Once populations have become adapted in the laboratory to a particular set of conditions, those populations may serve as effective controls for other studies of experimental evolution in the same laboratory. Some studies show little evidence of significant evolutionary change in well-established control populations for as many as fifty generations even at the molecular level (Teotónio et al. 2009). If a subset of flies from a long-established laboratory population is then challenged with a new environment, they almost always evolve rapidly, as they adapt to the new conditions. Adaptive evolution is evaluated as the differences found in traits of the experimental populations as compared with the control populations. Evolved differences in some traits are evident in experiments within less than ten generations (Gibbs 1999; Mueller and Joshi 2000).

Equally impressive are the wide range of environmental conditions that favor evolution and the great diversity of traits that evolve rapidly when populations are placed in novel environments. *Drosophila* populations have been shown to evolve, for example, under conditions of altered food, temperature, humidity, ethanol level in the medium, intraspecific competition, or interspecific interactions. Populations grown under drier conditions have evolved to be more tolerant to desiccation (Archer et al. 2007). Flies kept under low densities quickly evolve behavioral and life history traits that differ from flies kept under high densities (Mueller and Joshi 2000; Mueller 2009). Flies placed in competition with a second fly species in a novel physical environment evolve mechanisms to cope with the physical environment and with the competitors (Joshi and Thompson 1995, 1996). Flies subjected to pathogens evolve ways of resisting or tolerating these pathogens (Vijendravarma et al. 2009). And so on. These changes are often visible within a year, and sometimes more quickly.

Because these experiments generally start with populations that already have considerable genetic variation, some or much of the observed evolutionary change is due to natural selection acting on that variation rather than on new mutations. Continued refinement of molecular tools has made it possible to begin to assess how much of evolution is due to selection on standing genetic variation rather than on new mutations. In one experiment, replicate populations were established from a culture that had been maintained in the laboratory at population sizes of greater than a thousand individuals for almost thirty years under several experimental conditions: standard laboratory environment ("control environment"), selection for reproductive success late in life, selection for early reproductive success, and selection for resistance to starvation (Teotónio et al. 2009). Each of the noncontrol populations was subjected to fifty generations of reverse selection back to the ancestral state. The experimental populations evolved back to their ancestral state of adaptation in the control environment, but they did so mostly through selection on standing genetic variation rather than through new mutations. Selection occurred at several loci, and genetic diversity was retained throughout the process of adaptation. These results suggest that, in large populations, evolution over short time spans is indeed reversible, because populations tend to retain much of the genetic variation that allows adaptation in multiple directions.

Microbes and flies are just two of the groups now routinely used in laboratories to study rapid evolution, although they are the groups with which the greatest numbers of experiments and the longest-running experiments have been done. Flour beetles and other insects, *Daphnia* water fleas, annual plants, yeasts, and other organisms are all evolving in laboratories and greenhouses around the world this week. This is, of course, evolution under highly controlled conditions maintained generation after generation, and we have to be cautious about extending these observations directly to what we see in nature, where every year is at least a little different (Harshman and Hoffman 2000). Temperatures are warmer or colder in some years than usual. Parasites, predators, and competitors are abundant in some years but not in others. Rates of evolution in these laboratory microcosms are surely faster than we would find in the wild. Nevertheless, to dismiss these experiments as too artificial would be a mistake. The environments may be simple but the populations are real, and each individual has many genes that interact in complex ways. Despite all this genetic complexity, evolution proceeds generation after generation, continually shifting the genetic make-up of each population as natural selection favors some genetic combinations over others.

THE CHALLENGES AHEAD

Experimental microcosms have provided enormous insights into the genetic and ecological processes shaping evolution and coevolution. Their utility is increasing even more as it becomes possible to sequence many individuals within an experiment rather than just a few. We are at the point now where it is becoming possible to ask more precisely how many mutations occur within an experimental population in each generation. These studies are also beginning to allow assessments of how mutations are distributed throughout genomes, and how the distribution of mutations and the forms of selection change from one generation to another as selection continues to act on populations.

These studies will help us understand much more precisely how mutation, standing genetic variation, and natural selection continue to reshape the distributions of genotypes in evolving populations and coevolving species under different ecological conditions. Evolution in microcosms therefore helps us to understand the processes of both adaptation and diversification. That is important at a time when the relationship between adaptation and diversification is undergoing renewed scrutiny in evolutionary biology. It is to that relationship that we now turn.

Part 5

Diversification

14

Ecological Speciation

As we have come to appreciate the ongoing and sometimes rapid pace of adaptive evolution, we have had to reassess how new species form and the rates at which they form. If populations become locally adapted more quickly than we suspected a few decades ago, and if speciation is often a consequence of local adaptation and geographic mosaics, then we may have been underestimating the rates of speciation. Speciation may contribute more to the relentlessness of evolution than we previously thought. That is important for our understanding of the dynamics of evolutionary change, because the process of speciation gives us information about how adaptation is packaged. Speciation separates groups of locally adapted populations from other populations, which then influences the potential trajectories and dynamics of further adaptive evolution. Adaptation and speciation therefore must affect each other. If speciation is driven by natural selection, then the early stages of speciation should be visible on the same timescales as the process of adaptation (Hendry et al. 2007).

This chapter and the next consider why most speciation may be often an ongoing ecological process driven by geographic mosaics of local adaptation and coadaptation.

BACKGROUND: SPECIES AND SPECIATION

Speciation as an outcome of adaptation is an old view, going back to Darwin, but it is also a new one. The role of adaptation in speciation has had to be resurrected almost from first principles in recent decades, after a long period in which speciation was viewed primarily through the lens of genes and, specifically, genetic incompatibilities among populations. The resurgence of interest in the role of adaptation in speciation has brought with it an appreciation that the speciation process is probably much more dynamic than we have suspected until recently.

Species and *speciation* are notoriously difficult words to define in ways that apply equally well to all taxa. Every evolutionary biologist agrees that we need at least a general working definition of species for our discussions about how life is organized into genetically and ecologically distinct groups. Most evolutionary biologists, though, also agree that no single definition of a species can divide every diverging lineage in nature cleanly into a set of species that everyone would accept. That is not surprising. The impossibility of placing all of nature into clearly discrete species—and of placing all populations into either one species or another—is one of the most direct forms of evidence for evolution. Speciation is a continuous process, and the various definitions of species capture different stages of that process (Coyne and Orr 2004; Harrison 2010). We therefore should not be able to place every population cleanly into this species or that species. Estimates of the time it takes to form a fully self-sustaining species range from decades in some plants that have originated through hybrid speciation to tens of thousands or more than a million years in birds and mammals (Johnson and Cicero 2004; Weir and Schluter 2007; Price 2008; Abbott et al. 2009). Populations are commonly at intermediate stages of the speciation process.

Understanding species and speciation is made more complicated by the fact that there is no simple relationship between rates of adaptation and rates of speciation. Niles Eldredge and Stephen Jay Gould (1972) argued in their theory of punctuated equilibrium that species often diverge rapidly in morphology from their ancestors during speciation, but those species then look much the same for thousands or millions of years thereafter. That does not mean, though, that each species is not continuing to undergo adaptive evolution. Eldredge and Gould began with patterns found in the fossil record, which commonly show that the average sizes and shapes of species do not undergo continual directional change. Big changes sometimes happen quickly, followed by long periods in which no major directional changes seem to be happening at all, at least as observed within the resolution available in the fossil record.

The theory of punctuated equilibrium was immensely important as an impetus to the study of variation in the rates of net evolutionary change within and among species. The result has been studies showing every possible outcome over various timescales, from smooth divergence of traits to bursts of change followed by long periods of stasis (Elena et al. 1996; Eldredge et al. 2005; Toju and Sota 2009). At least for morphology, some lineages show patterns of divergence consistent with the arguments of punctuated equilibrium, whereas other lineages show evidence of more gradual divergence (Van Bocxlaer et al. 2008; Hunt 2010). Directional evolution can be gradual and sustained, but often it is not.

The theory of punctuated equilibrium does not argue that evolution always occurs in rare pulses, only that sustained directional evolution is often concentrated during speciation. Eldredge and Gould's arguments are important because they highlighted, through fossil evidence, that there is no reason to expect any simple relationship between rates of speciation and rates of adaptive evolutionary change within lineages. That decoupling has influenced our perceptions of the evolutionary process. It makes it clear that there is no inherent conflict between the often strong selection and rapid evolutionary rates found routinely by evolutionary ecologists studying populations year after year and the relatively unchanging morphologies of species found in the fossil record over millions of years (Eldredge et al. 2005). The highly dynamic evolution often shown in studies by evolutionary ecologists is crucial for keeping populations in the evolutionary game, but there is no reason to expect a general strong correlation between those rates and the rates of speciation, since much of adaptive evolution is nondirectional, driven by frequency dependence and other nonlinear forms of selection.

How we consider the relationship between the dynamics of adaptation and the process of speciation depends on how we view species. The classic definition of species as groups of populations reproductively isolated from other groups works well for sexual species that come into contact somewhere in their geographic ranges or can be mated in the laboratory. That captures only a relatively small number of species. Many recognized species are allospecies, which differ in phenotypic or molecular traits but are so geographically separated that they never encounter each other. When closely related species come into contact, hybridization at least sometimes occurs between populations that just about everyone would accept as separate species. Hence, reproductive isolation between populations is a state that develops to various degrees, sometimes slowly and sometimes quickly, among sexually reproducing populations as they become adapted to different ecological conditions and occupy different geographic regions.

In his masterful synthesis on speciation in birds, Trevor Price (2008) deals with the problem in a practical way by writing that he would use a definition of species based on reproductive isolation where it could be applied, but he would allow also for allospecies when populations are diagnostically different and show about the same degree of difference as found in sympatric pairs of species. The problem is even more difficult in other taxa such as plants, invertebrates, and microbial taxa that include asexual lineages in which the criterion of reproductive isolation cannot be applied.

Most current uses of the word *species* in evolutionary biology are based on the general view that species are groups of self-sustaining populations with traits, ecological attributes, and evolutionary trajectories that differ predictably from other groups, and that mixing the defining traits between these different groups often results in lower fitness in those individuals in the environments in which they live. Species are clouds of individuals whose fitnesses in various environments are distributed differently from other species. The distributions of the traits of each species generate the evolutionary dynamics found in nature. Understanding species and speciation therefore requires an understanding of how fitness among individuals is distributed within and among populations, and some definitions of species are based on fitness distributions (Wu 2001; Hausdorf 2011). By necessity, then, speciation research is increasingly becoming as much an ecological science as a genetic and molecular science.

A major meta-analysis of evolutionary change in the body size of mammals, birds, snakes, and lizards has suggested that divergence among populations is almost always highly restricted for up to a million years and then generally shows rapid divergence at longer timescales, creating a blunderbuss pattern of divergence (Uyeda et al. 2011). That pattern adds to the view that much generation-to-generation selection and evolution in nature often involves fluctuating changes over time rather than directional change. The possibility that short-term change is often not sustained in the long term is not surprising, but why divergence among populations of some taxa should increase so much after a million years is not intuitive and requires a better mechanistic understanding of the processes involved.

FROM GENETIC INCOMPATIBILITY TO ECOLOGICAL SPECIATION

Despite Darwin's clear argument that natural selection drove speciation as well as adaptation, it took more than a hundred years before rigorous studies began on exactly how adaptation of populations to different environments could drive the diversification of life. For some time following publication of *The Origin of Species*, the question seemed answered. The longer two populations are separated, the greater the number of differences they accumulate, as natural selection acts differently on different populations. Eventually, divergent populations would look sufficiently different to be regarded as varieties and, after more time, species. Speciation was about divergence over time as populations became adapted to different environments.

In the 1930s and 1940s, though, Theodosius Dobzhansky (1937) and Ernst Mayr (1942) argued that the major problem to solve in speciation was the evolution of reproductive isolation. Most species are sexual, and it was unclear whether the genes that favor local adaptation are the same as the genes that lead to reproductive isolation between populations. Research on speciation became focused on understanding the role of genetic incompatibilities. As Dolph Schluter (2009) has noted, this emphasis led to a long period in which there was almost no work on the role of adaptation in speciation.

In the genetic incompatibility model of speciation, populations diverge genetically due to natural selection or neutral genetic differences when they are isolated from each other. If the populations come back into contact, any hybrids between them are likely to suffer reduced viability or fertility due to epistatic effects among two or more loci. The reason is that the gene combinations found in hybrids will not have been subject to natural selection. Imagine an ancestor with two loci *a* and *b*, each with two alleles, *aabb*. If a new allele A evolves in one population and B evolves in another population, then the combination *AAbb* will occur in one population and *aaBB* in the other. *AaBb* hybrids between the populations would be combinations untested by natural selection and could therefore result in low fitness in these hybrids. This hypothesis, first proposed by Dobzhansky (1937) and Muller (1942), argues that speciation is more about gene interactions than about the genes themselves. It is inherently a genetic model of speciation that does not rely on ecological adaptation.

By this model of speciation, sterility or inviability increases faster than linearly as populations diverge genetically. The greater the number of genetic differences among populations, the disproportionately greater the degree or likelihood of reproductive incompatibility between a pair of populations (Orr 1995). This snowball effect on incompatibility, as genetic differences accumulate, has been difficult to test, but experimental studies in species as different as fruit flies and tomatoes have provided some strong evidence for it (Matute et al. 2010; Moyle and Nakazato 2010). Some genomic studies have suggested that these incompatibilities may be further augmented by molecular evolution among selfish genetic elements and the genes that evolve to suppress these parasitic genetic elements (Johnson 2010). The Dobzhansky-Muller model, the snowball effect, and evolving intragenomic interactions may all contribute to the process of speciation in situations in which speciation is driven by allelic substitutions (Presgraves 2011). It is, though, incomplete as a theory of speciation, because it fails to incorporate and articulate a clear role for the ecological processes that underlie selection on the genetic incompatibilities.

The rise of the field evolutionary ecology in the 1950s and 1960s brought a renewed focus on the role of adaptation in speciation. An expanding set of studies of local adaptation increased interest in the genetic basis of adaptive change. Those studies, in turn, increased interest in how adaptive changes may differ among populations, generating local adaptation and further genetic divergence of populations. These studies, then, fostered interest in how adaptation could drive speciation. In recent years, some major evolutionary biologists have been willing to argue that, although not all species arise through natural selection, most do (e.g., Schluter 2009; Via 2009). Price (2008) has argued that ecological factors affect the rate of speciation at every stage of the process. Ecological speciation is becoming the current paradigm.

THE FEATURES OF ECOLOGICAL SPECIATION

Ecological speciation is the origin of reproductive isolation between populations driven by divergent natural selection on populations in different environments (Schluter 2000). Under this view of speciation, the incompatible gene interactions among populations envisioned by Dobzhansky and Muller are ecologically incompatible combinations. These are often called environmentally based genetic incompatibilities, which is a mouthful. I shorten it here to ecological incompatibilities and contrast it with inherent genetic incompatibilities. Ecological incompatibilities are genetic combinations that result in lower fitness in a particular ecological setting, and inherent genetic incompatibilities are genetic combinations unfit in every environment.

The hypothesis of ecological speciation predicts that reproductive isolation will occur between populations if they adapt to different environments but not if they adapt to similar environments (Schluter 2009). Speciation occurs between populations because hybrids have gene combinations that are ecologically incompatible with either parental environment. As natural selection favors individuals better adapted than others to the local environment, it also reduces the chance that local individuals will encounter and mate with those from other populations (Schluter and Conte 2009). As selection continues, it may also act against individuals that move to the other environment, because those individuals will be less likely to survive than resident individuals (known as *immigrant inviability*). The genetics of immigrant inviability are therefore the same as the genetics of local adaptation. Because local adaptation can occur over a surprisingly small number of generations (see chaps. 1 and 10), this early stage of speciation probably occurs often in many species.

Ecological speciation is therefore population divergence driven by poor adaptation of hybrids to either parental environment. The critical tests of

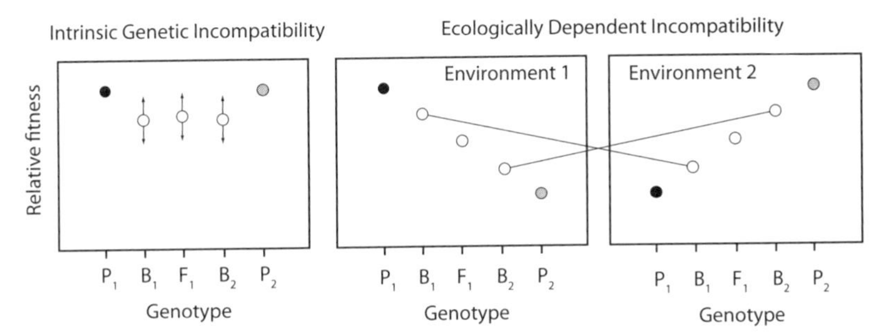

Fig. 14.1 Predictions for tests of speciation driven either by intrinsic genetic incompatibilities that are independent of the environment in which the genes are expressed or by ecological incompatibilities that depend on the environment in which individuals live. If the relative fitness of individuals depends on only the gene combinations themselves, then all hybrid genotypes will have lower fitness than parental genotypes. If the fitness of gene combinations depends on the environmental conditions in which those genes are expressed, then hybrids will show a graded series of fitnesses that depends on the proportion of genes suitable for that environment and that occur within an individual. The crucial generations are the backcrosses, which show the environmental dependence of gene combinations and are shown here as crossed lines. After Rundle (2002), based on arguments in Rundle and Whitlock (2001) and Rundle (2002).

ecological speciation are those in which multiple generations of two divergent populations and their hybrids can be evaluated in nature (fig. 14.1). It is insufficient to demonstrate that first-generation (F_1) or second-generation (F_2) hybrids have lower fitness than parental populations, because lower fitness could be due to inherent genetic incompatibilities rather than ecological incompatibilities. A full test requires creation of backcross populations (F_1 individuals crossed to the parental species) followed by tests of the performance of those individuals in the parental environments (Rundle and Whitlock 2001). Each type of backcross should show higher fitness in the environment of the parental type with which it shares the most genes.

This rigorous test has been applied so far only to a few species, but ecological incompatibilities have now been shown for taxa as different as threespine stickleback fish (Rundle 2002) and leaf beetles adapted to different host plant species (Egan and Funk 2009). The tests on leaf beetles have been on *Neochlamisus bebbianae*, which is a tree-feeding species with populations adapted to maples or willows. The backcross test shows that the fitness of hybrid individuals is environmentally dependent, favoring individuals on maple that have mostly maple-adapted genes and individuals on willow that have willow-adapted genes (fig. 14.2).

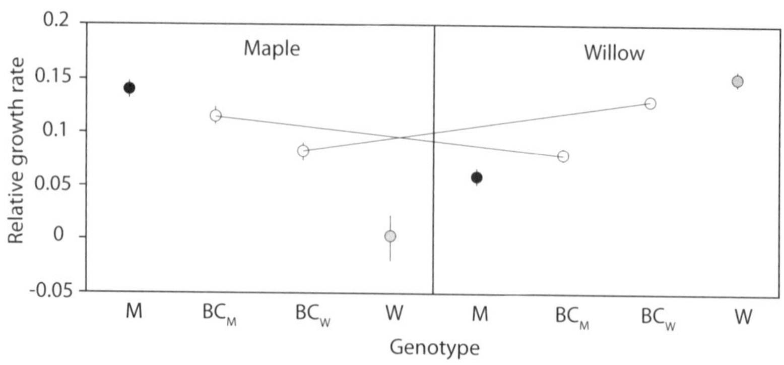

Fig. 14.2 Test of the environmental dependence of gene combinations of *Neochlamisus bebbianae* beetles that normally feed on maple (M) or willow (W) in eastern North America. The BC_M backcrosses are F1 hybrids backcrossed to maple, and the BC_W backcrosses are F1 hybrids backcrossed to willow. The crossed dotted lines between the backcrossed larvae grown on different plants indicate that the fitness of the gene combinations depends on the environment in which the individuals develop rather than on intrinsic genetic incompatibilities. The index of fitness used here is the relative growth rate of larvae, which takes into account initial mass and nonlinear growth rates. Values are means (±1 SE). After Egan and Funk (2009).

There is nothing in the theory of ecological speciation that requires the evolution of new mutations for the process to begin. Rather, speciation may often occur because natural selection acts on genetic variation already present in populations rather than on new mutations (Barrett and Schluter 2008; Schluter and Conte 2009). Under these circumstances, speciation proceeds by partitioning the standing genetic variation into populations adapted to different environments. Few studies, though, have evaluated the early stages of speciation in enough genetic detail to assess how often standing genetic variation rather than new mutations drives speciation (Schluter and Conte 2009).

Four results from the earlier chapters suggest that standing genetic variation is often important during the early stages of speciation: species harbor much more ecologically important genetic variation than previously suspected; the relative frequencies of genes and gene combinations are often distributed nonrandomly among populations; natural selection often acts in ways that maintain much genetic variation within and among populations; and genes are often expressed in different ways in different environments. Moreover, the standing variation found within species as they adapt to different physical environments can be amplified by selection mosaics, as interactions between species evolve through genotype-by-genotype-by-environment interactions and favor population divergence (Thompson 1994, 2005). More generally, variation in gene expression in all its different forms is likely to contribute to population divergence (West-Eberhard 2005; Nylin and Janz 2009; Pavey et al. 2010).

Most incipient species that arise through selection on divergent gene expression will not get beyond modest divergence from other populations. These early stages of speciation appear repeatedly among populations within any widely distributed species, but only a few lead to speciation. They are, though, part of the ongoing process of evolution in all species.

ECOLOGICAL SPECIATION AND GENE FLOW

In addition to the ecological incompatibility of hybrids, ecological speciation requires reduced gene flow. That part of speciation has long been clear. Populations adapted to different environments often differ in their geographic ranges and therefore rarely encounter each other, which is probably the most common starting point of speciation (Sobel et al. 2010). If most divergence involves relatively small peripheral populations that have budded off from other populations, then speciation may occur quickly in the isolated populations. If it involves large populations that become subdivided into separate but still large populations, then the process may be slower. In their broad synthesis of speciation theory and results, Jerry Coyne and Allen Orr (2004) considered speciation in completely allopatric species so ecologically and genetically simple a process that it hardly seems worth documenting.

In some lineages, speciation may be driven not by populations moving into new environments but rather by adaptations that subdivide the geographic range of a lineage in novel ways that happen to minimize gene flow. Macroevolutionary studies have shown that the geographic range limits within lineages have a strong phylogenetic signal up through the levels of genera and families (Roy et al. 2009). Hence, the geographic ranges that develop during speciation could often result from selection that partitions a species with a broad geographic distribution into two or more species with narrow distributions. The overall geographic range of the lineage may sometimes already be set earlier in the adaptive history of that lineage.

There is, though, nothing inherently binding about geographic range limits, if new ecological opportunities arise. The potential to colonize completely new regions has been shown time and again by the spread of thousands of species in recent centuries after their introduction onto new continents. Over longer periods of time, that flexibility is also apparent in the spread of taxa during the great intercontinental interchanges, such as occurred between North and South America following the rise of the Isthmus of Panama (Weir et al. 2009). These events are uncommon only because, historically, the opportunities for such long-distance movement have been few for many taxa.

The harder problem to solve is how populations might diverge when their range limits overlap rather than remain completely separate and some gene flow therefore persists. As Richard Harrison (2010) has noted, this problem has been the major recent focus of research on ecological speciation. When can speciation occur when divergent natural selection is initially the only major barrier to gene exchange? Harrison calls this process narrow-sense ecological speciation.

The long-standing view is that narrow-sense ecological speciation is rare in most vertebrates but broader-sense ecological speciation, involving divergent adaptation among populations with little to no gene flow, is common. Studies of some vertebrate taxa suggest that almost all speciation is geographic speciation. Initial divergence of populations occurs when matings cannot occur between them (Phillimore et al. 2008; Price 2008). For some taxa, though, such as fish and insects and plants, the recent focus has been on studies of ecological speciation that occur in the presence of some gene flow, and some of these studies have found evidence for divergence amid gene flow (Rieseberg et al. 2003; Ferrari et al. 2008; Via 2009; Harrison 2010; Schluter et al. 2010).

COMPETITION AND PREDATION

Competition in its various guises remains one of the favorite contenders among the major drivers of ecological speciation in large, mobile species. When competition drives population divergence, its signature should be clear: where diverging populations come into contact, traits with the greatest effect on competition between the populations will diverge more than others. Character displacement will occur within the hybrid zone but not elsewhere (Pfennig and Pfennig 2010). Either one or both populations will diverge in their traits in the region of sympatry. Divergence could be driven through multiple competitive mechanisms, including selection driven by competition for limiting food resources or through selection to escape from shared predators.

Among the most insightful experimental studies into the ecological mechanisms of speciation in vertebrates are those led by Dolph Schluter and colleagues on the divergence of threespine sticklebacks (*Gasterosteus aculeatus*). These fish have been undergoing speciation in coastal lakes in British Columbia over at least the past 12,000 years (e.g., Schluter 1994; Hatfield and Schluter 1999; Vamosi and Schluter 2004; Vines and Schluter 2006). Threespine sticklebacks are common in nearshore saltwater environments. At the end of the last ice age, some populations colonized nearby freshwater habitats and became trapped in lakes as the glaciers receded and the coastal land rebounded. The lake populations are subject to lower predation by predatory birds and fish, higher predation by invertebrates, greater availability of littoral habitat, and greater availability of invertebrate prey than ocean populations. Selection in these freshwater habitats has repeatedly favored reduced body armor, larger heads, greater body depth, reduced medial fins, and more downward-pointed jaws compared with their marine populations (Schluter et al. 2010).

As the glaciers receded and opportunities for colonization appeared and disappeared in various lakes, a second wave of stickleback colonizers occurred in some lakes. The colonists and the resident populations diverged into a benthic form and a limnetic form as selection favored divergence in traits (McPhail 1992; Schluter 1996). Sympatric populations of benthic and limnetic forms are known from five lakes, and each sympatric pair appears to have resulted from a separate sequential colonization followed by character displacement (Gow et al. 2008). That is quite a bit of replicated divergence among populations over a fairly short period of time.

The ancestral population in each lake is thought to have resembled the pelagic marine form, which is similar to the limnetic form found in the lakes. Limnetic fish have a slender body, extensive body armor, and many gill rakers (fig. 14.3). These fish feed on zooplankton and in the open water. In contrast, the benthic form is less slender, has reduced body armor, and fewer gill rakers. These fish feed on large invertebrates in the littoral zone. The forms now tend to form reproductively separate populations within lakes (Boughman et al. 2005; Vines and Schluter 2006).

Other studies have shown that these two forms are morphologically, genetically, and ecologically distinct in ways consistent with divergence from a common ancestor, followed by character displacement in the lakes where they have come into contact (Schluter and Conte 2009; Schluter et al. 2010). The sequence of adaptation to freshwater environments and subsequent divergence into different forms within freshwater has been driven by a combination of prey availability, predation, and competition. A move to a new physical environment set the stage for divergence, but selection imposed by

Fig. 14.3 Limnetic (*top*) and benthic (*bottom*) forms of threespine sticklebacks (*Gasterosteus aculeatus*) from Texada Island, British Columbia. The limnetic fish is a male from Paxton Lake, and the benthic fish is a male from Vananda Creek. Photographs copyright Gerrit J. Velema (www.gjvphoto.com), used with permission.

interspecific interactions drove the divergence. This sequence of divergence has occurred in multiple populations over thousands rather than millions of years.

THE ECOLOGICAL DRIVERS OF POPULATION DIVERGENCE

Identifying the major ecological drivers of speciation in threespine sticklebacks has required decades of observational and experimental studies on multiple populations. The problem is not that it is difficult to find ecological causes of divergent selection in natural populations. Rather, the problem is that it is difficult to study all the potentially important causes at the same time to sort out which are the most important. Any researcher studying speciation makes choices about what to include and exclude in any particular study. Some studies focus on how populations diverge during adaptation to different physical environments. Other studies focus on how they may diverge due to environmentally dependent sexual selection. Still others consider how populations can be driven apart through divergent interactions with other species.

These a priori decisions about what to study often differ among taxonomic groups. Studies of vertebrates have long probed the importance of competition in driving divergence among populations, and these studies have shown convincingly that competition can favor divergence of populations (Price 2008). Studies of insects have emphasized the importance of selection imposed by host plants, parasites, or predators in favoring divergent specialization among populations (Price et al. 1980; Murphy 2004; Nosil 2007; Ferrari et al. 2008; Wiklund and Friberg 2008). Studies of plants have varied in emphasis among researchers, leading some to suggest important roles for selection mediated by physical environments (e.g., soils, fire) and others to suggest important roles for competition, herbivores, or mutualistic interactions with pollinators (Kay and Sargent 2009; Givnish 2010; Schemske 2010; Anacker et al. 2011; Schnitzler et al. 2011).

The choices about what to study for each taxon are not random. They are often based on a deep natural history knowledge of different groups and the selective forces that researchers think may be important for any particular group. For example, studies of speciation in insects have often focused on selection imposed directly by food, predators, or parasites rather than competition, because there is a long history of studies showing that selection favors local adaptation of insect populations onto different resources (e.g., different plant species). Specialization allows a population to adapt better to that one resource, and it may also allow individuals to escape parasites or predators that may search more often in one environment than in others. Consistent with that view, a major survey of the pattern of attack by parasitic wasps and flies on forest moths and butterflies found that parasitism rates on larvae differ greatly among tree species (Lill et al. 2002). Related studies of temperate and tropical insects have suggested the diversification in tree-feeding insect species is associated with diversification of plant chemistry within and among plant species, which demands specialization by plant-feeding insects (Dyer et al. 2007; Becerra et al. 2009).

By studying different potential drivers of ecological speciation for different taxa, we have ended up with different kinds of ecological data for different groups. Complicating our interpretations are divergent views of just what constitutes an ecological interaction that is likely to result in speciation. Trevor Price (2008) has argued that coevolution with "resources" may be a less frequent cause of shifts to new adaptive peaks during population divergence than are some other ecological pressures such as competition. We have, though, few studies designed to ask when, in species such as birds, speciation is driven by coevolution with a prey or host species. Among the studies that have asked that question most directly—those involving interactions between crossbills and conifers (Benkman 2010)—speciation driven by selection for specialization on resources has been shown to be important. In that case, coevolution between crossbills and conifers drives divergent selection and reproductive isolation in the crossbills.

Crossbills and conifers may be a special extreme case, but there remain many unanswered questions about which kinds of interaction are most important in speciation. Some evolutionary biologists have argued that when trophic interactions are the major agents of selection, the pressures from above are greater than those from below. The "life–dinner principle" of Richard Dawkins and John Krebs (1979) and the process of escalation suggested by Geerat Vermeij (1987) suggest that predators impose greater selection on prey than do prey on their predators. If so, then the processes driving ecological speciation could differ among trophic levels. There is, though, no reason to suspect that any such asymmetry applies universally across taxa and ecosystems.

Hence, although the role of ecological incompatibilities seems to have now been solidly established as a major cause of speciation, there remains much to do. Sorting out which ecological drivers are most important in initiating divergence among populations and which are most important in furthering that divergence is among the current priorities. It could be that the ecological drivers do indeed differ among taxa and that different selective forces are important at different stages of speciation. What we can say for now is that accumulating evidence indicates that multiple kinds of ecological incompatibility are important in the speciation process.

ASSORTATIVE MATING AND ECOLOGICAL SPECIATION

As selection favors local adaptation, it could immediately or eventually also favor individuals that tend to mate only with other individuals from their local population. Selection may favor individuals that mate nonrandomly with individuals possessing similar traits. This assortative mating, as it is often called, is what has occurred in the sticklebacks. The two forms show a strong tendency to mate with others of the same morphological type (Vines and Schluter 2006), as individuals choose mates based on body size and nuptial color (Boughman et al. 2005). If, instead or in addition, individuals mate in the habitats to which they are best adapted, then assortative mating becomes a direct outcome of local adaptation.

Multiple examples of ecologically driven assortative mating have been found among plant-feeding insects that search for mates only on the plant species on which they lay their eggs (Via and Hawthorne 2002; Forbes et al. 2009; Craig and Itami 2010). In these cases, selection imposed primarily by host plants, parasites, or predators, rather than by competitors, is often the presumed driver of selection. Divergence has sometimes been rapid, and the major ecological cause can be something as simple as a difference in when the plants have sites available for mating and egg laying.

Rhagoletis fruit flies in North America are specialized to feed on fruits of different plant species that produce fruits at different times of the year. In *R. pomonella*, which feeds on native hawthorns, one or more populations became adapted to introduced domesticated apple plants in the 1800s. Apples often produce fruit three to four weeks earlier than do hawthorns in places where they occur together. That difference favors divergence in the annual dates of eclosion of adult flies from overwintering pupae, which further decreases the chance of mating between flies on these two plant species. These populations on hawthorn and apple now coexist essentially as separate species (Bush 1969; Feder et al. 2010). Hawthorn-adapted flies search for mates on hawthorn, and apple-adapted flies search for mates on apples.

Overall, hawthorn-feeding and apple-feeding populations have diverged as selection has acted on multiple ecologically important traits, including the length of time that the moths remain as overwintering pupae (i.e., diapause), and the interaction between the genes governing diapause and the pre-winter temperatures, which influences pupal energy reserves and choice of mating sites (Feder et al. 2010). These traits, though, are linked directly or indirectly to differences among plant species, such as when during the growing season they produce fruits. The complex interactions between genes and environments found in this fly species illustrate why phenological divergence of populations may be more likely to occur in some environments than in others throughout the range of a species. The ecological conditions are as important as the genes themselves in setting up the discontinuity between the populations.

ECOLOGICAL SPECIATION, SEXUAL SELECTION, AND SOCIAL SELECTION

In cases such as apple-feeding flies and hawthorn-feeding flies, it is easy to see how assortative mating and population divergence go hand in hand. Individuals choose mates on their host plants and population divergence follows. The links between assortative mating and population divergence become harder to disentangle when sexual selection seems divorced from ecological selection. Darwin was careful to separate natural selection from sexual selection, with natural selection shaping the adaptation of populations to their environments and sexual selection shaping competition for mates and choice of mates.

In the century and a half of research since Darwin's work, those distinctions remain useful as idealized differences, but we now know that these forms of selection influence each other. Sexual selection can favor brightly colored displays in local populations not under strong selection by predators to be cryptic, but it sometimes favors cryptic coloration or behaviors in regions in which selection by predators is strong (Endler 1980; Kemp et al. 2009). Similarly, sexual selection and parasite-mediated selection could also interact. Decades ago, William Hamilton (1980) suggested that the bright colors and vigorous courting displays sometimes favored by sexual selection may be signals that individuals can use to assess the degree to which potential mates are relatively free of parasites. Selection favoring ecological specialization also could affect mating preferences within populations, thereby affecting sexual selection (Hoskin and Higgie 2010). Ultimately, we would like to know how ecological selection and sexual selection act together to foster evolutionary change toward speciation.

Since differences in sexual selection are often evident once populations have already diverged ecologically, we want to know if divergence driven by sexual selection and ecological selection tends to occur simultaneously or if divergence in sexual selection starts later. Answering that question requires either being lucky enough to be able to track divergence right from its earliest stages or finding a set of populations under sexual selection that differ in other aspects of their evolutionary history. These latter studies rely on comparing populations that are all under sexual selection but differ in presence of a major ecological selection pressure that may also drive divergence in some of these populations but not others.

In the Bahamas mosquitofish (*Gambusia hubbsi*), for example, some populations occur in small lakes with predatory fish and others occur in lakes without these predators (Langerhans et al. 2007). Many of these ponds, called blue holes, sit atop now-submerged cave entrances. These freshwater ponds developed as rising seawater lifted freshwater to the surface about 15,000 years ago, flooding these voids. The blue holes have similar physical environments, and the major difference between them is the presence of predatory bigmouth sleepers (*Gobiomorus dormitor*). Mosquitofish from lakes with predators differ in shape from those in lakes without fish. The morphological differences among populations correlate with the presence or absence of predation but not with the pattern of mitochondrial genetic relatedness among populations, suggesting that the differences are due to ecological selection driven by the predators. Sexual selection follows from those morphological differences. Individuals choose mates based on their body shape. Assortative mating is twice as strong between populations with different-predator regimes than between populations with a same-predator regime. In this case, then, sexual selection reinforces the ecological divergence driven by predators (Langerhans et al. 2007).

Sexual selection acting through assortative mating affects so directly the distribution of genes within populations that it must often be involved in some stages of population divergence in many species. The results so far, though, suggest that sexual selection is more likely to augment ecological divergence rather than to drive the initial stages of divergence in at least some lineages (Ritchie 2007). Taxa with highly exaggerated characters do not seem to have more species than other taxa, but formal analyses have been completed on only a few major groups. In birds, lineages in which males and females differ strongly in color are not richer in species than other lineages, and ecological divergence is a better predictor of avian diversification (Phillimore et al. 2006; Kemp et al. 2009).

The results are more equivocal for other lineages. In cichlid fishes, which have been one of the model systems for trying to disentangle the roles of natural and sexual selection in speciation, sexual selection appears to have been an important driving force in diversification in some cichlid fish lineages (Wilson et al. 2000) but not in others (Day et al. 2008; Elmer et al. 2010). In some cases these processes may be too intertwined to completely disentangle (Salzburger 2009). Among insect lineages, *Enallagma* damselflies show evidence of diversification through a combination of ecological and sexual selection (McPeek and Brown 2000; McPeek 2008).

The relationships between ecological and sexual selection begin to blend even more when considering other aspects of social behavior that could affect speciation. Competition for nest sites and defense of food resources by social groups are among the aspects of social behavior that do not fit comfortably within the theory of sexual selection (West-Eberhard 1983; Price 2008). Highly ornamented adult males can be explained by competition for mates among males and by choice of mates by females, but characteristics such as brightly colored nestlings in birds cannot be explained directly by sexual selection theory. Rather, traits such as the bright orange colors of juveniles of avian species such as coots are more readily interpreted as a broader form of social selection involving competition among juveniles and female choice, as has been shown by manipulating the plumages of juveniles (Lyon et al. 1994). Other social characteristics such as intraspecific brood parasitism (Lyon and Eadie 2004), or bright plumages of females as well as males, are used in nonmating social signaling among conspecifics

(Filardi and Smith 2008). As Mary Jane West-Eberhard (1983) first noted, multiple forms of social selection could affect speciation.

In general, social selection appears to involve competition within species for resources rather than mates, and parental choice among offspring. Different environments are likely to favor different socially selected traits (Price 2008), which could amplify any ecologically based divergence among populations. Studies of speciation may therefore need to separate not only natural selection from sexual selection but also sexual selection from other forms of social selection.

MUTATION-ORDER SELECTION AND SPECIATION

Natural selection could also act directly on reproductive isolation in more subtle ways such as the order at which mutations occur in different populations. According to the hypothesis of mutation-order selection, reproductive isolation develops as populations become fixed for different mutations under similar selection pressures (Mani and Clarke 1990). Unlike the process of ecological speciation, the same alleles could be favored in each population, but the populations diverge because they differ in the order in which favorable mutations have become fixed.

Strong divergence through mutation-order selection requires low levels of gene flow among populations, because gene flow will allow favorable mutations to spread to other populations. Some models have suggested that mutation-order speciation is most likely to occur when some combination of several conditions occurs: low gene flow is accompanied by incompatible mutations that have similar beneficial effects; less fit mutations arise slightly earlier than more fit alternatives; or initial divergence occurs during a period in which the populations are allopatric (Nosil and Flaxman 2011). In the absence of those conditions, it could be difficult for selection to fix different alleles in different populations.

Cytoplasmic male sterility and other forms of intragenomic conflict (see chap. 6) are among the possible ways in which mutation-order selection could lead to speciation (Coyne and Orr 2004; Schluter 2009). In these situations a mutation could arise that allows the mutant to spread through a population by overrepresenting itself in gametes, even though it reduces the overall fitness of the individuals that carry it. In cases of incompatibility caused by symbionts, that would arise if selective death of male offspring occurs during crosses in which mating pairs do not have the same symbiont strains (Engelstädter and Hurst 2009). Similar incompatibilities also occur when only one population harbors the symbiont. Within a population, biased transmission of gametes favors other mutations that counter the selfish mutation and restore more equal genetic production of gametes. Populations, however, are likely to differ in their distorter and restorer genes, which would promote speciation. Incompatibilities caused by symbionts are therefore a special form of mutation-order speciation.

Ecological speciation and mutation-order speciation can become intermingled in situations in which speciation is driven by symbionts. The symbionts could simultaneously affect reproductive isolation among populations and adaptation to local environments. The possibility that speciation is sometimes driven by coevolution with symbionts seems increasingly likely, but, despite tremendous advances in recent years, there is much to learn about the ecological and genetic dynamics of these interactions and their roles in adaptation and divergence of host populations. We are probably only scratching the surface in understanding the complexity of these interactions, and the ongoing evolutionary dynamics as hosts and symbionts continue to adapt to each other. Bacteria in the genus *Wolbachia* were discovered just several decades ago. Since then, surveys have suggested that they occur in the majority of insect species and in a wide range of nematodes (Hilgenboecker et al. 2008). *Wolbachia* and some similar symbionts often cause partial reproductive isolation between host populations that normally have these bacteria and those that either lack the bacteria or have different strains of the bacteria. This partial reproductive isolation could serve as the initial stage of symbiont-induced speciation (Werren 1997).

Speciation through symbionts might be especially likely in situations in which the symbiont and host populations rapidly evolve in their ecological relationship, as has been observed in *Wolbachia*-infected *Drosophila simulans* in California (Weeks et al. 2007). Speciation could also occur in situations in which infection immediately changes multiple aspects of the life history of its hosts, as has been observed in the spread of a *Rickettsia* symbiont through populations of sweet potato whiteflies, *Bemisia tabaci* (Himler et al. 2011). In both cases, the symbiont has become, within some populations, a mutualist as well as a reproductive manipulator of its host. It has long been suggested that subsequent divergent coevolution involving symbionts could drive speciation among host populations. Through coevolution, host populations would rapidly favor genes that reduce the negative effects of their symbionts and magnify any positive effects (Thompson 1987; Wade and Goodnight 2006; Moran et al. 2008). Only recently, however, have studies begun to analyze how partial reproductive isolation caused by these symbionts could drive speciation among populations during the early stages of population divergence (Miller et al. 2010). It is likely that symbionts contribute to speciation in more ways than we have appreciated so far.

Ecological speciation and mutation-order speciation may also become intertwined during speciation between competitors. Speciation driven by competition could depend on the specific order in which adaptive mutations appear when competing populations come into contact. Even hybridization itself within the contact zone could contribute to the speciation process, as genes move from one species into the other and come under selection within their new genomes (Grant and Grant 2008b). Closely related species sometimes form complex mosaic hybrid zones in which populations overlap patchily across a region (Mallet et al. 2009; Harrison 2010). Which species is competitively ahead in different parts of a hybrid zone could depend on the order in which new mutants arose or spread into that part of the hybrid zone.

THE CHALLENGES AHEAD

Even with the renewed focus on ecological speciation in recent years, we are probably still underestimating how often divergent selection leads to speciation. Speciation probably contributes more to the dynamics of ongoing evolution than we expected until recently. We have come to realize that many species are barely distinguishable, if at all, to the human eye. Potentially cryptic species are being identified in taxon after taxon, ranging from amoeba (Douglas et al. 2011) to birds (Benkman et al. 2009). The number of microbial species in the world is anyone's guess (Piganeau et al. 2011; Zinger et al. 2011).

The estimated number of species continues to increase in all groups. Estimates of the number of bird species, which have been studied in more detail than any other group, have expanded from Ernst Mayr's (1946) estimate of 8,616 to more than 10,000. Some studies of plants have suggested that we may have underestimated plant speciation through failure to appreciate the importance of cryptic polyploid species (Soltis et al. 2007). Discovery of cryptic insect species is expanding fastest of all, lengthening the lead of insects in the estimated number of species worldwide (Hebert et al. 2004; Condon et al. 2008; Thompson 2008a; Janzen et al. 2009; Griffiths et al. 2011). Even within Europe, where some groups such as butterflies have been studied carefully for centuries, new cryptic species are being discovered that differ in ecologically important ways. The well-known common wood white butterfly *Leptidea sinapis* was shown two decades ago to be at least two species and more recently has been shown to be three genetically distinct species (Dincă et al. 2011). Detailed studies of two of these species have shown that they differ in the habitats in which they occur and in their life history traits, number of generations per year, and use of plant species (Friberg et al. 2008; Friberg and Wiklund 2010).

The many studies showing partial reproductive isolation among populations add to our increasing realization that the interplay between adaptation and speciation contributes more to the dynamics of evolution than we previously thought. Populations are continually diverging, then anastomosing, then diverging again. Those changes contribute to the many processes that drive continual adaptive change. One of the great next challenges, then, is to understand how natural selection juggles adaptation to physical environments, interactions with other species, sexual selection and, more generally, social selection at different stages of speciation.

Even the process of divergence itself can present new evolutionary possibilities, as divergent genomes hybridize and sometimes generate new adaptive, rather than maladaptive, genetic combinations. It is to these evolutionary opportunities, and others driven by major genomic change, that we now turn.

15

Reticulate Diversification

Speciation probably always proceeds at multiple rates, fast during some stages and slow during others. We are still trying to understand when punctuated change rather than smooth and directional evolutionary change accompanies different stages of speciation. The studies of the evolutionary ecology of speciation discussed in the previous chapter suggest that significant changes in morphology, behavior, physiology, and life history can progress rapidly during the early stages of speciation. Speciation is also sometimes accompanied by increased rates of molecular evolution (Venditti and Pagel 2010), although when during the speciation process the rates are particularly high is not yet known.

In some cases, rapid molecular changes may occur during the early stages of speciation through genomic processes such as hybridization between populations, sometimes accompanied by polyploidy (whole genome duplication). These events have the potential to produce, almost instantaneously, new species with novel adaptations and evolutionary trajectories. As genomic studies proceed, it is becoming clear that these are important processes contributing to the relentlessness of evolution.

In this chapter I explore the connections among hybridization, polyploidy, speciation, and adaptive radiation. We are learning that these processes may contribute not only to the diversification of lineages, but also to pulsed patterns of divergence over relatively short timescales.

BACKGROUND: SELF-SUSTAINING HYBRID POPULATIONS

Hybridization immediately increases genetic variation within a population, and that increased variation may sometimes disrupt local adaptation, including locally adapted interactions with other species (Dybdahl et al. 2008). Consequently, in most cases, hybrids do not form self-perpetuating populations. They rely for their persistence on the nearby parental species, which generate new hybrids each generation.

Occasionally, though, hybrid populations occur in environments that happen to be favorable to survival and reproduction. The environment may be favorable to the hybrid either in its physical characteristics or in interactions that the hybrids establish with other species (Whitham et al. 1999, 2006). For example, hybrid populations of *Daphnia* water fleas experience higher rates of protozoan infection in water at lower temperatures than at higher temperatures (Schoebel et al. 2011). When a hybrid population forms in a favorable environment, it has the potential to diverge, sometimes quickly, into a self-perpetuating species.

There are some tantalizing indications that hybridization and coevolution may work in concert during the divergence of some taxa. Hybridization and introgression of genes between populations have been detected in multiple coevolved interactions, such as those involving yuccas and their pollinating moths (Segraves et al. 2005), fescue grasses and their endophytic fungi (Wäli et al. 2007), and orchids and their pollinators and mycorrhizal fungi (Schatz et al. 2010). Even so, it is still unclear how hybridization and coevolution may together shape diversification in some lineages.

Hybridization can also affect speciation by producing asexual lineages that subsequently undergo their own separate evolutionary histories as separate species. The coevolving interaction between wild flax (*Linum marginale*) and flax rust (*Melampsora lini*) in Australia, two species that have provided insights into many aspects of the coevolutionary process (see chap. 3), provides some clues to this process. The rust is composed of two geographically overlapping lineages, one of which has a hybrid origin (Barrett et al. 2007). The rust lineages differ in geographic distribution, virulence, life history characteristics, and environmental tolerances. Additional differences occur among populations within each lineage. The hybrid lineage is asexual, whereas the other lineage is a mix of sexual and clonal populations (L. Barrett et al. 2008). The plants, too, are a mix of sexual and asexual populations. As the interactions have coevolved, flax, nonhybrid flax rust, and hybrid flax rust populations have diversified into populations that regularly exchange genes and those that never exchange genes. The formation of a hybrid asexual lineage has therefore contributed to the geographic diversification of this coevolving interaction.

Hybridization, then, is not just a temporary stage during the development of reproductive isolation among diverging populations. It is becoming increasingly clear that it is a common and important process that can contribute directly to speciation and reticulate evolution within lineages, and the diversification of interactions with other lineages.

HYBRID SPECIATION

Multiple examples of new hybrid species have been documented in the past 200 years (Ainouche et al. 2008; Abbott et al. 2009; Soltis and Soltis 2009). The possibility of hybrid speciation arises because hybrid offspring do not always have traits that are simply intermediate between their parents. Individual genes are expressed in different ways in different genetic backgrounds, and hybrids therefore can differ from their parents in their expression of multiple traits. Genes that were adaptively important in the parental species become recombined and expressed in novel and adaptively important ways through the process called transgressive segregation (Rieseberg et al. 1999, 2003). Some of the resulting phenotypes may fall outside the bounds of what has ever occurred in either parental species. Those traits could allow the hybrid population to expand into new habitats or form novel interactions with other species unavailable to the parents.

Hybridization, transgressive segregation, and adaptation to new habitats have been carefully documented in the origin of three hybrid sunflower species: *Helianthus anomalus* (sand sunflower), *H. deserticola* (desert sunflower), and *H. paradoxus* (paradox or pecos sunflower) (Rieseberg et al. 2003). These species originated by hybridization between *H. annuus* (common sunflower) and *H. petiolaris* (prairie sunflower) between 170,000 and 63,000 years ago (Gross et al. 2004). The hybrid species are restricted to drier environments than their widespread parental species, occurring only in desert habitats of the southwestern United States (fig. 15.1). The hybrids differ from their parents and each other in multiple ecological traits, and experimental studies have suggested that these beneficial traits arose from natural selection acting through transgressive segregation (Rieseberg et al. 2003; Donovan et al. 2010). Molecular studies have suggested that, following hybrid speciation, the genomes of these new species became stabilized within hundreds of generations (Buerkle and Rieseberg 2008). The detailed ecological and molecular studies of *Helianthus* speciation have been important for our understanding of the relationships among hybridization, adaptation, and speciation. They have demonstrated that new species can arise directly and repeatedly through natural selection acting on the novel phenotypic variation expressed in hybrid individuals.

Similar examples of hybrid speciation have been accumulating in animals (Mallet 2007; Mavárez and Linares 2008), and examples of incipient hybrid speciation linked to environmental change or altered geographic distributions of species are starting to appear (Ording et al. 2010; Stemshorn et al. 2011). These studies are changing the view that hybrid speciation

Fig. 15.1 *Left to right:* Two North American sunflower species, *Helianthus annuus* and *H. petiolaris*, and one of the three naturally occurring hybrid species, *H. anomalus*. This hybrid species occurs in much drier habitats than the parental species does, generally occupying sand dunes. Photographs by Jason Rick, courtesy of Loren Rieseberg.

is a phenomenon mostly restricted to plants, and they are refining our understanding of the mechanisms of this form of speciation (Gross and Rieseberg 2005; Nolte and Tautz 2010). Some studies suggest that the likelihood of hybrid speciation can depend strongly on the local environmental setting. For example, two species of *Lycaeides* butterflies, *L. idas* and *L. melissa*, occur throughout large parts of western North America and overlap in their distributions in multiple places. In the extreme alpine environments of the Sierra Nevada of California and Nevada near Lake Tahoe, they have formed a hybrid species that is genetically distinct from both nearby parental species (Gompert et al. 2006).

Data from nuclear and mitochondrial sequences, microsatellites, and arbitrary fragment length polymorphisms all indicate that the alpine species originated through hybridization of the two parental genomes. These molecular data also suggest that the hybrid populations are not maintained by continual hybridization between the parental species. Rather, the hybrids in the Sierra Nevada occur as a species genetically distinct from either parent. In contrast, in the Rocky Mountains hundreds of kilometers to the east, these same two species form a wide hybrid zone over a large geographic area and show no evidence of separate hybrid species (Gompert et al. 2010).

These results suggest that the greater the range of environmental settings in which two species co-occur, the greater the chance that hybrid species could form in at least one of those settings. Even subtle shifts in climate could change the number of places in which two species overlap, the number of individuals of each species within a hybrid zone, the width of the hybrid zone, and the relative fitnesses of hybrids in comparison with parental species. In some parts of eastern North America, eastern tiger swallowtails (*P. glaucus*) and Canadian tiger swallowtails (*P. canadensis*) hybridize along a narrow latitudinal band that is correlated with thermal environments (Scriber 2011). The hybrid zone corresponds approximately to the wavy band of overlap between the coniferous forests to the north and the deciduous forests to the south. The hybrids occur in a region known for hybridization among many species (Rissler and Smith 2010; Swenson 2010). For the swallowtails, it is a region that marks the transition between two life histories. Populations of eastern tiger swallowtails south of this band have a longer season for development and produce two generations a year, but populations of Canadian tiger swallowtails to the north produce only one generation a year (Scriber et al. 2008).

Amid the warming of climate in recent decades, there has been greater potential for phenological overlap between butterflies with these contrasting life histories. As these butterfly species have hybridized more extensively, some genes, but not others, have mixed (introgressed) between the parental species. For example, the ability of swallowtail populations to detoxify tulip tree (*Liriodendron tulipifera*) leaves has spread north, but the gene conferring the potential to produce two generations a year has moved little (Scriber 2011). In the Battenkill River valley of Vermont, hybrid individuals usually emerge over a month later than the parental species and are to a large extent reproductively isolated from their parents (Ording et al. 2010). In other parts of the geographic range of these two species, these butterflies overlap to varying degrees, creating multiple opportunities for hybridization, introgression, and incipient hybrid species over the more than a thousand kilometers in which this hybrid zone is spread.

HYBRIDIZATION AND ADAPTIVE RADIATION

Hybridization among closely related species also has the potential to drive adaptive radiation, which is repeated ecological speciation across space and time from a common ancestor. When George Gaylord Simpson (1944, 1953) made adaptive radiation the centerpiece of his view of how the diversity of life has evolved, he envisioned it as the divergence, almost simultaneously, of many evolutionary lineages from a single ancestral adaptive type. Since then the concept of adaptive radiation has loosened. It often connotes an unusually high rate of adaptive divergence into separate ecological niches within a lineage in comparison with closely related lineages that have not formed so many ecologically different species (Schluter 2000).

Adaptive radiations, then, have two hallmarks: an unusually high number of species arising from a common ancestor and exceptional adaptive diversity among those species. The adaptive diversity is often displayed through a wide range of morphologies, physiologies, behaviors, or life histories. Some methods of analyzing adaptive radiations focus in addition on the rates of divergence within lineages with a view toward understanding when during an adaptive radiation most of the species and adaptive diversity originated (Pinto et al. 2008; Harmon et al. 2010). That focus can help identify which processes, including hybridization, drive adaptive radiations.

Hybridization could contribute to adaptive radiations at any of three stages during the diversification of a lineage. It could be most important during the early stages of an adaptive radiation, when most species within the radiation are genetically and ecologically similar (Seehausen 2004). Alternatively, hybridization could be most important a little later in a radiation, when there are multiple species and some of them have come into secondary contact as their geographic ranges have expanded. At this stage the adaptive zone may not yet be filled with species, and multiple ecological opportunities may remain. Or hybridization could be most important even later in an adaptive radiation when multiple species within the adaptive zone differ ecologically only in small ways from other closely related species, because much of the adaptive zone is already filled with species. At this stage, hybrids may suffer relatively little fitness loss due to ecological incompatibility, because they are ecologically similar to their parents.

Among these possibilities, the potential for hybridization to contribute to adaptive radiation seems especially high during the early stages of diversification of lineages, and Ole Seehausen (2004) has formalized one mechanism as a hybrid swarm hypothesis. Early in the divergence of species, hybridization may often occur, creating a hybrid swarm that further increases the range of ecologically relevant variation, allowing the formation of an even greater range of ecologically different populations. Eventually, populations within the hybrid swarm undergo further divergence, leading to a decrease in hybridization events. By this hypothesis, the ecological diversification accompanying an adaptive radiation often depends on the opportunity for occasional hybridization among the diverging populations, rather than just on the initial genetic variation present within species early in a radiation.

Ecologically divergent species hybridize to varying degrees when they come into contact (Mallet 2007; Grant and Grant 2008b; Sturmbauer et al. 2010; Hudson et al. 2011), and mathematical models of speciation suggest that those events could be important in the speciation process (Duenez-Guzman et al. 2009). Crucial insights into how hybridization and introgression affect selection during an adaptive radiation have come from the work on Darwin's finches on the Gal·pagos Islands. Peter and Rosemary Grant (2008a, 2008b) have not only documented hybridization among the finch

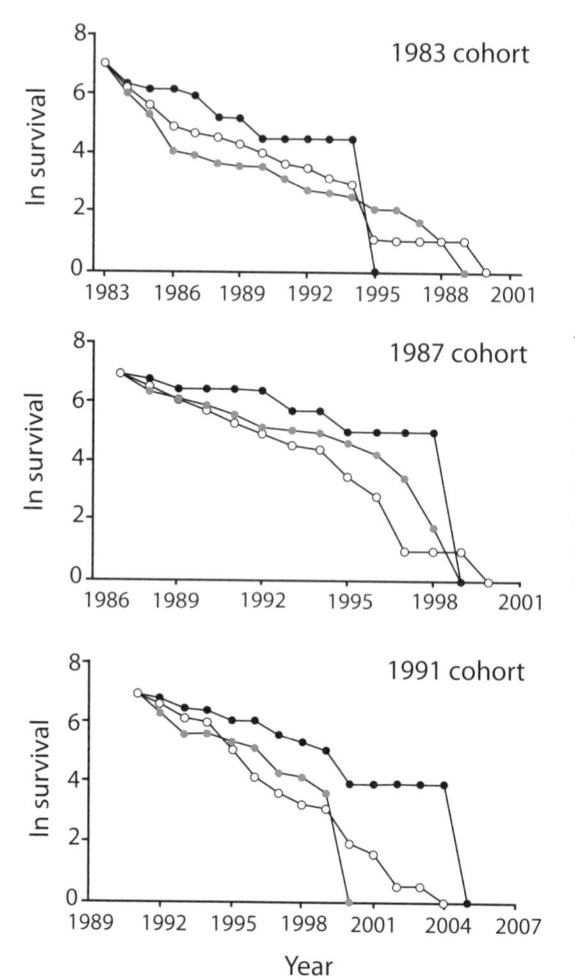

Fig. 15.2 Survival of three cohorts of Darwin's finches on the island of Daphne Major, Galápagos Islands. Gray circles are cactus finches (*Geospiza scandens*), open circles are medium ground finches (*Geospiza fortis*), and black circles are hybrids and backcrosses between these two species. Survival is on a natural log scale, and the initial numbers in each graph are scaled to 1,000. After Grant and Grant (2008a).

species on the island of Daphne Major, but they have also monitored in detail the introgression of genes from one species and the survival of these hybrids relative to their parental species (fig. 15.2). Less than 2 percent of the breeding pairs on the island are hybrids, but hybrid pairs have formed regularly across the years (Grant and Grant 2008a). These pairs are usually a result of a bird imprinting on the song of another species.

Hybrid survival varies among years, but it has been sufficiently high over time to allow introgression of genes as the hybrids backcross to parental species. Hybrids are able to feed on the soft seeds that are produced after the rains of El Niño years, but not on the large and hard seeds that are available in the seed bank in drought years. Over time, the finch species on Daphne

Major have fluctuated in their degree of difference in morphological traits as natural selection has acted on the parental species and the hybrids. At some moments in time on that island, the species have even shown convergence rather than divergence of traits (Grant et al. 2004).

In surveys of other islands of the archipelago, the Grants have found hybridization among multiple finch species in this relatively young adaptive radiation (Grant et al. 2005; Grant and Grant 2008a). From their long-term studies, they have concluded that hybridization and introgression have their greatest effect on adaptive radiations when the species have diverged in morphology but have not yet reached a point when genetic incompatibilities impose severe fitness costs on hybrid individuals. During these early stages of divergence, hybridization and introgression can expand the range and combination of traits on which natural selection can act (Grant and Grant 1994, 2008a). Archipelagos such as the Galápagos provide perhaps the ideal conditions for this kind of process, because populations become locally adapted on individual islands and hybridize with a small number of immigrants from other islands.

HYBRIDIZATION WITH GENOME DUPLICATION

Some other well-studied examples of speciation through hybridization involve not only hybridization but also genome duplication, producing new species called allopolyploids (Soltis and Soltis 2009). As with hybrid speciation, the process of species formation can be rapid and contribute to adaptive radiations. Polyploid offspring have a full set of chromosomes from both parental species and are to a large degree reproductively isolated from either parental species from the time of their formation. Matings with their parental species may produce offspring with uneven numbers of chromosomes, causing problems with chromosomal pairing during meiosis. In addition to the changes in gene expression that accompany hybridization, allopolyploid offspring also differ from their parents in gene dosage, because they have four alleles rather than two for each gene. The proteins produced by a gene or the regulation imposed by any gene can therefore be magnified by having these multiple copies.

The combination of hybridization and genome duplication therefore creates the opportunity for the formation of multiple novel phenotypes on which natural selection can act quickly by further altering gene expression, changing the dosage effects of some genes but not others, and even silencing some genes (Hegarty et al. 2008). To be sure, most instances of hybridization and genome duplication will produce maladapted offspring, but over the long term some novel genotypes may appear that are favored by selection in a particular environment. Most lineages of plants have genome duplication events somewhere in their past histories, and many lineages are composed of species that differ in ploidy levels (e.g., diploids, tetraploids, hexaploids, and so on). By one estimate, ploidy increase has accompanied 15 percent of speciation events in plants and 30 percent in ferns (Wood et al. 2009).

Over the past several hundred years, multiple new self-perpetuating populations have arisen through hybridization and genome duplication, and some of these are now formally recognized species. Well-studied examples in plants have shown that these forms of speciation often involve two or more events rather than a single event. Soon after diploid species of *Tragopogon* plants were introduced into the Palouse region of Washington State and Idaho, about eighty years ago, they repeatedly formed allopolyploid hybrids, which now co-occur as new species with their diploid parental species (Soltis et al. 1995, 2004; Symonds et al. 2010). Following polyploidization, the new tetraploids accumulated multiple additional chromosomal changes. These changes have been suggested by experiments that have compared the natural hybrids with synthetic laboratory-produced hybrids and the parent ancestors, using the technique of fluorescence in situ hybridization (Lim et al. 2008). This method detects the presence or loss of particular DNA sequences within genomes. By any standard, eighty years is a remarkably short length of time to produce genetically distinct populations that function as a new species. In addition to being genetically distinct, these polyploids are morphologically distinct (fig. 15.3).

Hybridization and polyploidy have combined in a different way to produce diversification in the groundsel genus *Senecio*. The initial hybridization took place between two *Senecio* species in their native habitats, and multiple new species evolved after the hybrids were introduced onto a new continent. The process began in Sicily, where two diploid species of *Senecio* that occur on Mount Etna naturally form hybrids that can cross freely with their parents (James and Abbott 2005). About 300 years ago, hybrid diploid individuals from Mount Etna were introduced into the United Kingdom and cultivated in the Oxford Botanical Garden (Abbott et al. 2009). The hybrids subsequently escaped into the wild and spread as far north as Scotland and as far west as Ireland, and diversified to the point where they were named as a separate species. The first recognized hybrid species was therefore a homoploid, rather than a polyploid, hybrid species.

Along the way, however, individuals of this hybrid species formed new hybrids with the native *Senecio vulgaris*, creating two types of allopolyploid—a tetraploid with four sets of chromosomes and a hexaploid with six sets.

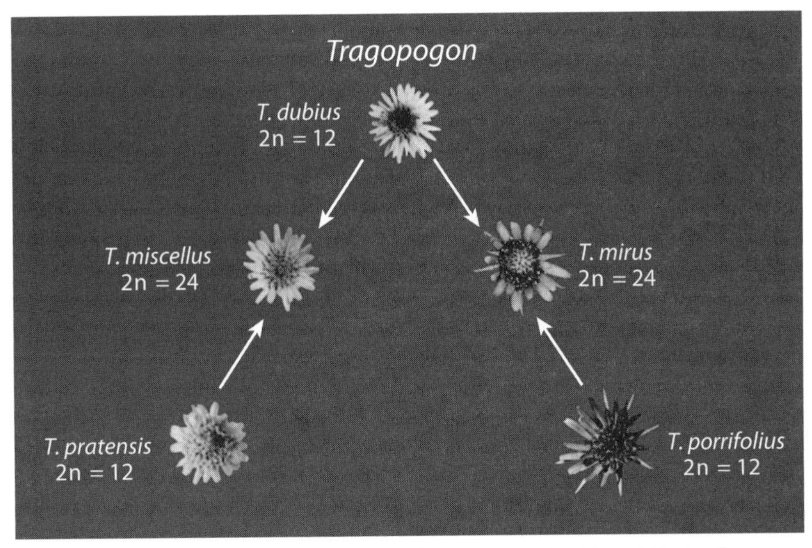

Fig. 15.3 Naturally occurring hybrid species of *Tragopogon*, *T. miris* and *T. miscellus*, formed through allopolyploidization during the past century in eastern Washington State. After Soltis and Soltis (2009).

These hybrids are also now each considered separate species (fig. 15.4). They differ in their patterns of gene expression and their morphological traits (Hegarty et al. 2008). Experiments re-creating the hybrids have shown that new patterns of gene expression occur immediately after hybridization of these species, therefore providing a source of genetic variation on which natural selection can act (Hegarty et al. 2008).

The rapid formation of yet other new plant species within the past 200 years has followed similar routes of hybridization and polyploidization. In some cases, the process has begun as populations have been introduced onto new continents (Soltis et al. 2004; Ainouche et al. 2008). In other cases it has begun as populations have evolved within continents following alteration of habitats from human activities (Ainouche et al. 2008). Some of these events involve initial formation of diploid hybrid species followed by formation of allopolyploid species, whereas others involve allopolyploidy from the start.

It seems likely that plant speciation through hybridization and genome duplication will increase as plant species continue to be moved among continents by our activities and as native species continue to change their geographic ranges as climates change and as we continue to alter habitats. That is, though, not a positive statement about how species

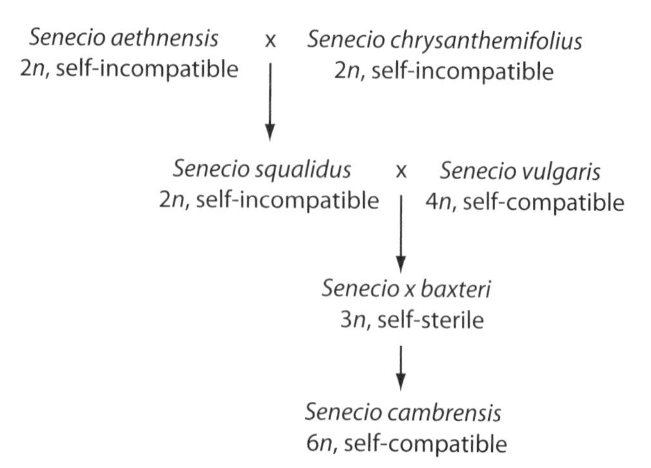

Fig. 15.4 The sequence of hybridization and genome duplication events that led to the formation and divergence of genetically novel populations of *Senecio* (Asteraceae) over the past 300 years. *Senecio aethnensis* and *S. chrysanthemifolius* are species that hybridize near Mt. Etna, Sicily. *Senecio squalidus* is the hybrid form that was introduced in the United Kingdom and rapidly diverged from the parental species in their absence. It hybridized with the UK native *S. vulgaris* and formed first *Senecio x baxteri* and then the allopolyploid *S. cambrensis* and another allopolyploid *S. eborcensis*, which is not shown here. After Hegarty et al. (2008).

introductions can lead to diversification. These hybrid species can sometimes have devastating effects on native species and their coevolved interactions with other species (Rhymer and Simberloff 1996; Brumfield 2010). The occurrence of these hybrid species highlights the fact that the introduction of species can affect not only patterns of adaptation within species but also how genetic variation becomes repackaged among species.

Whether newly self-sustaining hybrid populations survive for another few hundred years or a million years does not matter. There is no minimum time that a species must last to be considered a species. These are all ecologically and evolutionarily important events that contribute to ongoing dynamics of evolutionary change. The hybrid species have undergone evolutionary changes that have partitioned their adaptations into groups of populations with their own potential future evolutionary trajectories, which are sometimes very short. These hybrids differ from their parental species in their traits and their adaptive evolutionary changes, creating yet more opportunities for hybrid speciation in the future.

Finally, just as polyploidy is not a requirement for speciation involving hybridization, hybridization is not a requirement for speciation involving polyploidy. Genome duplication is more common in plants than in many animal groups, and it was once thought that most cases of polyploidy in plants resulted from a combination of hybridization and genome doubling. As more genomes have been studied, however, it has become evident that many polyploid plant populations originate through genome duplication within a population rather than through hybridization with other species. Autopolyploid populations have been found in many taxa, and the populations that have been studied in depth show differences from their parents in multiple ecological traits (see chap. 4). There is a growing view among those studying the role of genome duplication in plant speciation that we may have underrepresented the extent and importance of this mechanism by which plants adapt and sometimes speciate (Soltis et al. 2007). If so, then this is yet another way by which we have been underestimating the ongoing dynamics of evolutionary change.

Speciation through hybridization, genome duplication, or a combination of these events is inherently a process of sympatric speciation that can rapidly result in the divergence of populations. This is sympatric speciation in its most direct ecological sense of the divergence of populations whose individuals are capable of encountering each other with a high frequency (Futuyma and Mayer 1980; Mallet et al. 2009). Speciation by hybridization or genome duplication involves, by definition, the initial local overlap of the parental species and the hybrid or polyploid offspring. At least for plants, which account for several hundred thousand of the earth's species, these processes of sympatric speciation may be common. Even if many of these populations do not persist in the long term, their continued formation contributes to the dynamics of evolution as they occur repeatedly between species and sometimes repeatedly in the same species.

The sympatric speciation processes found in some plants contrast with the processes found in highly mobile animal taxa such as birds, in which sympatric speciation is rare (Phillimore et al. 2008; Price 2008). In the absence of genetic or ecological processes that sharply reduce gene flow between populations of these species, speciation is unlikely to occur (Gavrilets 2004). Sympatric speciation, though, remains a possibility among small, less mobile species such as some insects that partition environments very finely, because they specialize on particular resources and search for mates only during very restricted periods of a year. Much of the diversity of life is made up of plants and insects, and much more work remains before we can assess how much of the diversification of life has arisen through sympatric rather than allopatric divergence in animals other than vertebrates. It seems likely, though, that most speciation is probably allopatric in all taxa, but some taxa have the additional potential for speciating sympatrically, especially through genomic changes.

A CASE STUDY: CORAL BELLS, WOODLAND STARS, AND MITERWORTS

Diversification of the almost 90 herbaceous species within the *Heucherina* group of plants in the Saxifragaceae illustrates the genomic, morphological, life history, and interaction complexity that can occur even within relatively small radiations amid hybridization and genome duplication. This group of 9 genera includes horticultural plants such as coral bells and an ecologically diverse array of other plants distributed throughout western North America. A few genera also occur in eastern Asia or eastern North America. The plants grow in open coniferous forests, deciduous woodlands, talus slopes, wet cliff faces, dry cliff faces, and steppe. Taxonomically, they are divided into 3 relatively large genera—coral bells and their relatives (*Heuchera*), woodland stars (*Lithophragma*), and miterworts (*Mitella*)—and 6 small or monotypic genera (fig. 15.5). Two of the 3 large genera, coral bells and woodland stars, are clearly defined monophyletic lineages, but *Mitella* is a complex group of species, with some species more closely related to species in the other genera than to each other (Soltis and Kuzoff 1995).

The division into so many genera reflects appreciation by previous generations of biologists of the great genetic, morphological, and life history diversity found among these species. This has been one of the best-studied noncrop groups of plants with respect to patterns of relationship among species, including studies of hybridization, polyploidization, karyotype diversity, flavonoid diversity, allozyme diversity, chloroplast DNA divergence, multigene nuclear DNA divergence, allometric diversification in floral development, and divergence in use of pollinators. Among these genera, *Lithophragma* exhibits a particularly wide range of ovary positions within and among species (Kuzoff et al. 2001; Soltis and Hufford 2002), which is noteworthy because ovary position is generally considered a character that varies little at lower taxonomic levels. Other floral traits also vary widely within the *Heucherina* group (fig. 15.6).

The molecular complexity of this group results from the multiple genetic processes that have shaped the diversification. Hybrids occur within and between at least six of the genera (Soltis 1986), and some horticultural plants are intergeneric hybrids (e.g., *Heucherella*). Comparison of phylogenies constructed using chloroplast DNA sequences and nuclear DNA sequences shows evidence of chloroplast capture resulting from these hybridization events (Soltis and Kuzoff 1995; Kuzoff et al. 1999). The group includes

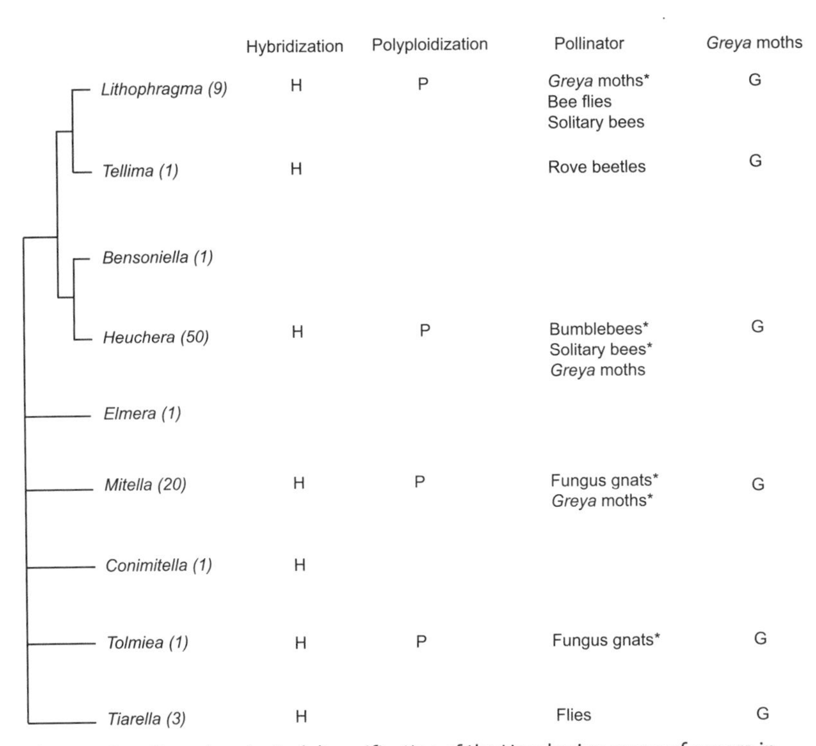

Fig. 15.5 Genetic and ecological diversification of the Heucherina group of genera in the plant family Saxifragaceae and the occurrence of hybridization (H), polyploidy (P), specialized pollination systems, and specialized herbivores (G) in these genera. Phylogeny is condensed from Soltis et al. (1993, 2001) and Soltis and Kuzoff (1995). *Mitella* is polyphyletic, with some species currently in that genus grouped with *Elmera* or *Conomitella*. Number of species, shown in parentheses, are from Soltis et al. (2001). Data for hybridization are from Soltis (1986, 2007), Soltis and Boehm (1984a, 1984b), Doyle et al. (1985), Soltis et al. (1991), and Kuzoff et al. (1999). Data for polyploidy, including autopolyploidy and allopolyploidy, are from Soltis (1988, 2007), Ness et al. (1989), Soltis and Soltis (1989), Thompson et al. (1997), and Kuzoff et al. (1999). Pollination data are summarized from Thompson and Pellmyr (1992), Pellmyr et al. (1996), Weiblen and Brehm (1996), Goldblatt et al. (2004), Okuyama et al. (2004, 2008), Thompson and Merg (2008), Cuautle and Thompson (2010), Thompson (2010), and Thompson et al. (2010). In most genera, only some of the species have been studied in detail. * = insects confirmed as the major pollinators of at least one species within the genus. Other species listed have been shown to contribute to pollination in populations of some species, except for *Tiarella*, for which the only data available are for the most the common floral visitors. Host associations for *Greya* are from Davis et al. (1992), Pellmyr et al. (1998), and Thompson (2010).

Fig. 15.6 Examples of the diversity of floral shapes found within the *Heucherina* group of plants in the Saxifragaceae. *Left to right: Heuchera grossulariifolia* (tetraploid form) with *Greya politella* moth, *Lithophragma affine*, and *Mitella stauropetala* with *Greya mitellae*. See chapter 9 for photographs of *Greya politella* moths on *Lithophragma*.

autopolyploids as well as allopolyploids (Soltis and Rieseberg 1986; Kuzoff et al. 1999). Moreover, autopolylploid populations have arisen multiple times from a diploid ancestor in at least one species (Segraves et al. 1999). The genera *Heuchera*, *Mitella*, and *Tolmiea* all have diploid ($2n = 14$) and tetraploid ($2n = 28$) populations or species (Soltis 2007), and the genus *Lithophragma* includes chromosome numbers of 14, 28, 35, and 42 (Taylor 1965).

The morphological and ecological complexity accompanying the diversification of the *Heucherina* group reflects the genetic and genomic complexity. Species differ greatly in floral morphology, leaf shape, life history characteristics (including self-compatibility in reproduction), the tendency to form asexual propagules in addition to sexual propagules, the pollinators they attract, and the specialist herbivores that attack them. Most species within the group are attacked by nonpollinating *Greya* species, and multiple species are pollinated by other *Greya* moths that lay their eggs in flowers or other plant parts before or after pollinating the plants (see chap. 9). Pollination in other species is distributed over a remarkably wide range of insect taxa, including fungus gnats, bee flies, beetles, bumblebees, and solitary bees. Multiple studies have separated the actual pollinators from other floral visitors, showing clearly that plant species within this group differ in their adaptation to different pollinator taxa (e.g., Pellmyr et al. 1996; Goldblatt et al. 2004; Thompson and Fernandez 2006; Thompson and Merg 2008; Thompson et al. 2010). Most species within the *Heucherina* group also are attacked by specialized species of *Puccinia* rust (Savile 1975), and a few are attacked by a geometrid moth species (Nuismer and Thompson 2001).

These ninety or so plant species are distributed among communities in combinations of up to about half a dozen sympatric species. The large genera include species that overlap to varying degrees, with several species co-occurring in some communities of western North America. *Tiarella* occurs as allospecies, with one species in eastern North America, another in western North America, and yet another in eastern Asia (Soltis and Bohm 1984b). Over a smaller geographic region, the complex of closely related species called *L. bolanderi*, *L. cymbalaria*, and *L. heterophyllum* occurs for the most part as a ring species around the central valley of California, with some populations of *L. bolanderi* and *L. heterophyllum* occurring as disjunct populations within the geographic range of the other. This ring species complex includes evidence of diversification through polyploidy, hybridization, and introgression (Taylor 1965; Kuzoff et al. 1999).

The *Heucherina* group of plants is therefore a microcosm of the diversity of genetic and ecological patterns and processes found in other adaptive radiations as species diverge, hybridize, undergo major genomic changes, colonize new habitats, and alter their interactions with other species, including, in some cases, co-opting antagonists into mutualistic roles (in this instance, *Greya* moth pollinators). No single genetic process or ecological interaction explains the radiation of species, traits, and interactions within this lineage. The radiation has undoubtedly been made possible because all these different mechanisms of genetic and ecological change have been available in different combinations in different populations.

ECOLOGICAL OPPORTUNITY, RETICULATION, AND ADAPTIVE RADIATION

Ecological speciation, often accompanied by hybridization or genome duplication, is likely at the center of all major adaptive radiations, as it favors speciation into new adaptive zones. It is difficult to imagine the radiation of entire lineages into different ecological niches without ecological speciation as the primary cause. Simple genetic incompatibilities among populations are unlikely to produce the remarkable radiations in the beaks of birds, shapes of flowers, and physiologies of insects that specialize on different plant species. Ecological speciation and the new ecological opportunities offered by events such as hybridization and polyploidization are likely often involved.

That does not mean, however, that every speciation event in an adaptive radiation is driven by strong ecological differentiation among populations or reticulate events during the speciation process. For example, among sawflies, the subfamily Nematinae includes a group of more than 700 species distributed over many plant taxa on multiple continents. Each species is specialized to a small group of host plants, but collectively they have evolved to cope with taxa as different as spruces, willows, roses, and blueberries (Nyman et al. 2010). Some feed as larvae externally on leaves or flowers, whereas others fold leaves, induce galls, or mine leaves or fruits. Shifts to use a different resource have accompanied at least 20 percent of speciation events in these insects. One interpretation of these results is that the remaining speciation events are non-ecological, resulting from additional speciation among allopatric populations (Nyman et al. 2010). There is, though, an alternative way to think of results such as these in which ecological divergence appears to be responsible for only a minority of speciation events. The 20 percent of speciation events attributable to shifts in resource use may have led to some of the remaining 80 percent, by allowing the new species to spread into new environments.

Adaptive radiations have generated many potential generalizations, including the view that divergence is most rapid in the early stages of radiations when new ecological opportunities are greatest (Gavrilets and Losos 2009). The appearance of a greater rate of diversification early in a radiation could arise either because speciation declines later in a radiation as ecological niches fill, extinction increases, or the role of reticulate evolution changes over time. Some analyses suggest that speciation rates are often high during an initial explosive stage of divergence and then decline over time (McPeek 2008; Rabosky and Lovette 2008). This suggests that the diversifying and reticulating processes that contribute most to the early stages of adaptive radiation also shape the eventual pattern of ecological divergence within a radiating lineage. Studies of some lineages, however, indicate little evidence of bursts of speciation or morphological evolution during the early stages of adaptive radiations or the difficulties of using extant taxa alone in estimating bursts (Harmon et al. 2010; Hulsey et al. 2010; Quental and Marshall 2010).

Mark McPeek (2008) has suggested that the distribution of speciation during diversification of lineages provides important clues to the role of ecological speciation in adaptive radiations. He has argued that lineages that have slowed over time in their accumulation of new species likely diversified through ecological speciation, whereas lineages that have accelerated their accumulation of new species likely diversified through little or no ecological diversification. Lineages that rapidly diversify ecologically as they speciate are more likely to decline over time in their rates of speciation than lineages that diversify slowly in ecologically important traits. How the commonness of reticulate evolution affects those rates remains unknown.

Many other generalizations about adaptive radiations concern the spatial scales, kinds of environments, and genetics of adaptation likely to favor multiple ecological speciation events. Each of these aspects of the evolutionary process also can influence the commonness of reticulate events during evolution. For example, one view is that the speciation and extinction rates within lineages depend on the geography of speciation, including the degree and rate of range expansion during diversification (Pigot et al. 2010). The degree and rate of range expansion, however, also affect the potential of divergent populations to hybridize at different stages of speciation and lineages to show reticulate evolution during the process of adaptive radiation. We therefore still need a much better understanding of how reticulate evolution shapes adaptive radiations.

THE CHALLENGES AHEAD

As we learn more about the roles of hybridization, genome duplication, and reticulate evolution in speciation, it is becoming clear that speciation is not just about the simple divergence of populations through selection on individual genes or suites of genes. It can also be a genomic process that has the potential to generate, within a few generations, populations that differ markedly from ancestors. As molecular genomic approaches expand, we are likely to find more evidence of reticulate diversification of lineages through genomic processes. The challenge is to understand the ecological conditions most likely to favor divergence through major genomic changes rather than other processes.

The other challenge is to gain a better understanding of the ecological conditions that favor continued adaptive diversification within lineages through all these genetic and genomic processes. We turn in the next chapter to those ecological conditions, and especially the role of species interactions in driving adaptive radiations.

16

Species Interactions and Adaptive Radiations

Adaptive radiations occur as the diversifying and reticulating processes of evolution amplify across environments. Many taxa probably have the potential for adaptive radiation but rarely the ecological opportunity to have that potential realized. From that view, evolution is relentless partly because lineages are often caught in a state that is evolutionarily highly dynamic but only rarely produces much net change either in traits or long-lasting species. Occasionally, though, a lineage undergoes sustained directional change in multiple directions, forming multiple new species adapted to different physical environments or different interactions with other species.

One common theme in studies of adaptive radiation is that they are often fueled by evolving interactions among species. The largest radiations—colonization of air, land, or freshwater, or later recolonization of oceans—began with exploitation of new physical environments. Subsequent diversification, however, has often largely been about the formation of new webs of interacting species. Interactions favor geographic mosaics of traits and ecological outcomes, which drive further diversification.

This chapter explores some of the ways in which evolving interactions among other species may contribute to the continued divergence of populations and species, thereby generating adaptive radiations.

BACKGROUND: MOVING BEYOND THE EARLY STAGES OF ADAPTIVE RADIATION

The arguments in the previous chapter suggest that hybridization and reticulate evolution are sometimes important during ecological speciation at least in the early stages of adaptive radiation. The hybrid swarm hypothesis of adaptive radiations suggests that hybrid swarms can be important because the swarms increase the range of ecological diversity within and among populations, and those populations eventually partition the environment into well-defined niches (Seehausen 2004). That ecological diversity is often due to interactions with other species. Multiple studies have shown that hybrid populations differ from their parental species in their ecological interactions with other species (Whitham et al. 2006; Schweitzer et al. 2008; Bailey et al. 2009).

Hybridization and reticulate evolution, though, are likely important only in some radiations. In a review of empirical and mathematical studies of adaptive radiation, Sergey Gavrilets and Jonathan Losos (2009) concluded that we continue to discover factors that contribute to adaptive radiations, but we still have no clear answers for why adaptive radiations have occurred in some places and groups but not in others, because there is likely no single answer. Adaptive radiations often seem to be about being in the right place at the right time with just the right genetic or genomic variation and population structure to favor the divergence of those particular populations. Some recent genomic studies have suggested that adaptive phenotypic divergence of geographically separated sublineages—such as cichlid fish in Africa and the Neotropics—may sometimes be driven by selection on very different suites of genes (Fan et al. 2012).

The coevolutionary process itself can set the stage for diversification. Selection mosaics in coevolving interactions generate multiple evolutionary outcomes among populations. Some of those locally adapted interactions can become the basis for adaptive radiations. As species diversify and come back together in different combinations in different regions, they can generate new and more complex selection mosaics that can favor further diversification. We are only now starting to understand how adaptive radiations can scale up directly from the geographic mosaics of coevolution found within many interactions.

THE INCREASING ECOLOGICAL COMPLEXITY OF ADAPTIVE RADIATIONS

The coevolutionary diversification of yucca plants and pollinating yucca moths has become one of the models of how coevolution can drive adaptive radiations. Yuccas have radiated into 34 recognized species (Pellmyr et al. 2007; Smith et al. 2008), with most occurring in arid parts of North and Central America. All these species are pollinated by yucca moths in the genus *Tegeticula*, which includes at least 20 species (Pellmyr et al. 2008). A closely related genus, *Hesperoyucca*, is pollinated by moths in the genus *Parategeticula*, which is a group of 5 species that are sister taxa to *Tegeticula* (Pellmyr et al. 2008). Most *Tegeticula* and *Parategeticula* moths pollinate the flowers in which they lay their eggs.

Two species of *Tegeticula* have evolved into cheaters that lay their eggs into reproductive tissues without pollinating their hosts. Molecular analyses suggest that the cheater moths hybridized with the pollinating moths early in their diversification (Segraves and Pellmyr 2004; Althoff et al. 2006). The cheater species rely on the other yucca moths to pollinate the flowers and occur only in some yucca populations and species. In addition, other prodoxid moth genera have radiated onto yuccas, feeding as larvae on various other tissues, including floral stalks (Althoff et al. 2001; Althoff and Pellmyr 2002).

The pollinating yucca moths are probably the strongest agents of selection on diversification in yuccas, because the plants depend completely on these moths for reproduction. Selection imposed by divergent physical environments or other herbivores could also favor speciation in yuccas, but studies designed to evaluate the relative roles of selection driven by differing physical environments rather than coevolution have suggested that coevolving yucca moths are the stronger agents of selection and divergence on these plant species (Godsoe et al. 2009). One way of disentangling the ecological factors driving divergence has been to compare divergence rates in reproductive characters and in vegetative characters. If coevolution with its obligate pollinators is the major factor driving divergence, then reproductive characters associated with pollination should have diverged more than vegetative characters associated with adaptation to physical environments. This kind of analysis cannot provide a definitive test of the role of coevolution in driving population divergence, but it is a pattern that should occur if coevolution with pollinators is driving the divergence of these plant populations.

Joshua trees (*Yucca brevifolia*) in the North American Mojave Desert provide a good model for this kind of study (fig. 16.1). These impressive plants occur in two morphologically distinct subspecies that are parapatric in part of their geographic ranges and sympatric in other regions (Smith et al. 2009). The subspecies differ in vegetative stature, the size and structure of the pistils, and the species of yucca moth that pollinates their flowers (fig. 16.2). In the zone of parapatry, the more western populations are pollinated by *T. synthetica* and the more eastern populations by *T. antithetica*. The plant and moth populations are thought to have begun diverging during a period starting 6.5 million years ago when the Sea of Cortez extended into the low-lying regions of the Mojave Desert, generating allopatric populations both in the plants and the moths (Pellmyr and Segraves 2003). Subsequent divergence of the populations could therefore have been due to other ecological factors, but coevolution with the moths is the likely explanation:

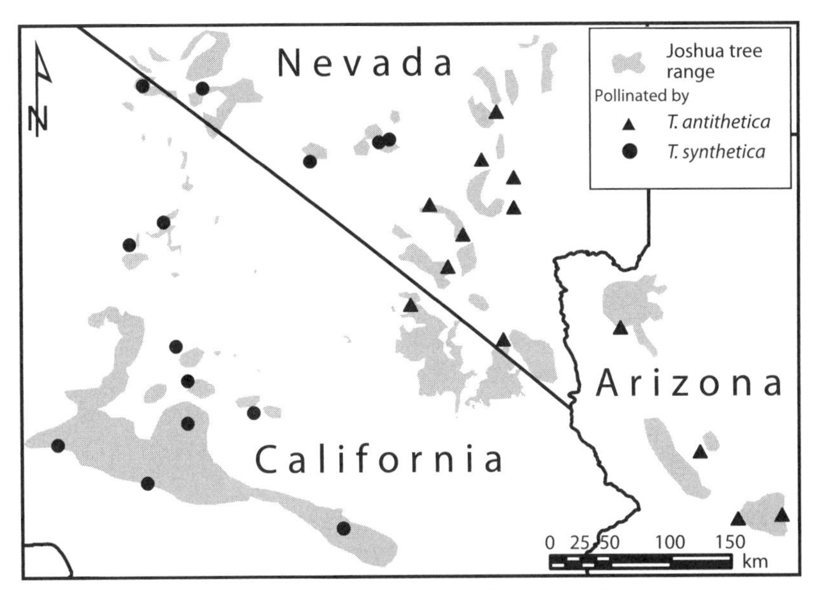

Fig. 16.1 Distribution of Joshua trees (*Yucca brevifolia*) in the North American Mojave Desert, where the subspecies pollinated by different yucca moth species occur parapatrically. The triangles and circles denote the study populations pollinated by *Tegeticula antithetica* and *T. synthetica*, respectively. After Godsoe et al. (2008).

Fig. 16.2 Differences in some of the morphological traits found in Joshua trees (*Yucca brevifolia*) pollinated by *Tegeticula synthetica* moths (*left*) and by *T. antithetica* (*right*). Photograph courtesy of Christopher Smith.

reproductive traits associated with pollination by yucca moths have diverged faster than vegetative traits (Godsoe et al. 2008).

Moth traits have also diverged, and they have done so in ways that match the divergence of the floral traits. The moths have diverged more in traits associated with divergence in floral traits than in body size (Pellmyr and Segraves 2003). Relaxed selection or random genetic drift could not have produced the disproportionate divergence in the reproductive traits of the plants and the moths relative to other traits, and it could not have produced the complementarity and codivergence of traits specifically associated with the mutualism. Coevolution between the plants and the moths therefore appears to have been responsible for the divergence of the plant populations and probably also the moth populations. In the zone of sympatry, each moth species has better reproductive success on its normal host population than on the other host population (Smith et al. 2009).

Yuccas and yucca moths, however, have not simply cospeciated (Smith et al. 2008). They arose at different times in the geological past, and they have spread and contracted multiple times over millions of years. Their changing geographic ranges have resulted in different sets of species interacting in different communities, generating novel selection pressures in different regions. Often only one yucca moth species pollinates a local yucca population, but some yucca species are pollinated by different yucca moths in different geographic regions, and some yucca moth species pollinate different yuccas in different regions (Pellmyr et al. 2008). Yuccas attract yucca moths by use of floral scents with unique chemical compounds, but the yucca species and populations that have been studied in detail for floral scents have similar floral scents even among populations pollinated by different yucca moths (G. Svensson et al. 2005, 2011). Hence, there seems to have been much potential for changing partners as these species have diversified and spread across North and Central America.

A yucca population may also be subject to attack by one or more other prodoxid species that either bore into stems or lay eggs in flowers without pollinating them. Some populations of *Hesperoyucca whipplei* in the southwestern United States have up to four prodoxid moth species, including a pollinating *Tegeticula* species and three nonpollinating *Prodoxus* species, two of which feed on fruits and two feed on different parts of the floral stalk (Althoff et al. 2007). In addition, several other insect taxa feed on yucca flowers, including leaf-footed bugs and beetles, adding to the complexity of selection on these interactions (Althoff et al. 2005). In a detailed study at one site, seed set in plants in one year resulted from a combination of the beneficial effects of pollination by yucca moths and the detrimental

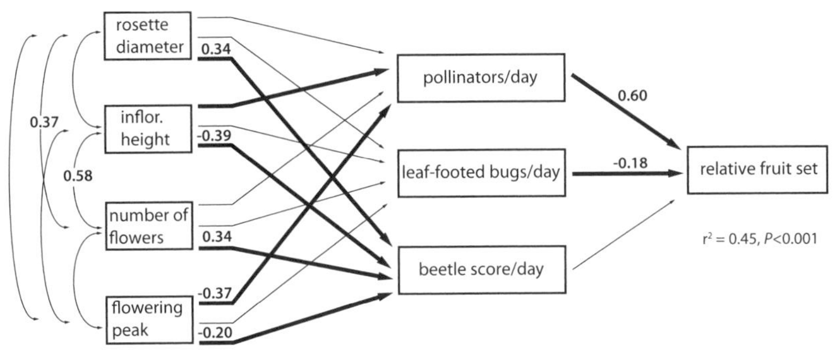

Fig. 16.3 The effects of pollinating yucca moths (pollinators/day) and flower-feeding insects (leaf-footed bugs and beetles) on the relative number of seeds produced by *Yucca filamentosa* plants in one year at one site in Florida. The analysis also shows how the effects of these three insect species depend, in turn, on the size of the plants, the height of the inflorescence, the number of flowers on a plant, and the phase when those flowers are produced during a season. Double-headed arrows show correlations between plant traits. Single-headed arrows show effects of plant traits on insect visitation rates and, in turn, effects of insect visitation rates on relative fruit set among yucca individuals. Numbers and arrows in boldface show statistically significant effects. Values are the path coefficients of the path analysis. After Althoff et al. (2005).

effects of herbivory by leaf-footed bugs (fig. 16.3). In the following year, the herbivores had no major effect on seed set. If these studies were repeated across the entire geographic range of this yucca species, the complexity of selection would surely be even greater.

Overall, the studies of yuccas and yucca moths suggest that coevolution may be a major driving force in some adaptive radiations as populations diversify, speciate, and even form small networks of interacting species. The descendent species come back together in various combinations in different parts of the range of the interaction. In other interactions, there could be even more complex relationships between geographic mosaics of coevolution and adaptive radiation. Robert Ricklefs (2010) has suggested that geographic mosaics of coevolution with parasites and predators could either foster or retard adaptive radiations. Population levels of species tend to be high in some regions and low in other regions. Predators and parasites could reduce population levels in regions where the numbers are high, thereby reducing competition and allowing a greater number of similar species to coexist than would otherwise be possible. Alternatively, predators and parasites could prevent some parts of niche space from being filled as a lineage diversifies. In this case, diversification would slow over time within a lineage because coevolving pathogens reduce population sizes and restrict geographic distributions. Possibly, then, the various stages of some adaptive radiations reflect geographic mosaics of coevolution driven by different forms of interaction, including competition, antagonistic trophic interactions, and mutualisms.

REPEATED ADAPTIVE RADIATIONS IN ECOLOGICAL NETWORKS

As geographic mosaics of coevolution extend across ecosystems, they seem to create opportunities for repeated mini-radiations. One version is the repeated pattern of competitive divergence found in the radiation of *Anolis* lizards (see chap. 11). Another version arises through the diversification of interactions between species across the continuum of antagonism, commensalism, and mutualism.

The diversification of the moth family Prodoxidae is among the best-studied examples. The family, which includes about a hundred species, includes not only the yucca moths but also the *Greya* moths. It is an ancient moth family thought to have arisen about 95 million years ago (Pellmyr 2003) and has therefore been present during much of the diversification of angiosperms over the past 130 million years (Soltis et al. 2005). During that time, the yucca moths have colonized monocotyledonous plants mostly throughout the drier parts of the North America, while *Greya* moths have colonized two families of eudicotyledonous plants in multiple habitats of western North America. These two moth taxa overlap in the drier parts of California and some other western regions.

The radiation of prodoxid moths has been driven to a large extent by shifts onto novel host plant species (Thompson 2010). In rare instances these host shifts have been onto plants only remotely related in angiosperm phylogeny. Following these jumps, each radiation has resulted in species specialized to feed on different plant parts. Among the monocot-feeding genera (*Tegeticula*, *Parategeticula*, *Prodoxus*) and the saxifrage-feeding genus *Greya*, the radiation has included pollinating floral feeders, nonpollinating floral and seed feeders, and stem feeders (fig. 16.4). In both lineages the interactions with plants have then diversified into multispecies networks that often include one or a few host species, a pollinating moth species, a nonpollinating or inefficiently pollinating moth species, and one or more other herbivores. The composition of these networks varies geographically. As a result, diversification of these interactions continues to foster further diversification, as selection on these interactions varies among ecosystems.

That diversification, though, is often bounded. As the moth and plant populations diverge into new species, they continue to interact in differ-

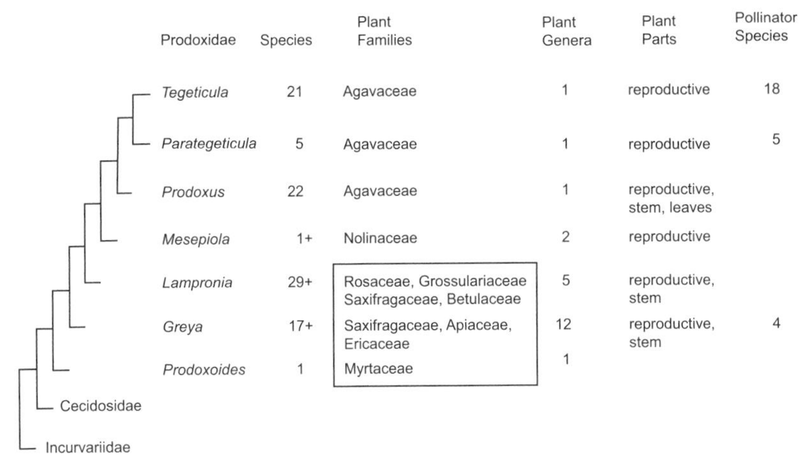

Fig. 16.4 Phylogenetic diversification of prodoxid moths as they have colonized multiple plant families to use as larval hosts. Phylogenetic branches are collapsed to equal length (the estimated lengths are shown in the references that follow). Eudicotyledonous plant families are shown enclosed in a box, and monocotyledonous species outside the box. Also shown are the number of plant genera used as hosts within each family, and the plant tissues into which females oviposit and in which early instars develop: reproductive parts (flowers or seeds), stems (floral scapes or, in one *Lampronia* species, twigs), and leaves (as a leafminer). *Pollinator species* refers to the number of moth species in that genus known to be a pollinator of its host plant(s). Compiled from Davis et al. (1992); Thompson (1997, unpub. data); Pellmyr et al. (1998, 2008); and Pellmyr and Leebens-Mack (1999). After Thompson (2010).

ent combinations in different ecosystems. That means that interactions among species and their descendants are often held together for thousands or even millions or years, as natural selection continues to act even as environments continue to change. The interacting species continue to evolve because they cannot get away. This creates both constraint and opportunity. It sets bounds on the divergence of a species, but it creates tremendous opportunity for continuing evolution and coevolution within those bounds.

Continued diversification of interacting lineages, however, is not inevitable. In some evolving interactions, much of the radiation of a lineage seems to have happened early during diversification and now shows little evidence for further directional change. That seems to be the case in *Enallagma*, *Ishneura*, and *Lestes* damselflies in lakes of North America. Species within these genera commonly coexist, and the genera show clear evidence of ecological partitioning (Siepielski et al. 2011b). The partitioning is consistent with the expectations generally found in theoretical models analyzing the conditions necessary for coexistence of potentially competing species (Chesson 2000; Chase and Leibold 2003; Leibold and McPeek 2006). Within *Enallagma*, some species have traits adapted to survival in lakes where they are subject to predation by fish, and others have traits adapted to survival in fishless lakes where they are subject to predation by large invertebrate predators such as dragonflies (McPeek and Brown 2000). But the situation is more complicated. Diversification within *Enallagma* appears to have involved ecological speciation in some species but not all (McPeek and Brown 2000; Stoks and McPeek 2006). Some species are ecological equivalents, showing few differences among them (Siepielski et al. 2010), and some speciation appears to have been driven primarily through differential sexual selection (McPeek et al. 2008).

ESCAPE-AND-RADIATE COEVOLUTION AND STARBURSTS OF SPECIATION

At even higher taxonomic levels, coevolution could foster multiple adaptive radiations. The first hypothesis on how coevolution between trophic levels may generate repeated starbursts of speciation in lineages was proposed by Paul Ehrlich and Peter Raven (1964) and is often called escape-and-radiate coevolution. It is a three-step process that they initially used to describe the adaptive radiation of plants and butterflies, but it could just as readily apply to any two or more lineages of interacting parasites and hosts or predators and prey. I describe it here using parasites and hosts as an example. The process begins with mutations in a host population that allow individuals to escape attack from enemies. The mutant population then expands its geographic range in the absence of the interaction and undergoes a starburst of speciation as it colonizes a wider range of environments. Eventually a mutant parasite population overcomes the new host defenses and radiates in species, with each new parasite species specializing on one or more of the many hosts now available to it. The process then repeats itself. The hypothesis assumes that there are defense syndromes (Agrawal 2007). Breaking free from enemies requires the evolution of a new defense syndrome that falls outside the arsenal of potential counterdefenses found in its normal enemies.

Escape-and-radiate coevolution would result in a temporal series of alternating starbursts of speciation on both sides of the interaction, forming entire clades with new defenses and counterdefenses. This view of coevolution makes clear predictions (Thompson 1994, 2005; Segraves 2010). Novel defenses occur among, rather than within, clades (i.e., within a sublineage produced by a starburst of speciation) because defenses accumulate starburst by starburst rather than species by species. Second, parasites do not colonize hosts within each host clade in any systematic fashion, because there is no pattern of accumulation of defenses from ancestors to descendants within each starburst of host speciation. Rather, the pattern of escape-and-radiate coevolution appears at higher taxonomic levels, where each starburst of species on one side of the interaction is matched later with a starburst of speciation on the other side. Different taxa may even be involved at different points in the radiation of defenses and counterdefenses in this form of parasite-mediated speciation.

Escape-and-radiate coevolution was the first hypothesis on how coevolution could affect not only speciation but also major patterns in the adaptive radiation of entire lineages. It has inspired a great deal of research on how interspecific interactions might drive speciation, and it remains a major framework for thinking about how the process of reciprocal selection could shape the web of life at multiple levels (Janz et al. 2006; Agrawal and Fishbein 2008; Futuyma and Agrawal 2009). It remains, however, a difficult hypothesis to test for any particular set of interacting lineages. It requires a detailed understanding of the phylogenetic histories of the interacting species and the distribution of the major traits under selection among these species. Escape-and-radiate is only one of many ways in which reciprocal evolutionary change could drive the diversification of interacting lineages. It has been, though, a highly motivating hypothesis, shaping ideas on how coevolution may affect the evolution of key innovations and adaptive radiations. The hypothesis suggests why speciation in coevolving species will usually not result in matched patterns of codivergence of interacting lineages.

Whether through the process of escape-and-radiate or through other processes, studies of adaptive radiation suggest that divergence is often pulsed rather than smooth (Currie et al. 2003; Toju and Sota 2009). Amid ongoing minor coadaptation, new traits and shifts of partners occasionally produce fundamental changes that alter the evolutionary trajectories of species. The pulsed origin of new species may sometimes arise as species continue to alter their degrees of specialization to other species. For example, Niklas Janz and colleagues (Janz et al. 2006; Janz and Nylin 2008; Slove and Janz 2011) have argued that plant-feeding insects may have diversified as selection has oscillated between favoring insects that feed on many plant species and attain large geographic ranges and favoring populations with narrower diets and are more restricted geographic ranges. The evolution of defenses is also likely to be pulsed as species accumulate enemies, escape to new adaptive zones where enemies are fewer, and then accumulate new enemies.

THE ECOLOGY OF COSPECIATION

One of the old arguments against the role of coevolution in the diversification of interacting taxa was that the lineages could not have coevolved because one of the lineages appeared before the other; another was that most of the diversification in one of the lineages occurred before the other. Those arguments confuse coevolution with cospeciation. There is no inherent relationship between the two. Coevolution is about reciprocal evolutionary change in the adaptations of interacting species, and there is no reason to expect that reciprocal adaptation would lead in lockstep to reciprocal speciation.

Cospeciation is the extreme alternative to escape-and-radiate coevolution. Cospeciation, or parallel cladogenesis, is a macroevolutionary pattern of speciation in which two or more co-occurring lineages undergo matched speciation events during their phylogenetic history. Each speciation event on one side of an interspecific interaction results in a speciation event on the other side of the interaction. This pattern is impossible with escape-and-radiate coevolution, because cospeciation at the species level could not occur during reciprocal starbursts of speciation.

Cospeciation can result for two completely different reasons. Either interacting species fully drive speciation in each other, as might occur in some interactions between symbionts and hosts, or interacting species codiverge simply because they live in the same environments and remain allopatric from other populations. At one side of the continuum, codivergence is driven directly by coevolutionary selection on the interaction, but at the other side codivergence is simply a result of population fragmentation or speciation in one species that forces speciation in the other.

The growing number of studies of the geographic mosaic of coevolution suggests that cospeciation should be uncommon in truly coevolving species. Within each species, populations differ geographically in their interactions and their traits. Multiple new evolutionary and coevolutionary directions can arise from a single ancestral species on one side of the interaction or the other. As coevolving species diversify and descendent populations come back into second contact, the possibilities increase for shifting among partners.

The same holds for most interactions, whether coevolving or not. There are often repeated opportunities over the history of a lineage for species to shift habitats and preferences in interactions with other species. For example, feather lice of birds have switched among avian orders at least twice during their evolutionary history (Johnson et al. 2011b), and the obligate marine worms that live within echinoderms have undergone occasional shifts onto new host lineages during the several hundred million years of their association (Lanterbecq et al. 2010).

There are, though, three situations that favor cospeciation in interacting species. One is in interactions between vertically transmitted symbionts and their hosts. If hosts transmit symbionts directly to their offspring during reproduction, then any divergence among host populations will result in divergence of the symbiont populations. Symbionts often become locally adapted to their hosts, thereby minimizing the opportunities to switch to new hosts and reinforcing cospeciation. Among the clearest examples are the cospeciation of hosts and maternally inherited symbionts that contribute directly to host nutrition, such as some bacterial symbionts that have cospeciated with insect hosts over millions or tens of millions of years (Degnan et al. 2004; Moran et al. 2008). In these situations, cospeciation can even extend to multiple species. These include, among others, the codiversification of pea aphids and *Buchnera* bacteria (Clark et al. 2000), *Camponotus* ants and their *Candidatus* bacteria (Degnan et al. 2004), and leafhoppers in the subfamily Cicadellinae and two bacterial genera (*Sulcia* and *Baumannia*) (Takiya et al. 2006). Some highly specialized symbiotic mutualisms, however, do not show evidence of codivergence. Light-producing symbionts of marine fish and squid show complex patterns of host switching and symbiont switching rather than codivergence (Dunlap et al. 2007).

The second situation that favors cospeciation is the evolutionary tracking of hosts by vertically inherited commensalistic species or by species specialized to the same habitats. Coevolutionary selection is not involved. The commensals diverge geographically along with their hosts or species diverged in tandem as their environments fragment over time (Page 2003; Segraves 2010). Most cospeciation probably results from this kind of shared habitat preference rather than from direct interactions among the species. It is an important process contributing continuity in the structure of species assemblages, but few lineages show sustained cospeciation with other lineages over long periods of time.

The third situation that sometimes favors cospeciation is the special case of pollinating floral parasites. Hundreds of plants worldwide are pollinated by insects that lay eggs in the same flowers that they pollinate. Unlike many other cases of pollination, this specialized form favors highly host-specific insect pollinators in which specialization is driven by the parasitic part of their lifestyle: a developing larva must be able to cope with the physiology of its hosts. Since the same insects also completely control the pattern of movement of pollen among plants, speciation in the plants follows from the pattern of specialization and speciation in the pollinators. Even so, host shifts sometimes occur, leading to cospeciation punctuated with occasional shifts of pollinators to distantly related host plants. Over 300 species of *Glochidion* plants are pollinated by *Epicephala* moths throughout eastern Asia, northeastern Australia, and the Pacific Ocean (fig. 16.5). These moths pollinate the flowers while laying eggs in them. Analysis of even a small subset of these lineages, however, shows evidence of some host shifts amid a more general pattern of cospeciation (fig. 16.6).

At higher taxonomic levels, diversification of these moths and their host plants shows even less codivergence. *Glochidion* is just one of multiple genera in the plant family Phyllanthaceae that have evolved pollination mutualisms with *Epicephala* moths. Members of this plant family may have evolved pollination mutualisms with the moths at least five times (Kawakita and Kato 2008), resulting in more than 500 plants in this family relying upon

Fig. 16.5 *Epicephala* moth on *Glochidion* flower. Photograph courtesy of Atsushi Kawakita.

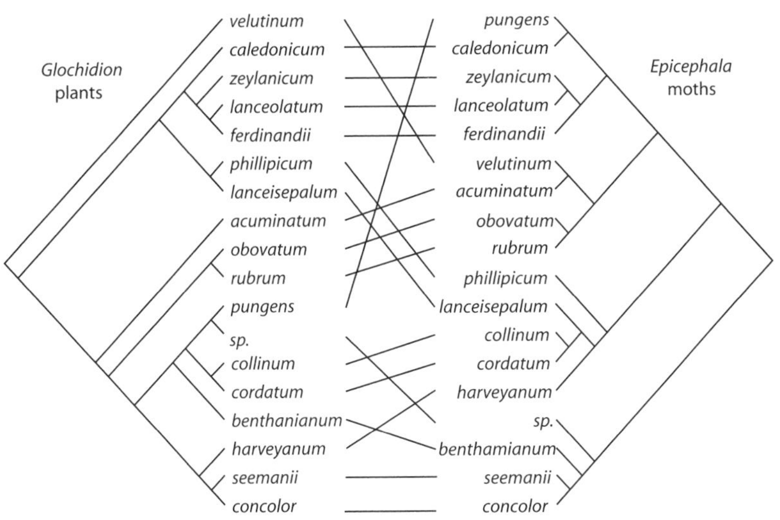

Fig. 16.6 Cospeciation in *Glochidion* plants and their coevolving pollinating floral parasitic moths in the genus *Epicephala*. In perfect cospeciation, each speciation event in either the plants or the moths would be matched by a speciation event in the other lineage. In this case, the plants and moths show an overall pattern of cospeciation with occasional shifts of the moths onto more distantly related plants. After Kawakita et al. (2004).

these moths for pollination (Kawakita 2010). Moreover, at least one clade of these moths has lost its pollinating behavior following colonization of a lineage of plants that is pollinated by ants (Kawakita and Kato 2008), and yet other members of this family are pollinated by bees or flies (Kawakita 2010). Codivergence between the Phyllanthaceae and the moths shows evidence of a combination of cospeciation and host switching at multiple taxonomic levels (Kawakita 2010). The same patterns are evident in some other coevolving interactions between plants and pollinating floral parasites, such as figs and fig wasps (Machado et al. 2005).

Most codivergence of lineages probably involves a mix of ecological and evolutionary processes. Some species sharing the same habitats will codiverge in some regions and not in others. Some coevolving populations will cospeciate, while other populations of one of the partners will switch interactions to other species. The result is constantly varying degrees of codivergence of interacting species at different geographic scales and timescales. A well-studied example is the divergence of coevolving leafcutter (attine) ants and the fungi that they cultivate as food in their fungus gardens. These fungi are directly transmitted by the ants to new colonies, generation after generation, creating the opportunity for codiversification of the ants and their symbiotic fungi. Although attine ant and fungal lineages show an overall pattern of codiversification, there have been multiple instances in which attine lineages have acquired new fungal species or cultivars during the millions of years of these associations (Mikheyev et al. 2010).

As interactions spread geographically and as different combinations of species come into contact in different regions, multiple new geographically restricted associations can develop. The complexity of the relationships between phylogeny, geography, and specialization is evident in the divergence of lice on doves, which has been intensively studied in recent decades as a model of the complexity of patterns of codivergence (Clayton et al. 2003; Harbison and Clayton 2011; Johnson et al. 2011a). The dove louse genus *Columbicola* includes eighty species, which form small groups restricted to particular groups of doves. These groups are, in turn, restricted in their geographic ranges, resulting in a mixture of phylogenetic and geographic patterns of codivergence in the lice and their hosts.

Mutualisms among free-living species show even less of a signal of codiversification, because selection favors the evolution of webs of interacting species through coevolution and through convergence of phylogenetically unrelated species. The interactions between plants and frugivores show a pattern that is common in many interactions between major lineages: one side of the interaction began to diversify earlier than the other side (fig. 16.7). Inferences about coevolution from this kind of pattern are not straightforward. Most importantly, the pattern cannot be used to suggest that two lineages have not undergone reciprocal adaptive change over time. This kind of asymmetry implies only that the species have not codiverged from the time of their first appearance.

The asymmetry could arise for any of multiple reasons. Lineages could converge on an interaction after their first appearance in the fossil record. The interacting groups could have inherently different rates of diversification. In addition, convergence of unrelated lineages on an interaction could have been greater on one side of the interaction than the other. Inferring back from current traits to those present at the first appearance of a lineage is therefore risky.

An asymmetry in diversification of two interacting lineages should therefore be interpreted only as an indication that the relationship between adaptation and speciation has differed between the two groups. Fleming and Kress (2011) suggest that the asymmetry in plant–frugivore interactions may fit a "long-fuse" model of diversification in evolving interactions. Although

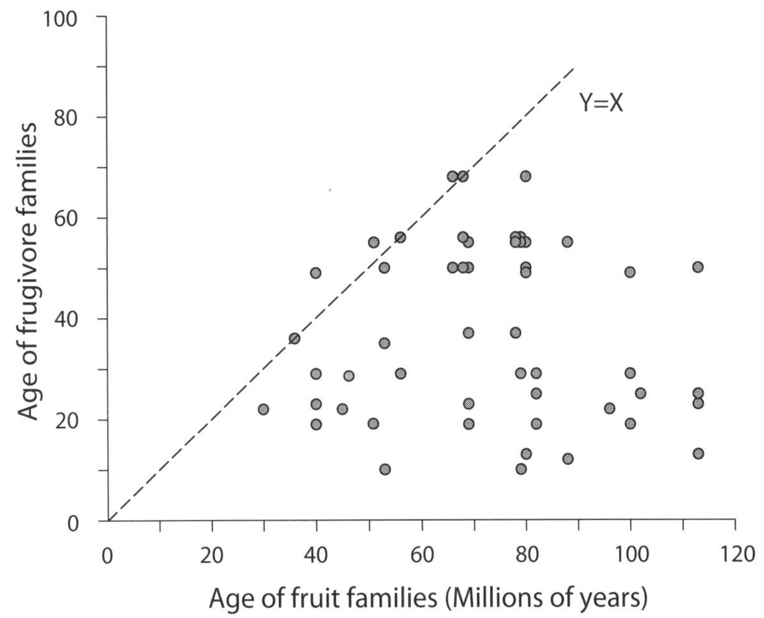

Fig. 16.7 Relative timing over the past 120 million years of the age of first appearance of angiosperm families with fleshy fruits eaten by birds and mammals, and avian and mammalian families that eat fleshy fruits. Families are included if at least some species have fleshy fruits or eat them. After Fleming and Kress (2011).

the plant families made their first appearance before most of the frugivore families, the major radiations in species within these plant families occurred much later, along with the radiations of frugivores and pollinators.

Strict cospeciation across multiple taxonomic levels, then, is rare, because so many genetic and ecological processes act in ways that prevent it. Even if codivergence occurs among some sublineages within interacting lineages, it often does not occur among other sublineages. Geographic mosaics of coevolution create mini-experiments in many different environments and often generate a richer set of outcomes than does cospeciation.

THE CHALLENGES AHEAD

We still have much to learn about which selection pressures drive wedges between populations and favor not just speciation but also unusually high rates of speciation and adaptive diversification in lineages. The view that competition between species is likely the most important force driving adaptive radiations is held most strongly by those who have studied divergence in vertebrates (Losos and Mahler 2010). Those who study other taxa often consider trophic interactions with predators, parasites, and mutualists to be more important than interspecific competition, or at least as important (Rintelen et al. 2004; Meyer and Kassen 2007).

We currently have conflicting views at even the deepest levels of the diversification of life. One view is that the upper limit of diversity is set by an overall carrying capacity (Rabosky 2009), implying competition as the great regulator of diversity, and but another argues that this assertion might hold for large animals but not for most of the diversity of life. Geerat Vermeij and Richard Grosberg (2010) have argued that the diversification of smaller species is often driven not by competition, but rather by the rarity and fragmentation of populations made possible by their sizes and their often limited dispersal and low metabolic rates. The implication of Vermeij and Grosberg's argument is that populations of small species often do not reach levels at which interspecific competition is the driving selective force. David Jablonski (2008) has argued that assessing more deeply the relative roles of competition, predation and parasitism, and mutualism remains one the major current challenges to our understanding of large-scale patterns in the evolution of biodiversity.

There is, then, still much to learn about which interactions drive adaptive radiations. We require studies designed specifically to evaluate how different forms of interaction drive diversification in different physical settings. It is a major challenge, because all species share a web of interactions that imposes selection pressures that sometimes conflict with each other and sometimes amplify each other. We need to understand how webs of interacting species assemble as lineages diversify. It is to that problem that we now turn.

17

The Web of Life

As the number of interacting species increases, the complexity of natural selection also increases to such an extent that it can sometimes seem hopeless to try to disentangle its effects. The links among species expand faster than the formation of new species, because each species often interacts with multiple other species. Species-rich biological communities therefore become a complex web of ever-changing interactions.

Ultimately, we want to know how evolution shapes large webs of interacting species. We are slowly feeling our way into this most complex of evolutionary problems. There have been glimmers of new insight in recent years as new approaches are developing. Part of the solution has been to find new ways of asking questions about how evolution shapes the web of life itself rather than just species, pairs of species, or small groups of species within the web.

This chapter explores some of these questions and approaches about evolving webs of species, and what the results may be telling us about why evolution is so relentless.

BACKGROUND: THE ASSEMBLY OF WEBS

The study of food webs has a long tradition within the field of ecology, as ecologists have searched for patterns in the links among species across three or more trophic levels (Paine 1966; May 1973; Cohen 1978; Pimm 1982; Wilson et al. 1996; Ings et al. 2009). The primary goal often has been to understand ecological stability, the resilience of webs following disturbance, and energy flow in ecological communities. Many of these studies focus on webs of interaction among plants, herbivores, and predators or parasites in terrestrial environments; among algae, herbivores, and predators in marine environments; or among plankton, planktivores, and predators in freshwater environments. The webs sometimes include complicated relationships among species, including species that feed at more than one trophic level and species that cannibalize conspecifics (Law and Rosenheim 2011; Long et al. 2011).

Another class of webs includes patterns of interaction among mutualists. Some of these webs describe interactions among free-living species such as plants and their pollinators or seed-dispersal agents, or cleaner fish and their host fish. Yet others describe mutualistic symbioses such as those between anemones and anemone fish. Complicated webs that include direct and indirect ecological outcomes, such as those involving plants and their mycorrhizal fungi and bacteria, are just starting to receive more attention now that molecular tools have made it possible to identify genetically distinct forms within these webs.

There are now hundreds of webs published in the scientific literature that can be used to search for the patterns and processes that shape the structure of ecological webs. Most of these studies treat species as if they were made up of genetically identical individuals, but that is changing. Some models have indicated that genetic, or at least phenotypic, variation among individuals can enhance the persistence of webs (Bolnick et al. 2011; Moya-Laraño 2011; Pires et al. 2011).

Webs of interacting species develop through adaptation and speciation, as they accumulate species that exploit webs in novel ways. The evolution of new lifestyles that accompanies expanding webs creates opportunities for further evolution. Selection can favor genetically based hierarchies in the preferences of predators for prey. It can favor polymorphisms in defense. It can favor lifestyles that rely on many species rather than a few species, and it can favor lifestyles in which individuals specialize their interactions to different species at different times of year. The evolutionary assembly of webs is fundamentally the process of adding new lifestyles that benefit from exploiting multiple species and adjusting the lifestyles of all species—generalists through specialists—to fit within webs.

The assembly of webs therefore involves both evolutionary and ecological diversification (fig. 17.1). Evolutionary diversification begins with local adaptation to physical environments (G × E) and geographic mosaics of interactions with other species (G × G × E interactions), which lead sometimes to speciation and occasionally to adaptive radiation. Ecological diversification begins as the evolutionary diversification generates small webs of interacting species that become linked together into large webs. The large webs, in turn, feed back to produce more small webs, as selection alters local adaptation, creates new multispecific selection mosaics, and fosters yet more speciation.

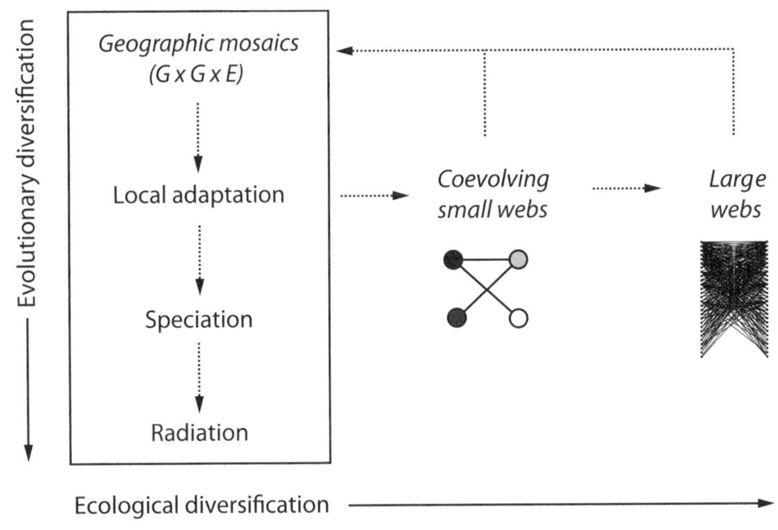

Fig. 17.1 The evolutionary and ecological diversification of webs of interacting species. As species diversify through adaptation and speciation, they also diversify in the ways in which they interact with other species and in the number of other species with which they interact, thereby forming larger interaction webs.

This process generates webs of interacting species that are structured hierarchically (Olesen et al. 2007; Bascompte 2010; Jordano 2010). Large webs are made up of multiple small webs, sometimes called compartments, modules, or motifs, which are connected to varying degrees. These small webs often include groups of closely related species and are a common unit in many coevolving interactions. They may include, for example, three insect species within a genus that feed only on three locally occurring plants within a genus. These small webs often also contain additional species that alter selection and coevolution among the interacting species. The webs are then linked by supergeneralists that interact with species in multiple small webs.

Large webs are therefore maps of patterns of specialization—who interacts with whom—that are due in part to the phylogenetic history of who is related to whom. The distribution of natural selection within large webs becomes complicated quickly as they expand in size. Daniel Janzen captured the problem in a paper titled "When Is It Coevolution" (1980), in which he suggested distinguishing between specific and diffuse coevolution. That distinction helped to generate greater rigor in studies purporting to show coevolution between pairs of species, but it had an unexpected effect on the study of multispecific coevolution.

During much of the 1980s, a general view developed that if pairwise coevolution could not be shown, then reciprocal selection probably was not important in that interaction. Even if coevolution was important, it was too complicated to study because selection was diffusely distributed among multiple interacting species. Nonetheless, any field biologist studying webs of interacting species in nature could see general patterns that suggested reciprocal evolution of traits, but these situations were often dealt with briefly by referring to the phrase diffuse coevolution and stopping there. The overuse of that phrase as a catchall, well beyond anything Janzen had intended, held back coevolutionary biology for many years. It hindered the development of testable hypotheses on multispecific coevolution and the role of natural selection in assembling large webs of interacting species.

At least six developments in recent years have fostered more specific hypotheses on the role of natural selection in shaping webs of interacting species. One is the development of a more populational approach to the study of coevolution, such as the geographic mosaic theory of coevolution discussed in earlier chapters. Another is the development of techniques to determine whether a local pairwise interaction is under selection primarily on that pair of species or if the interaction is shaped in important ways by yet other interacting species (Iwao and Rausher 1997; Gómez and Zamora 2000; Stinchcombe and Rausher 2002; Strauss et al. 2005). These informative studies often use the phrase diffuse coevolution, but their goal has been to evaluate the structure of selection among multiple interacting species. These and related approaches are helping us understand more precisely how traits may be shaped simultaneously by interactions with multiple species (Pieterse and Dicke 2007; Dicke and Baldwin 2010).

The third development is the formulation of specific hypotheses on the structure of selection within small webs (Davies and Brooke 1989; Thompson 2005; Nuismer and Thompson 2006; Gómez et al. 2009). These studies have explored how selection changes when small groups of species interact rather than pairs of species. The fourth development follows from the third and is the realization that coevolution often favors formation of webs of interacting species rather than pairwise interactions between species (Thompson 1982, 1994; Jordano 1987; Guimarães et al. 2007).

The fifth is the development of approaches to ask how the evolution of new lifestyles changes the structure of selection in small and large webs (Bascompte et al. 2003; Olesen et al. 2007, Jordano 2010; Guimarães et al. 2011). These approaches have drawn heavily on the broader theory of networks and its application to fields as different as neurobiology and patterns of connections within the Internet (Salathé et al. 2005; Alon 2007). Although Pedro Jordano's (1987) analysis of connectance and asymmetries in pollination and seed-dispersal webs introduced some formal aspects of network theory into evolutionary ecology and coevolutionary biology about a quarter century ago, it was not until the past decade that this approach has become a major way of thinking about evolution and coevolution within webs.

The sixth is the development of phylogenetic approaches and neutral approaches to the study of assemblages of species, both between and within trophic levels. Phylogenetic approaches explore the extent to which the evolutionary relatedness of species shapes the structure of assemblages (Bascompte and Jordano 2007; Rezende et al. 2007; Kraft et al. 2008). Neutral approaches evaluate the extent to which assemblages can be explained by the statistical properties of assemblages, independent of the idiosyncratic biology of species (Hubbell 2001; Krishna et al. 2008; Volkov et al. 2009; Rosindell et al. 2011)

Collectively, these developments are leading to a new understanding of how selection acts on webs to assemble, diversify, disassemble through species loss, and continually reassemble as environmental conditions change.

EVOLUTION AND COEVOLUTION IN SMALL WEBS

The simplest way to explore how evolution and coevolution can shape webs of interacting species is to ask how the range of evolutionary solutions expands as species interact in small groups rather than as pairs of species. Earlier chapters in this book considered some aspects of how the evolution of traits could be amplified or reduced as the number of interacting species expands into small webs of several species. As webs expand, completely novel evolutionary solutions become possible. I consider two of those solutions here as exemplars (see Thompson 2005 for a more extended discussion). These are the process of coevolutionary alternation in antagonistic interactions and the evolution of convergence and complementarity in mutualistic interactions.

The hypothesis of coevolutionary alternation suggests that groups of predators and prey coevolve through a cyclical process. The hypothesis assumes that predators have genetically based preference hierarchies for prey species and that these hierarchies can be altered by selection. It also assumes that defenses in prey are costly to maintain. Selection therefore favors escalated defenses in prey populations when they are under attack, but it also favors lowered defense levels when they are not under attack. During coevolutionary alternation, selection favors predators that preferentially attack the currently least defended prey species. These high levels of attack favor the evolution of increased defenses in the local population of that prey species. That, in turn, favors local predators that preferentially attack other, less defended prey species, and the cycle continues from there (Davies and Brooke 1989; Thompson 2005). When coevolutionary alternation is modeled mathematically, it produces cycles of increased and decreased specialization to particular prey in the predators and fluctuating levels of defense among potential prey species (Nuismer and Thompson 2006).

Models of coevolutionary alternation indicate that coevolutionary arms races are not an inevitable result of predator–prey (or parasite–host) interactions. Predators do not become increasingly specialized over time to any particular prey species by honing their abilities to attack just that one species. Instead, they cycle between periods of increased and decreased specificity to any one prey species. Meanwhile the prey species also cycle in their levels of their defenses. Predator–prey coevolution becomes a process of fluctuating specialization in one or more predator species and fluctuating defenses among a group of prey species.

Escalating arms races can sometimes occur in these interactions, if evolutionary changes in specialization in the predator occur faster than evolutionary changes in defenses in the prey. The escalation occurs because the defenses in the prey species never relax completely to their pre-attack states. Each cycle of predator alternation among prey species results in a higher average level of defenses among all the prey species. The models of coevolutionary alternation therefore show one way by which webs of interaction open up evolutionary solutions not possible between pairs of interacting species. Other scenarios of evolution in small webs of interacting predators and prey are possible, if prey are able to evolve adaptations effective against multiple predator species rather than evolve separate adaptations against each predator species (Ellner and Becks 2010).

As with antagonistic interactions, webs of mutualistic species also offer solutions beyond those found in pairs of interacting species. Many mutualisms involve short-term interactions between free-living species. Examples include the interactions between plants and their pollinators or seed-dispersal agents and the interactions between cleaner fish and host fish. In these mutualisms, selection often acts as a coevolutionary vortex that draws species into the interaction through two processes. Between trophic levels, mutualists evolve toward complementary traits (e.g., hummingbird bills and corolla tubes on flowers). Within trophic levels, species converge to exploit the same mutualists (e.g., flowers from multiple plant families converging on the traits that attract hummingbirds). Complementarity of traits and convergence of species together generate a coevolutionary process that leads to formation of mutualistic webs (Thompson 1994, 2005).

This point is crucial. Coevolution does not always lead to highly specific interactions between pairs of species. In some forms of interaction, it can lead directly to the formation of webs of interacting species.

There are multiple ways in which mutualistic webs can evolve. A pair of coevolving species can form the core of the interaction across all ecosystems, with additional species converging on that core. Alternatively, the central coevolving species could differ among ecosystems. Although an interaction would look highly diffuse when lumped across the entire geographic range of the interacting species, selection would show clear structure at the level of local communities.

Even seemingly extreme mutualisms among free-living species often generate small webs, rather than pairs, of coevolving species. A clear example is the interaction between long-tongued flies and about twenty species of plants with long-tubed flowers in several families in South Africa (Pauw et al. 2008). *Moegistorhynchus longirostris* flies have the longest proboscis relative to body size of any known pollinator (fig. 17.2). These flies and some other long-tongued flies occur in the lowlands of southwestern South Africa and are replaced inland at higher elevations with two species of unrelated long-tongued flies.

Plant species with long-tubed flowers adapted to these flies occur in different combinations among lowland and upland habitats. Flowers whose tube lengths are well matched to the flies are pollinated by the flies. Flowers with tubes shorter than the long tongues of the flies are visited but not pollinated, because the pollen on the heads of the flies never touches the floral reproductive parts. Plant species whose tubes are too long for the flies would be pollinated, but the flies would not be rewarded. Local communities differ in the number of plants that attract these flies and the number of long-tongued fly species that visit these plants. In some communities only some plant species visited by the flies have the potential to interact mutualistically with the flies (fig. 17.3). At most sites, one or more plant species fall outside the range at which mutualism is possible. That places selection on those plant species either to evolve in the direction of joining the assemblage or to evolve toward use of alternative pollinators.

There are certainly multiple impressive cases of exclusive pairwise interactions in mutualistic interactions between free-living species, but they are the exceptions. Detailed study of those special cases is important to our understanding of when coevolution favors reciprocal specialization. For example, the hundreds of plants pollinated by floral parasites often show

Fig. 17.2 A *Moegistorhynchus longirostris* fly visiting a *Lapeirousia anceps* (Iridaceae) flower in South Africa. Photograph courtesy of Anton Pauw.

reciprocal specialization, but specialization in those interactions is driven to a great degree by the parasitic part of the interaction. Closely related nonpollinating insect species that attack the same plants generally show the same high level of specialization to particular hosts as the pollinating species.

Among most other interactions between plants and pollinators, such exclusive relationships are rare (Fenster et al. 2004; Waser and Ollerton 2006). A minority of plant species have a pollination relationship with a single insect species; a small number of bird-pollinated plants have exclusive relationships with one bird species (Temeles and Kress 2003, 2010); only one

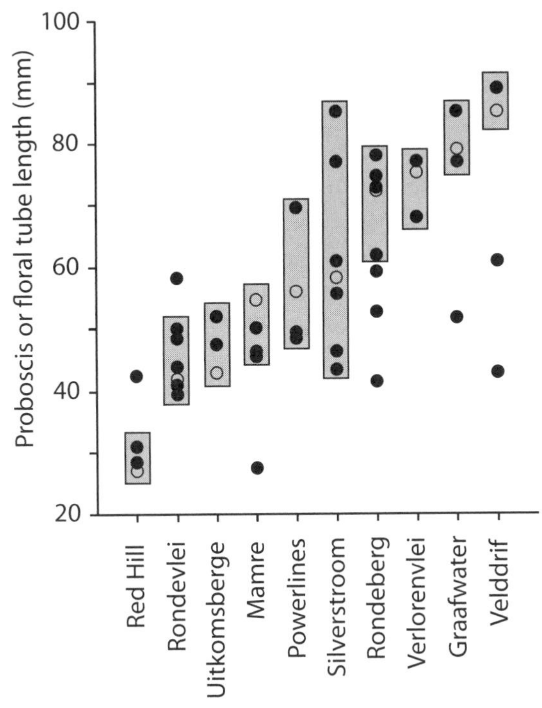

Fig. 17.3 The assemblage of interacting long-tongued fly species (*open circles*) and long-tubed flower species (*closed circles*) at multiple sites in South Africa. The regional assemblage includes three fly species and up to 20 plant species, but only subsets of species occur at each site. Individual species are not distinguished in the figure. Shaded boxes enclose the plant species whose tubes are within the range of variation in proboscis lengths found among the local flies and could therefore interact mutualistically with the flies. After Pauw et al. (2008).

bat-pollinated plant species is known to have an exclusive relationship with one bat species (Muchhala 2006); and there are no known reciprocally exclusive relationships between plants and frugivores (Jordano 1987, pers. comm.) The sheer fact that there are few examples of extreme reciprocal specialization among free-living mutualists suggests that these interactions generally tend to evolve through formation of webs.

Coevolution through complementarity and convergence in mutualisms and coevolutionary alternation between predators and prey (or between parasites and hosts) are two specific ways by which small webs of species may coevolve. Community-wide character displacement considered in earlier chapters is another. There are other ways as well (Thompson 2005). These hypotheses show that it is possible to replace general arguments about diffuse coevolution with specific hypotheses on the structure of selection on multispecific interactions.

THE STRUCTURE OF LARGE WEBS

Tracking selection and evolution in large webs requires new ways of asking questions about the evolution of interactions. Those ways are increasing quickly through incorporation of methods from network theory, because multiple mathematical properties of networks are informative for interpretation of biological webs (Bascompte et al. 2003; Jordano et al. 2003; Olesen et al. 2007; Guimarães et al. 2011). Much of network theory has been developed through work on the Internet and through analyses of social networks, neural networks, and physical processes. In recent years, these efforts have provided new ways of thinking about webs of interacting species. Even journals more traditionally focused on physical and mathematical principles, such as *Physical Review E* and *Physical Review Letters*, are rich sources of current studies attempting to apply network theory to webs of biological species (Lässig et al. 2001; Rosvall et al. 2006; Murase et al. 2010). As with all cross-disciplinary attempts, it is sometimes difficult to interpret the direct applicability of some of these models to biological interactions studied in the field. But the back-and-forth process of model development and application is generating new approaches on how to analyze large webs and the connections within them.

These new approaches have brought with them a specific jargon to describe patterns in links among species. Network studies tend to focus on the extent to which interactions among species are nested, asymmetric, and compartmentalized (Bascompte et al. 2003; Jordano et al. 2003; Olesen et al. 2007; Bascompte 2010). Nested webs occur if generalists interact with generalists, specialists interact with generalists, and specialists rarely interact with other specialists (fig. 17.4). That is, the interactions are asymmetric in the degrees of specialization between interacting species. Each set of interactions within a fully nested web is asymmetric and nested within an even larger set of interactions. Small subwebs, though, sometimes form identifiable compartments within the larger web. These are groups of species that tend to interact more with each other than with species in other groups. Most networks have generally focused on interactions between two trophic levels, such as plants and herbivores, pollinators, or seed dispersers.

The degrees of nestedness, asymmetry, and compartmentalization differ among forms of interaction in ways that follow from what we know so far from coevolutionary theory. Mutualistic webs among free-living species

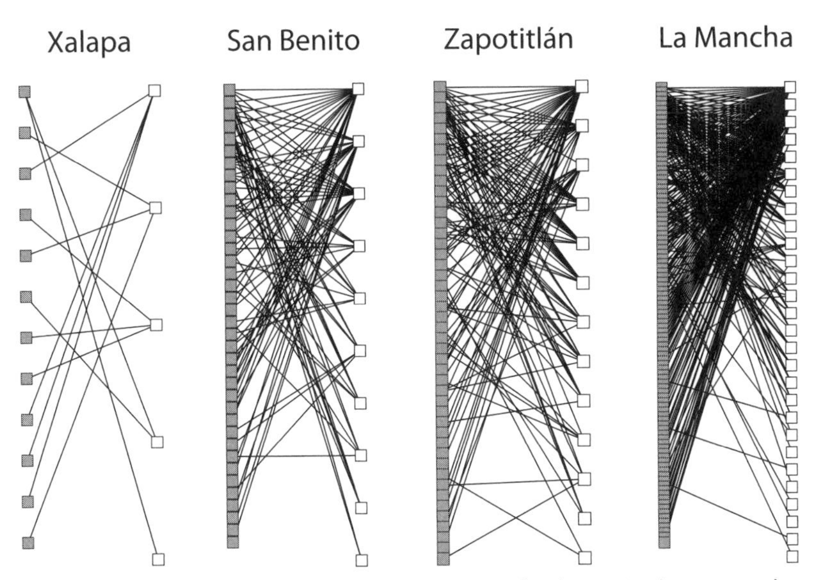

Fig. 17.4 Nested interactions between plants with extrafloral nectaries (*gray squares*) and ants (*open squares*) along an elevational gradient in Mexico. The species are ranked from the most generalist (i.e., the greatest number of links) at the top to the most specialized at the bottom. All but Xalapa show significant nestedness. Adapted from Guimarães et al. (2006).

are often more nested and less compartmentalized than are antagonistic webs between predators and prey species (Bascompte et al. 2003; Thébault and Fontaine 2010; Fontaine et al. 2011). We would expect this pattern if antagonistic interactions evolve through processes such as coevolutionary alternation, which can create small webs of interacting species. Mutualistic webs among free-living species also tend to be larger, more nested, and less compartmentalized than mutualistic webs among symbiotic species (Guimarães et al. 2007; Fontaine et al. 2011). This pattern follows from what we know about selection in these different forms of mutualism. Mutualisms among free-living species tend to draw in additional species, whereas mutualistic symbioses tend to favor reciprocal specialization.

There are, though, no discrete classes of webs among different forms of interaction. Rather there are gradations of structure, with some forms of interaction more likely to exhibit particular structures than other forms and with larger webs building up from smaller, sometimes nested, webs (Lewinsohn et al. 2006; Kondoh et al. 2010). Some degree of nestedness, asymmetry, and compartmentalization is expected by chance alone from the statistical properties of networks. Those effects need to be partitioned to find the biological influences on the structure of real webs.

Most interactions, not just mutualisms, tend to show some degree of nestedness. For example, the interactions among plants, herbivores, and large herbivores in the Serengeti show some nestedness (Dobson 2009). All biological webs also tend to form compartments. But the number of species in compartments seems to grow less over time in mutualisms between free-living species, resulting in fewer compartments than in interactions between predators and prey. Hence, the study of webs is about the evolutionary and ecological processes that generate differences in degrees of nestedness, asymmetry, and compartmentalization in different forms of interaction.

The structure of interaction webs also seems to depend on the overall size of the web. It is possible to see patterns forming as webs incorporate more species. For example, an analysis of over fifty pollination webs showed that those with more than 150 species were always compartmentalized, whereas those with less than 50 species never were (Olesen et al. 2007). The same holds for nestedness. Among four sites along an elevational gradient in Mexico, there is strong nestedness in interactions between ants and plants with extrafloral nectaries in sites with many species but little indication of nestedness at the site with fewest interacting species (fig. 17.4).

Within large webs, however, the structure of the web often seems to remain the same even as individual species change in their presence or relative abundance. At La Mancha, which is a Mexican site with many species of interacting ants and plants, the structure of the web has remained the same over a decade, even though the community had been invaded by new ant and plant species (Díaz-Castelzao et al. 2010). These are important results for our understanding of the ecological and evolutionary dynamics of webs. They suggest that the structure of webs does not depend simply on the relative abundance of species change. Although relative abundance affects asymmetries in patterns of interactions found in individual webs (Bascompte and Jordano 2007; Blüthgen et al. 2007; Vásquez et al. 2007), interactions tend to form webs with a somewhat predictable structure.

It therefore seems likely that the assembly of webs involves a combination of ecological sorting and the evolution of life histories that differ among different forms of interaction. Species are able to join a local web because they are able to fit into a particular position within the web. Over time, natural selection continues to shape the positions and traits of each species within each web. Some analyses suggest that compartmentalization within webs may be due mostly to coevolution and constraints, whereas nestedness may be due more to species abundances (Lewinsohn et al. 2006). The relative abundance of species, however, may be due in part to the evolutionary history of interactions within webs. Abundance of species is as much an outcome of the evolution of webs as it is an influence on the ecological interactions within webs.

PHYLOGENETIC STRUCTURE WITHIN WEBS

The other major influence on the structure of webs is the phylogenetic composition of the interacting species. The traits of each species are not independent of the traits of other closely related species. Close relatives are often similar in their ecological niches, as has long been noted by evolutionary biologists studying lineages at multiple timescales (Eldredge and Gould 1972; Harvey et al. 1995; Webb et al. 2002; Futuyma 2010; Wiens et al. 2010). Consequently, the phylogenetic relationships among species have the potential to shape the patterns of links among species within local webs.

A major analysis of 36 plant–pollinator webs and 23 plant–frugivore webs found that phylogenetic relationships among species explain the number of interactions found among species in about a third of the webs (Rezende et al. 2007). The deviation of real biological webs from random webs therefore results, in part, from the phylogenetic structure of the web of life and the traits found among species. Simulated extinctions within these webs generate cascading coextinction among related species (Rezende et al. 2007). The occasional phylogenetic disassembly of webs following extinctions may therefore be as important to the evolution of webs as their phylogenetic assembly.

Phylogenetic effects, however, may be less strong in mutualisms among free-living species than in some other forms of interaction, because selection in these interactions favors attraction of unrelated species into a coevolutionary vortex. Groups of unrelated species often converge on similar ecological characteristics. In the Pantanal of Brazil, the fleshy-fruited seeds dispersed by vertebrates fall into five webs (i.e., modules) that form a larger web through the interactions of a few species (fig. 17.5). Phylogenetic relationships contribute to the formation of these small webs, but most of the structure of the web has resulted from convergence of phylogenetically unrelated species.

Phylogenetic effects may also differ among trophic levels within a web. Studies of three trophic-level webs rather than two are still few, but some techniques are now in place (Ives and Godfray 2006). In a detailed analysis of interactions among plants, leaf-mining insects, and the parasitoids that attack the insects, host phylogeny was seen to have a strong effect on the

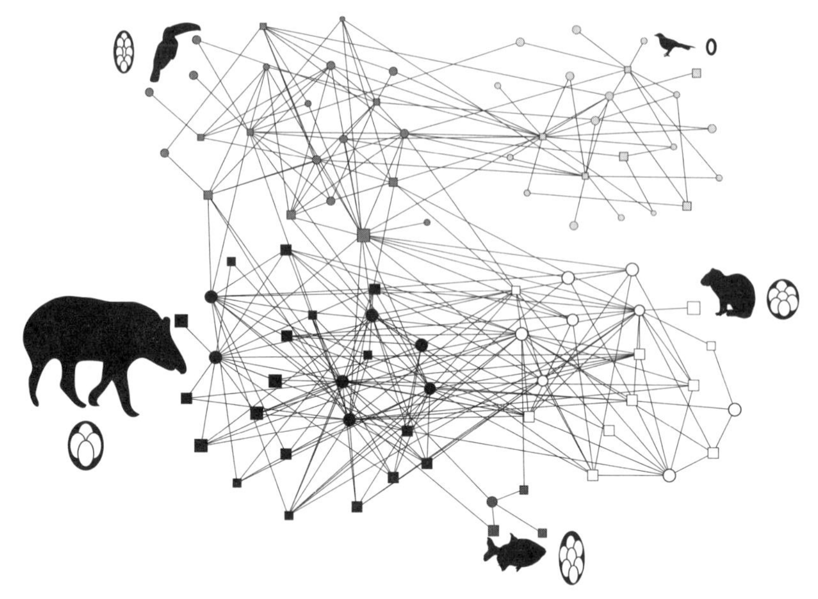

Fig. 17.5 Web of interaction between seeds and seed-dispersing vertebrates in the Pantanal of Brazil. The web involves 46 plant species (*squares*) and 46 animal species (circles). The interactions group into five smaller webs, represented here by differences in sizes and shading of the squares and circles and by a silhouette of the major seed disperser and relative average size of seeds in that module. After Donatti et al. (2011).

structure of the web but parasitoid phylogeny did not (Ives and Godfray 2006). Hence, phylogenetic effects at each higher level are not just a subset of the effects found at the lower levels.

Phylogenetic structure, however, can have important effects if it acts strongly at higher trophic levels. Predators can have powerful, and sometimes cascading, effects on lower trophic levels by altering either the diversity or abundance of prey species (Schmitz 2010; Estes et al. 2011). That could influence some aspects of how interactions are compartmentalized within ecosystems, thereby affecting the future evolution of interactions. A detailed study of a web of marine producer species, invertebrate species, and vertebrate species in the Caribbean showed that compartmentalization of predator–prey interactions is shaped by phylogenetic relationships among species, body sizes, and the spatial distribution of species (Rezende et al. 2009). In these communities, closely related top predators, usually sharks, occupy different compartments within the webs, and they are a major factor in organizing these webs into subwebs. Removal of these top predators, as is happening now through overfishing, influences the ecological architecture of these webs (Bascompte et al. 2005; Rezende et al. 2009). That change in web structure will likely have a domino effect on the future evolution of these webs. Similar trophic downgrading is occurring in most major ecosystems (Estes et al. 2011), imposing novel selection pressures on all species within interaction webs.

Phylogenetic methods are also becoming useful in asking whether selection favors predictable patterns of trait evolution within large assemblages. These methods can help to separate the roles of various ecological processes, the evolutionary divergence in traits among competing species, and the simple sorting of species by habitat during the formation of webs. These studies therefore can narrow the likely targets of selection and the forms of selection on species as they enter assemblages (Emerson and Gillespie 2008; Kraft et al. 2008). In some cases phylogenetic analyses can help determine if the morphological and other traits of species within large webs are constrained by the phylogenetic relatedness of species or if they are evolutionarily labile. Lability in traits can result in convergent lifestyles as species from different phylogenetic groups evolve in response to similar abiotic and biotic selection pressures.

The results of community-wide phylogenetic analyses are sometimes surprising, showing less phylogenetic constraint and more evolutionary convergence than expected. In the seasonal dry forest of Chamela, Mexico, the leaves of deciduous trees share a group of similar traits, as do the leaves of evergreen trees. Leaf traits, though, appear to have evolved under relatively little phylogenetic constraint (fig. 17.6). Trees from multiple lineages share similar traits, whereas some closely related species differ greatly in their traits (Pringle et al. 2010). The leaf traits included in this analysis have evolved in response to multiple abiotic and biotic selection pressures and therefore represent a composite picture of how natural selection mixes and matches traits as species evolve within complex webs.

Phylogenetic perspectives on community-wide patterns of assembly and selection are expanding in ways that link ecological and evolutionary processes. Peter Price (2003), for example, has argued that the great differences in population dynamics found within assemblages of insects can be grouped phylogenetically. Most species within lineages tend to exploit resources and interact with other species in similar ways, resulting in similar patterns of population dynamics. Species within some phylogenetic lineages tend to be more eruptive than others, quickly growing to high population levels and then crashing. Species in other lineages tend to fluctuate much less over time. Species in some lineages are also more dynamic in their geographic ranges than are species in other lineages. These and similar

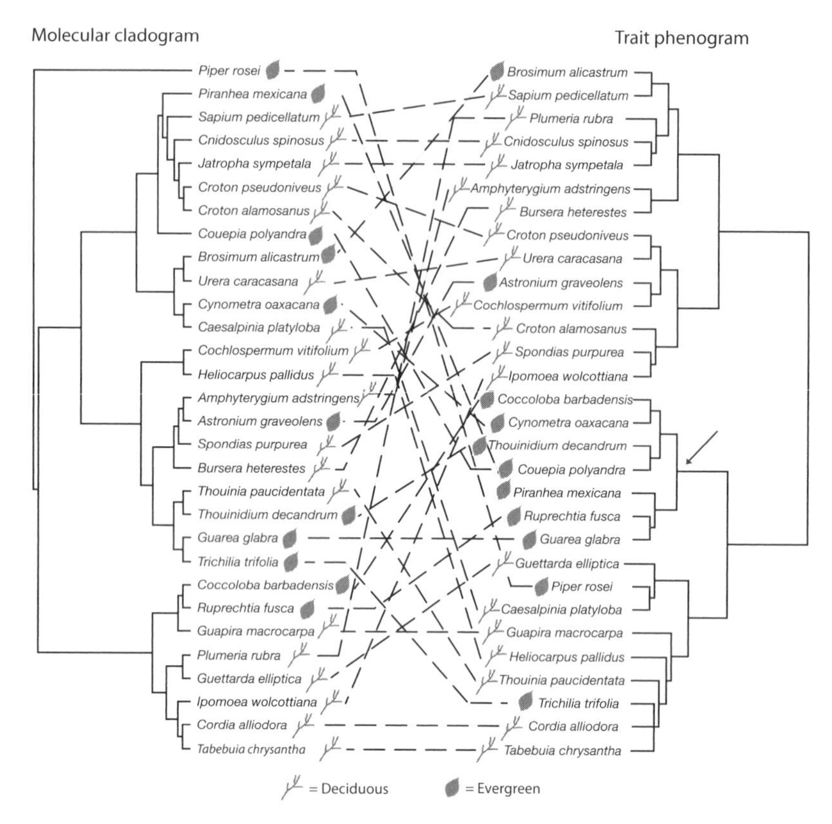

Fig. 17.6 Comparison of the pattern of molecular relatedness among 30 tree species in the seasonal dry forest of Chamela, Mexico, and the pattern of trait similarity among those species. The trait phenogram is based on a composite of five leaf characteristics, including toughness, amount of latex, ratio of carbon to nitrogen within the leaves, trichome density, and water content. The figure also shows that deciduous species (denoted with a branch) and evergreen species (denoted with a leaf) are distributed throughout the cladogram and phenogram, but most evergreen species tend to group in one part of the phenogram, as indicated by the arrow. After Pringle et al. (2010).

approaches should help to build the next generation of hypotheses on how webs are organized not just locally but also regionally and how selection on traits varies phylogenetically.

LARGE WEBS AND THE EVOLUTION OF NEW LIFESTYLES

The interplay between evolutionary and ecological diversification favors the evolution of fundamentally new lifestyles within large webs. The most generalist species within large webs are sometimes supergeneralists, which interact with species across the smaller webs (Jordano et al. 2003). The lifestyles of species such as honeybees and large tropical frugivores became possible only after webs have evolved into large sizes. These species rely on their ability to interact with a wide range of species within webs. They are one of the ways in which coevolution has produced the evolution of fundamentally new lifestyles as webs have increased in size over time (Thompson 1994, 2005). We are only now starting to explore how large webs assemble ecologically and evolutionarily through the evolution of novel lifestyles, and how each stage of that process affects the subsequent evolution of species and coevolution within webs.

Lifestyles that exploit mutualistic webs often appear repeatedly during the diversification of lineages. About 528 plant species are pollinated by bats, and these plants are distributed among 67 families and 28 orders (Fleming et al. 2009). Bat-pollinated plants have arisen erratically across almost the entire phylogeny of angiosperms, although they have arisen more commonly in more recently derived families. Similarly, dispersal of fruits by vertebrates has arisen multiple times throughout the angiosperms (Fleming and Kress 2011), as has seed dispersal by ants (Rico-Gray and Oliveira 2007). Within monocots alone, adaptations for dispersal by ants have arisen at least 24 times (Dunn et al. 2007). It seems, then, that webs assemble through adaptation within current lifestyles and occasional addition of new lifestyles.

The assembly and disassembly of large webs bring together all the different rates and trajectories of adaptation discussed in the previous chapters. Some recent mathematical models have begun to explore how evolution and coevolution might shape the traits of species within these large webs (Guimarães et al. 2007, 2011). These studies provide the first glimpses into questions such as how common and pervasive must coevolution be within a web to influence general patterns of trait evolution among species.

The models compare webs in which evolution and coevolution occur in different ways (fig. 17.7). The simplest simulations are those in which populations change only through the effects of evolution. That is, there is no coevolution between species, and evolution occurs only through the effects of one species on another species. Some additional complexity develops when evolution is allowed to cascade through several species: evolution of species A favors evolution in species B which favors evolution in species C. Another set of simulations allows coevolution but only between pairs of species. Still others allow coevolution between pairs of species to cascade into additional evolutionary events in yet other species. Finally, the most complicated simulations are those in which coevolution can also occur indirectly. The reciprocal effects flow through two or more other species before they affect evolution in the first species. The models have evaluated how the complementarity of traits evolves between trophic levels (e.g., corolla tube length and hummingbird bill length) and convergence of traits evolves within trophic levels (e.g., tube length among plant species). The models use, as a starting point, the structure of webs described in empirical studies (Guimarães et al. 2011).

These models suggest that coevolution has two major effects on species-rich mutualistic webs. Coevolution speeds up the overall rate of evolution within a web. Even occasional reciprocal evolution among some species increases the rate of evolutionary change in traits throughout the web. Webs

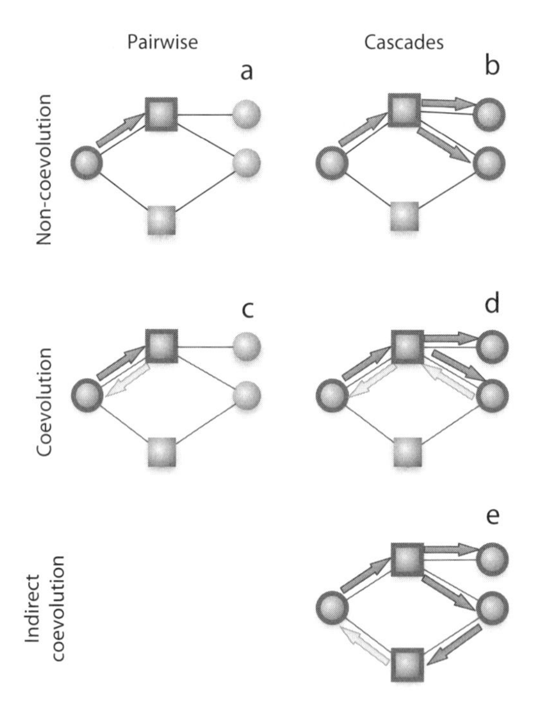

Fig. 17.7 Different levels of complexity of evolution and coevolution within multispecific webs. Squares represent one trophic level (e.g., animals) and circles represent another (e.g., plants). Arrows indicate whether natural selection acting on interacting species is unidirectional or reciprocal. Within networks, evolution and coevolution occur not just directly between pairs of species but also indirectly through cascades of evolutionary and coevolutionary change among species. After Guimarães et al. (2011).

with some coevolving interactions also reach greater levels of complementarity of traits between trophic levels and greater convergence of traits within trophic levels than webs lacking coevolution. The coevolutionary process therefore has the potential to have a strong effect on the evolution of traits even in large webs (Guimarães et al. 2011).

Webs with supergeneralists show even greater convergence of traits. These species act, directly and indirectly, as selection agents on species throughout a web. The models suggest that the introduction of supergeneralists, such as honeybees introduced onto new continents, or the loss of supergeneralists, as is occurring with large fruit-eating primates in many tropical environments, will have major effects on the evolution of mutualistic webs (Guimarães et al. 2011).

COMPETITION AND FACILITATION

As research continues on the evolution of webs between trophic levels, much remains to be understood also about the evolution of webs within trophic levels. The patterns of interaction among species within each trophic level have just as much complexity as those between trophic levels. The evolution of character divergence driven by competition provides some structure to the web of interactions within trophic levels (see chap. 11), but so can the evolution of facilitation among species (Valiente-Banuet and Verdc 2007; Verdú and Valiente-Banuet 2008). Facilitation occurs when individuals of one species ameliorate the local environments of other species, thereby increasing survival, growth rates, or reproduction of the facilitated individuals (Callaway 2007). We still do not know how competition and facilitation together shape the evolution of assemblages within trophic levels.

The null hypothesis is that the traits of species and the interactions do not matter for the structure of local assemblages within a trophic level. The pool of species can be viewed as equivalent. They have the same chance of reproduction and death. This view is formally captured in the unified neutral theory of biodiversity and biogeography by Stephen Hubbell and colleagues (Hubbell 2001; Rosindell et al. 2011). In that view, the structure of assemblages results mostly from the limited dispersal abilities of species, occasional speciation, and ecological drift, which is the stochastic loss of species from the assemblage over time. A local assemblage of, say, plant species results from continual colonization of individuals from a well-mixed regional species pool. The relative abundance of species in the regional pool varies stochastically over time due to ecological drift. The same process occurs again at larger regional scales, but speciation occasionally occurs.

The theory does not deny that species differ in their characteristics. Rather it argues that these differences do not have much of an effect on the overall summary statistics that ecologists often use to describe the relative abundance of species found within local assemblages. Classical models of neutral theory assume that, when an individual dies, it is immediately replaced by another individual. These models also assume that speciation is instantaneous. Any birth has a tiny probability of being a new species, which is a simplified way of modeling speciation from small, local populations.

Neutral theory is able to capture much of the local structure and dynamics of woody plant assemblages such as those found in the forest dynamics plots that have been established in permanent experimental plots of trees on multiple continents. That suggests that at these small spatial and temporal scales, limited dispersal together with occasional local loss or addition of species can capture the observed patterns in the number and relative abundances of species. A detailed analysis of the 50-hectare forest dynamics plot on Barro Colorado Island in Panama has suggested that intraspecific competition is more important than interspecific competition in structuring plant assemblages at that spatial scale (Volkov et al. 2009). The analysis, though, is restricted to the study of the relative abundances of plant species and the dynamics of their births and deaths within the plot. More generally, Volkov and colleagues have noted that the relatively weak effects of interspecific competition in comparison to intraspecific could themselves be a result of past selection, suggesting that it "is as though evolution has chosen weakly interacting species for proximal coexistence" (Volkov et al. 2009). That suggestion would be consistent with evolutionary character displacement among species.

Even as they compete, individuals of different species can facilitate the survival and reproduction of each other. Although ecological facilitation has been found among many taxa and in many ecological contexts (Bronstein 2009), it has most often been studied in interactions involving terrestrial plants or marine species (Bruno et al. 2003; Callaway 2007; Verdú et al. 2009; Adam 2011). Facilitation has often been suggested to occur among species in relatively harsh physical environments, but perhaps that is just where the effects are more readily apparent.

Some interactions may vary between facilitation and competition over the life of an individual. Early in life, the survival and growth of an individual may be enhanced by living on or near a "nurse" individual of another species, but later those same individuals may compete. As an extension of that view, facilitation has long been one of the processes invoked in the ecological process of succession within some communities, acting through its effect on some key species within assemblages.

More recently, studies have begun to ask how facilitation is distributed among species throughout large assemblages. Among the most interesting new observations are those suggesting that there are macroevolutionary patterns in the formation of assemblages that result from facilitation. These studies have relied on new techniques in the field of phylogenetic community ecology and new software (e.g., Phylocom) for evaluating how species assemble within communities with respect to their phylogenetic relatedness (Webb et al. 2002, 2008). Using the assumption that closely related species are more likely to compete than more distantly related species, Alfonso Valiente-Banuet, Miguel Verdú, and colleagues have hypothesized that phylogenetically distant plants should co-occur more often than by chance alone, and phylogenetically distant plants should more often be involved in facilitative interactions (Valiente-Banuet et al. 2006; Verdú and Valiente-Banuet 2008).

They have tested this hypothesis within plant assemblages in Mediterranean-like climates and more desert-like climates. Mediterranean climates have wet winters and dry summers, and they produce similar vegetation types in restricted regions of the Mediterranean Basin, North and South America, South Africa, and Australia. Studies of the arid and semiarid assemblages at seven sites in Mexico suggest that nurse plants and facilitated plants are more phylogenetically distant then predicted by a null model of random association (Valiente-Banuet and Verdú 2007).

Even successional patterns show a phylogenetic signal. At two sites studied in Spain, plants at the earliest stages of post-fire succession associate independent of their phylogenetic relatedness, but later in succession more distantly related plants tend to associate with each other (Verdú et al. 2009). Even later in succession, that phylogenetic pattern of association disappears, possibly because the balance between facilitative and competitive interactions changes.

The assemblages in Mediterranean-like climates tend to be nested in ways similar to that found in mutualistic interactions between trophic levels (Verdú and Valiente-Banuet 2008). Some species act ecologically as generalist nurse species that facilitate the establishment of many other plant species within an assemblage. There are, then, some predictable patterns in these plant assemblages, suggesting that selection acting within this trophic level is driven by a combination of facilitation and competition that follows from the phylogenetic relatedness of the interacting species.

Evolution within these assemblages is therefore likely molded, driven, and constrained by this broader macroevolutionary context.

WHY WEBS ARE SO ENTANGLED

There are at least six major reasons why the web of life has become so entangled, and we have only a rudimentary understanding of each of these reasons. First, the evolutionary processes that shape interactions within trophic levels are entangled with the processes that shape interactions among trophic levels. Even though we know that to be so, studies often concentrate on the structure of selection either within or among trophic levels. That is beginning to change as researchers grapple with how to study even more complex webs of interaction (e.g., Melián et al. 2009; Van Der Putten 2009; Fontaine et al. 2011). Some studies show that asymmetries and connectance in mutualistic and antagonistic webs change with increasing species diversity but in different ways (Thébault and Fontaine 2008). Antagonistic interactions may be better stabilized by a compartmentalized web in which the compartments are only weakly connected, and mutualistic interactions may be better stabilized by a nested and highly connected structure. These are the same structures that seem to be favored by selection on different forms of interaction, at least for interactions among free-living species. It seems, then, that natural selection and ecological stability may work in tandem during the evolution of large webs.

Second, species not only evolve but also coevolve, and some forms of coevolution favor the evolution of interaction webs. We have hypotheses on how selection shapes multispecific interactions, and network theory is helping to provide new approaches to studying multispecific coevolution. The challenge now is to develop better ways to evaluate the growing number of views of how coevolutionary and ecological processes change as species are added or removed from large webs. This is only one of the many possible questions to ask about the coevolutionary process in large webs, but it is a crucial one at a time when introduced species are altering communities worldwide.

The other major challenge in coevolution is to understand how the geographic mosaic of coevolution proceeds among large webs of interacting species. What happens when much of the gene flow into or out of a large web is due to movements of only a few species, as has been shown in some interactions between frugivores and plants (Jordano et al. 2007)? What happens when physically separated communities are connected by migrating species? What happens to the regional patterns of trait evolution if a supergeneralist is mutualistic with a crucial species in one community but antagonistic with that species in another community? These are not questions about which we will get detailed, quantitative answers. Rather the goal is to understand how evolution and coevolution in complex webs alter the general trajectories and rates of evolution in species, their traits, and their ecological interactions with other species.

Third, coevolving webs generate indirect effects on other species, and these effects can link seemingly different parts of webs. Even the web of species associated with a single plant species can generate highly complex relationships (fig. 17.8). Among the insects that feed on willow (*Salix miyabeana*) in northern Japan, a spittlebug species lays eggs in the distal parts of new shoots, which kills the terminal shoot but stimulates growth of new proximal shoots the next spring, which increases the density of 23 species of leaf-rolling caterpillars. After the larvae complete development, many of the abandoned rolled leaves are colonized by a species of aphid that specializes in using these structures. These aphids then attract 3 ant species that tend the aphids. Once attracted to the plants, the ants also kill large numbers of a leaf beetle, reducing beetle numbers by as much as 60 percent on some plants with high aphid numbers.

This web of 1 plant species and 29 insect species includes more indirect than direct interactions, about as many nontrophic interactions as trophic interactions, and a combination of direct and indirect mutualistic as well as antagonistic interactions (Ohgushi 2008). Similar indirect effects are found in all webs of interacting species and span all forms of interaction and spatial scales (Brown et al. 2001; Knight et al. 2005; Ohgushi 2005; Hartley and Gange 2009). No models or empirical studies have yet been able to explore most of these effects beyond small webs of interacting species.

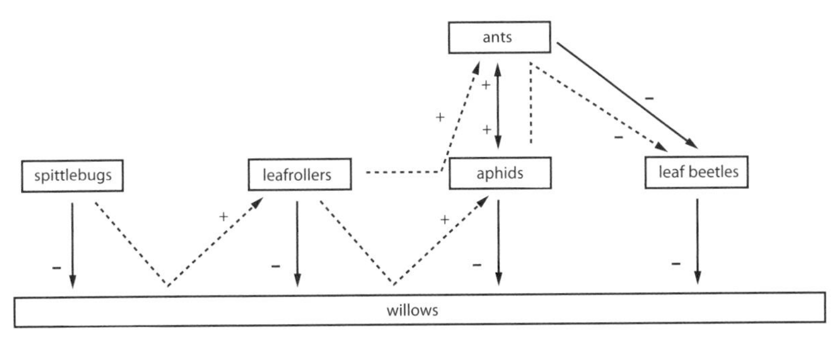

Fig. 17.8 Direct and indirect ecological effects among insects feeding on a willow species in northern Japan. Direct effects are indicated by solid lines and indirect effects by dashed lines. The ecological effects are indicated by plus and minus signs. After Ohgushi (2008).

Fourth, traits that evolved initially in response to one source of selection are often co-opted later to respond to new selective forces, thereby further tangling the web of life. Defenses become attractants, competitive behaviors grade into cooperative behaviors, and true signals become false signals. Our models of the evolution of webs generally assume that a trait could escalate or fluctuate within or among populations, but we know little about what happens to complex webs when selection favors completely new uses for old traits. Since each species interacts with multiple other species, each trait has the potential to be pulled in any of several directions, sometimes dragging with it multiple other traits. Darwin's book (1862) *On the Various Contrivances by Which British and Foreign Orchids Are Fertilized by Insects, and the Good Effects of Intercrossing*, which was published just three years after *The Origin of Species*, was explicitly about the reuse of floral structures for advertisement, transfer of pollen, or restriction of access to particular pollinators. In these orchids the same floral parts have been altered repeatedly by natural selection to attract different mutualists.

Selection on floral traits can even fundamentally alter the functions of the traits. In his worldwide studies of the evolution of floral traits in the diverse pantropical genus of vines called *Dalechampia*, Scott Armbruster has shown that the resin used as attractants for resin-collecting bees arose from compounds used as defenses against herbivores and that those attractants have been redeployed as defenses (Armbruster et al. 2009). All species are full of similar traits whose current use differs from the selection pressures that initially favored those traits, simply because much of the process of natural selection involves the development of new traits and new interactions by tinkering with old ones. The result can be completely unexpected new links among species within webs.

Fifth, although we know that the evolution of interaction webs allows the evolution of new lifestyles, we do not yet know in any systematic way how species such as supergeneralists, cheaters within mutualisms, and species that rely on facilitation by other species shape the evolution of webs. The effects of particular lifestyles on webs have been studied individually (Verdú and Valiente-Banuet 2008; Genini et al. 2010; Guimarães et al. 2011), but we do not know how webs accumulate species with "web-dependent" lifestyles and how these species generally shape the future evolution of webs.

Sixth, there is a great deal of ecologically important variation in nature beyond that which has traditionally been analyzed in most large webs. Cryptic species, ecological polymorphisms within species, complex life cycles in which species behave as ecologically different species at different life history stages, and migratory species that contribute to different webs at different times of year all add important dimensions to most large webs. The problem of cryptic species is being addressed through the use of molecular barcoding, which is helping to identify the actual structure of genetic diversity within and among communities (Janzen et al. 2009; Kress et al. 2009; Kesanakurti et al. 2011; Martiny et al. 2011). The question that remains, though, is how much deeper do we need to probe within the genetic diversity of species to capture the ecologically important genetic variation that structures webs within and among communities.

We need to understand better these six important aspects of the evolution of interaction webs if we are to understand how and why the entangled bank has become so entangled. We have come a long way, but there clearly is still much to learn.

THE CHALLENGES AHEAD

In a deeply biological sense, the webs of life *are* the history of life. The webs affect everything from the functioning of ecosystems to the functioning of the human gut. By extension, they are what our efforts to conserve life must be about. But gaining a deeper understanding of how natural selection shapes the webs of life is one of the most difficult challenges in evolutionary biology. Local patterns within webs are the result of processes happening at multiple temporal and spatial scales, some of which are beyond the range of any kind of direct experiments we can perform.

The developing molecular, phylogenetic, and ecological tools, however, are showing that we can go beyond what we already know. As we continue to disassemble and reassemble whole ecosystems through our activities, we should be able to apply these new tools to these inadvertent experiments on the composition of large assemblages. We can evaluate the evolutionary effects not just of adding new and sometimes invasive species, but also of losing crucial species. The loss of large predators from marine and terrestrial environments worldwide is having major effects on the diversity and relative abundance of species and on the patterns of interaction within webs of species (Dirzo et al. 2007; Estes et al. 2011). The ecological effects of these losses are as important as the spread of newly introduced and invasive species. Studies of the evolutionary effects of species loss and addition have not, until recently, kept pace with the ecological studies, but we now have the tools for better integration of ecological and evolutionary approaches.

Even more broadly, the evolutionary ecology of ecosystem dynamics and complexity remains an open frontier. A small but expanding number of studies, though, is showing that evolution and coevolution within webs can have major effects on nutrient dynamics, decomposition rates, and other ecosystem processes (Palkovacs et al. 2009; Post and Palkovacs 2009; Schmitz 2010). These ecosystem changes have the potential to affect future evolution within webs of interacting species.

The community and ecosystem effects of ongoing evolution are contributing to our changing perceptions of major patterns in the history of life. It is to those changing perceptions that we now turn.

Part 6

Synthesis

18

Our Changing Perceptions

Our perceptions of the relentlessness of evolution continue to change as we learn more about the pace and drivers of evolutionary change over many temporal scales and spatial scales. We are improving our ability to analyze evolutionary change in adaptive traits over short timescales, obtaining more information from molecular studies about adaptation and diversification over longer timescales, and finding new ways to extract information from fossils.

This chapter explores our shifting views on some of the fundamental patterns in the history of life and how those patterns fit with our growing understanding of the evolutionary process.

HAVE THE WEBS OF LIFE CHANGED DURING EVOLUTIONARY HISTORY?

We are getting better at inferring the processes that shape current webs of interacting species, but we still do not know whether these webs have changed in fundamental ways as the history of life has progressed. Are there larger patterns in how natural selection has shaped the web of life? David Jablonski (2008) has argued that we have so far mostly failed to exploit the rich fossil history of interactions among species in ways that would help us understand how current webs of interacting species fit within the history of past webs. Fossil webs are less complete than the webs we are able to study now, but fossils can provide important insights into some aspects of the history of specialization, the trajectories and dynamics of evolution in morphological traits, the patterns of interaction among species, and the evolution of new lifestyles.

In recent years paleobiologists have shown that it is possible to extract from fossils a remarkable level of detail about some species and their interactions. The study of fossils has provided insights into patterns of predation on hard-shelled species, the evolution of specialized pollinators and flowers, the variety of ways by which insect herbivores feed on leaves, and even changes in the levels of herbivory sustained by plants over millions of years (Vermeij 2002a; Alexander and Dietl 2003; Wilf et al. 2006; Ren et al. 2009; Wing et al. 2009). Studies of up to 25 plant species found in fossils from 53 million to 59 million years ago show evidence of 71 types of insect-feeding damage on their leaves, with increases in the diversity of damage during periods of warmer temperature (Currano et al. 2010).

It has even become possible to use fossilized leaves to assess the effects of parasitism on behavior in some animal species. Some fungi manipulate their ant hosts by causing them to bite into leaf tissue in a death grip. This grip puts the ant's body in a good position to disperse the fungal spores that develop as the hyphae grow out from the dead ant onto its body surface. Fossilized leaves from 48 million years ago show evidence of these death grips, suggesting that this manipulation of ant behavior has persisted at least that long (Hughes et al. 2011).

In marine environments, it has been possible to identify multiple forms of predation on molluscs by assessing patterns of shell damage (Vermeij 1987; Alexander and Dietl 2003; Dietl and Vega 2008). By studying changes in damage patterns in different layers of the fossil record, paleobiologists have been able to identify innovations in defense and counterdefense in species with shells and their shell-breaking predators, and then track the history of shell breaking over long periods of geological history.

One view of larger-scale patterns in the web of life, resulting from the study of marine fossils, is that the earth has become an inherently more dangerous place to live over many eons. Geerat Vermeij (1987, 2002b) has argued that enemies—competitors, predators, dangerous prey—are the major agents of natural selection, and has based this argument on the observation that so many characters of species are devoted to fighting off enemies. He has spent his professional career studying, among other things, how the astonishingly diverse sizes and shapes of the shells of fossil molluscs have changed over evolutionary time.

Shells are tremendously effective defenses against a range of enemies, but many of the enemies of molluscs have highly effective ways of breaking through these defenses. Some predators pry open shells, and others drill right through them. Marine species with shells seem to have responded by evolving more elaborate defenses, and those elaborations are evident in the fossil record. Using his work on shells as a starting point, Vermeij (1987, 1994) has argued that, with so many enemies about, coevolution has been supplanted over billions of years by a macroevolutionary trend that he calls escalation.

By this view, defenses and counterdefenses commonly evolve not through any identifiable process of reciprocal evolutionary change in which individual species impose strong selection pressures on one another. Instead, species evolve to ever higher levels of defense and counterdefense through the sum total of their increasingly dangerous enemies. It is the danger in the interaction that matters—the possibility of being killed, injured, or somehow having reproductive output depressed by the interaction. By this view, even the evolution of some mutualistic symbioses is often one of the elaborate ways of coping with ever more formidable enemies.

The crux of the argument for a worldwide escalation of defenses is that natural selection imposed by enemies is greater than natural selection imposed by food, essentially the life–dinner principle proposed by Dawkins and Krebs (1979) writ large. Selection acts to increase the defenses of populations against the sum total of their enemies. Only when the food itself is very dangerous to obtain does it approach dangerous enemies as a major agent of selection.

This kind of argument can be explored with mathematical models of the evolution of defense and counterdefense. You take the costs of failure and the benefits of success for predators and prey and evaluate what happens to their populations over time. As with all models, the answer depends on just how you fit the costs and benefits into the model and how much weight you put on each of the costs and each of the benefits. The models can then be analyzed to explore which ecological conditions favor greater asymmetry in costs and benefits between the predator and prey populations, and which conditions favor greater symmetry. The different ecological conditions can be anything from the relative densities of predators and prey to the number of meals required by the predator over its lifetime. The more symmetrical the costs and benefits shaped by a set of ecological conditions, the more equal the selection pressures and the greater the potential for coevolution.

In a classic set of models investigating the possibility of coevolutionary arms races, Peter Abrams (1986) evaluated 24 possible combinations of escalating costs and associated benefits. He found that prey responded to new adaptations in their predators in almost all cases, and predators responded to new antipredator adaptations in two-thirds of the combinations. Coevolutionary arms races were possible, but not inevitable, and occurred about 60 percent of the time. So, we are left with the glass half full and half empty in interpreting these results. Coevolution is not guaranteed, because asymmetries occur, but it is a likely outcome.

No one yet has tried a systematic revaluation of these results based on the newer generation of coevolutionary models. It would, though, be a worthwhile analysis. Abrams (pers. comm.) thinks that any reevaluation would likely reach similar conclusions: interactions between predators and prey can produce a wide range of possible evolutionary outcomes. That view is reinforced by models such as those for coevolutionary alternation in which changes in specialization and cycling of levels of defense are as likely as escalation (Nuismer and Thompson 2006). That does not mean escalation does not occur as a major macroevolutionary trend. Instead, it means that it could be one of multiple trends shaping the elaboration of webs of antagonistic species.

Antagonistic interactions are just one component of larger webs, and we do not yet know in any systematic way how other forms of interaction within webs have changed over time. It appears, though, that symbiotic mutualisms among species have not only expanded over geological time but that they also may have become more complex as species have accumulated new partnerships. Yet even this point must be tempered by our growing appreciation of the pervasiveness of horizontal gene transfer among prokaryotes and the interactions within microbial mats, which are themselves highly complex. We do not yet know how mutualistic webs form among prokaryotes, and how they may have changed over billions of years. Mutualisms among free-living species are also pervasive, but here too we have no clear idea of whether they have expanded over the eons in ways that have fundamentally restructured how selection acts within these webs. What we know is that mutualisms are now required by the majority of species on earth.

We cannot, then, go back to the fossil record and simply tally up the cases of apparent escalation of traits, or any other single pattern, as the basis of our assessment of which changes in webs have been most important during the diversification of life. The preserved record is bound to miss much of the continual but nondirectional coevolutionary change that may be just as common as, and sometimes more important than, directional change. It will certainly miss most of the mutualisms.

Nevertheless, there are indications that at least some major marine lineages have undergone long-term directional changes in traits consistent with selection imposed by enemies, either through non-coevolutionary escalation or through a process such as coevolutionary alternation. During the Mesozoic marine revolution in the middle to late Mesozoic period, predation on marine benthic invertebrates appears to have increased significantly (Vermeij 1977, 1987), as did disturbance of marine sediments by species searching for prey, a process called bioturbation or bulldozing (Thayer 1979). Molluscs from this period show evidence of increased defenses, such as shells more resistant to breakage by predators (Vermeij 1977). Other benthic marine lineages also changed in multiple ways associated with defense or avoidance of enemies. There was an increase in the proportion of mobile organisms, a decline in vulnerable organisms living on the substrate, and an increase in organisms living within the substrate rather than on the benthic surface (Aberhan et al. 2006).

These changes occurred in tandem across multiple marine lineages, and they were not correlated with any obvious changes in the abiotic environment such as seawater chemistry, climate, or global seawater level. Selection imposed by enemies is therefore the most likely cause of these changes in the web of benthic life. Just as interesting is the geographic component to the patterns. The observed faunal-wide trends were greater in the paleotropics than in other regions, suggesting that selection imposed by enemies may have varied among environments.

Our understanding of variation within and among fossil and living species would also benefit from better integration of approaches among the evolutionary disciplines. Gregory Dietl and Karl Flessa (2011) have argued that the long timeline of the fossil record provides a baseline for the patterns of phenotypic variation found currently within lineages. The range of variation sometimes increases or decreases during different periods of diversification of lineages. The fossil record of a lineage can provide insight into whether the current range of variation in morphological traits fills the range of morphological space that a lineage occupied over geological time. This is different from using the fossil record to ask how the average values of traits have shifted over time (Dietl and Flessa 2011).

New insights from fossils into the history of the web of life are currently limited more by the number of paleobiologists exploring these questions than by the availability of fossils. The insights are increasing as techniques from network theory and molecular biology are applied to fossil assemblages. Examples include analyses of the trophic structure of paleovertebrate assemblages (Roopnarine and Angielczyk 2012), studies of interactions based on stable carbon and nitrogen isotope signatures among species within communities (Fox-Dobbs et al. 2008), and the comparative phylogeography of Pleistocene mammals (Barnett et al. 2009; Shapiro et al. 2009; Enk et al. 2011).

Some of these studies suggest that interactions among species are dynamic over time partly because the geographic ranges of species are constantly changing. These changes are sometimes punctuated by wholesale turnover in assemblages, periodic cycles of change in diversity, and, rarely, intercontinental exchanges that greatly altered the webs of interacting species (Vrba 2005; Lieberman and Melott 2007; Weir et al. 2009). As studies of the ecological structure and dynamics of fossil assemblages continue, we will be in a better position to interpret the long-term implications of the continual evolution that we see in assemblages of living species.

METAPHORS OF EVOLUTIONARY CHANGE

As we have realized that evolution involves continual fine-scale change, we have also realized that those changes are not ecologically or evolutionarily trivial. Populations and species persist partly because they are constantly evolving in small ways. The high levels of genetic variation maintained within most species—through balancing selection, sexual reproduction, horizontal gene transfer, and differences in selection among populations—make those continual small changes possible.

As Francois Jacob (1977) once noted, evolution works like a tinkerer rather than like an engineer. Natural selection uses whatever is laying around in the morphologies, behaviors, physiologies, and life histories of organisms and remolds these characteristics to new uses. Occasionally a new mutation appears that provides something new to work with. Even without new mutations, an environmental change can alter how genes are expressed and, as a result, how species interact with each other, thereby opening new avenues of evolutionary change. It is even possible to evolve by continually altering in small ways the degrees to which populations specialize on other species within a web. As a result, all populations are jury-rigged rather than engineered, and sometimes that is accomplished by co-opting the fully functioning genomes of other species.

At the shortest timescales, then, most populations are essentially groups of boats sitting in the water but moored by a loose rope to a dock. The boats are constantly rising and falling as tides rise and fall and as storms batter the coast. Some boats are better at surviving those storms than others. The boats are not going anywhere, but most stay afloat because they are able to shift their position in the water relative to the constant changes in the surrounding water levels. If they could not continually shift up, down, and sideways in the restricted ecological space allowed by the rope, they would be swamped.

In a different version of this kind of metaphor, but one emphasizing the relationship between short- and long-term evolution, Niles Eldredge has thought of the continual back and forth of evolutionary change as a sloshing bucket (2001, 2008). Species are constantly sloshing around in the bucket, with much of the water first pushed up against one side and then another. Most of the change continues to occur within the bucket with only occasional spills. Big spills involving much water occur only following some large disturbance. In Eldredge's view these large disturbances occur mostly during periods of major physical catastrophe, producing the kinds of major turnover in biological communities that some paleobiologists have found in the fossil record (Vrba 2005).

The sloshing bucket view of evolution is an expansion of Eldredge and Gould's (1972) theory of punctuated equilibrium. In the "punkeek" view, most directional evolutionary change is concentrated at the time of formation of new species, with each species often remaining essentially unchanged with respect to sustained directional change for hundreds of thousands or millions of years. The sloshing bucket view argues that the speciation events themselves are often clustered around periods of major physical disturbance. Water in the buckets continues to slosh around as the applecart moves until, as Eldredge (2008) writes, "environmental change overturns the ecological applecart." The evolutionary hierarchy of life—organisms, populations, species, assemblages of species—is occasionally altered by these rare events, changing the distribution of organisms, populations, local ecosystems, regional ecosystems, and the state of the biosphere itself.

The essential point in all these metaphors, and evolutionary theory itself, is that evolution does not make populations inherently better at what they do. That idea was captured in Leigh Van Valen's (1973) argument that evolution is like the world described by the Red Queen in Lewis Carroll's *Through the Looking Glass*. In the book the Red Queen tells Alice, "Now *here*, you see, it takes all the running *you* can do, to keep in the same place." Van Valen's evolutionary version was that the environment of any species or lineage continually deteriorates at a fairly constant rate, because the physical world is always changing and enemies are always evolving. Van Valen argued that interactions among species—the biological environment rather the physical environment—were the driving force in this constant deterioration of the environment. In Van Valen's view, not only does coevolution produce diversity but it continually trims diversity as some species lose coevolutionary arms races while others win. The winners, however, are always only temporary. In making that argument, Van Valen was drawing on arguments going back to Darwin, who considered competition and other forms of interaction to be pervasive agents of natural selection.

Why the environments of species deteriorate has become clearer as we have come to understand the extent to which genes are expressed in different ways in different environments. As environments change, so do the effects of gene expression on fitness. Those effects are further magnified by the variable outcomes of interspecific interactions, which themselves vary with gene expression and community context. No set of gene combinations is best in all environments. Evolutionary tinkering, constantly shifting moored boats, sloshing buckets, and Red Queens are therefore all ways of expressing our increasing appreciation of the relentlessness of evolution.

SELECTION IMPOSED BY PHYSICAL AND BIOTIC ENVIRONMENTS

That tinkering, though, has been punctuated by periods of mass extinction. Given what we now know about how much of life has been repeatedly destroyed during periods of mass extinction, is it reasonable to argue that biotic interactions are as important as the physical environment in shaping the web of life? After all, there have been multiple periods of massive extinction throughout the history of life, and some of them have eliminated most species. At least one of these periods of mass destruction was caused by an asteroid striking the earth. That is about as physical as you can get.

The fine-tuned nuances of coevolutionary relationships do not matter much in the midst of a global disaster at that scale, unless the interactions among species determine who is likely to be among the remaining few percent that survive. Yet, surely the interactions must make a difference in who dies and who survives during mass extinctions. It has been difficult, though, to get data from the fossil record in a way that would answer that question directly. Instead, we often have to look for clues among living taxa. For example, adaptation to temperature is sometimes mediated by interactions with another species. A single mutation in the obligate bacterial symbionts (*Buchnera*) of pea aphids affects the tolerance of these aphids to warm or cool temperatures, affecting the age at which the aphids reproduce and their rate of reproduction (Dunbar et al. 2007).

Whether the physical environment or the biotic environment has been more important in shaping evolution therefore depends on the particular questions we are asking about the history of life. It is like many questions that have no simple answer. The physical environment and its cataclysms set the template on which evolution occurs. The massive physical disturbances continually reset that template during periods of mass extinction. The diversity of life, though, is the result of two processes: origination and extinction.

Massive changes in the physical environment impose heavy tolls, stripping the earth of its diversity through extinctions. These occasional events alter the mix of species on earth, creating huge turnovers in the structure of ecological communities. Before, during, and after each of these periods of mass extinction, natural selection takes the survivors, remolds them to fit into their new environments, and contributes to the formation of new species. If temperatures become higher than before, natural selection may favor individuals whose physiological enzymes function best at those raised temperatures. Selection culls the cold-adapted individuals while favoring the warm-adapted ones. It is easy to focus on these problems faced by organisms adapting to a changed physical landscape.

Meanwhile, the biotic template has changed as well, and that changed template appears often to be the major driver of the origination of new species. A large fraction of the studies showing ecological speciation suggests that differences in interactions among species in different physical environments are the focus of the divergent selection that drives speciation. During a mass extinction, some competitors, predators, parasites, and mutualists disappear, while others proliferate. As populations and species re-expand across environments, they form new alliances with other species, evolve defenses against new enemies, and exploit new opportunities to manipulate other species. Adaptation to the new physical and biotic landscapes proceeds hand in hand. The crucial difference in adapting to the physical and biotic environments is the coevolutionary feedback loop of the biotic environment.

Evolutionary responses to other species produce the counterresponses that are the basis of coevolution, local adaptation, and the divergence of populations into new species. Species track one another over evolutionary time, and that tracking links species together into ecological communities. Coevolution among the survivors of mass extinctions must often continue right through the periods of mass extinction. Some populations become extinct, but others squeeze through, still coevolving.

At the level of broadest brushstrokes, then, it seems that physical environments set the overall templates of adaptation and diversification and drive major patterns of extinction, while biotic environments often drive a great deal of the dynamics of adaptation, the process of speciation, and the organization of the web of life. The effects of physical and biotic environments, though, are intertwined so closely—through genotype-by-environment and genotype-by-genotype-by-environment interactions—that the dynamics of biodiversity cannot be said to be driven more by one or the other.

PROGRESS AND EVOLVABILITY

When evolution is directional, we might ask if it is progressive in any meaningful sense. Evolutionary biologists have long had a love–hate affair with the concept of progress, going back all the way to Darwin (Ruse 1993).

Vermeij's hypothesis (1987, 1994) that antagonistic interactions have escalated and Margulis's argument (Margulis and Fester 1991; Margulis and Sagan 2002) that the story of evolution is mostly about the evolution of symbioses can be viewed as arguments for progressive evolution in the restricted sense of directional change.

Like all things, though, our general perception depends on our definitions. Richard Dawkins (1986, 2004) has argued that evolution is commonly progressive in at least two senses. One is that natural selection favors mutations that make organisms better at performing particular functions. Hence, the human eye is better at visual acuity than the ancient light-sensitive cells found far back in the animal lineages that eventually gave rise to humans. Increased visual acuity developed step-by-step—that is, progressively—as natural selection favored in each generation those individuals with the most acute eyesight. In that view, evolutionary progress is adaptation: the accumulation of favorable mutations over time within evolutionary lineages.

Nothing in this argument suggests that the human eye or any other particular trait of organisms is an inevitable result of evolution. By this view, evolution is progressive only in the sense that natural selection works by accumulating favorable mutations while constantly eliminating deleterious mutations. Each gain or loss simply reflects selection in that particular environment. We can evaluate whether individuals within evolutionary lineages have progressed along particular pathways toward higher visual acuity, faster running speeds, or more aerodynamic flight by evaluating the accumulation of traits producing these results. Those changes, though, do not make organisms inherently better than their ancestors. Writing about interactions between species, Dawkins (1986) has summarized this point well: "There has been progress in design, but no progress in accomplishment, specifically because there has been equal progress in design on both sides of the arms race."

Evolution, though, can be progressive in a larger sense as well, as Stephen Jay Gould (1996, 2002) discussed repeatedly. Of course evolution looks to be progressive, he argued. Life started small and primitive and then diversified. Inevitably, evolution had to be toward increasing complexity in at least some organisms and larger size in others. Starting from small, single-celled organisms, there was no possibility of evolving to be simpler. So, if you track changes in the overall averages of traits over time, you come away with the biased impression that life has progressed. Yet what has changed over time is the range of complexity, the range of sizes of organisms, the range of shell thicknesses, and the range of running speeds of animals above that lower bound. The variance has increased. Meanwhile, many species are still doing just fine near the lower bounds of complexity, size, and other traits even after billions of years.

That brings us back to the second sense in which Dawkins has argued that evolution has been progressive. Despite the fact that there are still many bacteria faring just fine, each new major innovation in the structure of life on earth has opened up new vistas for subsequent evolution. The evolution of chromosomes, meiosis, sex, and segmentation in organisms, to name a few, created possibilities for subsequent evolution and diversification not present before these innovations. Dawkins (1989) has called this progressive accumulation of new evolutionary possibilities the "evolution of evolvability," and he has viewed these changes as evidence that evolution has been progressive in that sense. Gould had no argument with the importance of these new innovations in the diversification of life. He viewed them, though, as part of a change in the range of adaptation that has occurred in organisms over time.

Discussions on the evolution of evolvability continue (Wagner 2005; Draghi and Wagner 2008; Pavlicev et al. 2011), but the interpretation is becoming increasingly complex. As we learn more about the ability of bacteria to evolve through multiple genetic mechanisms, it has become clear that all life has evolved a wide range of abilities to evolve. Bacteria evolve through evolution of their own genome, through horizontal gene transfer, through development of cooperative behaviors, and, in myriad ways, through their use of the eukaryotes.

Has there, then, been progressive escalation of defenses against enemies or a progressive increase in the complexity of mutualisms in either Dawkins's or Gould's sense, or in any other sense of the word *progress*? The upper bound of complexity in defense, and counterdefense, has certainly increased during the history of life, creating a broader range of possibilities, either up or down, within the many traits of species. At the very least the evolutionary diversification of mutualistic and antagonistic interactions has made further evolution and diversification possible. If adaptive diversification itself is viewed as progress, then life has increased in the range of mechanisms that allow populations to explore an ever-wider range of ecological space over billions of years as the web of life has grown in complexity. It is the grand and continuing accomplishment of the process of natural selection.

THE CHALLENGES AHEAD

Even as the number of examples of rapid evolution continues to expand year after year, there remains a perception among nonbiologists that evolution

Table 18.1 *Euphemisms for the word* evolution *found in the scientific literature and in popular science writing*

Euphemism	*Example from one or more articles*
Accelerate	Pest resistance . . . accelerated
Acquire	Acquired new traits
Become	Weeds are becoming resistant
Change	Influenza virus . . . frequently changes
Come up with	Pests come up with
Create	Creates . . . new strains
Develop	Developed resistance
Devise	Devised novel ways
Emerge	Emergence of resistance
Get used to	Pathogens get used to
Grow	Pathogens have grown resistant
Overcome	Overcome their vulnerability
Sprout	Resistant [forms] . . . have begun sprouting and spreading
Thwart	Thwarted efforts to

Source: Adapted from Thompson (2008b).

must be something rare and unobservable. There are many reasons for this misconception, but a major one is that biologists and science writers often do not use the word *evolution* even when they write about it. There are so many synonyms used by writers when discussing evolution that some popular articles on the topic do not use the word *evolution* even once. Most of the time, it is not that writers are substituting other words specifically to avoid using the word *evolution* for a general audience that is leery of the word. Rather, its usage is just a consequence of the normal process that all writers go through as they try to avoid using the same word too many times—and end up sometimes not using it even once.

In writing about evolution, however, the synonyms have become so common, and they have become so clichéd, that they are now the default words or phrases when writing about some kinds of evolution. Consider the synonyms in table 18.1. You can mix and match these words to cover many of the general points about evolution that writers often try to make. It has become almost standard, for example, to use the phrase *developed resistance*, rather than *evolved resistance*, when writing about antibiotic or pesticide resistance. Biologists know that the word *developed*, when used in this context, means *evolved*, but many other readers do not.

These alternatives to the word *evolution* seem to be used most commonly to describe instances of very rapid evolution—the kinds of evolutionary change in pests and pathogens that make it into the news. At a time when it has become clear that evolution affects every aspect of human society from agriculture to medicine to conservation, we need to make it clearer to nonbiologists that all these euphemisms mean evolution. It is something that is happening now. We can do that just by using the word *evolution*. Otherwise, nonbiologists will never come to appreciate that evolution truly is relentless even on timescales observable during their own lifetimes.

19

Conclusions

The central conclusion from these chapters is that adaptive evolution is pervasive, relentless, and often surprisingly fast. That conclusion comes from long-term studies of natural populations; corroborating molecular studies; refined statistical, mathematical, and analytical assessments; and a new generation of studies in experimental evolution. Our growing appreciation of the relentlessness of evolution has been accompanied by an equally important appreciation of how the physical and biotic drivers of evolution continually reshape the web of life.

SUMMARY OF THE MAJOR ARGUMENTS

The ecological and genetic underpinnings of adaptive evolution can be summarized as nine major points:

1. Natural selection often acts strongly on populations and, despite the genetic complexity of many traits, populations often evolve quickly. Selection acts by favoring particular alleles, reorganizing the novel genomes formed by hybridization or polyploidization, and co-opting whole genomes or snippets of DNA of other species.
2. Much of adaptive evolution is about manipulating the genomes of other species and, simultaneously, evolving to minimize being manipulated by other species. Intergenomic conflicts and mutualisms therefore occur at every level in the hierarchy of life.
3. Consequently, interactions among species are often the strongest drivers of relentless evolution. Evidence for that view comes in part from the observation that the fastest evolving parts of genomes are often those under selection imposed by interactions with other species. Even social evolution, including sexual selection, is often shaped by interspecific interactions that influence which traits undergo selection and how they evolve.

4. Evolution driven by interspecific, and sometimes intraspecific, interactions is often subject to balancing selection, which results in the maintenance of high levels of adaptive variation within and among populations rather than directional arms races. Much of evolution is nondirectional in the long term but ecologically crucial in the short term to the persistence of adaptation in populations.
5. Because genes and genomes are expressed in different ways in different environments (genotype-by-environment interactions) and species interact in different ways in different environments (genotype-by-genotype-by-environment interactions), selection to local physical and biotic environments may often be linked.
6. The trajectories and dynamics of evolving interactions therefore differ among environments, creating geographic mosaics of adaptation and coadaptation through selection mosaics, coevolutionary hotspots and coldspots, and trait remixing. These mosaics generate much of the diversity of traits and life histories found within species.
7. These geographic mosaics sometimes produce new species through ecological speciation. Much of speciation appears to be ecological speciation.
8. The continued adaptive diversification of populations and species often favors webs rather than pairs of interacting species. It does so through multiple mechanisms of multispecific coevolution that we are only beginning to understand, such as coevolutionary alternation in antagonistic interactions and the evolution of complementarity of traits and convergence of traits in mutualisms.
9. As webs get larger, they favor the evolution of new lifestyles that exploit large webs (e.g., honeybees, large frugivorous primates), thereby further fueling the adaptive radiation of lineages.

The resulting entangled bank of life is constantly becoming disentangled and then reentangled as natural selection drives further evolutionary change. Everywhere on earth in every year, species undergo evolutionary change as physical environments change, webs of interacting species change, and selection acts anew on populations. It is a dynamic world of continually coevolving mutualistic and antagonistic interactions shifting about on the changing physical templates of land and sea. Although all species must evolve to survive in their physical environments, much of the diversity of life has come from the coevolutionary process that continually weaves species into intricate, ever-changing webs.

THE GREATEST CHALLENGE

The greatest challenge ahead in evolutionary biology is to understand how evolution continues to reshape the web of life itself at a time when we are altering biodiversity worldwide. If the past century and a half of studies has been mostly about the evolution of species, the next will be mostly about the evolution of interactions. That gradual shift in emphasis has become inevitable. Research in recent decades has shown that some of the fastest evolving genes in populations are those driven by selection on interactions.

The shift to a focus on interactions, though, is also being driven by the fact that many of the problems we face in medicine, agriculture, fisheries, forestry, and conservation are about how to manage interactions among species. It is a grand challenge that goes back to Darwin himself, who wrote in the last paragraph of *The Origin of Species* that it is "interesting to contemplate an entangled bank." That challenge has become increasingly important as we have come to appreciate that species undergo evolutionary change that is ecologically, agriculturally, and clinically relevant on the timescale of our lives.

Even with the great increase in studies of rapid evolution in recent years, we are undoubtedly underestimating the relentlessness of evolutionary change. We measure a few morphological traits, a few physiological or biochemical traits, or a few life history or behavioral traits. We make our best guess about which traits are important to measure. Once we have the molecular tools to measure the short-term dynamics of selection across whole genomes within populations, we will likely find that selection is even more pervasive than we thought.

Our biggest current gap, though, is our limited understanding of the dynamics of microbial evolution in nature—alone, or as microbial assemblages, or as symbionts associated with more complex organisms. As each year passes, our knowledge increases of the pervasiveness of microbes as mediators of ecological and evolutionary processes. Managing evolution, which is what our societies are increasingly trying to do, will often mean managing microbial evolution. Although that is obvious for problems such as rapid evolution of antibiotic resistance, it has, until recently, been less obvious for many other environmental changes. The invasions of tree-boring insects currently devastating forests worldwide are not simply insect invasions. They are invasions by assemblages of species, including fungi and bacteria, of which insects are only the most visible members (Klepzig et al. 2009; Adams et al. 2011).

In the coming decades, rapid and ongoing evolution will demand even more attention from us. Paul Ehrlich (2001) has argued that our ultimate goal in the manipulation and conservation of biodiversity should be toward maintaining the evolutionary options of species. We are changing the earth so quickly and in so many ways that we are imposing strong selection on a large proportion of the earth's biodiversity all at once, without knowing what those changes are doing collectively to the evolutionary options available to species. We are also imposing novel kinds of selection on species. Species have always shifted their geographic ranges over time as environments have changed, and individual species have occasionally colonized new regions through dispersal over long distances. Rarely, however, have so many new species invaded new continents or oceans all at once, as has been happening in recent centuries through our activities.

The communities of species that we manage in any particular locality will differ even in the near future from those that we are managing now. Populations continue to evolve. We still have much to learn about how we are altering evolution in the increasingly human-dominated communities worldwide, and how we can use our developing knowledge of the relentlessness of evolution to maintain the diverse web of life and our place within that web.

Literature Cited

Abbott, R. J., A. C. Brennan, J. K. James, D. G. Forbes, M. J. Hegarty, and S. J. Hiscock. 2009. Recent hybrid origin and invasion of the British Isles by a self-incompatible species, Oxford ragwort (*Senecio squalidus* L., Asteraceae). *Biological Invasions* 11:1145–1158.

Abbott, R. J., and A. J. Lowe. 2004. Origins, establishment and evolution of new polyploid species: *Senecio cambrensis* and *S. eboracensis* in the British Isles. *Biological Journal of the Linnean Society* 82:467–474.

Aberhan, M., W. Kiessling, and F. T. Fürsich. 2006. Testing the role of biological interactions in the evolution of mid-Mesozoic marine benthic ecosystems. *Paleobiology* 32:259–277.

Able, K. P., and J. R. Belthoff. 1998. Rapid "evolution" of migratory behaviour in the introduced house finch of eastern North America. *Proceedings of the Royal Society of London B* 265:2063–2071.

Abrahamson, W. G., and C. P. Blair. 2008. Sequential radiation through host-race formation: Herbivore diversity leads to diversity in natural enemies. Pages 188–202 in K. J. Tilmon, ed. *Specialization, speciation, and radiation: The evolutionary biology of herbivorous insects.* University of California Press, Berkeley.

Abrahamson, W. G., and A. E. Weis. 1997. *Evolutionary ecology across three trophic levels: Goldenrods, gallmakers, and natural enemies.* Princeton University Press, Princeton, NJ.

Abrams, P. A. 1986. Adaptive responses of predators to prey and prey to predators: The failure of the arms-race analogy. *Evolution* 40:1229–1247.

———. 2000. The evolution of predator–prey interactions: Theory and evidence. *Annual Review of Ecology and Systematics* 31:79–105.

Abrams, P. A., and H. Matsuda. 1997. Prey adaptation as a cause of predator–prey cycles. *Evolution* 51:1742–1750.

Abzhanov, A., M. Protas, B. Grant, P. Grant, and C. Tabin. 2004. Bmp4 and morphological variation of beaks in Darwin's finches. *Science* 305:1462–1465.

Abzhanov, A., W. P. Kuo, C. Hartmann, B. R. Grant, P. R. Grant, and C. J. Tabin. 2006. The calmodulin pathway and evolution of elongated beak morphology in Darwin's finches. *Nature* 442:563–567.

Achenbach, A., and S. Foitzik. 2009. First evidence for slave rebellion: Enslaved ant workers systematically kill the brood of their social parasite *Protomognathus americanus*. *Evolution* 63:1068–1075.

Adam, T. C. 2011. High-quality habitat and facilitation ameliorate competitive effects of prior residents on new settlers. *Oecologia* 166:121–130.

Adams, A. S., C. K. Boone, J. Bohlmann, and K. F. Raffa. 2011. Responses of bark beetle–associated bacteria to host monoterpenes and their relationship to insect life histories. *Journal of Chemical Ecology* 37:808–817.

Addison, J. A., B. S. Ort, K. A. Mesa, and G. H. Pogson. 2008. Range-wide genetic homogeneity in the California sea mussel (*Mytilus californianus*): A comparison of allozymes, nuclear DNA markers, and mitochondrial DNA sequences. *Molecular Ecology* 17:4222–4232.

Addison, J. A., and G. H. Pogson. 2009. Multiple gene genealogies reveal asymmetrical hybridization and introgression among strongylocentrotid sea urchins. *Molecular Ecology* 18:1239–1251.

Adler, L. S., and J. L. Bronstein. 2004. Attracting antagonists: Does floral nectar increase leaf herbivory? *Ecology* 85:1519–1526.

Agrawal, A. A. 2007. Macroevolution of plant defense strategies. *Trends in Ecology and Evolution* 22:103–109.

Agrawal, A. A., and M. Fishbein. 2008. Phylogenetic escalation and decline of plant defense strategies. *Proceedings of the National Academy of Sciences USA* 105:10057–10060.

Agrawal, A. F., E. D. Brodie, and L. H. Rieseberg. 2001. Possible consequences of genes of major effect: Transient changes in the G-matrix. *Genetica* 112:33–43.

Ainouche, M. L., P. M. Fortune, A. Salmon, C. Parisod, M.-A. Grandbastien, K. Fukunaga, M. Ricou, et al. 2008. Hybridization, polyploidy and invasion: Lessons from *Spartina* (Poaceae). *Biological Invasions* 11:1159–1173.

Akey, J. M., A. L. Ruhe, D. T. Akey, A. K. Wong, C. F. Connelly, J. Madeoy, T. J. Nicholas, et al. 2010. Tracking footprints of artificial selection in the dog genome. *Proceedings of the National Academy of Sciences USA* 107:1160–1165.

Albert, A. Y. K., S. Sawaya, T. H. Vines, A. K. Knecht, C. T. Miller, B. R. Summers, S. Balabhadra, et al. 2008. The genetics of adaptive shape shift in stickleback: Pleiotropy and effect size. *Evolution* 62:76–85.

Alberto, F. F., P. T. P. Raimondi, D. C. D. Reed, N. C. N. Coelho, R. R. Leblois, A. A. Whitmer, and E. A. E. Serrão. 2010. Habitat continuity and geographic distance predict population genetic differentiation in giant kelp. *Ecology* 91:49–56.

Albrechtsen, A., I. Moltke, and R. Nielsen. 2010. Natural selection and the distribution of identity-by-descent in the human genome. *Genetics* 186:295–308.

Alexander, R. R., and G. P. Dietl. 2003. The fossil record of shell-breaking predation on marine bivalves and gastropods. *Topics in Geobiology* 20:141–176.

Alford, R. A., G. P. Brown, L. Schwarzkopf, B. L. Phillips, and R. Shine. 2009. Comparisons through time and space suggest rapid evolution of dispersal behaviour in an invasive species. *Wildlife Research* 36:23–28.

Allee, W. C., O. Park, A. E. Emerson, and K. P. Schmidt. 1949. *Principles of animal ecology*. W. B. Saunders, Philadelphia.

Allen, C. E., P. Beldade, B. J. Zwaan, and P. M. Brakefield. 2008. Differences in the selection response of serially repeated color pattern characters: Standing variation, development, and evolution. *BMC Evolutionary Biology* 8:94.

Allen, H., K. Estrada, G. Lettre, S. Berndt, M. Weedon, F. Rivadeneira, C. J. Willer, et al. 2010. Hundreds of variants clustered in genomic loci and biological pathways affect human height. *Nature* 467:832–838.

Allen, M. F., W. Swenson, J. I. Querejeta, L. M. Egerton-Warburton, and K. K. Treseder. 2003. Ecology of mycorrhizae: A conceptual framework for complex interactions among plants and fungi. *Annual Review of Phytopathology* 41:271–303.

Allendorf, F. W., and J. J. Hard. 2009. Human-induced evolution caused by unnatural selection through harvest of wild animals. *Proceedings of the National Academy of Sciences USA* 106:9987–9994.

Allison, A. C. 1954. Protection offered by sickle-cell trait against subtertian malarial infection. *British Medical Journal* 1:290–294.

Alon, U. 2007. Network motifs: Theory and experimental approaches. *Nature Reviews Genetics* 8:450–461.

Alonso-Blanco, C., M. G. M. Aarts, L. Bentsink, J. J. B. Keurentjes, M. Reymond, D. Vreugdenhill, and M. Koornneef. 2009. What has natural variation taught us about plant development, physiology, and adaptation? *Plant Cell* 21:1877–1896.

Alonzo, S. H., and T. Pizzari. 2010. Male fecundity stimulation: Conflict and cooperation within and between the sexes: Model analyses and coevolutionary dynamics. *American Naturalist* 175:174–185.

Alstad, D. 1998. Population structure and the conundrum of local adaptation. Pages 3–21 in S. Mopper and S. Y. Strauss, eds., *Genetic variation and local adaptation in natural insect populations: Effects of ecology, life history, and behavior*. Chapman and Hall, New York.

Althoff, D. M., J. D. Groman, K. A. Segraves, and O. Pellmyr. 2001. Phylogeographic structure in the bogus yucca moth *Prodoxus quinquepunctellus* (Prodoxidae): Comparisons with coexisting pollinator yucca moths. *Molecular Phylogenetics and Evolution* 21:117–127.

Althoff, D. M., and O. Pellmyr. 2002. Examining genetic structure in a bogus yucca moth: A sequential approach to phylogeography. *Evolution* 56:1632–1643.

Althoff, D. M., K. A. Segraves, J. Leebens-Mack, and O. Pellmyr. 2006. Patterns of speciation in the yucca moths: Parallel species radiations within the *Tegeticula yuccasella* species complex. *Systematic Biology* 55:398–410.

Althoff, D. M., K. A. Segraves, and O. Pellmyr. 2005. Community context of an obligate mutualism: Pollinator and florivore effects on *Yucca filamentosa*. *Ecology* 86:905–913.

Althoff, D. M., G. P. Svensson, and O. Pellmyr. 2007. The influence of interaction type and feeding location on the phylogenetic structure of the yucca moth community associated with *Hesperoyucca whipplei*. *Molecular Phylogenetics and Evolution* 43:398–406.

Althoff, D. M., and J. N. Thompson. 1999. Comparative geographic structures of two parasitoid-host interactions. *Evolution* 53:818–825.

Altshuler, D., M. J. Daly, and E. S. Lander. 2008. Genetic mapping in human disease. *Science* 322:881–888.

Altshuler, D. L., R. M. Durbin, G. R. Abecasis, D. R. Bentley, A. Chakravarti, A. G. Clark, F. S. Collins, et al. 2010. A map of human genome variation from population-scale sequencing. *Nature* 467:1061–1073.

Alyokhin, A., M. Baker, D. Mota-Sanchez, G. Dively, and E. Grafius. 2008. Colorado potato beetle resistance to insecticides. *American Journal of Potato Research* 85:395–413.

Anacker, B. L., J. B. Whittall, E. E. Goldberg, and S. P. Harrison. 2011. Origins and consequences of serpentine endemism in the California flora. *Evolution* 65:365–376.

Andam, C. P., G. P. Fournier, and J. P. Gogarten. 2011. Multilevel populations and the evolution of antibiotic resistance through horizontal gene transfer. *FEMS Microbiology Reviews* 35:756–767.

Andam, C. P., and J. P. Gogarten. 2011. Biased gene transfer in microbial evolution. *Nature Reviews Microbiology* 9:543–555.

Andersen, S. B., S. Gerritsma, K. M. Yusah, D. Mayntz, N. L. H. Jones, J. Billen, J. J. Boomsma, et al. 2009. The life of a dead ant: The expression of an adaptive extended phenotype. *American Naturalist* 174:424–433.

Anderson, B., and S. D. Johnson. 2008. The geographic mosaic of coevolution in a plant-pollinator mutualism. *Evolution* 62:220–225.

———. 2009. Geographical covariation and local convergence of flower depth in a guild of fly-pollinated plants. *New Phytologist* 182:533–540.

Anderson, B., S. D. Johnson, and C. Carbutt. 2005. Exploitation of a specialized mutualism by a deceptive orchid. *American Journal of Botany* 92:1342–1349.

Anderson, B., J. S. Terblanche, and A. G. Ellis. 2010. Predictable patterns of trait mismatches between interacting plants and insects. *BMC Evolutionary Biology* 10:204.

Anderson, R. M., and R. M. May. 1979. Population biology of infectious disease. Part 1. *Nature* 280:361–367.

Andersson, J. O. 2009. Gene transfer and diversification of microbial eukaryotes. *Annual Review of Microbiology* 63:177–193.

Ando, Y., and T. Ohgushi. 2008. Ant- and plant-mediated indirect effects induced by aphid colonization on herbivorous insects on tall goldenrod. *Population Ecology* 50:181–189.

Andrés, A. M., M. J. Hubisz, A. Indap, D. G. Torgerson, J. D. Degenhardt, A. R. Boyko, R. N. Gutenkunst, et al. 2009. Targets of balancing selection in the human genome. *Molecular Biology and Evolution* 26:2755–2764.

Antonov, A., B. G. Stokke, J. R. Vikan, F. Fossøy, P. S. Ranke, E. Røskaft, A. Moksnes, et al. 2010. Egg phenotype differentiation in sympatric cuckoo *Cuculus canorus* gentes. *Journal of Evolutionary Biology* 23:1170–1182.

Antonovics, J., J. L. Abbate, C. H. Backer, D. Daley, M. E. Hood, C. E. Jenkins, L. J. Johnson, et al. 2007. Evolution by any other name: Antibiotic resistance and avoidance of the e-word. *PLoS Biology* 5:e30.

Antonovics, J., M. Boots, J. Abbate, C. Baker, Q. McFrederick, and V. Panjeti. 2011a. Biology and evolution of sexual transmission. *Annals of the New York Academy of Sciences* 1230:12–24.

Antonovics, J., P. H. Thrall, J. J. Burdon, and A.-L. Laine. 2011b. Partial resistance in the *Linum–Melampsora* host–pathogen system: Does partial resistance make the Red Queen run slower? *Evolution* 65:512–522.

Araguas, R. M., N. Sanz, R. Fernández, F. M. Utter, C. Pla, and J.-L. García-Marín. 2009. Role of genetic refuges in the restoration of native gene pools of brown trout. *Conservation Biology* 23:871–878.

Aravind, L., R. L. Tatusov, Y. I. Wolf, D. R. Walker, and E. V. Koonin. 1998. Evidence for massive gene exchange between archaeal and bacterial hyperthermophiles. *Trends in Genetics* 14:442–444.

Archer, M. A., T. J. Bradley, L. D. Mueller, and M. R. Rose. 2007. Using experimental evolution to study the physiological mechanisms of desiccation resistance in *Drosophila melanogaster*. *Physiological and Biochemical Zoology* 80:386–398.

Armbruster, W. S. 1990. Estimating and testing the shapes of adaptive surfaces: The morphology and pollination of *Dalechampia* blossoms. *American Naturalist* 135:14–31.

Armbruster, W. S., Y.-B. Gong, and S.-Q. Huang. 2011. Are pollination "syndromes" predictive? Asian *Dalechampia* fit neotropical models. *American Naturalist* 178:135–143.

Armbruster, W. S., J. Lee, and B. G. Baldwin. 2009. Macroevolutionary patterns of defense and pollination in *Dalechampia* vines: Adaptation, exaptation, and evolutionary novelty. *Proceedings of the National Academy of Sciences USA* 106:18085–18090.

Arnold, A. E., L. C. Mejía, D. Kyllo, E. I. Rojas, Z. Maynard, N. Robbins, and E. A. Herre. 2003. Fungal endophytes limit pathogen damage in a tropical tree. *Proceedings of the National Academy of Sciences USA* 100:15649–15654.

Arnold, S. J. 1981. Behavioral variation in natural populations. II. The inheritance of a feeding response in crosses between geographic races of the garter snake, *Thamnophis elegans*. *Evolution* 35:510–515.

Arnold, S. J., R. Bürger, P. A. Hohenlohe, B. C. Ajie, and A. G. Jones. 2008. Understanding the evolution and stability of the G-matrix. *Evolution* 62: 2451–2461.

Arnold, S. J., and M. J. Wade. 1984a. On the measurement of natural and sexual selection: Theory. *Evolution* 38:709–719.

———. 1984b. On the measurement of natural and sexual selection: Applications. *Evolution* 38:720–734.

Arslanyolu, M., and F. P. Doerder. 2000. Genetic and environmental factors affecting mating type frequency in natural isolates of *Tetraphymena thermophila*. *Journal of Eukaryotic Microbiology* 47:412–418.

Arvanitis, L., C. Wiklund, and J. Ehrlén. 2007. Butterfly seed predation: Effects of landscape characteristics, plant ploidy level and population structure. *Oecologia* 152:275–285.

Arvanitis, L., C. Wiklund, Z. Münzbergova, J. P. Dahlgren, and J. Ehrlén. 2010. Novel antagonistic interactions associated with plant polyploidization influence trait selection and habitat preference. *Ecology Letters* 13:330–337.

Ashelford, K. E., M. J. Day, and J. C. Fry. 2003. Elevated abundance of bacteriophage infecting bacteria in soil. *Applied and Environmental Microbiology* 69:285–289.

Auton, A., K. Bryc, A. R. Boyko, K. E. Lohmueller, J. Novembre, A. Reynolds, A. Indap, et al. 2009. Global distribution of genomic diversity underscores rich complex history of continental human populations. *Genome Research* 19: 795–803.

Avilés, J. M., J. R. Vikan, F. Fossøy, A. Antonov, A. Moksnes, E. Røskaft, J. A. Shykoff, et al. 2011. The common cuckoo *Cuculus canorus* is not locally adapted to its reed warbler *Acrocephalus scirpaceus* host. *Journal of Evolutionary Biology* 24:314–325.

Avise, J. C. 2000. *Phylogeography: The history and formation of species.* Harvard University Press, Cambridge, MA.

———. 2001. Evolving genomic metaphors: A new look at the language of DNA. *Science* 294:86–87.

Avise, J., W. Nelson, and H. Sugita. 1994. A speciational history of living fossils: Molecular evolutionary patterns in horseshoe crabs. *Evolution* 48:1986–2001.

Bäckhed, F., R. E. Ley, J. L. Sonnenburg, D. A. Peterson, and J. I. Gordon. 2005. Host-bacterial mutualism in the human intestine. *Science* 307:1915–1920.

Bailey, J. K., Z. A. Gu, R. Clark, K. Reinert, R. V. Samonte, S. Schwartz, M. D. Adams, et al. 2002. Recent segmental duplications in the human genome. *Science* 297:1003–1007.

Bailey, J. K., J. A. Schweitzer, M. D. Koricheva, C. J. LeRoy, B. J. Rehill, R. K. Bangert, D. G. Fisher, et al. 2009. From genes to ecosystems: Synthesizing the effects of plant genetic factors across systems. *Philosophical Transactions of the Royal Society B* 364:1607–1616.

Bailey, J. K., J. A. Schweitzer, B. J. Rehill, R. L. Lindroth, G. D. Martinsen, and T. G. Whitham. 2004. Beavers as molecular geneticists: A genetic basis to the foraging of an ecosystem engineer. *Ecology* 85:603–608.

Bailey, M. F., and L. Delph. 2007. A field guide to models of sex-ratio evolution in gynodioecious species. *Oikos* 116:1609–1617.

Bakker, E. G., C. Toomajian, and M. Kreitman. 2006. A genome-wide survey of *R* gene polymorphisms in *Arabidopsis*. *Plant Cell* 18:1803–1818.

Balanyá, J., R. B. Huey, G. W. Gilchrist, and L. Serra. 2009. The chromosomal polymorphism of *Drosophila subobscura*: A microevolutionary weapon to monitor global change. *Heredity* 103:364–367.

Balanyá, J., J. M. Oller, R. B. Huey, G. W. Gilchrist, and L. Serra. 2006. Global genetic change tracks global climate warming in *Drosophila subobscura*. *Science* 313:1773–1775.

Barbero, F., S. Bonelli, J. A. Thomas, E. Balletto, and K. Schönrogge. 2009. Acoustical mimicry in a predatory social parasite of ants. *Journal of Experimental Biology* 212:4084–4090.

Barnett, R., B. Shapiro, I. Barnes, S. Y. W. Ho, J. Burger, N. Yamaguchi, T. F. G. Higham, et al. 2009. Phylogeography of lions (*Panthera leo* ssp.) reveals three distinct taxa and a late Pleistocene reduction in genetic diversity. *Molecular Ecology* 18:1668–1677.

Barrett, L. G., P. H. Thrall, and J. J. Burdon. 2007. Evolutionary diversification through hybridization in a wild host-pathogen interaction. *Evolution* 61: 1613–1621.

Barrett, L. G., P. H. Thrall, J. J. Burdon, A. B. Nicotra, and C. C. Linde. 2008. Population structure and diversity in sexual and asexual populations of the pathogenic fungus *Melampsora lini*. *Molecular Ecology* 17:3401–3415.

Barrett, L. G., P. H. Thrall, P. N. Dodds, M. van der Merwe, C. C. Linde, G. J. Lawrence, and J. J. Burdon. 2009. Diversity and evolution of effector loci in natural populations of the plant pathogen *Melampsora lini*. *Molecular Biology and Evolution* 26:2499–2513.

Barrett, R. D. H., R. C. MacLean, and B. Bell. 2006. Mutations of intermediate effect are responsible for adaptation in evolving *Pseudomonas fluorescens* populations. *Biology Letters* 2:236–238.

Barrett, R. D. H., S. M. Rogers, and D. Schluter. 2008. Natural selection on a major armor gene in threespine stickleback. *Science* 322:255–257.

Barrett, R. D. H., and D. Schluter. 2008. Adaptation from standing genetic variation. *Trends in Ecology and Evolution* 23:38–44.

Barrett, S. C. H. 2010. Understanding plant reproductive diversity. *Philosophical Transactions of the Royal Society of London Series B* 365:99–109.

Barrick, J. E., and R. E. Lenski. 2009. Genome-wide mutational diversity in an evolving population of *Escherichia coli*. *Cold Spring Harbor Symposia on Quantitative Biology* 74:119–129.

Barrick, J. E., D. S. Yu, S. H. Yoon, H. Jeong, T. K. Oh, D. Schneider, R. E. Lenski, et al. 2009. Genome evolution and adaptation in a long-term experiment with *Escherichia coli*. *Nature* 461:1243–1247.

Bartel, R. A., K. S. Oberhauser, J. C. de Roode, and S. M. Altizer. 2011. Monarch butterfly migration and parasite transmission in eastern North America. *Ecology* 92:342–351.

Bartomeus, I., J. S. Ascher, D. Wagner, B. N. Danforth, S. Colla, S. Kornbluth, and R. Winfree. 2011. Climate-associated phenological advances in bee pollinators and bee-pollinated plants. *Proceedings of the National Academy of Sciences USA* 108:20645–20649.

Barton, N., and G. M. Hewitt. 1985. Analysis of hybrid zones. *Annual Review of Ecology and Systematics* 16:113–148.

Bascompte, J. 2010. Structure and dynamics of ecological networks. *Science* 329:765–766.

Bascompte, J., and P. Jordano. 2007. Plant-animal mutualistic networks: The architecture of biodiversity. *Annual Review of Ecology, Evolution, and Systematics* 38:567–593.

Bascompte, J., P. Jordano, C. J. Melián, and J. M. Olesen. 2003. The nested assembly of plant-animal mutualistic networks. *Proceedings of the National Academy of Sciences USA* 100:9383–9387.

Bascompte, J., C. J. Melián, and E. Sala. 2005. Interaction strength combinations and the overfishing of a marine food web. *Proceedings of the National Academy of Sciences USA* 102:5443–5447.

Baskett, M. L., and R. Gomulkiewicz. 2011. Introgressive hybridization as a mechanism for species rescue. *Theoretical Ecology* 4:223–239.

Bassar, R. D., M. C. Marshall, A. López-Sepulcre, E. Zandoná, S. K. Auer, J. Travis, C. M. Pringle, et al. 2010. Local adaptation in Trinidadian guppies alters ecosystem processes. *Proceedings of the National Academy of Sciences USA* 107:3616–3621.

Bates, H. W. 1862. XXXII. Contributions for an insect fauna of the Amazon valley. LEPIDOPTERA: HELICONIDAE. *Transactions of the Linnean Society of London* 23:495–566.

Bauer, S., V. Witte, M. Boehm, and S. Foitzik. 2009. Fight or flight? A geographic mosaic in host reaction and potency of a chemical weapon in the social parasite *Harpagoxenus sublaevis*. *Behavioral Ecology and Sociobiology* 64:45–56.

Baxter, L., S. Tripathy, N. Ishaque, N. Boot, A. Cabral, E. Kemen, M. Thines, et al. 2010. Signatures of adaptation to obligate biotrophy in the *Hyaloperonospora arabidopsidis* genome. *Science* 330:1549–1551.

Bearhop, S., W. Fiedler, R. W. Furness, S. C. Votier, S. Waldron, J. Newton, G. J. Bowen, et al. 2005. Assortative mating as a mechanism for rapid evolution of a migratory divide. *Science* 310:502–504.

Becerra, J. X., K. Noge, and D. L. Venable. 2009. Macroevolutionary chemical escalation in an ancient plant-herbivore arms race. *Proceedings of the National Academy of Sciences USA* 106:18062–18066.

Beibl, J., R. J. Stuart, J. Heinze, and S. Foitzik. 2005. Six origins of slavery in formicoxenine ants. *Insectes Sociaux* 52:291–297.

Beiko, R. G., T. J. Harlow, and M. A. Ragan. 2005. Highways of gene sharing in prokaryotes. *Proceedings of the National Academy of Sciences USA* 102:14332–14337.

Beldade, P., V. French, and P. M. Brakefield. 2008. Developmental and genetic mechanisms for evolutionary diversification of serial repeats: Eyespot size in *Bicyclus anynana* butterflies. *Journal of Experimental Zoology B* 310:191–201.

Beldade, P., K. Koops, and P. M. Brakefield. 2002. Developmental constraints *versus* flexibility in morphological evolution. *Nature* 416:844–847.

Bell, G. 1982. *The masterpiece of nature: The evolution and genetics of sexuality*. University of California Press, Berkeley.

———. 2008. *Selection: The mechanism of evolution*. Oxford University Press, Oxford.

———. 2010. Fluctuating selection: The perpetual renewal of adaptation in variable environments. *Philosophical Transactions of the Royal Society B* 365:87–97.

Bengtsson, B. O. 2009. Asex and evolution: A *very* large-scale overview. Pages 1–19 in I. Schön, K. Martens, and P. van Dijk, eds., *Lost sex: The evolutionary biology of parthenogenesis*. Springer, New York.

Benkman, C. W. 1993. Adaptation to single resources and the evolution of crossbill (*Loxia*) diversity. *Ecological Monographs* 63:305–325.

———. 1995. The impact of tree squirrels (*Tamiasciurus*) on limber pine seed dispersal adaptations. *Evolution* 49:585–592.

———. 1999. The selection mosaic and diversifying coevolution between crossbills and lodgepole pine. *American Naturalist* 153:S75–S91.

———. 2010. Diversifying coevolution between crossbills and conifers. *Evolution: Education and Outreach* 3:47–53.

Benkman, C. W., and R. P. Balda. 1984. Adaptations for seed dispersal and the compromises due to seed predation in limber pine. *Ecology* 65:632–642.

Benkman, C. W., and A. M. Siepielski. 2004. A keystone selective agent? Pine squirrels and the frequency of serotiny in lodgepole pine. *Ecology* 85:2082–2087.

Benkman, C. W., A. M. Siepelski, and T. L. Parchman. 2008. The local introduction of strongly interacting species and the loss of geographic variation in species and species interactions. *Molecular Ecology* 17:395–404.

Benkman, C. W., J. W. Smith, P. C. Keenan, T. L. Parchman, and L. Santisteban. 2009. A new species of the red crossbill (Fringillidae: *Loxia*) from Idaho. *Condor* 111:169–176.

Bennett, A. E., M. Thomsen, and S. Y. Strauss. 2011. Multiple mechanisms enable invasive species to suppress native species. *American Journal of Botany* 98:1086–1094.

Bentley, K. E., J. R. Mandel, and D. E. McCauley. 2010. Paternal leakage and heteroplasmy of mitochondrial genomes in *Silene vulgaris*: Evidence from experimental crosses. *Genetics* 185:961–968.

Berenbaum, M. R., and A. R. Zangerl. 1998. Chemical phenotype matching between a plant and its insect herbivore. *Proceedings of the National Academy of Sciences USA* 95:13743–13748.

———. 2006. Parsnip webworms and host plants at home and abroad: Trophic complexity in a geographic mosaic. *Ecology* 87:3070–3081.

Berenbaum, M. R., A. R. Zangerl, and J. K. Nitao. 1986. Constraints on chemical coevolution: Wild parsnips and the parsnip webworm. *Evolution* 40:1215–1228.

Bergelson, J., M. Kreitman, E. A. Stahl, and D. Tian. 2001. Evolutionary dynamics of plant R-genes. *Science* 292:2281–2285.

Bergh, O., K. Y. Borsheim, G. Bratbak, and M. Heldal. 1989. High abundance of viruses in aquatic environments. *Nature* 340:467–468.

Bernasconi, G., J. Antonovics, A. Biere, D. Charlesworth, L. F. Delph, D. Filatov, T. Giraud, et al. 2009. *Silene* as a model system in ecology and evolution. *Heredity* 103:5–14.

Best, A., S. Webb, J. Antonovics, and M. Boots. 2012. Local transmission processes and disease-driven host extinctions. *Theoretical Ecology* 5:211–217.

Best, A., S. Webb, A. White, and M. Boots. 2011. Host resistance and coevolution in spatially structured populations. *Proceedings of the Royal Society B* 278:2216–2222.

Bhatt, S., E. C. Holmes, and O. G. Pybus. 2011. The genomic rate of molecular adaptation of the human influenza A virus. *Molecular Biology and Evolution* 28:2443–2451.

Bidartondo, M. I., D. J. Read, J. M. Trappe, V. Merckx, R. Ligrone, and J. G. Duckett. 2011. The dawn of symbiosis between plants and fungi. *Biology Letters* 7:574–577.

Biek, R., A. J. Drummond, and M. Poss. 2006. A virus reveals population structure and recent demographic history of its carnivore host. *Science* 311:538–541.

Biémont, C. 2010. A brief history of the status of transposable elements: From junk DNA to major players in evolution. *Genetics* 186:1085–1093.

Billiard, S., M. Lopez-Villaciencio, B. Devier, M. E. Hood, C. Fairhead, and T. Giraud. 2011. Having sex, yes, but with whom? Inferences from fungi on the evolution of anisogamy and mating types. *Biological Reviews* 86: 421–442.

Birky, C. W., Jr. 2010. Positively negative evidence for asexuality. *Journal of Heredity* 101 (suppl. 1): S42–45.

Birky, C. W., Jr., and T. G. Barraclough. 2009. Asexual speciation. Pages 201–216 in I. Schön, K. Martens, and P. van Dijk, eds., *Lost sex: The evolutionary biology of parthenogenesis*. Springer, New York.

Black, W. C., C. F. Baer, M. F. Antolin, and N. M. DuTeau. 2001. Population genomics: Genome-wide sampling of insect populations. *Annual Review of Entomology* 46:441–469.

Blattner, F. R., G. Plunkett, C. A. Bloch, N. Perna, V. Burland, M. Riley, J. Collado-Vides, et al. 1997. The complete genome sequence of *Escherichia coli* K-12. *Science* 277:1453–1462.

Bleay, C., T. Comendant, and B. Sinervo. 2007. An experimental test of frequency-dependent selection on male mating strategy in the field. *Proceedings of the Royal Society B* 274:2019–2025.

Blossey, B., and R. Nötzold. 1995. Evolution of increased competitive ability in invasive nonindigenous plants: A hypothesis. *Journal of Ecology* 38:887–889.

Blount, Z. D., C. Z. Borland, and R. E. Lenski. 2008. Historical contingency and the evolution of a key innovation in an experimental population of *Escherichia coli*. *Proceedings of the National Academy of Sciences USA* 105:7899–7906.

Blüthgen, N., M. Florian, T. Hovestadt, B. Fiala, and N. Blüthgen. 2007. Specialization, constraints, and conflicting interests in mutualistic networks. *Current Biology* 17:341–346.

Bock, R. 2010. The give-and-take of DNA: Horizontal gene transfer in plants. *Trends in Plant Science* 15:11–22.

Boggs, C. L., and P. R. Ehrlich. 2008. Conservation of coevolved insect herbivores and plants. Pages 325–332 in K. J. Tilmon, ed. *Specialization, speciation, and radiation: The evolutionary biology of herbivorous insects.* University of Chicago Press, San Francisco.

Bollback, J. P., and J. P. Huelsenbeck. 2009. Parallel genetic evolution within and between bacteriophage species of varying degrees of divergence. *Genetics* 181:225–234.

Bolnick, D. I., P. Amarasekare, M. S. Araújo, R. Bürger, J. M. Levine, M. Novak, V. H. W. Rudolf, et al. 2011. Why intraspecific trait variation matters in community ecology. *Trends in Ecology and Evolution* 26:183–192.

Bone, E., and A. Farres. 2001. Trends and rates of microevolution in plants. *Genetica* 112–113:165–182.

Bongfen, S. E., A. Laroque, J. Berghout, and P. Gros. 2009. Genetic and genomic analyses of host-pathogen interactions in malaria. *Trends in Parasitology* 25:417–422.

Boomsma, J. J., and N. R. Franks. 2006. Social insects: From selfish genes to self organisation and beyond. *Trends in Ecology and Evolution* 21:303–308.

Boomsma, J. J., and F. L. W. Ratnieks. 1996. Paternity in eusocial Hymenoptera. *Philosophical Transactions of the Royal Society B* 351:947–975.

Boone, C. K., B. H. Aukema, J. Bohlmann, A. L. Carroll, and K. F. Raffa. 2011. Efficacy of tree defense physiology varies with bark beetle population density: A basis for positive feedback in eruptive species. *Canadian Journal of Forest Research* 41:1174–1188.

Boots, M. 2011. The evolution of resistance to a parasite is determined by resources. *American Naturalist* 178:214–220.

Boots, M., and A. Sasaki. 2003. Evolution in sexually and vector transmitted disease. *Ecology Letters* 6:176–182.

Bordenstein, S. R., M. L. Marshall, A. J. Fry, U. Kim, and J. J. Wernegreen. 2006. The tripartite associations between bacteriophage, *Wolbachia*, and arthropods. *PLoS Pathogens* 2:e43.

Bordenstein, S. R., and W. S. Reznikoff. 2005. Mobile DNA in obligate intracellular bacteria. *Nature Reviews Microbiology* 3:688–699.

Borevitz, J. O., S. P. Hazen, T. P. Michael, G. P. Morris, I. R. Baxter, T. T. Hu, H. Chen, et al. 2007. Genome-wide patterns of single-feature polymorphism in *Arabidopsis thaliana. Proceedings of the National Academy of Sciences USA* 104:12057–12062.

Bortolotto, E., A. Bucklin, M. Mezzavilla, L. Zane, and T. Patarnello. 2011. Gone with the currents: Lack of genetic differentiation at the circum-continental scale in the Antarctic krill *Euphausia superba. BMC Genetics* 12:232.

Bossart, J. L., and J. M. Scriber. 1995. Maintenance of ecologically significant genetic variation in the Tiger Swallowtail butterfly through differential selection and gene flow. *Evolution* 49:1163–1171.

Boto, L. 2010. Horizontal gene transfer in evolution: Facts and challenges. *Proceedings of the Royal Society B* 277:819–827.

Boughman, J., H. Rundle, and D. Schluter. 2005. Parallel evolution of sexual isolation in sticklebacks. *Evolution* 59:361–373.

Bourke, A. F. G. 2011. *Principles of Social Evolution*. Oxford University Press, Oxford.

Bourtzis, K., A. Nirgianaki, G. Markakis, and C. Savakis. 1996. *Wolbachia* infection and cytoplasmic incompatibility in *Drosophila* species. *Genetics* 144:1063–1073.

Boyko, A. R., P. Quignon, L. Li, J. J. Schoenebeck, J. D. Degenhardt, K. E. Lohmueller, K. Zhao, et al. 2010. A simple genetic architecture underlies morphological variation in dogs. *PLoS Biology* 8:e1000451.

Brachi, B., N. Faure, M. Horton, E. Flahauw, A. Vazquez, M. Nordborg, J. Bergelson, et al. 2010. Linkage and association mapping of *Arabidopsis thaliana* flowering time in nature. *PLoS Genetics* 6:e1000940.

Bradshaw, H. D., Jr., and D. W. Schemske. 2003. Allele substitution at a flower colour locus produces a pollinator shift in monkeyflowers. *Nature* 426:176–178.

Brakefield, P. M. 2010. Radiations of mycalesine butterflies and opening up their exploration of morphospace. *American Naturalist* 176:S77–87.

———. 2011. Evo-devo and accounting for Darwin's endless forms. *Philosophical Transactions of the Royal Society B* 366:2069–2075.

Brakefield, P. M., and T. G. Liebert. 1985. Studies of color polymorphism in some marginal populations of the aposematic jersey tiger moth *Callimorpha quadripunctaria*. *Biological Journal of the Linnean Society* 26:225–241.

———. 2000. Evolutionary dynamics of declining melanism in the peppered moth in The Netherlands. *Proceedings of the Royal Society Biological Sciences B* 267:1953–1957.

Brakefield, P. M., and J. C. Roskam. 2006. Exploring evolutionary constraints is a task for an integrative evolutionary biology. *American Naturalist* 168:S4–S13.

Brede, N., C. Sandrock, D. Straile, P. Spaak, T. Jankowski, B. Streit, and K. Schwenk. 2009. The impact of human-made ecological changes on the genetic architecture of *Daphnia* species. *Proceedings of the National Academy of Sciences USA* 106:4758–4763.

Brentlinger, K. L., S. Hafenstein, C. R. Novak, B. A. Fane, R. Borgon, R. McKenna, and M. Agbandje-McKenna. 2002. Microviridae, a family divided: Isolation, characterization, and genome sequence of phi MH2K, a bacteriophage of the obligate intracellular parasitic bacterium *Bdellovibrio bacteriovorus*. *Journal of Bacteriology* 184:1089–1094.

Brisson, J. A. 2010. Aphid wing dimorphisms: Linking environmental and genetic control of trait variation. *Philosophical Transactions of the Royal Society B* 365:605–616.

Brock, D. A., T. E. Douglas, D. C. Queller, and J. E. Strassmann. 2011. Primitive agriculture in a social amoeba. *Nature* 469:393–396.

Brockhurst, M. A., A. D. Morgan, P. B. Rainey, and A. Buckling. 2003. Population mixing accelerates coevolution. *Ecology Letters* 6:975–979.

Brodie, E. D., Jr., B. J. Ridenhour, and E. D. Brodie III. 2002. The evolutionary response of predators to dangerous prey: Hotspots and coldspots in the geographic mosaic of coevolution between newts and snakes. *Evolution* 56:2067–2082.

Brodie, E. D., III, and E. D. Brodie, Jr. 1999. Predator–prey arms races. *Bioscience* 49:557–568.

Brodie, E. D., III, C. R. Feldman, C. T. Hanifin, J. E. Motychak, D. G. Mulcahy, B. L. Williams, and E. D. Brodie, Jr. 2005. Parallel arms races between garter snakes and newts involving tetrodotoxin as the phenotypic interface of coevolution. *Journal of Chemical Ecology* 31:343–356.

Brodo, I. M., S. D. Sharnoff, and S. Sharnoff. 2001. *Lichens of North America*. Yale University Press, New Haven, CT.

Brody, A. K., D. E. Irwin, M. L. McCutcheon, and E. C. Parsons. 2008. Interactions between nectar robbers and seed predators mediated by a shared host plant, *Ipomopsis aggregata*. *Oecologia* 155:75–84.

Brody, A. K., T. M. Palmer, K. Fox-Dobbs, and D. F. Doak. 2010. Termites, vertebrate herbivores, and the fruiting success of *Acacia drepanolobium*. *Ecology* 91:399–407.

Bronstein, J. L. 2009. The evolution of facilitation and mutualism. *Journal of Ecology* 97:1160–1170.

Brown, A. 2001. Mating in mushrooms: Increasing the chances but prolonging the affair. *Trends in Genetics* 17:393–400.

Brown, C., A. Hobday, P. Ziegler, and D. Welsford. 2008. Darwinian fisheries science needs to consider realistic fishing pressures over evolutionary time scales. *Marine Ecology Progress Series* 369:257–266.

Brown, J. H., and G. B. West. 2000. *Scaling in Biology*. Oxford University Press, New York and Oxford.

Brown, J. H., T. G. Whitham, S. K. Morgan Ernest, and C. A. Gehring. 2001. Complex species interactions and the dynamics of ecological systems: Long-term experiments. *Science* 293:643–650.

Brown, J. K. M. J., and A. A. Tellier. 2011. Plant-parasite coevolution: Bridging the gap between genetics and ecology. *Annual Review of Phytopathology* 49:345–367.

Brown, J. M., J. H. Leebens-Mack, J. N. Thompson, O. Pellmyr, and R. G. Harrison. 1997. Phylogeography and host association in a pollinating seed parasite *Greya politella* (Lepidoptera: Prodoxidae). *Molecular Ecology* 6:215–224.

Brown, J. M., O. Pellmyr, J. N. Thompson, and R. G. Harrison. 1994. Phylogeny of *Greya* (Lepidoptera: Prodoxidae) based on nucleotide sequence variation in mitochondrial cytochrome oxidase I and II: Congruence with morphological data. *Molecular Biology and Evolution* 11:128–141.

Brown, M. J. F., and P. Schmid-Hempel. 2003. The evolution of female multiple mating in social Hymenoptera. *Evolution* 57:2067–2081.

Brown, W. L., and E. O. Wilson. 1956. Character displacement. *Systematic Zoology* 7:49–64.

Brumfield, R. T. 2010. Speciation genetics of biological invasions with hybridization. *Molecular Ecology* 19:5079–5083.

Bruno, J. F., J. J. Stachowicz, and M. D. Bertness. 2003. Inclusion of facilitation into ecological theory. *Trends in Ecology and Evolution* 18:119–125.

Bruns, T., and R. Shefferson. 2004. Evolutionary studies of ectomycorrhizal fungi: Recent advances and future directions. *Canadian Journal of Botany* 82:1122–1132.

Bryner, S. F., and D. Rigling. 2011. Temperature-dependent genotype-by-genotype interaction between a pathogenic fungus and its hyperparasitic virus. *American Naturalist* 177:65–74.

Buckler, E. S., J. B. Holland, P. J. Bradbury, C. B. Acharya, P. J. Brown, C. Browne, E. Ersoz, et al. 2009. The genetic architecture of maize flowering time. *Science* 325:714–718.

Buckling, A., and M. Brockhurst. 2005. RAMP resistance. *Nature* 438:170–171.

Buckling, A., and P. B. Rainey. 2002. Antagonistic coevolution between a bacterium and a bacteriophage. *Proceedings of the Royal Society of London B* 269:931–936.

Buerkle, C. A., and L. H. Rieseberg. 2008. The rate of genome stabilization in homoploid hybrid species. *Evolution* 62:266–275.

Buggs, R. J. A. 2007. Empirical study of hybrid zone movement. *Heredity* 99:301–312.

Bull, J. J. 1994. Virulence. *Evolution* 48:1423–1437.

Bull, J. J., M. R. Badgett, H. A. Wichman, J. P. Huelsenbeck, D. M. Hillis, A. Gulati, C. Ho, et al. 1997. Exceptional convergent evolution in a virus. *Genetics* 147:1497–1507.

Burdon, J. J. 1994. The distribution and origin of genes for race-specific resistance to *Melampsora lini* in *Linum marginale*. *Evolution* 48:1564–1575.

Burdon, J. J., and J. N. Thompson. 1995. Changed patterns of resistance in a population of *Linum marginale* attacked by the rust pathogen *Melampsora lini*. *Journal of Ecology* 83:199–206.

Burdon, J. J., and P. H. Thrall. 2000. Coevolution at multiple spatial scales: *Linum marginale-Melampsora lini*–from the individual to the species. *Evolutionary Ecology* 14:261–281.

———. 2009. Coevolution of plants and their pathogens in natural habitats. *Science* 324:755–756.

Burger, R. 1999. Evolution of genetic variability and the advantage of sex and recombination in changing environments. *Genetics* 153:1055–1069.

Burke, M. K., J. P. Dunham, P. Shahrestani, K. R. Thornton, M. R. Rose, and A. D. Long. 2010. Genome-wide analysis of a long-term evolution experiment with *Drosophila*. *Nature* 467:587–592.

Burke, M. K., and M. R. Rose. 2009. Experimental evolution with Drosophila. *American Journal of Physiology—Regulatory, Integrative and Comparative Physiology* 296:R1847–1854.

Burns, K. C. 2005. Is there limiting similarity in the phenology of fleshy fruits? *Journal of Vegetation Science* 16:617–624.

Burton, A., and R. Sinsheimer. 1963. Process of Infection with φX174: Effect of exonucleases on the replicative form. *Science* 142:962–963.

Bush, G. L. 1969. Sympatric host race formation and speciation in frugivorous flies of the genus *Rhagoletis* (Diptera: Tephritidae). *Evolution* 23:237–251.

Butlin, R. K. 2010. Population genomics and speciation. *Genetica* 138:409–418.

Buzbas, E. O., P. Joyce, and N. A. Rosenberg. 2011. Inference on the strength of balancing selection for epistatically interacting loci. *Theoretical Population Biology* 79:102–113.

Cafaro, M. J., M. Poulsen, A. E. F. Little, S. L. Price, N. M. Gerardo, B. Wong, A. E. Stuart, et al. 2011. Specificity in the symbiotic association between fungus-growing ants and protective *Pseudonocardia bacteria*. *Proceedings of the Royal Society B* 278:1814–1822.

Cain, A. J., and P. M. Sheppard. 1950. Selection in the polymorphic land snail *Cepaea nemoralis*. *Heredity* 4:275–294.

Callaway, R. M. 2007. *Positive interactions and interdependence in plant communities.* Springer, New York.

Calsbeek, R., and R. M. Cox. 2010. Experimentally assessing the relative importance of predation and competition as agents of selection. *Nature* 465:613–616.

Calsbeek, R., T. B. Smith, and C. Bardeleben. 2007. Intraspecific variation in *Anolis sagrei* mirrors the adaptive radiation in Greater Antillean anoles. *Biological Journal of the Linnean Society* 90:189–199.

Calsbeek, R., J. N. Thompson, and J. E. Richardson. 2003. Patterns of molecular evolution and diversification in a biodiversity hotspot: The California Floristic Province. *Molecular Ecology* 12:1021–1029.

Campbell, A. 1961. Conditions for the existence of bacteriophage. *Evolution* 15:153–165.

Campbell, D. R., and N. M. Waser. 2007. Genotype-by-environment interaction and the fitness of plant hybrids in the wild. *Evolution* 55:669–676.

Carrington, L., J. Lipkowitz, and A. A. Hoffmann. 2011. A re-examination of *Wolbachia*-induced cytoplasmic incompatibility in California *Drosophila simulans*. *PLoS One* 6:e22565.

Carroll, S. B. 2005. *Endless forms most beautiful: The new science of evo-devo.* Norton, New York.

Carroll, S. P., and C. Boyd. 1992. Host race radiation in the soapberry bug: Natural history with the history. *Evolution* 46:1052–1069.

Carroll, S. P., H. Dingle, T. R. Famula, and C. W. Fox. 2001. Genetic architecture of adaptive differentiation in evolving host races of the soapberry bug, *Jadera haematoloma*. *Genetica* 112–113:257–272.

Carroll, S. P., A. P. Hendry, D. N. Reznick, and C. W. Fox. 2007. Evolution on ecological time-scales. *Functional Ecology* 21:387–393.

Carroll, S. P., S. P. Klassen, and H. Dingle. 1998. Rapidly evolving adaptations to host ecology and nutrition in the soapberry bug. *Evolutionary Ecology* 12:955–968.

Carroll, S. P., J. Loye, H. Dingle, M. Mathieson, and M. Zalucki. 2005. Ecology of

Leptocoris Hahn (Hemiptera: Rhopalidae) soapberry bugs in Australia. *Australian Journal of Entomology* 44:344–353.

Carson, E. W., M. Tobler, W. L. Minckley, R. J. Ainsworth, and T. E. Dowling. 2012. Relationships between spatio-temporal environmental and genetic variation reveal an important influence of exogenous selection in a pupfish hybrid zone. *Molecular Ecology* 21:1209–1222.

Casacci, L. P., M. Witek, F. Barbero, D. Patricelli, G. Solazzo, E. Balletto, and S. Bonelli. 2011. Habitat preferences of *Maculinea arion* and its *Myrmica* host ants: Implications for habitat management in Italian Alps. *Journal of Insect Conservation* 15:103–110.

Case, A. L., and T.-L. Ashman. 2007. An experimental test of the effects of resources and sex ratio on maternal fitness and phenotypic selection in gynodioecious *Fragaria virginiana*. *Evolution* 61:1900–1911.

Casida, J. E. 2009. Pest toxicology: The primary mechanisms of pesticide action. *Chemical Research in Toxicology* 22:609–619.

Cavalier-Smith, T. 2009. Deep phylogeny, ancestral groups and the four ages of life. *Philosophical Transactions of the Royal Society B* 365:111–132.

Cavicchioli, R., P. M. G. Curmi, N. Saunders, and T. Thomas. 2003. Pathogenic archaea: Do they exist? *BioEssays* 25:1119–1128.

Chafee, M. E., D. J. Funk, R. G. Harrison, and S. R. Bordenstein. 2010. Lateral phage transfer in obligate intracellular bacteria (*Wolbachia*): Verification from natural populations. *Molecular Biology and Evolution* 27:501–505.

Chafee, M. E., C. N. Zecher, M. L. Gourley, V. T. Schmidt, J. H. Chen, S. R. Bordenstein, M. E. Clark, et al. 2011. Decoupling of host-symbiont-phage coadaptations following transfer between insect species. *Genetics* 187:203–215.

Chan, Y. F., M. E. Marks, F. C. Jones, G. Villarreal, M. D. Shapiro, S. D. Brady, A. M. Southwick, et al. 2010. Adaptive evolution of pelvic reduction in sticklebacks by recurrent deletion of a Pitx1 enhancer. *Science* 327:302–305.

Chao, L., B. R. Levin, and F. M. Stewart. 1977. Complex community in a simple habitat: Experimental study with bacteria and phage. *Ecology* 58:369–378.

Chapman, A. D. 2009. *Numbers of living species in Australia and the world*. 2nd ed. Report for the Australian Biological Resources Study, Canberra. Australian Information Services, Department of the Environment, Water, Heritage, and the Arts.

Charbonnel, N., and J. Pemberton. 2005. A long-term genetic survey of an ungulate population reveals balancing selection acting on MHC through spatial and temporal fluctuations in selection. *Heredity* 95:377–388.

Charlat, S., J. Engelstädter, E. A. Dyson, E. A. Hornett, A. Duplouy, P. Tortosa, N. Davies, et al. 2006. Competing selfish genetic elements in the butterfly *Hypolimnas bolina*. *Current Biology* 16:2453–2458.

Charlat, S., E. A. Hornett, J. H. Fullard, N. Davies, G. K. Roderick, N. Wedell, and G. D. D. Hurst. 2007. Extraordinary flux in sex ratio. *Science* 317:214.

Charlesworth, B. 2012. The effects of deleterious mutations on evolution at linked sites. *Genetics* 190:5–22.

Charlesworth, D. 2006a. Balancing selection and its effects on sequences in nearby genome regions. *PLoS Genetics* 2:e64.

———. 2006b. Evolution of plant breeding systems. *Current Biology* 16:R726–R735.

Charlesworth, D., and B. Charlesworth. 1987. Inbreeding depression and its evolutionary consequences. *Annual Review of Ecology and Evolution* 18:237–268.

Charlesworth, D., and J. H. Willis. 2009. The genetics of inbreeding depression. *Nature Reviews Genetics* 10:783–796.

Charnov, E. L. 1982. *The theory of sex allocation*. Princeton University Press, Princeton, NJ.

Chase, J. M., and M. A. Leibold. 2003. *Ecological niches: Linking classical and contemporary approaches*. University of Chicago Press, Chicago.

Chatfield, M. W. H., K. H. Kozak, B. M. Fitzpatrick, and P. K. Tucker. 2010. Patterns of differential introgression in a salamander hybrid zone: Inferences from genetic data and ecological niche modelling. *Molecular Ecology* 19:4265–4282.

Chen, I.-C., J. K. Hill, R. Ohlemüller, D. B. Roy, and C. D. Thomas. 2011. Rapid range shifts of species associated with high levels of climate warming. *Science* 333:1024–1026.

Chen, J.-M., Y.-X. Sun, J.-W. Chen, S. Liu, J.-M. Yu, C.-J. Shen, X.-D. Sun, et al. 2009. Panorama phylogenetic diversity and distribution of type A influenza viruses based on their six internal gene sequences. *Virology Journal* 6:137.

Chesson, P. 2000. Mechanisms of maintenance of species diversity. *Annual Review of Ecology and Systematics* 31:343–366.

Cirrimotich, C. M., Y. Dong, A. M. Clayton, S. L. Sandiford, J. A. Souza-Neto, M. Mulenga, and G. Dimopoulos. 2011. Natural microbe-mediated refractoriness to *Plasmodium* infection in *Anopheles gambiae*. *Science* 332:855–858.

Clamp, M., B. Fry, M. Kamal, X. Xie, J. Cuff, M. F. Lin, M. Kellis, et al. 2007. Distinguishing protein-coding and noncoding genes in the human genome. *Proceedings of the National Academy of Sciences USA* 104:19428–19433.

Clark, M. A., N. A. Moran, P. Baumann, and J. J. Wernegreen. 2000. Cospeciation between bacterial endosymbionts (*Buchnera*) and a recent radiation of aphids (*Uroleucon*) and pitfalls of testing for phylogenetic congruence. *Evolution* 2000:517–525.

Clark, R. M., G. Schweikert, C. Toomajian, S. Ossowski, G. Zeller, P. Shinn, N. Warthmann, et al. 2007. Common sequence polymorphisms shaping genetic diversity in *Arabidopsis thaliana*. *Science* 317:338–342.

Clarke, B. 1962. Natural selection in mixed populations of two polymorphic snails. *Heredity* 17:319–345.

Clarke, C. A., and P. M. Sheppard. 1959. The genetics of *Papilio dardanus*, Brown. I. Race *cenea* from South-Africa. *Genetics* 44:1347–1358.

———. 1963. Interactions between major genes and polygenes in determination of mimetic patterns of *Papilio dardanus*. *Evolution* 17:404–413.

Clausen, J., and W. M. Hiesey. 1958. Experimental studies on the nature of species IV. Genetic structure of ecological races. *Carnegie Institution of Washington Publication* 615:1–312.

Clausen, J., D. D. Keck, and W. M. Hiesey. 1941. Regional differentiation in plant species. *American Naturalist* 75:231–250.

———. 1947. Heredity of geographically and ecologically isolated races. *American Naturalist* 81:114–133.

Clay, K. 2009. Defensive mutualism and grass endophytes: Still valid after all these years. Pages 9–20 in J. F. White, Jr., and M. S. Torres, eds., *Defensive mutualism in microbial symbiosis*. CRC Press, Boca Raton, FL.

Clay, K., J. Holah, and J. A. Rudgers. 2005. Herbivores cause a rapid increase in hereditary symbiosis and alter plant community composition. *Proceedings of the National Academy of Sciences USA* 102:12465–12470.

Clayton, D. H., S. E. Bush, B. M. Goates, and K. P. Johnson. 2003. Host defense reinforces host–parasite cospeciation. *Proceedings of the National Academy of Sciences USA* 100:15694–15699.

Cockram, J., H. Jones, F. J. Leigh, D. O'Sullivan, W. Powell, D. A. Laurie, and A. J. Greenland. 2007. Control of flowering time in temperate cereals: Genes, domestication, and sustainable productivity. *Journal of Experimental Botany* 58:1231–1244.

Coelho, S. 2009. European pesticide rules promote resistance, researchers warn. *Science* 323:450.

Coffey, K., C. W. Benkman, and B. G. Milligan. 1999. The adaptive significance of spines on pine cones. *Ecology* 80:1221–1229.

Cohen, J. E. 1978. *Food webs and niche space*. Princeton University Press, Princeton, NJ.

Cohen, M. S., N. Hellmann, J. A. Levy, K. DeCock, and J. Lange. 2008. The spread, treatment, and prevention of HIV-1: Evolution of a global pandemic. *Journal of Clinical Investigation* 118:1244.

Colautti, R. I., J. L. Maron, and S. H. Barrett 2009. Common garden comparisons of native and introduced plant populations: Latitudinal clines can obscure evolutionary inferences. *Evolutionary Applications* 2:187–199.

Coleman, M. L., and S. W. Chisholm. 2010. Ecosystem-specific selection pressures revealed through comparative population genomics. *Proceedings of the National Academy of Sciences USA* 107:18634–18639.

Colosimo, P. F., K. E. Hosemann, S. Balabhadra, G. Villarreal, Jr., M. Dickson, J. Grimwood, J. Schmutz, et al. 2000. Widespread parallel evolution in sticklebacks by repeated fixation of *ectodyplasin* alleles. *Science* 307:1927–1933.

Condon, M., D. C. Adams, D. Bann, K. Flaherty, J. Gammons, J. Johnson, M. L. Lewis, et al. 2008. Uncovering tropical diversity: Six sympatric species of *Blepharoneura* (Diptera: Tephritidae) in flowers of *Gurania spinulosa*

(Cucurbitaceae) in eastern Ecuador. *Biological Journal of the Linnean Society* 93:779–797.

Conover, D. O, S. A. Arnott, M. R. Walsh, and S. B. Munsch. 2005. Darwinian fishery science: Lessons from the Atlantic silverside (*Menidia menidia*). *Canadian Journal of Fisheries and Aquatic Sciences* 62:730–737.

Conover, D. O., L. M. Clarke, S. B. Munch, and G. N. Wagner. 2006. Spatial and temporal scales of adaptive divergence in marine fishes and the implications for conservation. *Journal of Fish Biology* 69:21–47.

Conover, D. O. , S. B. Munch, and S. A. Arnott. 2009. Reversal of evolutionary downsizing caused by selective harvest of large fish. *Proceedings of the Royal Society B* 276:2015–2020.

Cook, L. M. 2003. The rise and fall of the Carbonaria form of the peppered moth. *Quarterly Review of Biology* 78:399–417.

———. 2008. Variation with habitat in *Cepaea nemoralis*: The Cain and Sheppard diagram. *Journal of Molluscan Studies* 74:239–243.

Cook, L. M., and C. W. A. Pettitt. 2008. Morph frequencies in the snail *Cepaea nemoralis*: Changes with time and their interpretation. *Biological Journal of the Linnean Society* 64:137–150.

Cook, L. M., and J. R. G. Turner. 2008. Decline of melanism in two British moths: Spatial, temporal and inter-specific variation. *Heredity* 101:483–489.

Cook-Patton, S. C., S. H. McArt, A. L. Parachnowitsch, J. S. Thaler, and A. A. Agrawal. 2011. A direct comparison of the consequences of plant genotypic and species diversity on communities and ecosystem function. *Ecology* 92:915–923.

Cooper, M. D., and M. N. Alder. 2006. The evolution of adaptive immune systems. *Cell* 124:815–822.

Cooper, V., S., and R. Lenski, E. 2000. The population genetics of ecological specialization in evolving *Escherichia coli* populations. *Nature* 407:736–739.

Corl, A., A. R. Davis, S. R. Kuchta, T. Comendant, and B. Sinervo. 2010a. Alternative mating strategies and the evolution of sexual size dimorphism in the side-blotched lizard, *Uta stansburiana*: A population-level comparative analysis. *Evolution* 64:79–96.

Corl, A., A. R. Davis, S. R. Kuchta, and B. Sinervo. 2010b. Selective loss of polymorphic mating types is associated with rapid phenotypic evolution during morphic speciation. *Proceedings of the National Academy of Sciences USA* 107:4254–4259.

Coventry, A., L. M. Bull-Otterson, X. Liu, A. G. Clark, T. J. Maxwell, J. Crosby, J. E. Hixson, et al. 2010. Deep resequencing reveals excess rare recent variants consistent with explosive population growth. *Nature Communications* 1:131.

Coyne, J. A., and H. A. Orr. 2004. *Speciation*. Sinauer Associates, Sunderland, MA.

Craig, T. P., and J. K. Itami. 2010. Divergence of *Eurosta solidaginis* in response to host plant variation and natural enemies. *Evolution* 65:802–817.

Craig, T. P., J. K. Itami, and J. D. Horner. 2007. Geographic variation in the evolution and coevolution of a tritrophic interaction. *Evolution* 61:1137–1152.

Crespi, B. J. 2010. The origins and evolution of genetic disease risk in modern humans. *Annals of the New York Academy of Sciences* 1206:80–109.

Crisci, J. L., A. Wong, J. M. Good, and J. D. Jensen. 2011. On characterizing adaptive events unique to modern humans. *Genome Biology and Evolution* 3:791–798.

Cuautle, M., and J. N. Thompson. 2010. Evaluating the co-pollinator network structure of two sympatric *Lithophragma* species with different morphology. *Oecologia* 162:71–80.

Cubillos, F. A., E. Billi, E. Zörgö, L. Parts, P. Fargier, S. Omholt, A. Blomberg, et al. 2011. Assessing the complex architecture of polygenic traits in diverged yeast populations. *Molecular Ecology* 20:1401–1413.

Currano, E. D., C. C. Labandeira, and P. Wilf. 2010. Fossil insect folivory tracks paleotemperature for six million years. *Ecological Monographs* 80:547–567.

Currie, C. R., M. Poulsen, J. Mendenhall, J. J. Boomsma, and J. Billen. 2006. Coevolved crypts and exocrine glands support mutualistic bacteria in fungus-growing ants. *Science* 311:81–83.

Currie, C. R., B. Wong, A. E. Stuart, T. R. Schultz, S. A. Rehner, U. G. Mueller, G.-H. Sung, et al. 2003. Ancient tripartite coevolution in the attine ant-microbe symbiosis. *Science* 299:386–388.

Danchin, E., J. Flot, L. Perfus-Barbeoch, and K. Van Doninck. 2011. Genomic perspectives on the long-term absence of sexual reproduction in animals. Pages 223–242 in P. Pontarotti, ed., *Evolutionary biology: Concepts, biodiversity, macroevolution, and genome evolution*. Springer-Verlag, Berlin.

Danovaro, R., A. Dell'Anno, C. Corinaldesi, M. Magagnini, R. Noble, C. Tamburini, and M. Weinbauer. 2008. Major viral impact on the functioning of benthic deep-sea ecosystems. *Nature* 454:1084–1087.

Darimont, C. T., S. M. Carlson, M. T. Kinnison, P. C. Paquet, T. E. Rieimchen, and C. C. Wilmers. 2009. Human predators outpace other agents of trait change in the wild. *Proceedings of the National Academy of Sciences USA* 106:952–954.

Darwin, C. 1862. *On the various contrivances by which British and foreign orchids are fertilised by insects, and the good effects of intercrossing*. John Murray, London.

Dasmahapatra, K. K., G. Lamas, F. Simpson, and J. Mallet. 2010. The anatomy of a "suture zone" in Amazonian butterflies: A coalescent-based test for vicariant geographic divergence and speciation. *Molecular Ecology* 19:4283–4301.

Davey, N. E., G. Travé, and T. J. Gibson. 2011. How viruses hijack cell regulation. *Trends in Biochemical Sciences* 36:159–169.

Davies, N. B. 2000. *Cuckoos, cowbirds and other cheats*. T. and A. D. Poyser, London.

Davies, N. B., and M. d. L. Brooke. 1989. An experimental study of co-evolution between the cuckoo, *Cuculus canorus*, and its hosts. II. Host egg markings, chick discrimination and general discussion. *Journal of Animal Ecology* 58:225–236.

Davis, D. R., O. Pellmyr, and J. N. Thompson. 1992. Biology and systematics of *Greya* Busck and *Tetragma*, new genus (Lepidoptera: Prodoxidae). *Smithsonian Contributions to Zoology* 524:1–88.

Dawkins, R. 1976. *The selfish gene*. Paladin, London.

———. 1982. *The extended phenotype*. Oxford University Press, Oxford.

———. 1986. *The blind watchmaker: Why the evidence of evolution reveals a universe without design*. Norton, New York.

———. 1989. The evolution of evolvability. Pages 201–220 in C. Langdon, ed., *Artificial life*. Addison-Wesley, Boston.

———. 2004. *The ancestor's tale: A pilgrimage to the dawn of evolution*. Mariner Books, New York.

Dawkins, R., and J. R. Krebs. 1979. Arms races between and within species. *Proceedings of the Royal Society B* 205:489–511.

Day, J. J., J. A. Cotton, and T. G. Barraclough. 2008. Tempo and mode of diversification of Lake Tanganyika cichlid fishes. *PLoS ONE* 3:e1730.

Dayan, T., and D. Simberloff. 2005. Ecological and community-wide character displacement: The next generation. *Ecology Letters* 8:875–894.

De Bruin, A., B. W. Ibelings, M. Kagami, W. M. Mooij, and D. Van. 2008. Adaptation of the fungal parasite *Zygorhizidium planktonicum* during 200 generations of growth on homogeneous and heterogeneous populations of its host, the diatom *Asterionella formosa*. *Journal of Eukaryotic Microbiology* 55:69–74.

De Fine Licht, H. H., M. Schiøtt, U. G. Mueller, and J. J. Boomsma. 2010. Evolutionary transitions in enzyme activity of ant fungus gardens. *Evolution* 64:2055–2069.

de Jong, P. W., C. J. Breuker, H. d. Vos, K. M. C. A. Vermeer, K. Oku, P. Verbaarschot, J. K. Nielsen, et al. 2009. Genetic differentiation between resistance phenotypes in the phytophagous flea beetle, *Phyllotreta nemorum*. *Journal of Insect Science* 9:1–8.

Dearnaley, J. D. W. 2007. Further advances in orchid mycorrhizal research. *Mycorrhiza* 17:475–486.

Decaestecker, E., S. Gaba, J. A. M. Raeymaekers, R. Stoks, L. Van Kerckhoven, D. Ebert, and L. De Meester. 2007. Host–parasite "Red Queen" dynamics archived in pond sediment. *Nature* 450:870–873.

Degnan, P., A. Lazarus, C. Brock, and J. Wernegreen. 2004. Host–symbiont stability and fast evolutionary rates in an ant-bacterium association: Cospeciation of *Camponotus* species and their endosymbionts, *Candidatus biochmannia*. *Systematic Biology* 53:95–110.

DeLong, E. F. 2002. Microbial population genomics and ecology. *Current Opinion in Microbiology* 5:520–524.

———. 2009. The microbial ocean from genomes to biomes. *Nature* 459:200–206.

DeLong, E. F., and N. R. Pace. 2001. Environmental diversity of bacteria and archaea. *Systematic Biology* 50:470–478.

Delph, L. F. 2003. Sexual dimorphism in gender plasticity and its consequences for breeding system evolution. *Evolution and Development* 5:34–39.

Delph, L. F., P. Touzet, and M. F. Bailey. 2007. Merging theory and mechanism in studies of gynodioecy. *Trends in Ecology and Evolution* 22:17–24.

Delph, L. F., and D. E. Wolf. 2005. Evolutionary consequences of gender plasticity in genetically dimorphic breeding systems. *New Phytologist* 166:119–128.

Després, L., and N. Jaeger. 1999. Evolution of oviposition strategies and speciation in the globeflower flies *Chiastocheta* spp. (Anthomyiidae). *Journal of Evolutionary Biology* 12:822–831.

Díaz-Castelzao, C., P. R. Guimarães, Jr., P. Jordano, J. N. Thompson, R. J. Marquis, and V. Rico-Gray. 2010. Changes of a mutualistic network over time: Reanalysis over a 10-year period. *Ecology* 91:793–801.

Dicke, M., and I. T. Baldwin. 2010. The evolutionary context for herbivore-induced plant volatiles: Beyond the "cry for help." *Trends in Plant Science* 15:167–175.

Dietl, G. P., and K. W. Flessa. 2011. Conservation paleobiology: Putting the dead to work. *Trends in Ecology and Evolution* 26:30–37.

Dietl, G. P., and F. J. Vega. 2008. Specialized shell-breaking crab claws in Cretaceous seas. *Biology Letters* 4:290–293.

Dincă, V., V. A. Lukhtanov, G. Talavera, and R. Vila. 2011. Unexpected layers of cryptic diversity in wood white *Leptidea* butterflies. *Nature Communications* 2:324.

Dingle, H. 2007. What is migration? *BioScience* 57:113–121.

Dingle, H., S. P. Carroll, and T. R. Famula. 2009. Influence of genetic architecture on contemporary local evolution in the soapberry bug, *Jadera haematoloma*: Artificial selection on beak length. *Journal of Evolutionary Biology* 22:2031–2040.

Dionne, M., K. M. Miller, J. J. Dodson, and L. Bernatchez. 2009. MHC standing genetic variation and pathogen resistance in wild Atlantic salmon. *Philosophical Transactions of the Royal Society B: Biological Sciences* 364:1555–1565.

Dirzo, R., and M. Loreau. 2005. Biodiversity science evolves. *Science* 310:943.

Dirzo, R., E. Mendoza, and P. Ortiz. 2007. Size-related differential seed predation in a heavily defaunated neotropical rain forest. *Biotropica* 39:355–362.

Dlugosch, K. M., and I. M. Parker. 2008. Invading populations of an ornamental shrub show rapid life history evolution despite genetic bottlenecks. *Ecology Letters* 11:701–709.

Dobrindt, U., M. G. Chowdary, G. Krumbholz, and J. Hacker. 2010. Genome dynamics and its impact on evolution of *Escherichia coli*. *Medical Microbiology and Immunology* 199:145–154.

Dobson, A. 2009. Food-web structure and ecosystem services: Insights from the Serengeti. *Philosophical Transactions of the Royal Society B* 364:1665–1682.

Dobzhansky, T. 1937. *Genetics and the origin of species*. Columbia University Press, New York.

Dodds, P., and P. Thrall. 2009. Recognition events and host–pathogen co-evolution in gene-for-gene resistance to flax rust. *Functional Plant Biology* 36:395–408.

Domingo-Gil, E., R. Toribio, J. Nájera, and M. Esteban. 2011. Diversity in viral anti-PKR mechanisms: A remarkable case of evolutionary convergence. *PLoS Biology* 6:e16711.

Donatti, C. I., P. R. Guimarães, M. Galetti, M. A. Pizo, F. Marquitti, and R. Dirzo. 2011. Analysis of a hyper-diverse seed dispersal network: Modularity and underlying mechanisms. *Ecology Letters* 14:773–781.

Donohue, K., ed. 2011. *Darwin's finches: Readings in the evolution of a scientific paradigm.* University of Chicago Press, Chicago.

Donohue, K., L. Dorn, C. Griffith, E. Kim, A. Aguilera, C. R. Polisetty, and J. Schmitt. 2005. The evolutionary ecology of seed germination of *Arabidopsis thaliana*: Variable natural selection on germination timing. *Evolution* 59:758–770.

Donovan, L. A., D. R. Rosenthal, M. Sanchez-Velenosi, L. H. Rieseberg, and F. Ludwig. 2010. Are hybrid species more fit than ancestral parent species in the current hybrid species habitats? *Journal of Evolutionary Biology* 23:805–816.

Doolittle, W. F. 2010. The attempt on the life of the Tree of Life: Science, philosophy and politics. *Biology and Philosophy* 25:455–473.

Doolittle, W. F., and C. Sapienza. 1980. Selfish genes, the phenotype paradigm and genome evolution. *Nature* 284:601–603.

Doorduin, L. J., and K. Vrieling. 2011. A review of the phytochemical support for the shifting defence hypothesis. *Phytochemistry Reviews* 10:99–106.

Douglas, A. E. 2010. *The symbiotic habit.* Princeton University Press, Princeton, NJ.

Douglas, A. E., and J. A. Raven. 2003. Genomes at the interface between bacteria and organelles. *Philosophical Transactions of the Royal Society B* 358:5–18.

Douglas, T. E., M. R. Kronforst, D. C. Queller, and J. E. Strassman. 2011. Genetic diversity in the social amoeba *Dictyostelium discoideum*: Population differentiation and cryptic species. *Molecular Phylogenetics and Evolution* 60:455–462.

Doyle, J. J., D. Soltis, and P. Soltis. 1985. An intergeneric hybrid in the Saxifragaceae: Evidence from ribosomal-RNA genes. *American Journal of Botany* 72:1388–1391.

Draghi, J., and G. P. Wagner. 2008. Evolution of evolvability in a developmental model. *Evolution* 62:301–315.

Dridi, B., D. Raoult, and M. Drancourt. 2011. Archaea as emerging organisms in complex human microbiomes. *Anaerobe* 17:56–63.

Dubuffet, A., S. Dupas, F. Frey, J.-M. Drezen, M. Poirie, and Y. Carton. 2007. Genetic interactions between the parasitoid wasp *Leptopilina boulardi* and its *Drosophila* hosts. *Heredity* 98:21–27.

Duenez-Guzman, E. A., J. Mavárez, M. D. Vose, and S. Gavrilets. 2009. Case studies and mathematical models of ecological speciation. 4. Hybrid speciation in butterflies in a jungle. *Evolution* 63:2611–2626.

Dufay, M., and J. R. Pannell. 2010. The effect of pollen versus seed flow on the maintenance of nuclear-cytoplasmic gynodioecy. *Evolution* 64:772–784.

Duffy, M. A., C. Brassil, S. Hall, A. Tessier, C. E. Cáceres, and J. K. Conner. 2008. Parasite-mediated disruptive selection in a natural *Daphnia* population. *BMC Evolutionary Biology* 8:80.

Duffy, M. A., S. R. Hall, C. E. Cáceres, and A. R. Ives. 2009. Rapid evolution, seasonality, and the termination of parasite epidemics. *Ecology* 90:1441–1448.

Duffy, M. A., and L. Sivars-Becker. 2007. Rapid evolution and ecological host–parasite dynamics. *Ecology Letters* 10:44–53.

Dunbar, H. E., A. C. C. Wilson, N. R. Ferguson, and N. A. Moran. 2007. Aphid thermal tolerance is governed by a point mutation in bacterial symbionts. *PLoS Biology* 5:e96.

Duncan, R. P., P. Cassey, and T. M. Blackburn. 2009. Do climate envelope models transfer? A manipulative test using dung beetle introductions. *Proceedings of the Royal Society B* 276:1449–1457.

Dunlap, P. V., J. C. Ast, S. Kimura, A. Fukui, T. Yoshino, and H. Endo. 2007. Phylogenetic analysis of host-symbiont specificity and codivergence in bioluminescent symbioses. *Cladistics* 23:507–532.

Dunn, R. R., A. D. Gove, T. G. Barraclough, T. J. Givnish, and J. D. Majer. 2007. Convergent evolution of an ant–plant mutualism across plant families, continents, and time. *Evolutionary Ecology Research* 9:1349–1362.

Duperron, S., M. Sibuet, B. J. MacGregor, M. M. M. Kuypers, C. R. Fisher, and N. Dubilier. 2007. Diversity, relative abundance and metabolic potential of bacterial endosymbionts in three *Bathymodiolus* mussel species from cold seeps in the Gulf of Mexico. *Environmental Microbiology* 9:1423–1438.

Duplouy, A., G. D. D. Hurst, S. L. O:Neill, and S. Charlat. 2010. Rapid spread of male-killing *Wolbachia* in the butterfly *Hypolimnas bolina*. *Journal of Evolutionary Biology* 23:231–235.

Duvaux, L., K. Belkhir, M. Boulesteix, and P. Boursot. 2011. Isolation and gene flow: Inferring the speciation history of European house mice. *Molecular Ecology* 20:5248–5264.

Dybdahl, M. F., J. Jokela, L. F. Delph, B. Koskella, and C. M. Lively. 2008. Hybrid fitness in a locally adapted parasite. *American Naturalist* 172:772–782.

Dyer, L. A., M. S. Singer, J. Lill, J. Stireman, G. Gentry, R. Marquis, R. E. Ricklefs, et al. 2007. Host specificity of Lepidoptera in tropical and temperate forests *Nature* 448:696–699.

Ebert, D. 2004. Conceptual issues in local adaptation. *Ecology Letters* 7:1225–1241.

———. 2008. Host–parasite coevolution: Insights from the *Daphnia*-parasite model system. *Current Opinion in Microbiology* 11:290–301.

Edelaar, P. 2008. Assortative mating also indicates that common crossbill *Loxia curvirostra* vocal types are species. *Journal of Avian Biology* 39:9–12.

Edelaar, P., and C. W. Benkman. 2006. Replicated population divergence caused by localized coevolution? A test of three hypotheses in the red crossbill–lodgepole pine system. *Journal of Evolutionary Biology* 19:1652–1659.

Edeline, E., S. M. Carlson, L. C. Stige, I. J. Winfield, J. M. Fletcher, J. B. James, T. O. Haugen, et al. 2007. Trait changes in a harvested population are driven by a dynamic tug-of-war between natural and harvest selection. *Proceedings of the National Academy of Sciences USA* 104:15799–15804.

Edmunds, G. F., and D. N. Alstad. 1978. Coevolution in insect herbivores and conifers. *Science* 199:941–945.

Edwards, S., and S. Bensch. 2009. Looking forwards or looking backwards in avian phylogeography? A comment on Zink and Barrowclough 2008. *Molecular Ecology* 18:2930–2933.

Egan, S. P., and D. J. Funk. 2009. Ecologically dependent postmating isolation between sympatric host forms of *Neochlamisus bebbianae* leaf beetles. *Proceedings of the National Academy of Sciences USA* 106:19426–19431.

Egan, S. P., and J. R. Ott. 2007. Host plant quality and local adaptation determine the distribution of a gall-forming herbivore. *Ecology* 88:2868–2879.

Ehrlén, J., and Z. Münzbergova. 2009. Timing of flowering: Opposed selection on different fitness components and trait covariation. *American Naturalist* 173:819–830.

Ehrlich, P. R. 1965. The population biology of the checkerspot butterfly, *Euphydryas editha*. II. The structure of the Jasper Ridge colony. *Evolution* 19:327–336.

———. 2001. Intervening in evolution: Ethics and actions. *Proceedings of the National Academy of Sciences USA* 98:5477–5480.

Ehrlich, P. R., and A. H. Ehrlich. 2008. *The dominant animal: Human evolution and the environment*. Island Press, Washington, DC.

Ehrlich, P. R., and I. Hanski. 2004. *On the wings of checkerspots: A model system for population biology*. Oxford University Press, New York.

Ehrlich, P. R., and L. G. Mason. 1966. Population biology of the butterfly *Euphydryas editha*. 3. Selection and phenetics of the Jasper Ridge Colony. *Evolution* 20:165–173.

Ehrlich, P. R., and P. H. Raven. 1964. Butterflies and plants: A study in coevolution. *Evolution* 18:586–608.

Ehrlich, P. R., R. R. White, M. C. Singer, S. W. McKechnie, and L. E. Gilbert. 1975. Checkerspot butterflies: A historical perspective. *Science* 188:221–228.

Ehrlich, S. D., and K. E. Nelson. 2010. MetaHIT: The European Union project on metagenomics of the human intestinal tract. Pages 307–316 in K. E. Nelson, ed., *Metagenomics of the human body*. Springer, New York.

Elde, N. C., S. J. Child, A. P. Geballe, and H. S. Malik. 2009. Protein kinase R reveals an evolutionary model for defeating viral mimicry. *Nature* 457: 485–489.

Elde, N. C., and H. S. Malik. 2009. The evolutionary conundrum of pathogen mimicry. *Nature Reviews Microbiology* 7:787–797.

Eldredge, N. 2001. The sloshing bucket: How the physical realm controls evolution. Pages 3–32 in J. Crutchfield and P. Schuster, eds., *Evolutionary dynamics: Exploring the interplay of selection, neutrality, accident, and function*. Oxford University Press, New York.

———. 2008. Hierarchies and the sloshing bucket: Toward the unification of evolutionary biology. *Evolution: Education and Outreach* 1:10–15.

Eldredge, N., and S. J. Gould. 1972. Punctuated equilibrium: An alternative to phyletic gradualism. Pages 82–115 in T. J. M. Schopf, ed., *Models in paleobiology*. Freeman, Cooper, and Co., San Francisco.

Eldredge, N., J. N. Thompson, P. M. Brakefield, S. Gavrilets, D. Jablonski, J. B. C. Jackson, R. E. Lenski, et al. 2005. The dynamics of evolutionary stasis. *Paleobiology* 31:133–145.

Elena, S. F., V. S. Cooper, and R. E. Lenski. 1996. Punctuated evolution caused by selection of rare beneficial mutations. *Science* 272:1802–1804.

Elena, S. F., P. Agudelo-Romero, P. Carrasco, F. Codoñer, S. Martin, C. Torres-Barceló, and R. Sanjuán. 2008. Experimental evolution of plant RNA viruses. *Heredity* 100:478–483.

Ellner, S. P., and L. Becks. 2010. Rapid prey evolution and the dynamics of two-predator food webs. *Theoretical Ecology* 4:133–152.

Ellner, S. P., M. A. Geber, and N. G. J. Hairston. 2011. Does rapid evolution matter? Measuring the rate of contemporary evolution and its impacts on ecological dynamics. *Ecology Letters* 14:603–614.

Ellner, S. P., N. G. Hairston, C. M Kearns, and D. Babai. 1999. The roles of fluctuating selection and long-term diapause in microevolution of diapause timing in a freshwater copepod. *Evolution* 53:111–122.

Elmer, K. R., T. K. Lehtonen, A. F. Kautt, C. Harrod, and A. Meyer. 2010. Rapid sympatric ecological differentiation of crater lake cichlid fishes within historic times. *BMC Biology* 8:60.

Elmer, K. R., and A. Meyer. 2011. Adaptation in the age of ecological genomics: Insights from parallelism and convergence. *Trends in Ecology and Evolution* 26:298–306.

Elmes, G. W., T. Akino, J. A. Thomas, R. T. Clarke, and J. J. Knapp. 2002. Interspecific differences in cuticular hydrocarbon profiles of *Myrmica* ants are sufficiently consistent to explain host specificity by *Maculinea* (Large Blue) butterflies. *Oecologia* 130:525–535.

Elton, C. 1927. *Animal ecology*. Sidgwick and Jackson, London.

Elzinga, J., A. Atlan, A. Biere, and L. Gigord. 2007. Time after time: Flowering phenology and biotic interactions. *Trends in Ecology and Evolution* 8:432–439.

Emerson, B., and R. G. Gillespie. 2008. Phylogenetic analysis of community assembly and structure over space and time. *Trends in Ecology and Evolution* 23:619–630.

Emerson, J. J., M. Cardoso-Moreira, J. O. Borevitz, and M. Long. 2008. Natural selection shapes genome-wide patterns of copy-number polymorphism in *Drosophila melanogaster*. *Science* 320:1629–1631.

Endler, J. A. 1977. *Geographic variation, speciation, and clines*. Princeton University Press, Princeton, NJ.

———. 1980. Natural selection on color patterns in *Poecilia reticulata*. *Evolution* 34:76–91.

———. 1986. *Natural selection in the wild*. Princeton University Press, Princeton, NJ.

Engelstädter, J., and G. D. D. Hurst. 2009. The ecology and evolution of microbes that manipulate host reproduction. *Annual Review of Ecology, Evolution, and Systematics* 40:127–149.

Engh, I. B., I. Skrede, G.-P. Saetre, and H. Kauserud. 2010. High variability in a mating type linked region in the dry rot fungus *Serpula lacrymans* caused by frequency-dependent selection? *BMC Genetics* 11:64.

Enk, J., A. Devault, R. Debruyne, C. E. King, T. Treangen, D. O'Rourke, S. L. Salzberg, et al. 2011. Complete Columbian mammoth mitogenome suggests interbreeding with woolly mammoths. *Genome Biology* 12:R51.

Epling, C., and T. Dobzhansky. 1942. Genetics of natural populations. VI. Microgeographical races of *Linanthus parryi*. *Genetics* 27:317–332.

Erbilgin, N., N. E. Gillette, S. R. Mori, J. D. Stein, D. R. Owen, and D. L. Wood. 2007. Acetophenone as an anti-attractant for the western pine beetle, *Dendroctonus brevicomis* LeConte (Coleoptera: Scolytidae). *Journal of Chemical Ecology* 33:817–823.

Ercolin, F., and D. Reinhardt. 2011. Successful joint ventures of plants: Arbuscular mycorrhiza and beyond. *Trends in Plant Science* 16:356–362.

Erken, M., M. Weitere, S. Kjelleberg, and D. McDougald. 2011. *In situ* grazing resistance of *Vibrio cholerae* in the marine environment. *FEMS Microbiology Ecology* 76:504–512.

Estes, J. A., J. Terborgh, J. S. Brashares, M. E. Power, J. Berger, W. J. Bond, S. R. Carpenter, et al. 2011. Trophic downgrading of planet Earth. *Science* 333:301–306.

Estes, S., and S. J. Arnold. 2007. Resolving the paradox of stasis: Models with stabilizing selection explain evolutionary divergence on all timescales. *American Naturalist* 169:227–244.

Evans, L. M., G. J. Allan, S. M. Shuster, S. A. Woolbright, and T. G. Whitham. 2008. Tree hybridization and genotypic variation drive cryptic speciation of a specialist mite herbivore. *Evolution* 62:3027–3040.

Ewald, P. W. 1994. *Evolution of infectious disease*. Oxford University Press, Oxford.

Faeth, S. H. 2009. Asexual fungal symbionts alter reproductive allocation and herbivory over time in their native perennial grass hosts. *American Naturalist* 173:554–565.

Fan, S., K. R. Elmer, and A. Meyer. 2012. Genomics of adaptation and speciation in cichlid fishes: Recent advance and analyses in African and Neotropical lineages. *Philosophical Transactions of the Royal Society B* 367:385–394.

Feder, J. L., and A. A. Forbes. 2008. Host fruit-odor discrimination and sympatric host-race formation. Pages 101–116 in K. J. Tilmon, ed., *Specialization, speciation and radiation: The evolutionary biology of herbivorous insects*. University of California Press, Berkeley.

Feder, J. L., T. H. Q. Powell, K. Filchak, and B. Leung. 2010. The diapause response of *Rhagoletis pomonella* to varying environmental conditions and its significance for geographic and host plant-related adaptation. *Entomologia Experimentalis et Applicata* 136:31–44.

Feldman, C. R., E. D. J. Brodie, E. D. I. Brodie, and M. E. Pfrender. 2010. Genetic architecture of a feeding adaptation: Garter snake (*Thamnophis*) resistance to tetrodotoxin bearing prey. *Proceedings of the Royal Society B* 277:3317–3325.

Felsenstein, J. 1974. The evolutionary advantage of recombination. *Genetics* 78:737–756.

Fenster, C. B., W. S. Armbruster, P. Wilson, M. R. Dudash, and J. D. Thomson. 2004. Pollination syndromes and floral specialization. *Annual Review of Ecology and Systematics* 35:375–403.

Ferrari, J., S. Via, and H. C. J. Godfray. 2008. Population differentiation and genetic variation in performance on eight hosts in the pea aphid complex. *Evolution* 62:2508–2524.

Fidock, D., R. Eastman, S. Ward, and S. Meshnick. 2008. Recent highlights in antimalarial drug resistance and chemotherapy research. *Trends in Parasitology* 24:537–544.

Filardi, C. E., and C. E. Smith. 2008. Social selection and geographic variation in two monarch flycatchers from the Solomon islands. *Condor* 110:24–34.

Fiore, C., J. Jarett, and N. Olson. 2010. Nitrogen fixation and nitrogen transformations in marine symbioses. *Trends in Microbiology* 18:455–463.

Fischbach, M. A, and C. T. Walsh. 2009. Antibiotics for emerging pathogens. *Science* 325:1089–1093.

Fischer, B., and S. Foitzik. 2004. Local co-adaptation leading to a geographic mosaic of coevolution in a social parasite system. *Journal of Evolutionary Biology* 17:1026–1034.

Fischer-Blass, B., J. Heinze, and S. Foitzik. 2006. Microsatellite analysis reveals strong but differential impact of a social parasite on its two host species. *Molecular Ecology* 15:863–872.

Fisher, R. A. 1930. *The genetical theory of natural selection*. Oxford University Press, Oxford.

Fisher, R. A., and E. B. Ford. 1947. The spread of a gene in natural conditions in a colony of the moth *Panaxia dominula* L. *Heredity* 1:143–174.

Fleming, T. H., C. Geiselman, and W. J. Kress. 2009. The evolution of bat pollination: A phylogenetic perspective. *Annals of Botany* 104:1017–1043.

Fleming, T. H., and W. J. Kress. 2011. A brief history of fruits and frugivores. *Acta Oecologica* 37:521–530.

Flint, J., and T. F. C. Mackay. 2009. Genetic architecture of quantitative traits in mice, flies, and humans. *Genome Research* 19:723–733.

Floate, K. D., G. D. Martinsen, and T. G. Whitham. 1997. Cottonwood hybrid zones as centres of abundance for gall aphids in western North America: Importance of relative habitat size. *Journal of Animal Ecology* 66:179–188.

Flor, H. H. 1942. Inheritance of pathogenicity in *Melampsora lini*. *Phytopathology* 32:653–669.

———. 1955. Host–parasite interaction in flax rust: Its genetics and other implications. *Phytopathology* 45:680–685.

———. 1956. The complementary genic systems of flax and flax rust. *Advances in Genetics* 8:29–54.

Foitzik, S., A. Achenbach, and M. Brandt. 2009a. Locally adapted social parasite affects density, social structure, and life history of its ant hosts. *Ecology* 90:1195–1206.
Foitzik, S., S. Bauer, S. Laurent, and P. S. Pennings. 2009b. Genetic diversity, population structure and sex-biased dispersal in three co-evolving species. *Journal of Evolutionary Biology* 22:2470–2480.
Fontaine, C., P. R. Guimarães, Jr., S. Kéfi, N. Loeuille, J. Memmott, W. H. van der Putten, F. J. F. van Veen, et al. 2011. The ecological and evolutionary implications of merging different types of networks. *Ecology Letters* 14: 1170–1181.
Forbes, A. A., T. H. Q. Powell, L. L. Stelinski, J. J. Smith, and J. L. Feder. 2009. Sequential sympatric speciation across trophic levels. *Science* 323:776–779.
Ford, E. B. 1945. Polymorphism. *Biological Reviews* 20:73–88.
———. 1964. *Ecological genetics*. Methuen, London.
Forde, S. E., R. E. Beardmore, I. Gudelj, S. S. Arkin, J. N. Thompson, and L. D. Hurst. 2008a. Understanding the limits to generalisability of experimental evolutionary models. *Nature* 455:220–223.
Forde, S. E., J. N. Thompson, and B. J. M. Bohannan. 2004. Adaptation varies through space and time in a coevolving host–parasitoid interaction. *Nature* 431:841–844.
———. 2007. Gene flow reverses an adaptive cline in a coevolving host–parasitoid interaction. *American Naturalist* 169:794–801.
Forde, S. E., J. N. Thompson, R. D. Holt, and B. J. M. Bohannan. 2008b. Coevolution drives temporal changes in fitness and diversity across environments in a bacteria-bacteriophage interaction. *Evolution* 62:1830–1839.
Forsman, A., M. Karlsson, L. Wennersten, J. Johansson, and E. Karpestam. 2011. Rapid evolution of fire melanism in replicated populations of pygmy grasshoppers. *Evolution* 65:2530–2540.
Fossøy, F., A. Antonov, A. Moksenes, E. Røskaft, J. R. Vikan, A. P. Møller, J. A. Shykoff, and B. G. Stokke. 2011. Genetic differentiation among sympatric cuckoo host races: Males matter. *Proceedings of the Royal Society B* 278:1639–1645.
Foster, K. R. 2011. The sociobiology of molecular systems. *Nature Reviews Genetics* 12:193–203.
Foster, K. R., and H. Kokko. 2006. Cheating can stabilize cooperation in mutualisms. *Proceedings of the Royal Society B* 273:2233–2239.
Fox, L. R., and P. A. Morrow. 1981. Specialization: Species property or local phenomenon? *Science* 211:887–893.
Foxcroft, L. C., S. T. A. Pickett, and M. L. Cadenasso. 2011. Expanding the conceptual frameworks of plant invasion ecology. *Perspectives in Plant Ecology Evolution and Systematics* 13:89–100.
Fox-Dobbs, K., J. A. Leonard, and P. L. Koch. 2008. Pleistocene megafauna from eastern Beringia: Paleoecological and paleoenvironmental interpretations of

stable carbon and nitrogen isotope and radiocarbon records. *Palaeogeography, Palaeoclimatology, Palaeoecology* 261:30–46.

Franks, S. J., P. D. Pratt, F. A. Dray, and E. L. Simms. 2008. Selection on herbivory resistance and growth rate in an invasive plant. *American Naturalist* 171:678–691.

Franks, S. J., S. Sim, and A. E. Weis. 2007. Rapid evolution of flowering time by an annual plant in response to a climate fluctuation. *Proceedings of the National Academy of Sciences USA* 104:1278–1282.

Franks, S. J., and A. E. Weis. 2008. A change in climate causes rapid evolution of multiple life-history traits and their interactions in an annual plant. *Journal of Evolutionary Biology* 21:1321–1334.

Fraser, J. A., Y.-P. Hsueh, K. M. Findley, and J. Heitman. 2007. Evolution of the mating-type locus: The Basidiomycetes. Pages 19–34 in J. Heitman, J. W. Kronstat, J. W. Taylor, and L. A. Casselton, eds., *Sex in fungi: Molecular determination and evolutionary implications.* ASM Press, Washington, DC.

Frederickson, M. E. 2009. Conflict over reproduction in an ant–plant symbiosis: Why *Allomerus octoarticulatus* ants sterilize *Cordia nodosa* trees. *American Naturalist* 173:675–681.

Frey-Klett, P., J. Garbaye, and M. Tarkka. 2007. The mycorrhiza helper bacteria revisited. *New Phytologist* 176:22–36.

Friberg, M., M. Bergman, J. Kullberg, and N. Wahlberg. 2008. Niche separation in space and time between two sympatric sister species: A case of ecological pleiotropy. *Evolutionary Ecology* 22:1–18.

Friberg, M., I. M. A. Haugen, J. Dahlerus, K. Gotthard, and C. Wiklund. 2011. Asymmetric life-history decision-making in butterfly larvae. *Oecologia* 165:301–310.

Friberg, M., and C. Wiklund. 2010. Host-plant-induced larval decision-making in a habitat/host-plant generalist butterfly. *Ecology* 91:15–21.

Fukami, T., H. J. E. Beaumont, X.-X. Zhang, and P. B. Rainey. 2007. Immigration history controls diversification in experimental adaptive radiation. *Nature* 446:436–439.

Fumagalli, M., R. Cagliani, U. Pozzoli, S. Riva, G. P. Comi, G. Menozzi, N. Bresolin, et al. 2009. Widespread balancing selection and pathogen-driven selection at blood group antigen genes. *Genome Research* 19:199–212.

Funk, D. J., J. J. Wernegreen, and N. A. Moran. 2001. Intraspecific variation in symbiont genomes: Bottlenecks and the aphid-Buchnera association. *Genetics* 157:477–489.

Fürst, M. A., M. Durey, and D. R. Nash. 2011. Testing the adjustable threshold model for intruder recognition on *Myrmica* ants in the context of a social parasite. *Proceedings of the Royal Society B* 279:516–522.

Fussman, G. F., M. Loreau, and P. A. Abrams. 2007. Eco-evolutionary dynamics of communities and ecosystems. *Functional Ecology* 21:465–477.

Futuyma, D. J. 2010. Evolutionary constraint and ecological consequences. *Evolution* 64:1865–1884.

Futuyma, D. J., and A. A. Agrawal. 2009. Macroevolution and the biological diversity of plants and herbivores. *Proceedings of the National Academy of Sciences USA* 106:18054–18061.

Futuyma, D. J., and G. C. Mayer. 1980. Non-allopatric speciation in animals. *Systematic Zoology* 29:254–271.

Gagneux, S., C. Long, P. Small, T. Van, G. Schoolnik, and B. Bohannan. 2006. The competitive cost of antibiotic resistance in *Mycobacterium tuberculosis*. *Science* 312:1944–1946.

Galen, C., R. Kaczorowski, S. L. Todd, J. C. Geib, and R. Raguso. 2011. Dosage-dependent impacts of a floral volatile compound on pollinators, larcenists, and the potential for floral evolution in the alpine skypilot *Polemonium viscosum*. *American Naturalist* 177:258–272.

Gall, B. G., A. N. Stokes, S. S. French, E. A. Schlepphorst, E. D. Brodie III, and E. D. Brodie, Jr. 2011. Tetrodotoxin levels in larval and metamorphosed newts (*Taricha granulosa*) and palatability to predatory dragonflies. *Toxicon* 57:978–983.

Gandon, S., and T. Day. 2009. Evolutionary epidemiology and the dynamics of adaptation. *Evolution* 63:826–838.

Gandon, S., and S. L. Nuismer. 2009. Interactions between genetic drift, gene flow, and selection mosaics drive parasite local adaptation. *American Naturalist* 173:212–224.

Garland, T., Jr., and M. R. Rose, eds. 2009. *Experimental evolution: Concepts, methods, and applications of selection experiments*. University of California Press, Berkeley.

Garraud, C., B. Brachi, M. Dufay, P. Touzet, and J. A. Shykoff. 2011. Genetic determination of male sterility in gynodioecious *Silene nutans*. *Heredity* 106: 757–764.

Gavrilets, S. 1997. Evolution and speciation on holey adaptive landscapes. *Trends in Ecology and Evolution* 13:307–312.

———. 2004. *Fitness landscapes and the origin of species*. Princeton University Press, Princeton, NJ.

Gavrilets, S., and J. B. Losos. 2009. Adaptive radiation: Contrasting theory with data. *Science* 323:732–737.

Geffeney, S. L., E. Fujimoto, E. D. Brodie III, E. D. Brodie, Jr., and P. C. Ruben. 2005. Evolutionary diversification of TTX-resistant sodium channels in a predator–prey interaction. *Nature* 434:759–763.

Genini, J., L. P. C. Morellato, J. P. R. Guimaraes, Jr., and J. M. Olesen. 2010. Cheaters in mutualism networks. *Biology Letters* 6:494–497.

Genung, M. A., G. M. Crutsinger, J. K. Bailey, J. A. Schweitzer, and N. J. Sanders. 2012. Aphid and ladybird beetle abundance depend on the interaction of spatial effects and genotypic diversity. *Oecologia* 168:167–174.

Ghalambor, C. K., J. K. McKay, S. P. Carroll, and D. N. Reznick. 2007. Adaptive versus non-adaptive phenotypic plasticity and the potential for contemporary adaptation in new environments. *Functional Ecology* 21:394–407.

Ghedin, E., N. A. Sengamalay, M. Shumway, J. Zaborsky, T. Feldblyum, V. Subbu, D. J. Spiro, et al. 2005. Large-scale sequencing of human influenza reveals the dynamic nature of viral genome evolution. *Nature* 437:1162–1166.

Gibbs, A. 1999. Laboratory selection for the comparative physiologist. *Journal of Experimental Biology* 202:2709–2718.

Gienapp, P., R. Leimu, and J. Merilä. 2007a. Responses to climate change in avian migration time: Microevolution versus phenotypic plasticity. *Climate Research* 35:25–35.

Gienapp, P., C. Teplitsky, J. S. Alho, J. A. Mills, and J. Merilä. 2007b. Climate change and evolution: Disentangling environmental and genetic responses. *Molecular Ecology* 17:167–178.

Gil, R., B. Sabater-Muñoz, A. Latorre, F. J. Silva, and A. Moya. 2002. Extreme genome reduction in *Buchnera* spp.: Toward the minimal genome needed for symbiotic life. *Proceedings of the National Academy of Sciences USA* 99:4454–4458.

Gilbert, L. E. 2003. Adaptive novelty through introgression in *Heliconius* wing patterns: Evidence for a shared genetic "toolbox" from synthetic hybrid zones and a theory of diversification. Pages 281–318 in C. L. Boggs, W. B. Watt, and P. R. Ehrlich, eds., *Butterflies: Ecology and evolution taking flight*. University of Chicago Press, Chicago.

Gilchrist, G. W., R. B. Huey, J. Balanya, M. Pascual, and L. Serra. 2004. A time series of evolution in action: A latitudinal cline in wing size in South American *Drosophila subobscura*. *Evolution* 58:768–780.

Gilchrist, G. W., L. M. Jeffers, B. West, D. G. Folk, J. Suess, and R. B. Huey. 2008. Clinal patterns of desiccation and starvation resistance in ancestral and invading populations of *Drosophila subobscura*. *Evolutionary Applications* 1:513–523.

Gill, E. E., and F. S. L. Brinkman. 2011. The proportional lack of archaeal pathogens: Do viruses/phages hold the key? *BioEssays* 33:248–254.

Gingerich, P. D. 1983. Rates of evolution: Effects of time and temporal scaling. *Science* 222:159–161.

———. 1993. Quantification and comparison of evolutionary rates. *American Journal of Science* 293A:453–478.

———. 2000. Arithmetic or geometric normality of biological variation: An empirical test of theory. *Journal of Theoretical Biology* 204:201–221.

———. 2009. Rates of evolution. *Annual Review of Ecology, Evolution, and Systematics* 40:657–675.

Givnish, T. J. 2010. Ecology of plant speciation. *Taxon* 59:1326–1366.

Gjerde, B., B. F. Terjesen, Y. Barr, I. Lein, and I. Thorland. 2004. Genetic variation for juvenile growth and survival in Atlantic cod (*Gadus morhua*). *Aquaculture* 236:167–177.

Gladyshev, E. A., M. Meselson, and I. R. Arkhipova. 2008. Massive horizontal gene transfer in bdelloid rotifers. *Science* 320:1210–1213.

Glor, R. E., J. J. Kolbe, R. Powell, A. Larson, and J. B. Losos. 2001. Phylogenetic analysis of ecological and morphological diversification in Hispaniolan trunk-ground anoles (*Anolis cybotes* group). *Evolution* 57:2383–2397.

Glor, R. E., J. B. Losos, and A. Larson. 2005. Out of Cuba: Overwater dispersal and speciation among lizards in the *Anolis carolinensis* subgroup. *Molecular Ecology* 14:2419–2432.

Godsoe, W., E. Strand, C. I Smith, J. B. Yoder, T. C. Esque, and O. Pellmyr. 2009. Divergence in an obligate mutualism is not explained by divergent climatic factors. *New Phytologist* 183:589–599.

Godsoe, W., J. B. Yoder, C. I. Smith, and O. Pellmyr. 2008. Coevolution and divergence in the Joshua tree/yucca moth mutualism. *American Naturalist* 171:816–824.

Goecke, F., A. Labes, J. Wiese, and J. F. Imhoff. 2010. Chemical interactions between marine macroalgae and bacteria. *Marine Ecology Progress Series* 409: 267–299.

Goetz, T. D. 2010. The global problem of antibiotic resistance 30:79–93.

Gojobori, J., H. Tang, J. M. Akey, and C.-I. Wu. 2007. Adaptive evolution in humans revealed by the negative correlation between the polymorphism and fixation phases of evolution. *Proceedings of the National Academy of Sciences USA* 104:3907–3912.

Goldberg, K., E. Eltzov, M. Schnit-Orland, R. S. Marks, and A. Kushmaro. 2011. Characterization of quorum sensing signals in coral-associated bacteria. *Microbial Ecology* 61:783–792.

Goldblatt, P., P. Bernhardt, P. Vogan, and J. C. Manning. 2004. Pollination by fungus gnats (Diptera: Mycetophilidae) and self-recognition sites in *Tolmiea menziesii* (Saxifragaceae). *Plant Systematics and Evolution* 244:55–67.

Goldenfield, N., and C. Woese. 2007. Biology's next revolution. *Nature* 445:369.

Gómez, J. M., F. Perfectti, J. Bosch, and J. P. M. Camacho. 2009. A geographic selection mosaic in a generalized plant–pollinator–herbivore system. *Ecological Monographs* 79:245–263.

Gómez, J. M., and R. Zamora. 2000. Spatial variation in the selective scenarios of *Hormathophylla spinosa* (Cruciferae). *American Naturalist* 155:657–668.

Gompert, Z., J. A. Fordyce, M. L. Forister, A. M. Shapiro, and C. C. Nice. 2006. Homoploid hybrid speciation in an extreme habitat. *Science* 314:1923–1925.

Gompert, Z., L. K. Lucas, J. A. Fordyce, M. L. Forister, and C. C. Nice. 2010. Secondary contact between *Lycaeides idas* and *L. melissa* in the Rocky Mountains: Extensive admixture and a patchy hybrid zone. *Molecular Ecology* 19:3171–3192.

Gomulkiewicz, R., D. M. Drown, M. F. Dybdahl, W. Godsoe, S. L. Nuismer, K. M. Pepin, B. J. Ridenhour, et al. 2007. Dos and don'ts for testing the geographic mosaic theory of coevolution. *Heredity* 2007:1–10.

Gomulkiewicz, R., and R. D. Holt. 1995. When does evolution by natural selection prevent extinction? *Evolution* 49:201–207.

Gomulkiewicz, R., R. D. Holt, M. Barfield, and S. L. Nuismer. 2010. Genetics, adaptation, and invasion in harsh environments. *Evolutionary Applications* 3:97–108.

Gomulkiewicz, R., and D. Houle. 2009. Demographic and genetic constraints on evolution. *American Naturalist* 174:E218–E229.

Gomulkiewicz, R., J. N. Thompson, R. D. Holt, S. L. Nuismer, and M. E. Hochberg. 2000. Hot spots, cold spots, and the geographic mosaic theory of coevolution. *American Naturalist* 156:156–174.

Gonzalez, E., H. Kulkarni, H. Bolivar, A. Mangano, R. Sanchez, G. Catano, R. Nibbs, et al. 2005. The influence of CCL3L1 gene-containing segmental duplications on HIV-1/AIDS susceptibility. *Science* 307:1434–1440.

González-Teuber, M., and M. Heil. 2010. *Pseudomyrmex* ants and *Acacia* host plants join efforts to protect their mutualism from microbial threats. *Plant Signaling and Behavior* 5:890–892.

Gordon, D. M. 2011. The fusion of behavioral ecology and ecology. *Behavioral Ecology* 22:225–230.

Gould, S. J. 1996. *Full house: The spread of excellence from Plato to Darwin*. Three Rivers Press, New York.

———. 2002. *The structure of evolutionary theory*. Harvard University Press, Cambridge, MA.

Gow, J. L., S. M. Rogers, M. Jackson, and D. Schluter. 2008. Ecological predictions lead to the discovery of a benthic-limnetic sympatric species pair of threespine stickleback in Little Quarry Lake, British Columbia. *Canadian Journal of Zoology* 86:564–571.

Grant, B. R., and P. R. Grant. 1996a. Cultural inheritance of song and its role in the evolution of Darwin's finches. *Evolution* 50:2471–2487.

———. 2008a. Fission and fusion of Darwin's finches populations. *Philosophical Transactions of the Royal Society B* 363:2821–2829.

———. 2010a. Songs of Darwin's finches diverge when a new species enters the community. *Proceedings of the National Academy of Sciences USA* 107:20156–20163.

Grant, P. R., and B. R. Grant. 1994. Phenotypic and genetic effects of hybridization in Darwin's finches. *Evolution* 48:297–316.

———. 1996b. Speciation and hybridization of island birds. *Philosophical Transactions of the Royal Society* 351:765–772.

———. 2002. Unpredictable evolution in a 30-year study of Darwin's finches. *Science* 296:707–711.

———. 2008b. *How and why species multiply: The radiation of Darwin's finches*. Princeton University Press, Princeton, NJ.

———. 2008c. Pedigrees, assortative mating and speciation in Darwin's finches. *Proceedings of the Royal Society B* 275:661–668.

———. 2010b. Conspecific versus heterospecific gene exchange between populations of Darwin's finches. *Philosophical Transactions of the Royal Society of London B* 365:1065–1076.

Grant, P. R, B. R. Grant, and A. Abzhanov. 2006. A developing paradigm for the development of bird beaks. *Biological Journal of the Linnean Society* 88:17–22.

Grant, P. R., B. R. Grant, J. A. Markert, L. F. Keller, and K. Petren. 2004. Convergent evolution of Darwin's finches caused by introgressive hybridization and selection. *Evolution* 58:1588–1599.

Grant, P. R., B. R. Grant, and K. Petren. 2005. Hybridization in the recent past. *American Naturalist* 166:56–67.

Greenwood, B. M, D. A. Fidock, D. E. Kyle, S. H. Kappe, P. L. Alonso, F. H. Collins, and P. E. Duffy. 2008. Malaria: Progress, perils, and prospects for eradication. *Journal of Clinical Investigation* 118:1266.

Greischar, M. A., and B. Koskella. 2007. A synthesis of experimental work on parasite local adaptation. *Ecology Letters* 10:418–434.

Gribaldo, S., A. M. Poole, V. Daubin, P. Forterre, and C. Brochier-Armanet. 2010. The origin of eukaryotes and their relationship with the Archaea: Are we at a phylogenomic impasse? *Nature Reviews Microbiology* 8:743–752.

Griffiths, K. E., J. W. H. Trueman, G. R. Brown, and R. Peakall. 2011. Molecular genetic analysis and ecological evidence reveals multiple cryptic species among thynnine wasp pollinators of sexually deceptive orchids. *Molecular Phylogenetics and Evolution* 211:195–205.

Grimsley, N., B. Péquin, C. Bachy, and H. Moreau. 2010. Cryptic sex in the smallest eukaryotic marine green alga. *Molecular Biology and Evolution* 27:47–54.

Gripenberg, S., E. Morriën, A. Cudmore, J.-P. Salminen, and T. Roslin. 2007. Resource selection by female moths in a heterogeneous environment: What is a poor girl to do? *Journal of Animal Ecology* 76:854–865.

Gross, B. L., and L. H. Rieseberg. 2005. The ecological genetics of homoploid hybrid speciation. *Journal of Heredity* 96:241–252.

Gross, B. L., A. E. Schwarzbach, and L. H. Rieseberg. 2004. Origin(s) of the diploid hybrid species *Helianthus deserticola* (Asteraceae). *American Journal of Botany* 90:1708–1719.

Grubisha, L. C., S. E. Bergemann, and T. D. Bruns. 2007. Host islands within the California Northern Channel Islands create fine-scale genetic structure in two sympatric species of the symbiotic ectomycorrhizal fungus *Rhizopogon*. *Molecular Ecology* 16:1811–1822.

Guimarães, P. R., Jr., P. Jordano, and J. N. Thompson. 2011. Evolution and coevolution in mutualistic networks. *Ecology Letters* 14:887–885.

Guimarães, P. R., Jr., V. Rico-Gray, S. Furtado dos Reis, and J. N. Thompson. 2006. Asymmetries in specialization in ant-plant mutualistic networks. *Proceedings of the Royal Society of London B* 273:2041–2047.

Guimarães, P. R., Jr., V. Rico-Gray, P. S. Oliveira, T. J. Izzo, S. F. dos Reis, and J. N. Thompson. 2007. Interaction intimacy affects structure and coevolutionary dynamics in mutualistic networks. *Current Biology* 17:1797–1803.

Guinane, C. M., J. R. Penadés, and J. R. Fitzgerald. 2011. The role of horizontal gene transfer in *Staphylococcus aureus* host adaptation. *Virulence* 2:241–243.

Guitian, P. 1998. Latitudinal variation in the fruiting phenology of a bird-dispersed plant (*Crataegus monogyna*) in western Europe. *Plant Ecology* 137:139–142.

Gullberg, E., S. Cao, O. G. Berg, C. Ilbäck, L. Sandegren, D. Hughes, and D. I. Andersson. 2011. Selection of resistant bacteria at very low antibiotic concentrations. *PLoS Pathogens* 7:e1002158.

Haag, C. R., M. Saastamoinen, J. H. Marden, and I. Hanski. 2005. A candidate locus for variation in dispersal rate in a butterfly metapopulation. *Proceedings of the Royal Society B* 272:2449–2456.

Haas, B. J., S. Kamoun, M. C. Zody, R. H. Y. Jiang, R. E. Handsaker, L. M. Cano, M. Grabherr, et al. 2009. Genome sequence and analysis of the Irish potato famine pathogen *Phytophthora infestans*. *Nature* 461:393–398.

Haegeman, A., J. T. Jones, and E. G. J. Danchin. 2011. Horizontal gene transfer in nematodes: A catalyst for plant parasitism? *Molecular Plant-Microbe Interactions* 24:879–887.

Hairston, N.G., Jr., and T. Dillon. 1990. Fluctuating selection and response in a population of fresh-water copepods. *Evolution* 44:1796–1805.

Hairston, N. G., Jr., S. P. Ellner, M. A. Geber, T. Yoshida, and J. A. Fox. 2005. Rapid evolution and the convergence of ecological and evolutionary time. *Ecology Letters* 8:1114–1127.

Hairston, N. G., Jr., W. Lampert, C. E. Cáceres, C. L. Holtmeier, L. J. Weider, U. Gaedke, J. M. Fischer, et al. 1999. Lake ecosystems: Rapid evolution revealed by dormant eggs. *Nature* 401:446.

Haldane, J. B. S. 1932. *The causes of evolution*. Longmans, Green, New York.

———. 1949. Suggestions as to quantitative measurement of rates of evolution. *Evolution* 3:51–56.

———. 1964. A defense of beanbag genetics. *Perspectives in Biology and Medicine* 7:343–359.

Haldane, J. B. S., and S. Jayakar. 1963. Polymorphism due to selection of varying direction. *Journal of Genetics* 58:237–242.

Hall, A. R., P. D. Scanlan, A. D. Morgan, and A. Buckling. 2011. Host–parasite coevolutionary arms races give way to fluctuating selection. *Ecology Letters* 14:282–288.

Hallatschek, O., P. Hersen, S. Ramanathan, and D. R. Nelson. 2007. Genetic drift at expanding frontiers promotes gene segregation. *Proceedings of the National Academy of Sciences USA* 104:19926–19930.

Halverson, K., S. B. Heard, J. D. Nason, and J. O. I. Stireman. 2008. Differential attack on diploid, tetraploid, and hexaploid *Solidago altissima* L. by five insect gallmakers. *Oecologia* 154:755–761.

Hamilton, W. D. 1967. Extraordinary sex ratios. *Science* 156:477–488.

———. 1980. Sex versus non-sex versus parasite. *Oikos* 35:282–290.

———. 1987. Kinship, recognition, disease, and intelligence: Constraints on social evolution. Pages 81–102 in Y. Ito, J. L. Brown, and J. Kikkawa, eds., *Animal societies: Theories and facts*. Japan Scientific Societies Press, Tokyo.

Hampe, A., and F. Bairlein. 2000. Modified dispersal-related traits in disjunct populations of bird-dispersed *Frangula alnus* (Rhamnaceae): A result of its Quaternary distribution shifts? *Ecography* 23:603–613.

Handley, R. J., T. Steinger, U. A. Treier, and H. Mueller-Schaerer. 2008. Testing the evolution of increased competitive ability (EICA) hypothesis in a novel framework. *Ecology* 89:407–417.

Hanifin, C. T., E. D. Brodie, Jr., and E. D. III. Brodie. 2008. Phenotypic mismatches reveal escape from arms-race coevolution. *PLOS Biology* 6:e60.

Hanski, I. A. 2011. Eco-evolutionary spatial dynamics in the Glanville fritillary butterfly. *Proceedings of the National Academy of Sciences USA* 108:14397–14404.

Hanski, I., M. Kuussaari, and M. Nieminen. 1994. Metapopulation structure and migration in the butterfly *Melitaea cinxia*. *Ecology* 75:747–762.

Hanski, I., and E. Meyke. 2005. Large-scale dynamics of the Glanville fritillary butterfly: Landscape structure, population processes, and weather. *Annales Zoologici Fennici* 42:379–395.

Hanski, I., T. Mononen, and O. Ovaskainen. 2011. Eco evolutionary metapopulation dynamics and the spatial scale of adaptation. *American Naturalist* 177:29–43.

Hanski, I., and I. Saccheri. 2006. Molecular-level variation affects population growth in a butterfly metapopulation. *PLOS Biology* 4:e129.

Harbison, C. W., and D. H. Clayton. 2011. Community interactions govern host-switching with implications for host–parasite coevolutionary history. *Proceedings of the National Academy of Sciences USA* 108:9525–9529.

Harmon, L. J., J. B. Losos, T. Jonathan Davies, R. G. Gillespie, J. L. Gittleman, W. Bryan Jennings, K. H. Kozak, et al. 2010. Early bursts of body size and shape evolution are rare in comparative data. *Evolution* 64:2385–2396.

Harper, J. L. 1967. A Darwinian approach to plant ecology. *Journal of Ecology* 55:247–270.

Harrison, R. G., ed. 1993. *Hybrid zones and the evolutionary process*. New York, NY, Oxford University Press.

Harrison, R. G. 2010. Understanding the origin of species: Where have we been? Where are we going? Pages 319–348 in M. A. Bell, D. J. Futuyma, W. F. Eanes, and J. S. Levinton, eds., *Evolution since Darwin: The first 150 years*. Sinauer Associates, Sunderland, MA.

Harrison, S., D. D. Murphy, and P. R. Ehrlich. 1988. Distribution of the bay checkerspot butterfly, *Euphydryas editha bayensis*: Evidence for a metapopulation model. *American Naturalist* 132:360–382.

Harshman, L. G., and A. A. Hoffman. 2000. Laboratory selection experiments using *Drosophila*: What do they really tell us? *Trends in Ecology and Evolution* 15:32–36.

Hartley, S. E., and A. C. Gange. 2009. Impacts of plant symbiotic fungi on insect herbivores: Mutualism in a multitrophic context. *Annual Review of Entomology* 54:323–342.

Harvey, P. H., A. F. Read, and S. Nee. 1995. Why ecologists need to be phylogenetically challenged. *Ecology* 83:535–536.

Hatfield, T., and D. Schluter. 1999. Ecological speciation in sticklebacks: Environment-dependent hybrid fitness. *Evolution* 53:866–873.

Hatfull, G. F. 2008. Bacteriophage genomics. *Current Opinion in Microbiology* 11:447–453.

Hausdorf, B. 2011. Progress toward a general species concept. *Evolution* 64: 923–931.

Hawks, J., E. T. Wang, G. M. Cochran, H. C. Harpending, and R. K. Moyzis. 2007. Recent acceleration of human adaptive evolution. *Proceedings of the National Academy of Sciences USA* 104:20753–20758.

Hazarika, L. K., M. Bhuyan, and B. N. Hazarika. 2009. Insect pests of tea and their management. *Annual Review of Entomology* 54:267–284.

Heap, I. 2011. The international survey of herbicide resistant weeds; *http://www.weedscience.org*.

Heath, K. D. 2010. Intergenomic epistasis and coevolutionary constraint in plants and rhizobia. *Evolution* 64:1446–1458.

Heath, K. D., A. J. Stock, and J. R. Stinchcombe. 2010. Mutualism variation in the nodulation response to nitrate. *Journal of Evolutionary Biology* 23:2494–2500.

Hebert, P. D. N., E. H. Penton, J. M. Burns, D. H. Janzen, and W. Hallwachs. 2004. Ten species in one: DNA barcoding reveals cryptic species in the neotropical skipper butterfly *Astraptes fulgerator*. *Proceedings of the National Academy of Sciences USA* 101:14812–14817.

Hechinger, R. F., K. D. Lafferty, A. P. Dobson, J. H. Brown, and A. M. Kuris. 2011. A common scaling rule for abundance, energetics, and production of parasitic and free-living species. *Science* 333:445–448.

Hedges, L. M., J. C. Brownlie, S. L. O'Neill, and K. N. Johnson. 2008. *Wolbachia* and virus protection in insects. *Science* 322:702.

Hedrick, P. W. 2007. Balancing selection. *Current Biology* 17:R230–R231.

Hegarty, M. J., G. L. Barker, A. C. Brennan, K. J. Edwards, R. J. Abbott, and S. J. Hiscock. 2008. Changes in gene expression associated with speciation in plants: Further insights from transcriptomic studies of *Senecio*. *Philosophical Transactions of the Royal Society B* 363:3055–3069.

Hegreness, M., N. Shoresh, D. Damian, D. Hartl, and R. Kishony. 2008. Accelerated evolution of resistance in multidrug environments. *Proceedings of the National Academy of Sciences USA* 105:13977–13981.

Heitman, J. 2010. Evolution of eukaryotic microbial pathogens via covert sexual reproduction. *Cell Host and Microbe* 8:86–99.

Hendrikse, J. L., T. E. Parsons, and B. Hallgrímsson. 2007. Evolvability as the proper focus of evolutionary developmental biology. *Evolution and Development* 9:393–401.

Hendrix, S. D. 1979. Compensatory reproduction in a biennial herb *Pastinaca sativa* following insect defloration. *Oecologia* 42:107–118.

Hendry, A. P., T. J. Farrugia, and M. T. Kinnison. 2008. Human influences on the rates of phenotypic change in wild animal populations. *Molecular Ecology* 17:20–29.

Hendry, A. P., and M. T. Kinnison. 1999. The pace of modern life: Measuring rates of contemporary microevolution. *Evolution* 53:1637–1653.

Hendry, A. P., P. Nosil, and L. H. Rieseberg. 2007. The speed of ecological speciation. *Functional Ecology* 21:455–464.

Herbers, J. M., and S. Foitzik. 2002. The ecology of slavemaking ants and their hosts in north temperate forests. *Ecology* 83:148–163.

Hereford, J. 2009. A quantitative survey of local adaptation and fitness trade-offs. *American Naturalist* 173:579–588.

Hereford, J., T. F. Hansen, and D. Houle. 2004. Comparing strengths of directional selection: How strong is strong? *Evolution* 58:2133–2143.

Hermansen, J. S., S. A. Saether, T. O. Elgvin, T. Borge, E. Hjelle, and G.-P. Saetre. 2011. Hybrid speciation in sparrows I: Phenotypic intermediacy, genetic admixture and barriers to gene flow. *Molecular Ecology* 20:3812–3822.

Herre, E. A., and K. C. Jandér. 2008. Evolutionary ecology of figs and their associates: Recent progress and outstanding puzzles. *Annual Review of Ecology, Evolution and Systematics* 39:439–458.

Herrera, C. M. 2009. *Multiplicity in unity: Plant subindividual variation and interactions with animals*. University of Chicago Press, Chicago.

Herrera, C. M., C. de Vega, A. Canto, and M. I. Pozo. 2009. Yeasts in floral nectar: A quantitative survey. *Annals of Botany* 103:1415–1423.

Hewitt, G. M. 2011. Quaternary phylogeography: The roots of hybrid zones. *Genetica* 139:617–638.

Hilgenboecker, K., P. Hammerstein, P. Schlattmann, A. Telschow, and J. H. Werren. 2008. How many species are infected with *Wolbachia*?–A statistical analysis of current data. *FEMS Microbiology Letters* 281:215–220.

Hillesland, K. L., and D. A. Stahl. 2010. Rapid evolution of stability and productivity at the origin of a microbial mutualism. *Proceedings of the National Academy of Sciences USA* 107:2124–2129.

Himler, A. G., T. Adachi-Hagimori, J. E. Bergen, A. Kozuch, S. E. Kelly, B. E. Tabashnik, E. Chiel, et al. 2011. Rapid spread of a bacterial symbiont in an invasive whitefly is driven by fitness benefits and female bias. *Science* 332: 254–256.

Hindorff, L. A., P. Sethupathy, H. A. Junkins, E. M. Ramos, J. P. Mehta, F. S. Collins, and T. A. Manolio. 2009. Potential etiologic and functional implications of genome-wide association loci for human diseases and traits. *Proceedings of the National Academy of Sciences USA* 106:9362–9367.

Hobbs, J.-P. A., A. J. Frisch, G. R. Allen, and L. Van Herwerden. 2009. Marine hybrid hotspot at Indo-Pacific biogeographic border. *Biology Letters* 5:258–261.

Hochberg, M. E., and M. van Baalen. 1998. Antagonistic coevolution over productivity gradients. *American Naturalist* 152:620–634.

Hochberg, M. E., R. Gomulkiewicz, R. D. Holt, and J. N. Thompson. 2000. Weak sinks could cradle mutualistic symbioses—Strong sources should harbour parasitic symbioses. *Journal of Evolutionary Biology* 13:213–222.

Hodkinson, T. R., M. B. Jones, S. Waldren, and J. A. N. Parnell. 2011. *Climate change, ecology and systematics*. Cambridge University Press, Cambridge.

Hoeksema, J. D. 2010. Ongoing coevolution in mycorrhizal interactions. *New Phytologist* 187:286–300.

Hoeksema, J. D., and S. E. Forde. 2008. A meta-analysis of factors affecting local adaptation between interacting species. *American Naturalist* 171:275–290.

Hoekstra, H. E., R. J. Hirschmann, R. A. Bundey, P. A. Insel, and J. P. Crossland. 2006. A single amino acid mutation contributes to adaptive beach mouse color pattern. *Science* 313:101–104.

Hoekstra, H. E., J. M. Hoekstra, D. Berrigan, S. N. Vignieri, A. Hoang, C. E. Hill, P. Beerli, et al. 2001. Strength and tempo of directional selection in the wild. *Proceedings of the National Academy of Sciences USA* 98:9157–9160.

Hoeksema, J. D., and J. N. Thompson. 2007. Geographic structure in a widespread plant-mycorrhizal interaction: Pines and false truffles. *Journal of Evolutionary Biology* 20:1148–1163.

Hoffman, A. A., M. Turelli, and G. M. Simmon. 1986. Unidirectional incompatibility between populations of *Drosophila simulans*. *Evolution* 40: 692–701.

Hoffman, M. T., and A. E. Arnold. 2010. Diverse bacteria inhabit living hyphae of phylogenetically diverse fungal endophytes. *Applied and Environmental Microbiology* 76:4063–4075.

Hofreiter, M., and T. Schoeneberg. 2010. The genetic and evolutionary basis of colour variation in vertebrates. *Cellular and Molecular Life Sciences* 67:2591–2603.

Holland, J. N., and D. L. DeAngelis. 2010. A consumer-resource approach to the density-dependent population dynamics of mutualism. *Ecology* 91:1286–1295.

Hölldobler, B., and E. O. Wilson. 1990. *The ants*. Harvard University Press, Cambridge, MA.

Holmes, E. 2009. The evolutionary genetics of emerging viruses. *Annual Review of Ecology, Evolution, and Systematics* 40:353–372.

Holmes, R. 2008. Bacteria make major evolutionary shift in the lab. *New Scientist*, June 9; http://www.newscientist.com/article/dn14094.

Holt, K. E., J. Parkhill, C. J. Mazzoni, P. Roumagnac, F.-X. Weill, I. Goodhead, R. Rance, et al. 2008. High-throughput sequencing provides insights into genome variation and evolution in *Salmonella typhi*. *Nature Genetics* 40:987–993.

Holt, R. D. 2009. Bringing the Hutchinsonian niche into the 21st century: Ecological and evolutionary perspectives. *Proceedings of the National Academy of Sciences USA* 106:19659–19665.

Hornett, E. A., S. Charlat, A. M. R. Duplouy, N. Davies, G. K. Roderick, N. Wedell, and G. D. D. Hurst. 2006. Evolution of male-killer suppression in a natural population. *PLoS Biology* 4:e283.

Hornett, E. A., S. Charlat, N. Wedell, C. D. Jiggins, and G. D. D. Hurst. 2009. Rapidly shifting sex ratio across a species range. *Current Biology* 19:1628–1631.

Hörvandl, E. 2009. Geographical parthenogenesis: Opportunities for asexuality. Pages 161–186 in I. Schön, K. Martens, and P. van Dijk, eds., *Lost sex: The evolutionary biology of parthenogensis*. Springer, New York.

Horvitz, C. C., T. Coulson, S. Tuljapurkar, and D. W. Schemske. 2010. A new way to integrate selection when both demography and selection gradients vary over time. *International Journal of Plant Sciences* 171:945–959.

Hoskin, C. J., and M. Higgie. 2010. Speciation via species interactions: The divergence of mating traits within species. *Ecology Letters* 13:409–420.

Hosokawa, T., R. Koga, Y. Kikuchi, X. Y. Meng, and T. Fukatsu. 2010. *Wolbachia* as a bacteriocyte-associated nutritional mutualist. *Proceedings of the National Academy of Sciences USA* 107:769–774.

Howard, R. S., and C. M. Lively. 2002. The ratchet and the Red Queen: The maintenance of sex in parasites. *Journal of Evolutionary Biology* 52:648–656.

Huang, X., J. Schmitt, L. Dorn, C. Griffith, S. Effgen, S. Takao, M. Koornneef, et al. 2010. The earliest stages of adaptation in an experimental plant population: Strong selection on QTLS for seed dormancy. *Molecular Ecology* 19:1335–1351.

Hubbell, S. P. 2001. *The unified theory of biodiversity and biogeography*. Princeton University Press, Princeton, NJ.

Huchard, E., L. A. Knapp, J. Wang, M. Raymond, and G. Cowlishaw. 2010. MHC, mate choice and heterozygote advantage in a wild social primate. *Molecular Ecology* 19:2545–2561.

Hudson, A. G., P. Vonlanthen, and O. Seehausen. 2011. Rapid parallel adaptive radiations from a single hybridogenic ancestral population. *Proceedings of the Royal Society B* 278:58–66.

Huey, R. B., G. W. Gilchrist, M. L. Carlson, D. Berrigan, and L. Serra. 2000. Rapid evolution of a geographic cline in size in an introduced fly. *Science* 287:308–309.

Huey, R. B., G. W. Gilchrist, and A. P. Hendry. 2005. Using invasive species to study evolution: Case studies of *Drosophila* and salmon. Pages 139–164 in D. F. Sax, J. J. Stachowicz, and S. D. Gaines, eds., *Species invasions: Insights into ecology, evolution, and biogeography*. Sinauer Associates, Sunderland, MA.

Huey, R. B., and M. Pascual. 2009. Partial thermoregulatory compensation by a rapidly evolving invasive species along a latitudinal cline. *Ecology* 90:1715–1720.

Hughes, D. P., T. Wappler, and C. C. Labandeira. 2011. Ancient death-grip leaf scars reveal ant-fungal parasitism. *Biology Letters* 7:67–70.

Hughes, J. B., G. C. Daily, and P. R. Ehrlich. 1997. Population diversity: Its extent and extinction. *Science* 278:689–692.

Hughes, J., and A. P. Vogler. 2004. The phylogeny of acorn weevils (genus *Curculio*) from mitochondrial and nuclear DNA sequences: The problem of incomplete data. *Molecular Phylogenetics and Evolution* 32:601–615.

Hulsey, C. D., P. R. J. Hollingsworth, and J. A. Fordyce. 2010. Temporal diversification of Central American cichlids. *BMC Evolutionary Biology* 10:279.

Hunt, B. G., L. Ometto, Y. Wurm, D. Shoemaker, S. V. Yi, L. Keller, and M. A. D. Goodisman. 2011. Relaxed selection is a precursor to the evolution of phenotypic plasticity. *Proceedings of the National Academy of Sciences USA* 108:15936–15941.

Hunt, G. 2010. Evolution in fossil lineages: Paleontology and *The origin of species*. *American Naturalist* 176:S61–76.

Hunt, G., M. A. Bell, and M. P. Travis. 2008. Evolution toward a new adaptive optimum: Phenotypic evolution in a fossil stickleback lineages. *Evolution* 62:700–710.

Huntzinger, M., R. Karban, T. P. Young, and T. M. Palmer. 2004. Relaxation of induced indirect defenses of acacias following exclusion of mammalian herbivores. *Ecology* 85:609–614.

Husband, B. C., and D. W. Schemske. 2000. Ecological mechanisms of reproductive isolation between diploid and tetraploid *Chamerion angustifolium*. *Journal of Ecology* 88:689–701.

Hutchinson, G. E. 1959. Homage to Santa Rosalia, or why are there so many kinds of animals? *American Naturalist* 93:145–159.

Ings, T. C., J. M. Montoya, J. Bascompte, N. Bluthgen, L. Brown, C. F. Dormann, F. Edwards, et al. 2009. Ecological networks: Beyond food webs. *Journal of Animal Ecology* 78:253–269.

Ionita-Laza, I., A. J. Rogers, C. Lange, B. A. Raby, and C. Lee. 2009. Genetic association analysis of copy-number variation (CNV) in human disease pathogenesis. *Genomics* 93:22–26.

Irwin, R. E. 2009. Realized tolerance to nectar robbing: Compensation to floral enemies in *Ipomopsis aggregata*. *Annals of Botany* 103:1425–1433.

Ishizuka, M., T. Tanikawa, K. Tanaka, M. Heewon, F. Okajima, K. Sakamoto, and S. Fujita. 2008. Pesticide resistance in wild mammals: Mechanisms of anticoagulant resistance in wild rodents. *Journal of Toxicological Sciences* 33: 283–291.

Ishmael, N., J. C. Dunning Hotopp, P. Ioannidis, S. Biber, J. Sakamoto, S. Siozios, V. Nene, et al. 2009. Extensive genomic diversity of closely related *Wolbachia* strains. *Microbiology* 155:2211–2222.

Ives, A. R., and H. C. J. Godfray. 2006. Phylogenetic analysis of trophic associations. *American Naturalist* 168:E1–E14.

Iwao, K., and M. D. Rausher. 1997. Evolution of plant resistance to multiple herbivores: Quantifying diffuse coevolution. *American Naturalist* 149:316–335.

Iwasaki, A., and R. Medzhitov. 2010. Regulation of adaptive immunity by the innate immune system. *Science* 327:291–295.

Jablonski, D. 2008. Biotic interactions and macroevolution: Extensions and mismatches across scales and levels. *Evolution* 62:715–739.

Jacob, F. 1977. Evolution and tinkering. *Science* 196:1161–1166.

Jaenike, J. 2007. Spontaneous emergence of a new *Wolbachia* phenotype. *Evolution* 61:2244–2252.

Jaenike, J., J. K. Stahlhut, L. M. Boelio, and R. L. Unckless. 2010a. Association between *Wolbachia* and *Spiroplasma* within *Drosophila neotestacea*: An emerging symbiotic mutualism. *Molecular Ecology* 19:414–425.

Jaenike, J., R. Unckless, S. N. Cockburn, L. M. Boelio, and S. J. Perlman. 2010b. Adaptation via symbiosis: Recent spread of a *Drosophila* defensive symbiont. *Science* 329:212–215.

Jain, S. K., and A. D. Bradshaw. 1966. Evolution in closely adjacent plant populations. I. The evidence and its theoretical analysis. *Heredity* 21:407–441.

James, A. K., and R. J. Abbott. 2005. Recent, allopatric, homoploid hybrid speciation: The origin of *Senecio squalidus* (Asteraceae) in the British Isles from a hybrid zone on Mount Etna, Sicily. *Evolution* 59:2533–2547.

Jandér, K. C., and E. A. Herre. 2010. Host sanctions and pollinator cheating in fig tree-fig-wasp mutualism. *Proceedings of the Royal Society B* 277:1481–1488.

Janova, E., J. Matiasovic, J. Vahala, R. Vodicka, E. Van Dyk, and P. Horin. 2009. Polymorphism and selection in the major histocompatibility complex DRA and DQA genes in the family Equidae. *Immunogenetics* 61:513–527.

Janz, N. 2003. Sex linkage of host plant use in butterflies. Pages 229–239 in C. L. Boggs, W. B. Watt, and P. R. Ehrlich, eds., *Butterflies: Ecology and evolution taking flight*. University of Chicago Press, Chicago.

Janz, N., and S. Nylin. 2008. The oscillation hypothesis of host-plant range and speciation. Pages 230–215 in K. J. Tilmon, ed., *Specialization, speciation, and radiation: The evolutionary biology of herbivorous insects*. University of California Press, Berkeley.

Janz, N., S. Nylin, and N. Wahlberg. 2006. Diversity begets diversity: Host expansions and the diversification of plant-feeding insects. *BMC Evolutionary Biology* 6:4.

Janz, N., and J. N. Thompson. 2002. Plant polyploidy and host expansion in an insect herbivore. *Oecologia* 130:570–575.

Janzen, D. H. 1966. Coevolution of mutualism between ants and acacias in Central America. *Evolution* 20:249–275.

———. 1980. When is it coevolution? *Evolution* 34:611–612.

Janzen, D. H., W. Hallwachs, P. Blandin, J. M. Burns, J.-M. Cadiou, I. Chacon, T. Dapkey, et al. 2009. Integration of DNA barcoding into an ongoing inventory of complex tropical biodiversity. *Molecular Ecology Resources* 9:1–26.

Janzen, D., W. Hallwachs, and J. M. Burns. 2010. A tropical horde of counterfeit predator eyes. *Proceedings of the National Academy of Sciences USA* 107:11659–11665.

Janzen, F. J., J. G. Krenz, T. S. Haselkorn, E. D. Brodie, Jr., and E. D. Brodie III. 2002. Molecular phylogeography of common garter snakes (*Thamnophis sirtalis*) in western North America: Implications for regional historical forces. *Molecular Ecology* 11:1739–1751.

Jarosz, A. M., and J. J. Burdon. 1992. Host–pathogen interactions in natural populations of *Linum marginale* and *Melampsora lini*. III. Influence of pathogen epidemics on host survivorship and flower production. *Oecologia* 89:53–61.

Jarrell, K. F., A. D. Walters, C. Bochiwal, J. M. Borgia, T. Dickinson, and J. P. J. Chong. 2011. Major players on the microbial stage: Why archaea are important. *Microbiology* 157:919–936.

Jeong, H., V. Barbe, C. H. Lee, D. Vallenet, D. S. Yu, S.-H. Choi, A. Couloux, et al. 2009. Genome sequences of *Escherichia coli* B strains REL606 and BL21(DE3). *Journal of Molecular Biology* 394:644–652.

Jiao, Y., N. J. Wickett, S. Ayyampalayam, A. S. Chanderbali, L. Landherr, P. E. Ralph, L. P. Tomsho, et al. 2011. Ancestral polyploidy in seed plants and angiosperms. *Nature* 473:97–100.

Johnsen, P. J., J. P. Townsend, T. Bøhn, G. S. Simonsen, A. Sundsfjord, and K. M. Nielsen. 2009. Factors affecting the reversal of antimicrobial-drug resistance. *Lancet Infectious Diseases* 9:357–364.

Johnson, C. A., and J. M. Herbers. 2006. Impact of parasite sympatry on the geographic mosaic of coevolution. *Ecology* 87:382–394.

Johnson, J. B., S. M. Peat, and B. J. Adams. 2009. Where's the ecology in molecular ecology? *Oikos* 118:1601–1609.

Johnson, K. P., J. D. Weckstein, S. E. Bush, and D. H. Clayton. 2011a. The evolution of host specificity in dove body lice. *Parasitology* 138:1730–1736.

Johnson, K. P., J. D. Weckstein, M. J. Meyer, and D. H. Clayton. 2011b. There and back again: Switching between host orders by avian body lice (Ischnocera: Goniodidae). *Biological Journal of the Linnean Society* 102:614–625.

Johnson, L. J. 2007. The genome strikes back: The evolutionary importance of defence against mobile elements. *Evolutionary Biology* 34:121–129.

Johnson, M. A., L. J. Revell, and J. B. Losos. 2010. Behavioral convergence and adaptive radiation: Effects of habitat use on territorial behavior in *Anolis* lizards. *Evolution* 64:1151–1159.

Johnson, M. T. J. 2008. Bottom-up effects of plant genotype on aphids, ants, and predators. *Ecology* 89:145–154.

Johnson, N. A. 2010. Hybrid incompatibility genes: Remnants of a genomic battlefield? *Trends in Genetics* 26:317–325.

Johnson, N. K., and C. Cicero. 2004. New mitochondrial DNA data affirm the importance of Pleistocene speciation in North American birds. *Evolution* 58:1122–1130.

Johnson, S. D., and B. Anderson. 2010. Coevolution between food-rewarding flowers and their pollinators. *Evolution: Education and Outreach* 3:32–39.

Johnson, T., and N. Barton. 2005. Theoretical models of selection and mutation on quantitative traits. *Philosophical Transactions of the Royal Society B* 360:1411–1425.

Johnson, Z. I,, E. R, Zinser, A. Coe, N. P. McNulty, E. M. Woodward, and S. W. Chisholm. 2006. Niche partitioning among *Prochlorococcus* ecotypes along ocean-scale environmental gradients. *Science* 311:1737–1740.

Jokela, J., M. F. Dybdahl, and C. M. Lively. 2009. The maintenance of sex, clonal dynamics, and host–parasite coevolution in a mixed population of sexual and asexual snails. *American Naturalist* 174:S43–S53.

Jones, E. I., R. Ferrière, and J. L. Bronstein. 2009. Eco-evolutionary dynamics of mutualists and exploiters. *American Naturalist* 174:780–794.

Jones, E. O., A. White, and M. Boots. 2010. The evolutionary implications of conflict between parasites with different transmission modes. *Evolution* 64:2408–2416.

Jordano, P. 1987. Patterns of mutualistic interactions in pollination and see dispersal: Connectance, dependence asymmetries, and coevolution. *American Naturalist* 129:657–677.

———. 2010. Coevolution in multi-specific interactions among free-living species. *Evolution: Education and Outreach* 3:40–46.

Jordano, P., J. Bascompte, and J. M. Olesen. 2003. Invariant properties in coevolutionary networks of plant–animal interactions. *Ecology Letters* 6:69–81.

Jordano, P., C. Garcia, J. A. Godoy, and J. L. Garcia-Castano. 2007. Differential contribution of frugivores to complex seed dispersal patterns. *Proceedings of the National Academy of Sciences USA* 104:3278–3282.

Jose, C., and F. Dufresne. 2010. Differential survival among genotypes of *Daphnia pulex* differing in reproductive mode, ploidy level, and geographic origin. *Evolutionary Ecology* 24:413–421.

Joshi, A., and J. N. Thompson. 1995. Alternative routes to the evolution of competitive ability in two competing species of *Drosophila*. *Evolution* 49:616–625.

———. 1996. Evolution of broad and specific competitive ability in novel versus familiar environments in *Drosophila* species. *Evolution* 50:188–194.

Joshi, J., and K. Vrieling. 2005. The enemy release and EICA hypothesis revisited: Incorporating the fundamental difference between specialist and generalist herbivores. *Ecology Letters* 8:704–714.

Judson, O. P., and B. B. Normack. 1996. Ancient asexual scandals. *Trends in Ecology and Evolution* 11:41–46.

Kaiser, T. S., and D. G. Heckel. 2012. Genetic architecture of local adaptation in lunar and diurnal emergence times of the marine midge *Clunia marinus* (Chironomidae, Diptera). *PLoS One* 7:e32092.

Kaiwa, N., T. Hosokawa, Y. Kikuchi, N. Nikoh, X. Y. Meng, N. Kimura, M. Ito, et al. 2010. Primary gut symbiont and secondary, *Sodalis*-allied symbiont of the scutellerid stinkbug *Cantao ocellatus*. *Applied and Environmental Microbiology* 76:3486–3494.

Kaltenpoth, M., S. A. Winter, and A. Kleinhammer. 2009. Localization and transmission route of *Coriobacterium glomerans*, the endosymbiont of pyrrhocorid bugs. *FEMS Microbiology Ecology* 69:373–383.

Karban, R. 2011. The ecology and evolution of induced resistance against herbivores. *Functional Ecology* 25:339–347.

Kawakita, A. 2010. Evolution of obligate pollination mutualism in the tribe Phyllantheae (Phyllanthaceae). *Plant Species Biology* 25:3–19.

Kawakita, A., and M. Kato. 2008. Repeated independent evolution of obligate pollination mutualism in the *Phyllantheae–Epicephala* association. *Proceedings of the Royal Society B* 276:417–426.

Kawakita, A., A. Takimura, T. Terachi, T. Sota, and M. Kato. 2004. Cospeciation analysis of an obligate pollination mutualism: Have *Glochidion* trees (Euphorbiaceae) and pollinating *Epicephala* moths (Gracillariidae) diversified in parallel? *Evolution* 58:2201–2214.

Kawecki, T. J. 2008. Adaptation to marginal habitats. *Annual Review of Ecology, Evolution, and Systematics* 39:321–342.

Kay, K. M., and R. D. Sargent. 2009. The role of animal pollination in plant speciation: Integrating ecology, geography, and genetics. *Annual Review of Ecology, Evolution, and Systematics* 40:637–656.

Kays, R., A. Curtis, and J. J. Kirchman. 2010. Rapid adaptive evolution of northeastern coyotes via hybridization with wolves. *Biology Letters* 6:89–93.

Keeling, P. J. 2010. The endosymbiotic origin, diversification and fate of plastids. *Philosophical Transactions of the Royal Society B* 365:729–748.

Keeling, P. J., and J. D. Palmer. 2008. Horizontal gene transfer in eukaryotic evolution. *Nature Reviews Genetics* 9:605–618.

Keith, A. R., J. K. Bailey, and T. G. Whitham. 2010. A genetic basis to community repeatability and stability. *Ecology* 91:3398–3406.

Keller, M. J., and H. C. Gerhardt. 2001. Polyploidy alters advertisement call structure in gray treefrogs. *Proceedings of the Royal Society B* 268:341–345.

Kembel, S. W., and S. Hubbell. 2006. The phylogenetic structure of a neotropical forest tree community. *Ecology* 87:S86–99.

Kemp, D. J., D. N. Reznick, G. F. Grether, and J. A. Endler. 2009. Predicting the direction of ornament evolution in Trinidadian guppies (*Poecilia reticulata*). *Proceedings of the Royal Society B* 276:4335–4343.

Kennedy, B. F., H. A. Sabara, D. Haydon, and B. C. Husband. 2006. Pollinator-mediated assortative mating in mixed ploidy populations of *Chamerion angustifolium* (Onagraceae). *Oecologia* 150:398–408.

Kennedy, P. G., S. E. Bergemann, S. Hortal, and T. D. Bruns. 2007. Determining the outcome of field-based competition between two *Rhizopogon* species using real-time PCR. *Molecular Ecology* 16:881–890.

Kent, B. N., and S. R. Bordenstein. 2010. Phage WO of *Wolbachia*: Lambda of the endosymbiont world. *Trends in Microbiology* 18:173–181.

Kent, B. N., L. Salichos, J. G. Gibbons, A. Rokas, I. L. G. Newton, M. E. Clark, and S. R. Bordenstein. 2011. Complete bacteriophage transfer in a bacterial endosymbiont (*Wolbachia*) determined by targeted genome capture. *Genome Biology and Evolution* 3:209–218.

Kerfoot, W. C., and L. J. Weider. 2004. Experimental paleoecology (resurrection ecology): Chasing Van Valen's Red Queen hypothesis. *Limnology and Oceanography* 49:1300–1316.

Kesanakurti, P. R., A. J. Fazekas, K. S. Burgesss, D. M. Percy, S. G. Newmaster, S. W. Graham, S. C. H. Barrett, et al. 2011. Spatial patterns of plant diversity below-ground as revealed by DNA barcoding. *Molecular Ecology* 20:1289–1302.

Kessin, R. H. 2001. *Dictyostelium: Evolution, cell biology and the development of multicellularity*. Cambridge University Press, Cambridge.

Kettlewell, H. 1959. New aspects of the genetic control of industrial melanism in the Lepidoptera. *Nature* 183:918–921.

———. 1973. *The evolution of melanism: The study of a recurring necessity*. Oxford University Press, Oxford.

Keys, D. N., D. L. Lewis, J. E. Selegue, B. J. Pearson, L. V. Goodrich, R. L. Johnson, J. Gates, et al. 1999. Recruitment of a hedgehog regulatory circuit in butterfly eyespot evolution. *Science* 283:532–534.

Khan, A. I., D. M. Dinh, D. Schneider, R. E. Lenski, and T. F. Cooper. 2011. Negative epistasis between beneficial mutations in an evolving bacterial population. *Science* 332:1193–1196.

Kilner, R. M., and N. E. Langmore. 2011. Cuckoos versus hosts in insects and birds: Adaptations, counter-adaptations and outcomes. *Biological Reviews of the Cambridge Philosophical Society* 86:836–852.

Kimura, M. 1977. Preponderance of synonymous changes as evidence for the neutral theory of molecular evolution. *Nature* 267:275–276.

———. 1983. *The neutral theory of molecular evolution*. Cambridge University Press, Cambridge.

King, K. C., L. F. Delph, J. Jokela, and C. M. Lively. 2009. The geographic mosaic of sex and the Red Queen. *Current Biology* 19:1438–1441.

———. 2011. Coevolutionary hotspots and coldspots for host sex and parasite local adaptation in a snail–trematode interaction. *Oikos* 120:1335–1340.

King, K. C., and C. M. Lively. 2009. Geographic variation in sterilizing parasite species and the Red Queen. *Oikos* 118:1416–1420.

King, T. L., M. S. Eackles, A. P. Spidle, and H. J. Brockmann. 2005. Regional differentiation and sex-biased dispersal among populations of the horseshoe crab Limulus polyphemus. *Transaction of the American Fisheries Society* 134:441–465.

Kingsolver, J. G., and S. E. Diamond. 2011. Phenotypic selection in natural populations: What limits directional selection? *American Naturalist* 177:346–357.

Kingsolver, J. G., H. E. Hoekstra, J. M. Hoekstra, D. Berrigan, S. N. Vignieri, C. E. Hill, A. Hoang, et al. 2001. The strength of phenotypic selection in natural populations. *American Naturalist* 157:245–261.

Kingsolver, J. G., and D. W. Pfennig. 2008. Patterns and power of phenotypic selection in nature. *Bioscience* 57:561–572.

Kinnison, M. T., and N. G. Hairston. 2007. Eco-evolutionary conservation biology: Contemporary evolution and the dynamics of persistence. *Functional Ecology* 21:444–454.

Kinnison, M. T., and A. P. Hendry. 2001. The pace of modern life II: From rates of contemporary microevolution to pattern and process. *Genetica* 112–113:145–164.

Kirkness, E. F., B. J. Haas, W. Sun, H. R. Braig, M. A. Perotti, J. M. Clark, S. H. Lee, et al. 2011. Genome sequences of the human body louse and its primary endosymbiont provide insights into the permanent parasitic lifestyle. *Proceedings of the National Academy of Sciences USA* 108:6335–6336.

Klemme, I., and I. Hanski. 2009. Heritability of and strong single gene (*Pgi*) effects on life-history traits in the Glanville fritillary butterfly. *Journal of Evolutionary Biology* 22:1944–1953.

Klepzig, K. D., A. S. Adams, J. Handelsman, and K. F. Raffa. 2009. Symbioses: A key driver of insect physiological processes, ecological interactions, evolutionary diversification, and impacts on humans. *Environmental Entomology* 38:67–77.

Knapp, C. W., J. Dolfing, P. A. I. Ehlert, and D. W. Graham. 2010. Evidence of increasing antibiotic resistance gene abundances in archived soils since 1940. *Environmental Science and Technology* 44:580–587.

Knight, T. M., M. W. McCoy, J. M. Chase, K. A. McCoy, and R. D. Holt. 2005. Trophic cascades across ecosystems. *Nature* 437:880–883.

Kniskern, J. M., and M. D. Rausher. 2007. Natural selection on a polymorphic disease-resistance locus in *Ipomoea purpurea*. *Evolution* 61:377–387.

Kohn, M. H., H. J. Pelz, and R. K. Wayne. 2000. Natural selection mapping of the warfarin-resistance gene. *Proceedings of the National Academy of Sciences* 97: 7911–7915.

Kondoh, M., S. Kato, and Y. Sakato. 2010. Food webs are built up with nested subwebs. *Ecology* 91:3123–3130.

Kondrashov, A. S., and L. Y. Yampolsky. 1996. High genetic variability under the balance between symmetric mutation and fluctuating stabilizing selection. *Genetical Research* 68:157–164.

Koonin, E. 2010. The origin and early evolution of eukaryotes in the light of phylogenomics. *Genome Biology* 11:209.

Koskella, B., and C. M. Lively. 2007. Advise of the rose: Experimental coevolution of a trematode and its snail host. *Evolution* 61:152–159.

———. 2009. Evidence for negative frequency-dependent selection during experimental coevolution of a freshwater snail and a sterilizing trematode. *Evolution* 63:2213–2221.

Koskella, B., J. N. Thompson, G. M. Preston, and A. Buckling. 2011. Local biotic environment shapes spatial bacteriophage adaptation to bacteria. *American Naturalist* 177:440–451.

Kraaijeveld, A. R., and H. C. J. Godfray. 1997. Trade-off between parasitoid resistance and larval competitive ability in *Drosophila melanogaster*. *Nature* 389:278–280.

Kraaijeveld, A. R., E. C. Limentani, and H. C. J. Godfray. 2001. Basis of the trade-off between parasitoid resistance and larval competitive ability in *Drosophila melanogaster*. *Proceedings of the Royal Society of London B* 268:259–261.

Kraft, N. J. B., W. K. Cornwell, C. O. Webb, and D. D. Ackerly. 2008. Trait evolution, community assembly, and the phylogenetic structure of ecological communities. *American Naturalist* 170:271–283.

Kress, W. J., D. L. Erickson, F. A. Jones, N. G. Swenson, R. Perez, O. Sanjur, and E. Bermingham. 2009. Plant DNA barcodes and a community phylogeny of a tropical forest dynamics plot in Panama. *Proceedings of the National Academy of Sciences USA* 106:18621–18626.

Krishna, A., P. R. Guimarães, Jr., and J. Bascompte. 2008. A neutral-niche theory of nestedness in mutualistic networks. *Oikos* 117:1609–1618.

Kryazhimskiy, S., G. A. Bazykin, and J. Dushoff. 2008. Natural selection for nucleotide usage at synonymous and nonsynonymous sites in influenza A virus genes. *Journal of Virology* 82:4938–4945.

Kryazhimskiy, S., and J. B. Plotkin. 2008. The population genetics of dN/dS. *PLoS Genetics* 4:e1000304.

Kuo, C.-H., and H. Ochman. 2009. The fate of new bacterial genes. *FEMS Microbiology Reviews* 33:38–43.

Kuussaari, M., M. Nieminen, and I. Hanski. 1996. An experimental study of migration in the Glanville fritillary butterfly *Melitaea cinxia*. *Journal of Animal Ecology* 65:791–801.

Kuzoff, R. K., L. Hufford, and D. E. Soltis. 2001. Structural homology and developmental transformations associated with ovary diversification in *Lithophragma* (Saxifragaceae). *American Journal of Botany* 88:196–205.

Kuzoff, R. K., D. E. Soltis, L. Hufford, and P. S. Soltis. 1999. Phylogenetic relationships with *Lithophragma* (Saxifragaceae): Hybridization, allopolyploidy and ovary diversification. *Systematic Botany* 24:598–615.

Labandeira, C. C. 2010. The pollination of mid Mesozoic seed plants and the early history of long-proboscid insects. *Annals of the Missouri Botanical Garden* 97:469–513.

Lack, D. 1947. *Darwin's finches*. Cambridge University Press, Cambridge.

Ladle, R., R. Johnstone, and O. Judson. 1993. Coevolutionary dynamics of sex in a metapopulation: Escaping the Red Queen. *Proceedings of the Royal Society London B* 253:155–160.

Laine, A.-L. 2005. Spatial scale of local adaptation in a plant-pathogen metapopulation. *Journal of Evolutionary Biology* 18:930–938.

———. 2006. Evolution of host resistance: Looking for coevolutionary hotspots at small spatial scales. *Proceedings of the Royal Society of London B* 273:267–273.

———. 2007. Detecting local adaptation in a natural plant-pathogen metapopulation: A laboratory vs. field transplant approach. *Journal of Evolutionary Biology* 20:1665–1673.

———. 2008. Temperature-mediated patterns of local adaptation in a natural plant-pathogen metapopulation. *Ecology Letters* 11:327–337.

———. 2009. Role of coevolution in generating biological diversity: Spatially divergent selection trajectories. *Journal of Experimental Botany* 60:2957–2970.

Laine, A.-L., J. J. Burdon, P. N. Dodds, and P. H. Thrall. 2011. Spatial variation in disease resistance: From molecules to metapopulations. *Journal of Ecology* 99:96–112.

Lake, J. A. 2009. Evidence for an early prokaryotic endosymbiosis. *Nature* 460: 967–971.

Lambrechts, L., J. Halbert, P. Durand, L. C. Gouagna, and J. C. Koella. 2005. Host genotype by parasite genotype interactions underlying the resistance of anopheline mosquitoes to *Plasmodium falciparum*. *Malaria Journal* 4:3.

Lambrinos, J. G. 2004. How interactions between ecology and evolution influence contemporary invasion dynamics. *Ecology* 85:2061–2070.

Lande, R. 1983. The response to selection on major and minor mutations affecting a metrical trait. *Heredity* 50:47–65.

Lande, R., and S. J. Arnold. 1983. The measurement of selection on correlated characters. *Evolution* 36:1210–1226.

Langerhans, R. B., M. E. Gifford, and E. O. Joseph. 2007. Ecological speciation in *Gambusia* fishes. *Evolution* 61:2056–2074.

Langerhans, R. B., J. H. Knouft, and J. B. Losos. 2006. Shared and unique features of diversification in Greater Antillean *Anolis* ecomorphs. *Evolution* 60:362–369.

Lanterbecq, D., G. W. Rouse, and I. Eeckhaut. 2010. Evidence for cospeciation events in the host-symbiont system involving crinoids (Echinodermata) and their obligate associates, the myzostomids (Myzostomida, Annelida). *Molecular Phylogenetics and Evolution* 54:357–371.

Larsdotter Mellstrom, H., M. Friberg, A.-K. Borg-Karlson, R. Murtazina, M. Palm, and C. Wiklund. 2010. Seasonal polyphenism in life history traits: Time costs of direct development in a butterfly. *Behavioral Ecology and Sociobiology* 64:1377–1383.

Lässig, M., U. Bastolla, S. Manrubia, and A. Valleriani. 2001. Shape of ecological networks. *Physical Review Letters* 86:4418–4421.

Law, Y.-H., and J. A. Rosenheim. 2011. Effects of combining an intraguild predator with a cannibalistic intermediate predator on a species-level trophic cascade. *Ecology* 92:333–341.

Lederberg, J., and E. L. Tatum. 1946. Gene recombination in *E. coli*. *Nature* 158:558.

Lee, S. C., M. Ni, W. Li, C. Shertz, and J. Heitman. 2010. The evolution of sex: A perspective from the fungal kingdom. *Microbiology and Molecular Biology Reviews* 74:298–340.

Lee, Y. K., and S. K. Mazmanian. 2010. Has the microbiota played a critical role in the evolution of the adaptive immune system? *Science* 330:1768–1773.

Lefevre, T., S. A. Adamo, D. G. Biron, D. Misse, D. Hughes, and F. Thomas. 2009. Invasion of the body snatchers: The diversity and evolution of manipulative strategies in host–parasite interactions. *Advances in Parasitology* 68:45–83.

Leibold, M. A., and M. McPeek. 2006. Coexistence of the niche and neutral perspectives in community ecology. *Ecology* 87:1399–1410.

Leimu, R., and M. Fischer. 2008. A meta-analysis of local adaptation in plants. *PLoS ONE* 3:e4010.

Leitch, A., and I. Leitch. 2008. Genomic plasticity and the diversity of polyploid plants. *Science* 320:481–483.

Lenski, R. E. 1984. 2-step resistance by *Escherichia coli* B to bacteriophage T2. *Genetics* 107:1–7.

———. 1988. Experimental studies of pleiotropy and epistasis in *Escherichia coli*. II. Compensation for maladaptive effects associated with resistance to virus T4. *Evolution* 42:433–440.

———. 2004. Phenotypic and genomic evolution during a 20,000-generation experiment with the bacterium *Escherichia coli*. *Plant Breeding Reviews* 24:225–265.

Lenski, R. E., M. R. Rose, S. C. Simpson, and S. C. Tadler. 1991. Long-term experimental evolution in *Escherichia coli*. I. Adaptation and divergence during 2,000 generations. *American Naturalist* 138:1315–1341.

Lenski, R. E., and M. Travisano. 1994. Dynamics of adaptation and diversification: A 10,000 generation experiment with bacterial populations. *Proceedings of the National Academy of Sciences USA* 91:6808–6814.

Lenski, R. E., C. L. Winkworth, and M. A. Riley. 2003. Rates of DNA sequence evolution in experimental populations of *Escherichia coli* during 20,000 generations. *Journal of Molecular Evolution* 56:498–506.

Letarov, A., and E. Kulikov. 2009. The bacteriophages in human and animal body-associated microbial communities. *Journal of Applied Microbiology* 107:1–13.

Levin, B. R., and O. E. Cornejo. 2009. The population and evolutionary dynamics of homologous gene recombination in bacteria. *PloS Genetics* 5:e1000601.

Levin, B. R., B. S. Stewart, and L. Chao. 1977. Resource limited growth, competition, and predation: A model and experimental studies with bacteria and bacteriophage. *American Naturalist* 111:3–24.

Levins, R. 1969. Some demographic and genetic consequences of environmental heterogeneity for biological control. *Bulletin of the Entomological Society of America* 15:237–240.

Lewinsohn, T., P. I. Prado, P. Jordano, J. Bascompte, and J. M. Olesen. 2006. Structure in plant–animal interaction assemblages. *Oikos* 113:174–184.

Lewontin, R. C., and J. L. Hubby. 1966. A molecular approach to study of genic heterozygosity in natural populations. II. Amount of variation and degree of heterozygosity in natural populations of *Drosophila pseudoobscura*. *Genetics* 54:595–609.

Ley, R. E., C. A. Lozupone, M. Hamady, R. Knight, and J. I. Gordon. 2008. Worlds within worlds: Evolution of the vertebrate gut microbiota. *Nature Reviews Microbiology* 6:776–788.

Ley, R. E., D. A. Peterson, and J. I. Gordon. 2006. Ecological and evolutionary forces shaping microbial diversity in the human intestine. *Cell* 124:837–848.

Lieberman, B. S., and A. L. Melott. 2007. Considering the case for biodiversity cycles: Re-examining the evidence for periodicity in the fossil record. *PLoS ONE* 2:e759.

Lill, J. T., R. J. Marguis, and R. E. Ricklefs. 2002. Host plants influence parasitism of forest caterpillars. *Nature* 417:170–173.

Lim, K. Y., D. E. Soltis, P. S. Soltis, J. Tate, R. Matyasek, H. Srubarova, A. Kovarik, et al. 2008. Rapid chromosome evolution in recently formed polyploids in *Tragopogon* (Asteraceae). *PloS One* 3:e3353.

Lin, X., Y. Zhou, J. Zhang, X. Lu, F. Zhang, Q. Shen, S. Wu, et al. 2011. Enhancement of artemisinin content in tetraploid *Artemisia annua* plants by modulating the expression of genes in artemisinin biosynthetic pathway. *Biotechnology and Applied Biochemistry* 58:50–57.

Lindell, D., J. D. Jaffee, M. L. Coleman, M. E. Futschik, I. M. Axmann, T. Rector, G. Kettler, et al. 2007. Genome-wide expression dynamics of a marine virus and host reveal features of co-evolution. *Nature* 449:83–86.

Lindell, D., M. B. Sullivan, Z. I. Johnson, A. C. Tolonen, F. Rohwer, and S. W. Chisholm. 2004. Transfer of photosynthesis genes to and from *Prochlorococcus* viruses. *Proceedings of the National Academy of Sciences USA* 101:11013–11018.

Linnen, C. R., E. P. Kingsley, J. D. Jensen, and H. E. Hoekstra. 2009. On the origin and spread of an adaptive allele in deer mice. *Science* 325:1095–1098.

Linnenbrink, M., J. M. Johnsen, I. Montero, C. R. Brzezinski, B. Harr, and J. F. Baines. 2011. Long-term balancing selection at the blood group-related gene B4galnt2 in the genus *Mus* (Rodentia; Muridae). *Molecular Biology and Evolution* 28:2999–3003.

Litman, G. W., L. J. Dishaw, J. P. Cannon, R. N. Haire, and J. P. Rast. 2007. Alternative mechanisms of immune receptor diversity. *Current Opinion in Immunology* 19:526–534.

Little, A. E. F., and C. R. Currie. 2007. Symbiotic complexity: Discovery of a fifth symbiont in the attine ant-microbe symbiosis. *Biology Letters* 3:501–504.

Little, A. E. F., C. J. Robinson, S. B. Peterson, K. F. Raffa, and J. Handelsman. 2008. Rules of engagement: Interspecies interactions that regulate microbial communities. *Annual Review of Microbiology* 62:375–401.

Little, T. J., K. Watt, and D. Ebert. 2006. Parasite–host specificity: Experimental studies on the basis of parasite adaptation. *Evolution* 60:31–38.

Lively, C. M. 1987. Evidence from a New Zealand snail for the maintenance of sex by parasitism. *Nature* 328:519–521.

———. 2010a. Antagonistic coevolution and sex. *Evolution: Education and Outreach* 3:19–25.

———. 2010b. A review of Red Queen models for the persistence of obligate sexual reproduction. *Journal of Heredity* 101:S13–20.

Lively, C. M., and M. F. Dybdahl. 2000. Parasite adaptation to locally common host genotypes. *Nature* 405:679–681.

Loik, M. E. 2008. The effect of cactus spines on light interception and photosystem II for three sympatric species of *Opuntia* from the Mojave Desert. *Physiologia Plantarum* 134:87–98.

Lombardo, M. P. 2008. Access to endosymbiotic mutualistic microbes: An under-appreciated benefit of group living. *Behavioral Ecology and Sociobiology* 62:479–497.

Long, Z. T., J. F. Bruno, and J. E. Duffy. 2011. Food chain length and omnivory determine the stability of a marine subtidal food web. *Journal of Animal Ecology* 80:586–594.

Lopez-Pascua, L. D. C., M. A. Brockhurst, and A. Buckling. 2010. Antagonistic coevolution across productivity gradients: An experimental test of the effects of dispersal. *Journal of Evolutionary Biology* 23:207–211.

Lopez-Pascua, L. D. C., and A. Buckling. 2008. Increasing productivity accelerates host–parasite coevolution. *Journal of Evolutionary Biology* 21:853–860.

Lorenzi, M. C., R. Cervo, and S. Turillazzi. 1992. Effects of social parasitism of *Polistes atrimandibularis* on the colony cycle and brood production of *Polistes biglumis* bimaculatus (Hymenoptera, Vespidae). *Bollettino di Zoologia* 59: 267–271.

Lorenzi, M. C., and J. N. Thompson. 2011. The geographic structure of selection on a coevolving interaction between social parasitic wasps and their hosts hampers social evolution. *Evolution* 65:3527–3542.

Losos, J. B. 1990. Ecomorphology, performance capability, and scaling of West Indian *Anolis* lizards: An evolutionary analysis. *Ecological Monographs* 60:369–388.

———. 2009. *Lizards in an evolutionary tree: Ecology and adaptive radiation of anoles.* University of California Press, Berkeley.

Losos, J. B., T. R. Jackman, A. Larson, K. de Queiroz, and L. Rodriguez-Schettino. 1998. Contingency and determinism in replicated adaptive radiations of island lizards. *Science* 279:2115–2118.

Losos, J. B., and D. L. Mahler. 2010. Adaptive radiation: The interaction of ecological opportunity, adaptation, and speciation. Pages 381–420 in M. A. Bell, D. J. Futuyma, W. F. Eanes, and J. S. Levinton, eds., *Evolution since Darwin: The first 150 years.* Sinauer Associates, Sunderland, MA.

Losos, J. B., and R. E. Ricklefs. 2009. Adaptation and diversification on islands. *Nature* 457:830–836.

Losos, J. B., and C. J. Schneider. 2009. Anolis lizards. *Current Biology* 19:R316–R318.

Louthan, A. M., and K. M. Kay. 2011. Comparing the adaptive landscape across trait types: Larger QTL effect size in traits under biotic selection. *BMC Evolutionary Biology* 11:60.

Lowder, B. V., C. M. Guinane, N. L. Ben Zakour, L. A. Weinert, A. Conway-Morris, R. A. Cartwright, A. J. Simpson, et al. 2009. Recent human-to-poultry host jump, adaptation, and pandemic spread of *Staphylococcus aureus*. *Proceedings of the National Academy of Sciences USA* 106:19545–19550.

Lozovsky, E. R., T. Chookajorn, K. Brown, M. Imwong, P. J. Shaw, S. Kamchonwongpaisan, D. E. Neafsey, et al. 2009. Stepwise acquisition of pyrimethamine resistance in the malaria parasite. *Proceedings of the National Academy of Sciences USA* 106:12025.

Luciani, F., S. A. Sisson, H. Jiang, A. R. Francis, and M. M. Tanaka. 2009. The epidemiological fitness cost of drug resistance in *Mycobacterium tuberculosis*. *Proceedings of the National Academy of Sciences USA* 106:14711–14715.

Ludin, P., D. Nilsson, and P. Maeser. 2011. Genome-wide identification of molecular mimicry candidates in parasites. *PLoS ONE* 6:e17546.

Lumaret, R., J.-L. Guillerm, J. Delay, A. Ait Lhaj Loutfi, J. Izco, and M. Jay. 1987. Polyploidy and habitat differentiation in *Dactylis glomerata* L. from Galicia (Spain). *Oecologia* 73:436–446.

Luria, S. 1945. Genetics of bacterium–bacterial virus relationship. *Annals of the Missouri Botanical Garden* 32:235–242.

Luria, S., and M. Delbrück. 1943. Mutations of bacteria from virus sensitivity to virus resistance. *Genetics* 28:491–511.

Luyten, Y. A., J. R. Thompson, W. Morrill, M. F. Polz, and D. L. Distel. 2006. Extensive variation in intracellular symbiont community composition among members of a single population of the wood-boring bivalve *Lyrodus pedicellatus* (Bivalvia: Teredinidae). *Applied and Environmental Microbiology* 72:412–417.

Lyell, C. 1830. *Principles of geology, being an attempt to explain the former changes of the earth's surface by references to causes now in operation.* Vol. 1. Murray, London. (Reprinted by the University of Chicago Press, 1990.)

———. 1832. *Principles of geology, being an attempt to explain the former changes of the earth's surface by references to causes now in operation.* Vol. 2. Murray, London. (Reprinted by the University of Chicago Press, 1991.)

———. 1833. *Principles of geology, being an attempt to explain the former changes of the earth's surface by references to causes now in operation.* Vol. 3. Murray, London. (Reprinted by the University of Chicago Press, 1992.)

Lynch, M. 1990. The rate of morphological evolution in mammals from the standpoint of the neutral expectation. *American Naturalist* 136:727–741.

Lynch, M., and B. Walsh. 1998. *Genetics and analysis of quantitative traits.* Sinauer Associates, Sunderland, MA.

Lynch, R. M., T. Shen, S. Gnanakaran, and C. A. Derdeyn. 2009. Appreciating HIV type 1 diversity: Subtype differences in Env. *AIDS Research and Human Retroviruses* 25:237–248.

Lyon, B. E., and J. M. Eadie. 2004. An obligate brood parasite trapped in the intraspecific arms race of its hosts. *Nature* 432:390–393.

———. 2008. Conspecific brood parasitism in birds: A life-history perspective. *Annual Review of Ecology, Evolution, and Systematics* 39:343–363.

Lyon, B. E., J. M. Eadie, and L. D. Hamilton. 1994. Parental choice selects for ornamental plumage in American coot chicks. *Nature* 371:240–243.

Lyytinen, A., P. M. Brakefield, L. Lindström, and J. Mappes. 2004. Does predation maintain eyespot plasticity in *Bicyclus anynana*. *Proceedings of the Royal Society B* 271:279–283.

Machado, C. A., N. Robbins, M. T. P. Gilbert, and E. A. Herre. 2005. Critical review of host specificity and its coevolutionary implications in the fig/fig-wasp mutualism. *Proceedings of the National Academy of Sciences USA* 102: 6558–6565.

Magnello, M. E. 2009. Karl Pearson and the establishment of mathematical statistics. *International Statistical Review* 77:3–29.

Magurran, A. E., B. H. Seghers, G. R. Carvalho, and P. W. Shaw. 1992. Behavioral consequences of an artificial introduction of guppies (*Poecilia reticulata*) in N. Trinidad: Evidence for the evolution of antipredator behavior in the wild. *Proceedings of the Royal Society of London B* 248:117–122.

Mahler, D. L., L. J. Revell, R. E. Glor, and J. B. Losos. 2010. Ecological opportunity and the rate of morphological evolution in the diversification of greater Antillean anoles. *Evolution* 64:2731–2745.

Majerus, M. E. N. 1998. *Melanism: Evolution in action*. Oxford University Press, Oxford.

Majerus, T. M. O., and M. E. N. Majerus. 2010. Intergenomic arms races: Detection of a nuclear rescue gene of male-killing in a ladybird. *PLoS Pathogens* 6:e1000987.

Makinen, H. S., M. Cano, and J. Merila. 2008. Identifying footprints of directional and balancing selection in marine and freshwater three-spined stickleback (*Gasterosteus aculeatus*) populations. *Molecular Ecology* 17:3565–3582.

Mallet, J. 2005. Hybridization as an invasion of the genome. *Trends in Ecology and Evolution* 20:229–237.

———. 2007. Hybrid speciation. *Nature* 446:279–283.

———. 2008. Hybridization, ecological races and the nature of species: Empirical evidence for the ease of speciation. *Philosophical Transactions of the Royal Society of London Series B* 363:2971–2986.

Mallet, J., M. Beltran, W. Neukirchen, and M. Linares. 2007. Natural hybridization in heliconiine butterflies: The species boundary as a continuum. *BMC Evolutionary Biology* 7:28.

Mallet, J., A. Meyer, P. Nosil, and J. L. Feder. 2009. Space, sympatry, and speciation. *Journal of Evolutionary Biology* 95:3–16.

Mani, G. S., and B. C. Clarke. 1990. Mutational order: A major stochastic process in evolution. *Proceedings of the Royal Society B* 240:29–37.

Manier, M. K., and S. J. Arnold. 2005. Population genetic analysis identified source-sink dynamics for two sympatric garter snake species (*Thamnophis elegans* and *Thamnophis sirtalis*). *Molecular Ecology* 14:3965–3976.

Manolio, T. 2010. Genomewide association studies and assessment of the risk of disease. *New England Journal of Medicine* 363:166–176.

Margulis, L., and R. Fester. 1991. *Symbiosis as a source of evolutionary innovation: Speciation and morphogenesis*. MIT Press, Cambridge, MA.

Margulis, L., and D. Sagan. 2002. *Acquiring genomes: A theory of the origins of species*. Basic Books, New York.

Martén-Rodríguez, S., W. John Kress, E. J. Temeles, and E. Meléndez-Ackerman. 2011. Plant–pollinator interactions and floral convergence in two species of *Heliconia* from the Caribbean Islands. *Oecologia* 167:1075–1083.

Martín-Gálvez, D., J. J. Soler, J. G. Martínez, A. P. Krupa, M. Soler, and T. Burke. 2007. Cuckoo parasitism and productivity in different magpie subpopulations predict frequencies of the 457bp allele: A mosaic of coevolution at a small geographic scale. *Evolution* 61:2340–2348.

Martinez, J. L. 2008. Antibiotics and antibiotic resistance genes in natural environments. *Science* 321:365–367.

Martinez del Rio, C., A. Silva, R. Medel, and M. Hourdequin. 1996. Seed dispersers as disease vectors: Bird transmission of mistletoe seeds to plant hosts. *Ecology* 77:912–921.

Martiny, J. B. H., J. A. Eisen, K. Penn, S. D. Allison, and M. C. Horner-Devine. 2011. Drivers of bacterial beta-diversity depend on spatial scale. *Proceedings of the National Academy of Sciences USA* 108:7850–7854.

Maslowski, K. M., A. T. Vieira, A. Ng, J. Kranich, F. Sierro, D. Yu, H. C. Schilter, et al. 2009. Regulation of inflammatory responses by gut microbiota and chemoattractant receptor GPR43. *Nature* 461:1282–1286.

Matessi, C., and K. A. Schneider. 2009. Optimization under frequency-dependent selection. *Theoretical Population Biology* 76:1–12.

Matsuoka, Y. 2011. Evolution of polyploid *triticum* wheats under cultivation: The role of domestication, natural hybridization and allopolyploid speciation in their diversification. *Plant and Cell Physiology* 52:750–764.

Matsuura, K., T. Yashiro, K. Shimizu, S. Tatsumi, and T. Tamura. 2009. Cuckoo fungus mimics termite eggs by producing the cellulose-digesting enzyme β-glucosidase. *Current Biology* 19:1–7.

Matute, D. R., I. A. Butler, D. A. Turissini, and J. A. Coyne. 2010. A test of the snowball theory for the rate of evolution of hybrid incompatibilities. *Science* 329:1518–1521.

Mavárez, J., and M. Linares. 2008. Homoploid hybrid speciation in animals. *Molecular Ecology* 17:4181–4185.

May, G., F. Shaw, H. Badrane, and X. Vekemans. 1999. The signature of balancing selection: Fungal mating compatibility gene evolution. *Proceedings of the National Academy of Sciences USA* 96:9172–9177.

May, R. M. 1973. *Stability and complexity in model ecosystems*. Princeton University Press, Princeton, NJ.

May, R. M., and R. M. Anderson. 1983. Epidemiology and genetics in the coevolution of parasites and hosts. *Proceedings of the Royal Society of London B* 219:281–313.

Maynard Smith, J. 1978. *The evolution of sex*. Cambridge University Press, Cambridge.

Mayr, E. 1942. *Systematics and the origin of species*. Columbia University Press, New York.

———. 1946. The number of species of birds. *Auk* 63:64–69.

———. 1947. Ecological factors in speciation. *Evolution* 1:263–288.

———. 1963. *Animal species and evolution*. Harvard University Press, Cambridge, MA.

———. 1992. Darwin's principle of divergence. *Journal of the History of Biology* 25:343–359.

McBride, R., D. Greig, and M. Travisano. 2008. Fungal viral mutualism moderated by ploidy. *Evolution* 62:2372–2380.
McCauley, D. E., and M. S. Olson. 2008. Do recent findings in plant mitochondrial molecular and population genetics have implications for the study of gynodioecy and cytonuclear conflict? *Evolution* 62:1013–1025.
McCutcheon, J. P., and N. A. Moran. 2007. Parallel genomic evolution and metabolic interdependence in an ancient symbiosis. *Proceedings of the National Academy of Sciences USA* 104:19392–19397.
McCutcheon, J. P., B. R. McDonald, and N. A. Moran. 2009. Convergent evolution of metabolic roles in bacterial co-symbionts of insects. *Proceedings of the National Academy of Sciences USA* 106:15394–15399.
McDaniel, L. D., E. Yound, J. Delaney, F. Ruhnau, K. B. Ritchie, and J. H. Paul. 2010. High frequency of horizontal gene transfer in the oceans. *Science* 330:50.
McDonald, M. J., T. F. Cooper, H. J. E. Beaumont, and P. B. Rainey. 2011. The distribution of fitness effects of new beneficial mutations in *Pseudomonas fluorescens*. *Biology Letters* 7:98–100.
McDonald, M. J., S. M. Gehrig, P. L. Meintjes, X.-X. Zhang, and P. B. Rainey. 2009. Adaptive divergence in experimental populations of *Pseudomonas fluorescens*. IV. Genetic constraints guide evolutionary trajectories in a parallel adaptive radiation. *Genetics* 183:1041–1053.
McEvoy, B. P., G. W. Montgomery, A. F. McRae, S. Ripatti, M. Perola, T. D. Spector, L. Cherkas, et al. 2009. Geographical structure and differential natural selection among North European populations. *Genome Research* 19:804–814.
McFall-Ngai, M. 2002. Unseen forces: The influence of bacteria on animal development. *Developmental Biology* 242:1–14.
———. 2007. Adaptive immunity: Care for the community. *Nature* 445:153.
———. 2008. Hawaiian bobtail squid. *Current Biology* 18:R1043–R1044.
———. 2011. Origins of the immune system. Pages 199–205 in L. Margulis, C. A. Asikainen, and W. E. Krumbein, eds., *Chimeras and consciousness: Evolution of the sensory self*. MIT Press, Cambridge, MA.
McFall-Ngai, M., S. V. Nyholm, and M. G. Castillo. 2010. The role of the immune system in the initiation and persistence of the *Euprymna scolopes–Vibrio fischeri* symbiosis. *Seminars in Immunology* 22:48–53.
McPeek, M. A. 2008. The ecological dynamics of clade diversification and community assembly. *American Naturalist* 172:E270–E284.
McPeek, M. A., and J. M. Brown. 2000. Building a regional species pool: Diversification of the *Enallagma* damselflies of eastern North American waters. *Ecology* 81:904–920.
McPeek, M. A., A. L. Shen, and H. Farid. 2008. The correlated evolution of 3-dimensional reproductive structures between male and female damselflies. *Evolution* 63:73–83.

McPhail, J. D. 1992. Ecology and evolution of sympatric sticklebacks (*Gasterosteus*): Evidence for genetically divergent populations in Paxton Lake, Texada Island, British Columbia. *Canadian Journal of Zoology* 70:361–369.

Medel, R. 2000. Assessment of parasite-mediated selection in a host–parasite system in plants. *Ecology* 81:1554–1564.

Medel, R., M. A. Mendez, C. G. Ossa, and C. Botto-Mahan. 2010. Arms race coevolution: The local and geographic structure of a host–parasite interaction. *Evolution: Education and Outreach* 3:26–31.

Melián, C. J., J. Bascompte, P. Jordano, and V. Krivan. 2009. Diversity in a complex ecological network with two interaction types. *Oikos* 118:122–130.

Mercader, R. J., and J. M. Scriber. 2008. Divergence in the ovipositional behavior of the *Papilio glaucus* group. *Insect Science* 15:361–367.

Merhej, V., and D. Raoult. 2011. Rickettsial evolution in the light of comparative genomics. *Biological Reviews of the Cambridge Philosophical Society* 86:379–405.

Meyer, D., and G. Thomson. 2001. How selection shapes variation of the human major histocompatibility complex: A review. *Annals of Human Genetics* 65:1–26.

Meyer, J. R., A. A. Agrawal, R. T. Quick, D. T. Dobias, D. Schneider, and R. E. Lenski. 2010. Parallel changes in host resistance to viral infection during 45,000 generations of relaxed selection. *Evolution* 64:3024–3034.

Meyer, J. R., D. T. Dobias, J. S. Weitz, J. E. Barrick, R. T. Quick, and R. E. Lenski. 2012. Repeatability and contingency in the evolution of a key innovation in phage lambda. *Science* 335:428–432.

Meyer, J. R., and R. Kassen. 2007. The effects of competition and predation on diversification in a model adaptive radiation. *Nature* 446:432–435.

Mezquida, E. T., and C. W. Benkman. 2005. The geographic selection mosaic of squirrels, crossbills, and Aleppo pine. *Journal of Evolutionary Biology* 18:348–357.

Mikheyev, A. S., U. G. Mueller, and P. Abbot. 2010. Comparative dating of attine ant and lepiotaceous cultivar phylogenies reveals coevolutionary synchrony and discord. *American Naturalist* 175:E126–E133.

Miller, E., E. Kutter, G. Mosig, F. Arisaka, T. Kunisawa, and W. Ruger. 2003. Bacteriophage T4 genome. *Microbiology and Molecular Biology Reviews* 67:86–156.

Miller, W. J., L. Ehrman, D. Schneider, and C. Parrish. 2010. Infectious speciation revisited: Impact of symbiont-depletion on female fitness and mating behavior of *Drosophila paulistorum*. *PLoS Pathogens* 6:71–125.

Ming, R., A. Bendahmane, and S. S. Renner. 2011. Sex chromosomes in land plants. 62:485–514.

Misevic, D., C. Ofria, and R. E. Lenski. 2010. Experiments with digital organisms on the origin and maintenance of sex in changing environments. *Journal of Heredity* 101:S46–S54.

Mita, T., K. Tanabe, and K. Kita. 2009. Spread and evolution of *Plasmodium falciparum* drug resistance. *Parasitology International* 58:201–209.

Mitchell-Olds, T., and J. Schmitt. 2006. Genetic mechanisms and evolutionary significance of natural variation in *Arabidopsis*. *Nature* 441:947–952.

Mitchell-Olds, T., and R. G. Shaw. 1987. Regression analysis of natural selection: Statistical inference and biological interpretation. *Evolution* 1149–1161.

Modak, S. G., K. M. Satish, J. Mohan, S. Dey, N. Raghavendra, M. Shakarad, and A. Joshi. 2009. A possible tradeoff between developmental rate and pathogen resistance in *Drosophila melanogaster*. *Journal of Genetics* 88:253–256.

Mode, C. J. 1958. A mathematical model for the co-evolution of obligate parasites of their hosts. *Evolution* 12:158–165.

Moermond, T. C. 1981. Prey-attack behavior of *Anolis* lizards. *Zeitschriftüfür Tierpsychologie* 56:128–136.

Møller, A. P., N. Saino, P. Adamík, R. Ambrosini, A. Antonov, D. Campobello, B. G. Stokke, et al. 2011. Rapid change in host use of the common cuckoo *Cuculus canorus* linked to climate change. *Proceedings of the Royal Society B* 278:733–738.

Montague, J. L., S. C. H. Barrett, and C. G. Eckert. 2008. Re-establishment of clinal variation in flowering time among introduced populations of purple loosestrife (*Lythrum salicaria*, Lythraceae). *Journal of Evolutionary Biology* 21:234–245.

Monteiro, A., B. Chen, L. C. Scott, L. Vedder, H. J. Prijs, A. Belicha-Villanueva, and P. M. Brakefield. 2007. The combined effect of two mutations that alter serially homologous color pattern elements on the fore and hindwings of a butterfly. *BMC Genetics* 8:22.

Montesinos, A., S. J. Tonsor, C. Alonso-Blanco, and F. X. Picó. 2009. Demographic and genetic patterns of variation among populations of *Arabidopsis thaliana* from contrasting native environments. *PLoS ONE* 4:e7213.

Mooney, K. A., and A. A. Agrawal. 2008. Plant genotype shapes ant-aphid interactions: Implications for community structure and indirect plant defense. *American Naturalist* 171:E195–E205.

Mopper, S., and S. Y. Strauss, eds. 1998. *Genetic structure and local adaptation in natural insect populations*. Chapman and Hall, London.

Moran, N. A. 2002. The ubiquitous and varied role of infection in the lives of animals and plants. *American Naturalist* 160:S1–8.

———. 2007. Symbiosis as an adaptive process and source of phenotypic complexity. *Proceedings of the National Academy of Sciences USA* 104:8627–8633.

Moran, N. A., P. H. Degnan, S. R. Santos, H. E. Dunbar, and H. Ochman. 2005. The players in a mutualistic symbiosis: Insects, bacteria, viruses, and virulence genes. *Proceedings of the National Academy of Sciences USA* 102:16919–16926.

Moran, N. A., and T. Jarvik. 2010. Lateral transfer of genes from fungi underlies carotenoid production in aphids. *Science* 328:624–627.

Moran, N. A., J. P. McCutcheon, and A. Nakabachi. 2008. Genomics and evolution of heritable bacterial symbionts. *Annual Review of Genetics* 42:165–190.

Moreau, S., E. Huguet, and J.-M. Drezen. 2009. Polydnaviruses as tools to deliver wasp virulence factors to impair lepidopteran host immunity. Pages 137–158 in J. Rolff and S. E. Reynolds, eds., *Insect infection and immunity: Evolution, ecology, and mechanisms*. Oxford University Press, Oxford.

Moreira, L. A., I. Iturbe-Ormaetxe, J. A. Jeffery, G. Lu, A. T. Pyke, L. M. Hedges, B. C. Rocha, et al. 2009. A *Wolbachia* symbiont in *Aedes aegypti* limits infection with dengue, Chikungunya, and *Plasmodium*. *Cell* 139:1268–1278.

Morgan, A. D., M. B. Bonsall, and A. Buckling. 2010. Impact of bacterial mutation rate on coevolutionary dynamics between bacteria and phages. *Evolution* 64:2980–2987.

Morgan, A. D., S. Gandon, and A. Buckling. 2005. The effects of migration on local adaptation in a coevolving host–parasite system. *Nature* 437:253–256.

Morgan, T., M. Evans, T. Garland, J. G. Swallow, and P. A. Carter. 2005. Molecular and quantitative genetic divergence among populations of house mice with known evolutionary histories. *Heredity* 94:518–525.

Morgan, T. H. 1926. *The theory of the gene*. Yale University Press, New Haven, CT.

Morgan-Richards, M., R. D. Smissen, L. D. Shepherd, G. P. Wallis, J. J. Hayward, C.-H. Chan, G. K. Chambers, et al. 2009. A review of genetic analyses of hybridization in New Zealand. *Journal of the Royal Society of New Zealand* 39:15–34.

Morran, L. T., O. G. Schmidt, I. A. Gelarden, R. C. Parrish, and C. M. Lively. 2011. Running with the Red Queen: Host–parasite coevolution selects for biparental sex. *Science* 333:216–218.

Morris, D. W. 2011. Adaptation and habitat selection in the eco-evolutionary process. *Proceedings of the Royal Society B* 278:2401–2411.

Morris, J. J., Z. I. Johnson, M. J. Szul, M. Keller, and E. R. Zinser. 2011. Dependence of the cyanobacterium *Prochlorococcus* on hydrogen peroxide scavenging microbes for growth at the ocean's surface. *PLoS ONE* 6:e16805.

Morrissey, M. B., and J. D. Hadfield. 2012. Directional selection in temporally replicated studies is remarkably consistent. *Evolution* 66:435–442.

Mougi, A., and Y. Iwasa. 2011. Unique coevolutionary dynamics in a predator–prey system. *Journal of Theoretical Biology* 277:83–89.

Moya-Laraño, J. 2011. Genetic variation, predator–prey interactions and food web structure. *Philosophical Transactions of the Royal Society B* 366:1425–1437.

Moyle, L. C., and T. Nakazato. 2010. Hybrid incompatibility "snowballs" between *Solanum* species. *Science* 329:1521–1523.

Muchhala, N. 2006. Nectar bat stows huge tongue in its rib cage. *Nature* 444:701–702.

Mueller, L. D. 2009. Fitness, demography, and population dynamics in laboratory experiments. Pages 551–584 in T. J. Garland, and M. R. Rose, eds., *Experimental evolution: Concepts, methods, and applications of selection experiments*. University of California Press, Berkeley.

Mueller, L. D., and A. Joshi. 2000. *Stability in model populations*. Princeton University Press, Princeton, NJ.

Mueller, U. G., D. Dash, C. Rabeling, and A. Rodrigues. 2008. Coevolution between attine ants and actinomycete bacteria: A reevaluation. *Evolution* 62:2894–2912.

Mueller, U., N. M. Gerardo, D. K. Aanen, D. L. Six, and T. R. Schultz. 2005. The evolution of agriculture in insects. *Annual Review of Ecology and Systematics* 36:563–595.
Mueller, U. G., A. S. Mikheyev, S. E. Solomon, and M. Cooper. 2011. Frontier mutualism: Coevolutionary patterns at the northern range limit of the leaf-cutter ant-fungus symbiosis. *Proceedings of the Royal Society B* 278:3050–3059.
Mulet, M., J. Lalucat, and E. Garcia-Valdes. 2010. DNA sequence-based analysis of the *Pseudomonas* species. *Environmental Microbiology* 12:1513–1530.
Mullen, L. M., S. N. Vignieri, J. A. Gore, and H. E. Hoekstra. 2009. Adaptive basis of geographic variation: Genetic, phenotypic and environmental differences among beach mouse populations. *Proceedings of the Royal Society B* 276:3809–3818.
Mullen, S. P., E. B. Dopman, and R. G. Harrison. 2008. Hybrid zone origins, species boundaries, and the evolution of wing-pattern diversity in a polytypic species complex of North American admiral butterflies (Nymphalidae: *Limenitis*). *Evolution* 62:1400–1417.
Müller, F. 1879. *Ituna* and *Thyridia*: A remarkable case of mimicry in butterflies. *Proceedings of the Entomological Society of London* 1879:xx–xxix.
Müller, H. 1873. On the fertilization of flowers and on the reciprocal adaptations of both. *Nature* 8:187–189.
———. 1883. *The fertilization of flowers*. Translated by D'Arcy Thompson. Macmillan, London.
Muller, H. J. 1932. Some genetic aspects of sex. *American Naturalist* 66:118–138.
———. 1942. Isolating mechanisms, evolution, and temperature. Pages 71–125 in T. Dobzhansky, ed. *Temperature, evolution, development. Biological symposia: A series of volumes devoted to current symposia in the field of biology*, vol. 6. Jaques Cattell Press, Lancaster, PA.
———. 1964. The relation of recombination to mutational advance. *Mutation Research* 1:2–9.
Müller-Schärer, H., U. Schaffner, and T. Steinger. 2004. Evolution in invasive plants: Implication for biological control. *Trends in Ecology and Evolution* 19:417–422.
Münzbergová, Z. 2007. Population dynamics of diploid and hexaploid populations of a perennial herb. *Annals of Botany* 100:1259–1270.
Muola, A., P. Mutikainen, M. Lilley, L. Laukkanen, J.-P. Salminen, and R. Leimu. 2010. Associations of plant fitness, leaf chemistry, and damage suggest selection mosaic in plant-herbivore interactions. *Ecology* 91:2650–2659.
Murase, Y., T. Shimada, N. Ito, and P. A. Rikvold. 2010. Effects of demographic stochasticity on biological community assembly on evolutionary time scales. *Physical Review E* 81:041908.
Murphy, S. M. 2004. Enemy-free space maintains swallowtail butterfly host shift. *Proceedings of the National Academy of Sciences USA* 101:18048–18052.

Murphy, S. M., and P. Feeny. 2006. Chemical facilitation of a naturally occurring host shift by *Papilio machaon* butterflies (Papilionidae). *Ecological Monographs* 76:399–414.

Murphy, S. M., and Y. B. Linhart. 1999. Comparative morphology of the gastrointestinal tract of the feeding specialist *Sciurus aberti* and several generalist congeners. *Journal of Mammalogy* 80:1325–1330.

Murray, C., E. Huerta-Sanchez, F. Casey, and D. G. Bradley. 2010. Cattle demographic history modelled from autosomal sequence variation. *Philosophical Transactions of the Royal Society B* 365:2531–2539.

Nadeau, N. J., and C. D. Jiggins. 2010. A golden age for evolutionary genetics? Genomic studies of adaptation in natural populations. *Trends in Genetics* 26:484–492.

Nahmias, A., and D. Danielsson. 2011. Introduction to the evolution of infectious agents in relation to sex. *Annals of the New York Academy of Sciences* 1230:xiii–xix.

Nair, S., B. Miller, M. Barends, A. Jaidee, J. Patel, M. Mayxay, P. Newton, et al. 2008. Adaptive copy number evolution in malaria parasites. *PLoS Genetics* 4:e1000243.

Nakagawa, S., and K. Takai. 2008. Deep-sea vent chemoautotrophs: Diversity, biochemistry and ecological significance. *FEMS Microbiology Ecology* 65:1–14.

Nash, D. R., T. D. Als, R. Maile, G. R. Jones, and J. J. Boomsma. 2008. A mosaic of chemical coevolution in a large blue butterfly. *Science* 319:88–90.

Naumann, M., A. Schüssler, and P. Bonfante. 2010. The obligate endobacteria of arbuscular mycorrhizal fungi are ancient heritable components related to the mollicutes. *ISME Journal* 4:862–871.

Nealson, K. H., T. Platt, and J. W. Hastings. 1970. Cellular control of the synthesis and activity of the bacterial luminescent system. *Journal of Bacteriology* 104:313–322.

Neiman, M., and B. Koskella. 2009. Sex and the Red Queen. Pages 133–159 in I. Schön, K. Martens, and P. van Dijk, eds., *Lost sex: The evolutionary biology of parthenogensis*. Springer, New York.

Nelson, M. I., and E. C. Holmes. 2007. The evolution of epidemic influenza. *Nature Reviews Genetics* 8:196–205.

Ness, B. D., D. E. Soltis, and P. E. Soltis. 1989. Autopolyploidy in *Heuchera micrantha* (Saxifragaceae). *American Journal of Botany* 76:614–626.

Nesse, R. M., and S. C. Stearns. 2008. The great opportunity: Evolutionary applications to medicine and public health. *Evolutionary Applications* 1:28–48.

Newton, I. L. G., and S. R. Bordenstein. 2011. Correlations between bacterial ecology and mobile DNA. *Current Microbiology* 62:198–208.

Nguyen, D.-Q., C. Webber, J. Hehir-Kwa, R. Pfundt, J. Veltman, and C. P. Ponting. 2008. Reduced purifying selection prevails over positive selection in human copy number variant evolution. *Genome Research* 18:1711–1723.

Niaré, O., K. Markianos, J. Volz, F. Oduol, A. Touré, M. Bagayoko, D. Sangaré, et al. 2002. Genetic loci affecting resistance to human malaria parasites in a West African mosquito vector population. *Science* 298:213–216.

Nicholson, K. E., R. E. Glor, J. J. Kolbe, A. Larson, S. B. Hedges, and J. B. Losos. 2005. Mainland colonization by island lizards. *Journal of Biogeography* 32:929–938.

Nielsen, J. K., T. Nagao, H. Okabe, and T. Shinoda. 2010. Resistance in the plant, *Barbarea vulgaris*, and counter-adaptations in flea beetles mediated by saponins. *Journal of Chemical Ecology* 36:277–285.

Niitepõld, K., A. D. Smith, J. L. Osborne, D. R. Reynolds, N. L. Carreck, A. P. Martin, J. H. Marden, et al. 2009. Flight metabolic rate and *Pgi* genotype influence butterfly dispersal rate in the field. *Ecology* 90:2223–2232.

Noedl, H., Y. Se, K. Schaecher, B. L. Smith, D. Socheat, and M. M. Fukuda. 2008. Evidence of artemisinin-resistant malaria in western Cambodia. *New England Journal of Medicine* 359:2619.

Nolte, A. W., and D. Tautz. 2010. Understanding the onset of hybrid speciation. *Trends in Genetics* 26:54–58.

North, A., J. Pennanen, O. Ovaskainen, and A.-L. Laine. 2011. Local adaptation in a changing world: The roles of gene-flow, mutation, and sexual reproduction. *Evolution* 65:79–89.

Nosil, P. 2007. Divergent host plant adaptation and reproductive isolation between ecotypes of *Timema cristinae* walking sticks. *American Naturalist* 169:151–162.

Nosil, P., and S. M. Flaxman. 2011. Conditions for mutation-order speciation. *Proceedings of the Royal Society B* 278:399–407.

Novak, M., T. Pfeiffer, R. E. Lenski, U. Sauer, and S. Bonhoeffer. 2006. Experimental tests for an evolutionary trade-off between growth rate and yield in *E. coli*. *American Naturalist* 168:242–251.

Nuismer, S. L., and B. M. Cunningham. 2005. Selection for phenotypic divergence between diploid and autotetraploid *Heuchera grossulariifolia*. *Evolution* 59:1928–1935.

Nuismer, S. L., and S. Gandon. 2008. Moving beyond common-garden and transplant designs: Insight into the causes of local adaptation in species interactions. *American Naturalist* 171:658–668.

Nuismer, S. L., R. Gomulkiewicz, and B. J. Ridenhour. 2010. When is correlation coevolution? *American Naturalist* 175:525–537.

Nuismer, S. L., and B. J. Ridenhour. 2008. The contribution of parasitism to selection on floral traits in *Heuchera grossulariifolia*. *Journal of Evolutionary Biology* 21:958–965.

Nuismer, S. L., B. J. Ridenhour, and B. P. Oswald. 2007. Antagonistic coevolution mediated by phenotypic differences between quantitative traits. *Evolution* 61:1823–1834.

Nuismer, S. L., and J. N. Thompson. 2001. Plant polyploidy and non-uniform effects on insect herbivores. *Proceedings of the Royal Society of London B* 268: 1937–1940.

———. 2006. Coevolutionary alternation in antagonistic interactions. *Evolution* 60:2207–2217.

Nuismer, S. L., J. N. Thompson, and R. Gomulkiewicz. 1999. Gene flow and geographically structured coevolution. *Proceedings of the Royal Society of London B* 266:605–609.

———. 2000. Coevolutionary clines across selection mosaics. *Evolution* 54:1102–1115.

———. 2003. Coevolution between species with partially overlapping geographic ranges. *Journal of Evolutionary Biology* 16:1337–1345.

Nyholm, S. V., and M. J. McFall-Ngai. 2004. The winnowing: Establishing the squid-*Vibrio* symbiosis. *Nature Reviews Microbiology* 2:632–642.

Nylin, S., and N. Janz. 2009. Butterfly host plant range: An example of plasticity as a promoter of speciation? *Evolutionary Ecology* 23:137–146.

Nylin, S., G. H. Nygren, J. J. Windig, N. Janz, and A. Bergstrom. 2005. Genetics of host-plant preference in the comma butterfly *Polygonia c-album* (Nymphalidae), and evolutionary implications. *Biological Journal of the Linnean Society* 84:755–765.

Nyman, T., V. Vikberg, D. R. Smith, and J.-L. Boeve. 2010. How common is ecological speciation in plant-feeding insects? A "higher" Nematinae perspective. *BMC Evolutionary Biology* 10:266.

Obbard, D. J., S. A. Harris, R. J. A. Buggs, and J. R. Pannell. 2006. Hybridization, polyploidy, and the evolution of sexual systems in *Mercurialis* (Euphorbiaceae). *Evolution* 60:1801–1815.

Obbard, D. J., J. J. Welch, K.-W. Kim, and F. M. Jiggins. 2009. Quantifying adaptive evolution in the *Drosophila* immune system. *PLoS Genetics* 5:e1000698.

Ochman, H., M. Worobey, C.-H. Kuo, J.-B. N. Ndjango, M. Peeters, B. H. Hahn, and P. Hugenholtz. 2010. Evolutionary relationships of wild hominids recapitulated by gut microbial communities. *PLoS Biology* 8:e1000546.

Oduor, A. M. O., R. A. Lankau, S. Y. Strauss, and J. M. Gómez. 2011. Introduced *Brassica nigra* populations exhibit greater growth and herbivore resistance but less tolerance than native populations in the native range. *New Phytologist* 191:536–544.

O'Fallon, B. D., J. Seger, and F. R. Adler. 2010. A continuous-state coalescent and the impact of weak selection on the structure of gene genealogies. *Molecular Biology and Evolution* 27:1162–1172.

O'Gorman, C. M., H. T. Fuller, and P. S. Dyer. 2009. Discovery of a sexual cycle in the opportunistic fungal pathogen *Aspergillus fumigatus*. *Nature* 457:471–474.

O'Hara, R. 2005. Comparing the effects of genetic drift and fluctuating selection on genotype frequency changes in the scarlet tiger moth. *Proceedings of the Royal Society B* 272:211–217.

Ohgushi, T. 2005. Indirect interaction webs: Herbivore-induced effects through trait change in plants. *Annual Review of Ecology, Evolution, and Systematics* 36:81–105.

———. 2008. Herbivore-induced indirect interaction webs on terrestrial plants: The importance of non-trophic, indirect, and facilitative interactions *Entomologia Experimentalis et Applicata* 128:217–229.

O'Keefe, K. J., O. K. Silander, H. McCreery, D. Weinreich, K. M. Wright, L. Chao, S. V. Edwards, et al. 2010. Geographic differences in sexual reassortment in RNA phage. *Evolution* 64:3010–3023.

Okuyama, Y., M. Kato, and M. Murakami. 2004. Pollination by fungus gnats in four species of the genus *Mitella* (Saxifragaceae). *Botanical Journal of the Linnean Society* 144:449–460.

Okuyama, Y., O. Pellmyr, and M. Kato. 2008. Parallel floral adaptations to pollination by fungus gnats within the genus *Mitella* (Saxifragaceae). *Molecular Phylogenetics and Evolution* 46:560–575.

Olesen, J. M., J. Bascompte, Y. L. Dupont, and P. Jordano. 2007. The modularity of pollination networks. *Proceedings of the National Academy of Sciences USA* 104:19891–19896.

Oliver, K. M., P. H. Degnan, M. S. Hunter, and N. A. Moran. 2009. Bacteriophages encode factors required for protection in a symbiotic mutualism. *Science* 325:992–994.

Ollerton, J., R. Winfree, and S. Tarrant. 2011. How many flowering plants are pollinated by animals? *Oikos* 120:321–326.

Olofsson, M., A. Valin, S. Jakobsson, and C. Wiklund. 2010. Marginal eyespots on butterfly wings deflect bird attacks under low light intensities with UV wavelengths. *PLoS ONE* 5:e10798.

Olsen, E. M., S. M. Carlson, J. Gjøsaeter, and N. C. Stenseth. 2009. Nine decades of decreasing phenotypic variability in Atlantic cod. *Ecology Letters* 12:622–631.

Omsland, A., and R. A. Heinzen. 2011. Life on the outside: The rescue of *Coxiella burnetii* from its host cell. *Annual Review of Microbiology*.

Oostra, V., M. A. de Jong, B. M. Invergo, F. Kesbeke, F. Wende, P. M. Brakefield, and B. J. Zwaan. 2011. Translating environmental gradients into discontinuous reaction norms via hormone signalling in a polyphenic butterfly. *Proceedings of the Royal Society B* 278:789–797.

Orcutt, B. N., J. B. Sylvan, N. J. Knab, and K. J. Edwards. 2011. Microbial ecology of the dark ocean above, at, and below the seafloor. *Microbiology and Molecular Biology Reviews* 75:361–422.

Ording, G. J., R. J. Mercader, M. L. Aardema, and J. M. Scriber. 2010. Allochronic isolation and incipient hybrid speciation in tiger swallowtail butterflies. *Oecologia* 162:523–531.

Orgel, L. E., and F. H. Crick. 1980. Selfish DNA: The ultimate parasite. *Nature* 284: 604–607.

Orians, C. M., and D. Ward. 2010. Evolution of plant defenses in nonindigenous environments. *Annual Review of Entomology* 55:439–459.

Orians, G. H. 1962. Natural selection and ecological theory. *American Naturalist* 96:257–263.

Orr, H. A. 1995. The population genetics of speciation: The evolution of hybrid incompatibilities. *Genetics* 139:1805–1813.

———. 1998. The population genetics of adaptation: The distribution of factors fixed during adaptive evolution. *Evolution* 52:935–949.
———. 1999. The evolutionary genetics of adaptation: A simulation study. *Genetical Research* 74:207–214.
———. 2005. The genetic theory of adaptation: A brief history. *Nature Reviews Genetics* 6:119–127.
———. 2009. Fitness and its role in evolutionary genetics. *Nature Reviews Genetics* 10:531–539.
———. 2010. The population genetics of beneficial mutations. *Philosophical Transactions of the Royal Society B* 365:1195–1201.
Orr, H. A., and R. L. Unckless. 2008. Population extinction and the genetics of adaptation. *American Naturalist* 172:160–169.
Orsini, L., J. Corander, A. Alasentie, and I. Hanski. 2008. Genetic spatial structure in a butterfly metapopulation correlates better with past than present demographic structure. *Molecular Ecology* 17:2629–2642.
Orsini, L., C. W. Wheat, C. R. Haag, J. Kvist, M. J. Frilander, and I. Hanski. 2009. Fitness differences associated with *Pgi* SNP genotypes in the Glanville fritillary butterfly (*Melitaea cinxia*). *Journal of Evolutionary Biology* 22:367–375.
Osnas, E., and C. M. Lively. 2006. Host ploidy, parasitism and immune defence in a coevolutionary snail-trematode system. *Journal of Evolutionary Biology* 19: 42–48.
O'Steen, S., A. J. Cullum, and A. F. Bennett. 2002. Rapid evolution of escape ability in Trinidadian guppies (*Poecilia reticulata*). *Evolution* 56:776–784.
Oswald, B. P., and S. L. Nuismer. 2007. Neopolyploidy and pathogen resistance. *Proceedings of the Royal Society of London B* 274:2393–2397.
———. 2011. Neopolyploidy and diversification in *Heuchera grossulariifolia*. *Evolution* 65:1667–1679.
Otto, S. P. 2009. The evolutionary enigma of sex. *American Naturalist* 174:S1–S14.
Otto, S. P., and S. L. Nuismer. 2004. Species interactions and the evolution of sex. *Science* 304:1018–1020.
Ouanes, K., L. Bahri-Sfar, O. K. Ben Hassine, and F. Bonhomme. 2011. Expanding hybrid zone between *Solea aegyptiaca* and *Solea senegalensis*: Genetic evidence over two decades. *Molecular Ecology* 20:1717–1728.
Ozgo, M. 2011. Rapid evolution in unstable habitats: A success story of the polymorphic land snail *Cepaea nemoralis* (Gastropoda: Pulmonata). *Biological Journal of the Linnean Society* 102:251–262.
Page, R. D. M. 2003. *Tangled trees: Phylogeny, cospeciation, and coevolution*. University of Chicago Press, Chicago.
Paine, R. T. 1966. Food web complexity and species diversity. *American Naturalist* 100:65–75.
Paixão, T., S. S. Phadke, R. B. R. Azevedo, and R. A. Zufall. 2011. Sex ratio evolution under probabilistic sex determination. *Evolution* 65:2050–2060.

Palkovacs, E. P., M. C. Marshall, B. A. Lamphere, B. R. Lynch, D. J. Weese, D. F. Fraser, D. N. Reznick, et al. 2009. Experimental evaluation of evolution and coevolution as agents of ecosystem change in Trinidadian streams. *Philosophical Transactions of the Royal Society B* 364:1617–1628.

Palmer, T. M., D. F. Doak, M. L. Stanton, J. L. Bronstein, E. T. Kiers, T. P. Young, J. R. Goheen, et al. 2010. Synergy of multiple partners, including freeloaders, increases host fitness in a multispecies mutualism. *Proceedings of the National Academy of Sciences USA* 107:17234–17239.

Palumbi, S. R. 2001a. *The evolution explosion: How humans cause rapid evolutionary change*. Norton, New York.

———. 2001b. Humans are the world's greatest evolutionary force. *Science* 293:1786–1790.

Parchman, T. L., and C. W. Benkman. 2002. Diversifying coevolution between crossbills and black spruce on Newfoundland. *Evolution* 56:1663–1672.

———. 2008. The geographic selection mosaic for ponderosa pine and crossbills: A tale of two squirrels. *Evolution* 62:348–360.

Parchman, T., C. Benkman, and S. Britch. 2006. Patterns of genetic variation in the adaptive radiation of New World crossbills (Aves: *Loxia*). *Molecular Ecology* 15:1873–1887.

Parchman, T. L., C. W. Benkman, and E. T. Mezquida. 2007. Coevolution between hispaniolan crossbills and pine: Does more time allow for greater phenotypic escalation at lower latitude ? *Evolution* 61:2142–2153.

Parniske, M. 2008. Arbuscular mycorrhiza: The mother of plant root endosymbioses. *Nature Reviews Microbiology* 6:763–775.

Partensky, F., and L. Garczarek. 2010. *Prochlorococcus*: Advantages and limits of minimalism. *Annual Review of Marine Science* 2:305–331.

Pascual, M., M. P. Chapuis, F. Mestres, J. Balanya, R. B. Huey, G. W. Gilchrist, L. Serra, et al. 2007. Introduction history of *Drosophila subobscura* in the New World: A microsatellite-based survey using ABC methods. *Molecular Ecology* 16:3069–3083.

Paterson, S., T. Vogwill, A. Buckling, R. Benmayor, A. J. Spiers, N. R. Thomson, M. Quail, et al. 2010. Antagonistic coevolution accelerates molecular evolution. *Nature* 464:275–278.

Pauw, A., J. Stofberg, and R. J. Waterman. 2008. Flies and flowers in Darwin's race. *Evolution* 63:268–279.

Pavey, S. A., H. Collin, P. Nosil, and S. M. Rogers. 2010. The role of gene expression in ecological speciation. *Annals of the New York Academy of Sciences* 1206:110–129.

Pavlicev, M., J. M. Cheverud, and G. P. Wagner. 2011. Evolution of adaptive phenotypic variation patterns by direct selection for evolvability. *Proceedings of the Royal Society B* 278:1903–1912.

Peakall, R., D. Ebert, J. Poldy, R. Barrow, W. Francke, C. C. Bower, and F. P. Schiestl. 2010. Pollinator specificity, floral odour chemistry and the phylogeny of Australian sexually deceptive *Chiloglottis* orchids: Implications for pollinator-driven speciation. *New Phytologist* 188:437–450.

Pearse, D., S. Hayes, and M. Bond. 2009. Over the falls? Rapid evolution of ecotypic differentiation in steelhead/rainbow trout (*Oncorhynchus mykiss*). *Journal of Heredity* 100:515–525.

Peay, K. G., T. D. Bruns, and M. Garbelotto. 2010. Testing the ecological stability of ectomycorrhizal symbiosis: Effects of heat, ash and mycorrhizal colonization on *Pinus muricata* seedling performance. *Plant and Soil* 330: 291–302.

Peay, K. G., P. G. Kennedy, and T. D. Bruns. 2008. Fungal community ecology: A hybrid beast with a molecular master. *BioScience* 58:799–810.

Peccoud, J., J.-C. Simon, H. J. McLaughlin, and N. A. Moran. 2009. Post-Pleistocene radiation of the pea aphid complex revealed by rapidly evolving endosymbionts. *Proceedings of the National Academy of Sciences USA* 106:16315–16320.

Pélabon, C., T. F. Hansen, A. J. R. Carter, and D. Houle. 2010. Evolution of variation and variability under fluctuating, stabilizing, and disruptive selection. *Evolution* 64:1912–1925.

Pelletier, F., D. Garant, and A. P. Hendry. 2009. Eco-evolutionary dynamics. *Philosophical Transactions of the Royal Society B* 364:1483–1490.

Pellmyr, O. 2003. Yuccas, yucca moths, and coevolution: A review. *Annals of the Missouri Botanical Garden* 90:35–55.

Pellmyr, O., M. Balcázar-Lara, K. A. Segraves, D. M. Althoff, and R. J. Littlefleld. 2008. Phylogeny of the pollinating yucca moths, with revision of Mexican species (*Tegeticula* and *Parategeticula*; Lepidoptera, Prodoxidae). *Zoological Journal of the Linnean Society* 12:297–314.

Pellmyr, O., and J. Leebens-Mack. 1999. Forty million years of mutualism: Evidence for Eocene origin of the yucca–yucca moth association. *Proceedings of the National Academy of Sciences USA* 96:9178–9183.

Pellmyr, O., J. H. Leebens-Mack, and J. N. Thompson. 1998. Herbivores and molecular clocks as tools in plant biogeography. *Biological Journal of the Linnean Society* 63:367–378.

Pellmyr, O., and K. Segraves. 2003. Pollinator divergence within an obligation mutualism: Two yucca moth species (Lepidoptera: Prodoxidae: *Tegeticula*) on the Joshua tree (*Yucca brevifolia*; Agavaceae). *Annals of the Entomological Society of America* 96:716–722.

Pellmyr, O., K. Segraves, D. M. Althoff, M. Balcázar-Lara, and J. Leebens-Mack. 2007. The phylogeny of yuccas. *Molecular Phylogenetics and Evolution* 43:493–501.

Pellmyr, O., J. N. Thompson, J. M. Brown, and R. G. Harrison. 1996. Evolution of pollination and mutualism in the yucca moth lineage. *American Naturalist* 148:827–847.

Penman, B. S., O. G. Pybus, D. J. Weatherall, and S. Gupta. 2009. Epistatic interactions between genetic disorders of hemoglobin can explain why the sickle-cell gene is uncommon in the Mediterranean. *Proceedings of the National Academy of Sciences USA* 106:21242–21246.

Pennings, P. S., A. Achenbach, and S. Foitzik. 2011. Similar evolutionary potentials in an obligate ant parasite and its two host species. *Journal of Evolutionary Biology* 24:871–886.

Pernice, M., S. Wetzel, O. Gros, R. Boucher-Rodoni, and N. Dubilier. 2007. Enigmatic dual symbiosis in the excretory organ of *Nautilus macromphalus* (Cephalopoda: Nautiloidea). *Proceedings of the Royal Society B* 274:1143–1152.

Peters, A. D., and C. Lively. 2007. Short- and long-term benefits and detriments to recombination under antagonistic coevolution. *Journal of Evolutionary Biology* 20:1206–1217.

Petit, R. J., and A. Hampe 2006. Some evolutionary consequences of being a tree. *Annual Review of Ecology, Evolution, and Systematics* 37:187–214.

Petrusek, A., R. Tollrian, K. Schwenk, A. Haas, and C. Laforsch. 2009. A "crown of thorns" is an inducible defense that protects *Daphnia* against an ancient predator. *Proceedings of the National Academy of Sciences USA* 106:2248–2252.

Pfennig, D. W., and K. S. Pfennig. 2010. Character displacement and the origins of diversity. *American Naturalist* 176:S26–S44.

Pfennig, D. W., A. M. Rice, and R. A. Martin. 2007. Field and experimental evidence for competition's role in phenotypic divergence. *Evolution* 61:257–271.

Pfennig, D. W., M. A. Wund, E. C. Snell-Rood, T. Cruickshank, C. D. Schlichting, and A. P. Moczek. 2010. Phenotypic plasticity's impacts on diversification and speciation. *Trends in Ecology and Evolution* 25:459–467.

Phadke, S. S., and R. A. Zufall. 2009. Rapid diversification of mating systems in ciliates. *Biological Journal of the Linnean Society* 98:187–197.

Phillimore, A. B., R. P. Freckleton, C. D. L. Orme, and I. P. F. Owens. 2006. Ecology predicts large-scale patterns of diversification in birds. *American Naturalist* 168:220–229.

Phillimore, A. B., C. D. L. Orme, G. H. Thomas, T. M. Blackburn, P. M. Bennett, K. J. Gaston, and I. P. F. Owens. 2008. Sympatric speciation in birds is rare: Insights from range data and simulations. *American Naturalist* 171:646–657.

Phillips, B. L. 2009. The evolution of growth rates on an expanding range edge. *Biology Letters* 5:802–804.

Piculell, B. J., J. D. Hoeksema, and J. N. Thompson. 2008. Interactions of biotic and abiotic environmental factors on an ectomycorrhizal symbiosis, and the potential for selection mosaics. *BMC Biology* 6:23.

Pieterse, C. M. J., and M. Dicke. 2007. Plant interactions with microbes and insects: From molecular mechanisms to ecology. *Trends in Plant Science* 12: 564–569.

Piganeau, G., A. Eyre-Walker, N. Grimsley, and H. Moreau. 2011. How and why DNA barcodes underestimate the diversity of microbial eukaryotes. *PLoS ONE* 6:e16342.

Pigliucci, M. 2008. Is evolvability evolvable? *Nature Reviews Genetics* 9:75–82.

Pigot, A. L., A. B. Phillimore, I. P. F. Owens, and C. D. L. Orme. 2010. The shape and temporal dynamics of phylogenetic trees arising from geographic speciation. *Systematic Biology* 59:660–673.

Pimm, S. L. 1982. *Food webs*. Chapman and Hall, London.

Pinto, G., D. L. Mahler, L. J. Harmon, and J. B. Losos. 2008. Testing the island effect in adaptive radiation: Rates and patterns of morphological diversification in Caribbean and mainland *Anolis* lizards. *Proceedings of the Royal Society B* 275:2749–2757.

Pires, M. M., P. R. Guimarães, M. S. Araújo, A. A. Giaretta, J. C. L. Costa, and S. F. dos Reis. 2011. The nested assembly of individual-resource networks. *Journal of Animal Ecology* 80:896–903.

Platt, A., M. Horton, Y. S. Huang, Y. Li, A. E. Anastasio, N. W. Mulyati, J. Agren, et al. 2010. The scale of population structure in *Arabidopsis thaliana*. *PLoS Genetics* 6:e1000843.

Pogson, G. H., C. T. Taggart, K. A. Mesa, and R. G. Boutilier. 2001. Isolation by distance in the Atlantic cod, *Gadus morhua*, at large and small geographic scales. 55:131–146.

Poisot, T., P. H. Thrall, and M. E. Hochberg. 2011. Trophic network structure emerges through antagonistic coevolution in temporally varying environments. *Proceedings of the Royal Society B* 279:299–308.

Posfai, G., G. Plunkett III, T. Feher, D. Frisch, G. Keil, K. Umenhoffer, V. Kolisnychenko, et al. 2006. Emergent properties of reduced-genome *Escherichia coli*. *Science* 312:1044–1046.

Post, D. M., and E. P. Palkovacs. 2009. Eco-evolutionary feedbacks in community and ecosystem ecology: Interactions between the ecological theatre and the evolutionary play. *Philosophical Transactions of the Royal Society B* 364:1629–1640.

Poulin, B., S. J. Wright, G. Lefebvre, and O. Calderon. 1999. Interspecific synchrony and asynchrony in the fruiting phenologies of congeneric bird-dispersed plants in Panama. *Journal of Tropical Ecology* 15:213–227.

Poulin, R. 2010. Parasite manipulation of host behavior: An update and frequently asked questions. *Advances in the Study of Behavior* 41:151–186.

Powles, S. B., and Q. Yu. 2010. Evolution in action: Plants resistant to herbicides. *Annual Review of Plant Biology* 61:317–347.

Poxleitner, M. K., M. L. Carpenter, J. J. Mancuso, C. J. Wang, S. C. Dawson, and W. Z. Cande. 2008. Evidence for karyogamy and exchange of genetic material in the binucleate parasite *Giardia intestinalis*. *Science* 319:1530–1533.

Prado, F., A. Sheih, J. D. West, and B. Kerr. 2009. Coevolutionary cycling of host sociality and pathogen virulence in contact networks. *Journal of Theoretical Biology* 261:561–569.

Prangishvili, D., P. Forterre, and R. A. Garrett. 2006. Viruses of the Archaea: A unifying view. *Nature Reviews Microbiology* 4:837–848.

Prentis, P., J. Wilson, E. E. Dormontt, D. M. Richardson, and A. J. Lowe. 2008. Adaptive evolution in invasive species. *Trends in Plant Science* 13:288–294.

Presgraves, D. C. 2011. Speciation genetics: Search for the missing snowball. *Current Biology* 20:R1073–R1074.

Prevosti, A., G. Ribo, L. Serra, M. Aguade, J. Balañá, M. Monclus, and F. Mestres. 1988. Colonization of America by *Drosophila subobscura*: Experiment in natural populations that supports the adaptive role of chromosomal-inversion polymorphism. *Proceedings of the National Academy of Sciences USA* 85:5597–5600.

Prevosti, A., L. Serra, C. Segarra, M. Aguade, G. Ribo, and M. Monclus. 1990. Clines of chromosomal arrangements of *Drosophila subobscura* in South America evolve closer to Old World patterns. *Evolution* 44:218–221.

Price, P. W. 1980. *Evolutionary biology of parasites*. Princeton University Press, Princeton, NJ.

———. 2003. *Macroevolutionary theory on macroecological patterns*. Cambridge University Press, Cambridge.

Price, P. W., C. E. Bouton, P. Gross, B. A. McPheron, J. N. Thompson, and A. E. Weis. 1980. Interactions among three trophic levels: Influence of plants on interactions between insect herbivores and natural enemies. *Annual Review of Ecology and Systematics* 11:41–65.

Price, T. 2008. *Speciation in birds*. Roberts and Co., Greenwood Village, CO.

Price, T. D., A. Qvarnstrom, and D. E. Irwin. 2003. The role of phenotypic plasticity in driving genetic evolution. *Proceedings of the Royal Society of London B* 270:1433–1440.

Pringle, E. G., R. I. Adams, E. Broadbent, P. E. Busby, C. I. Donatti, E. L. Kurten, K. Renton, et al. 2010. Distinct leaf-trait syndromes of evergreen and deciduous trees in a seasonally dry tropical forest. *Biotropica* 43:299–308.

Proctor, M., B. McLellan, C. Strobeck, and R. Barclay. 2005. Genetic analysis reveals demographic fragmentation of grizzly bears yielding vulnerably small populations. *Proceedings of the Royal Society B* 272:2409–2416.

Provine, W. B. 1971. *The origins of theoretical population genetics*. University of Chicago Press, Chicago.

Prudic, K. L., C. Jeon, H. Cao, and A. Monteiro. 2011. Developmental plasticity in sexual roles of butterfly species drives mutual sexual ornamentation. *Science* 331:73–75.

Pujol, B., and J. R. Pannell. 2008. Reduced responses to selection after species range expansion. *Science* 321:96.

Pulido, F., and P. Berthold. 2010. Current selection for lower migratory activity will drive the evolution of residency in a migratory bird population. *Proceedings of the National Academy of Sciences USA* 107:7341–7346.

Pureswaran, D. S., B. T. Sullivan, and M. P. Ayres. 2006. Fitness consequences of pheromone production and host selection strategies in a tree-killing bark beetle (Coleoptera: Curculionidae: Scolytinae). *Oecologia* 148:720–728.

Pybus, O. G., and A. Rambaut. 2009. Evolutionary analysis of the dynamics of viral infectious disease. *Nature Reviews Genetics* 10:540–550.

Quek, S.-P., B. A. Counterman, P. Albuquerque de Moura, M. Z. Cardoso, C. R. Marshall, W. O. McMillan, and M. R. Kronforst. 2010. Dissecting comimetic radiations in *Heliconius* reveals divergent histories of convergent butterflies. *Proceedings of the National Academy of Sciences USA* 107:7365–7370.
Queller, D. C., and J. E. Strassmann. 2009. Beyond society: The evolution of organismality. *Philosophical Transactions of the Royal Society B* 364:3143–3155.
Quental, T. B., and C. R. Marshall. 2010. Diversity dynamics: Molecular phylogenies need the fossil record. *Trends in Ecology and Evolution* 25:434–441.
Rabeling, C., O. Gonzales, T. R. Schultz, M. Bacci, M. V. B. Garcia, M. Verhaagh, H. D. Ishak, et al. 2011. Cryptic sexual populations account for genetic diversity and ecological success in a widely distributed, asexual fungus-growing ant. *Proceedings of the National Academy of Sciences USA* 108:12366–12371.
Rabosky, D. L. 2009. Ecological limits and diversification rate: Alternative paradigms to explain the variation in species richness among clades and regions. *Ecology Letters* 12:735–743.
Rabosky, D. L., and I. J. Lovette. 2008. Explosive evolutionary radiations: Decreasing speciation or increasing extinction through time? *Evolution* 62:1866–1875.
Raghoebarsing, A. A., A. Pol, K. T. van de Pas-Schoonen, A. J. Smolders, K. F. Ettwig, W. I. Rijpstra, S. Schouten, et al. 2006. A microbial consortium couples anaerobic methane oxidation to denitrification. *Nature* 440:918–921.
Raguso, R. A. 2008. Wake up and smell the roses: The ecology and evolution of floral scent. *Annual Review of Ecology, Evolution, and Systematics* 39:549–569.
Rainey, P. B., and M. Travisano. 1998. Adaptive radiation in a heterogeneous environment. *Nature* 394:69–72.
Ranson, H., R. N'Guessan, J. Lines, N. Moiroux, Z. Nkuni, and V. Corbel. 2011. Pyrethroid resistance in African anopheline mosquitoes: What are the implications for malaria control? *Trends in Parasitology* 27:91–98.
Rausher, M. D., Y. Lu, and K. Meyer. 2008. Variation in constraint versus positive selection as an explanation for evolutionary rate variation among anthocyanin genes. *Journal of Molecular Evolution* 67:137–144.
Ravensdale, M., A. Nemri, P. H. Thrall, J. G. Ellis, and P. N. Dodds. 2011. Co-evolutionary interactions between host resistance and pathogen effector genes in flax rust disease. *Molecular Plant Pathology* 12:93–102.
Rawls, J. F., M. A. Mahowald, R. E. Ley, and J. I. Gordon. 2006. Reciprocal gut microbiota transplants from zebrafish and mice to germ-free recipients reveal host habitat selection. *Cell* 127:423–433.
Reimchen, T. E. 2000. Predator handling failures of lateral plate morphs in *Gasterosteus aculeatus*: Functional implications for the ancestral plate condition. *Behaviour* 137:1081–1094.
Reiss, J. O. 2009. *Not by design: Retiring Darwin's watchmaker*. University of California Press, San Francisco.

Ren, D., C. C. Labandeira, J. A. Santiago-Blay, A. Rasnitsyn, C. Shih, A. Bashkuev, M. A. V. Logan, et al. 2009. A probable pollination mode before angiosperms: Eurasian, long-proboscid scorpionflies. *Science* 326:840–847.
Restif, O. 2009. Evolutionary epidemiology 20 years on: Challenges and prospects. *Infection, Genetics and Evolution* 9:108–123.
Rezende, E., E. Albert, and M. Fortuna. 2009. Compartments in a marine food web associated with phylogeny, body mass, and habitat structure. *Ecology* 12:779–788.
Rezende, E. L., J. E. Lavabre, P. R. Guimarães, P. Jordano, and J. Bascompte. 2007. Non-random coextinctions in phylogenetically structured mutualistic networks. *Nature* 448:925–928.
Reznick, D. N. 2011. Guppies and the empirical study of adaptation. Pages 205–232 in J. B. Losos, ed., *In the light of evolution: Essays from the laboratory and field*. Roberts and Co., Greenwood Village, CO.
Reznick, D. N., H. Bryga, and J. A. Endler. 1990. Experimental induced life history evolution in a natural population. *Nature* 346:357–359.
Reznick, D. N., and C. K. Ghalambor. 2001. The population ecology of contemporaneous adaptations: What empirical studies reveal about the conditions that promote adaptive radiation. *Genetica* 112–113:183–198.
———. 2005. Selection in nature: Experimental manipulations of natural populations. *Integrative and Comparative Biology* 45:456–462.
Reznick, D. N., C. K. Ghalambor, and K. Crooks. 2008. Experimental studies of evolution in guppies: A model for understanding the evolutionary consequences of predator removal in natural communities. *Molecular Ecology* 17:97–107.
Reznick, D. N., F. H. Shaw, F. H. Rodd, and R. G. Shaw. 1997. Evaluation of the rate of evolution in natural populations of guppies (*Poecilia reticulata*). *Science* 275:1934–1936.
Rhone, B., C. Remoue, N. Galic, I. Goldringer, and I. Bonnin. 2008. Insight into the genetic bases of climatic adaptation in experimentally evolving wheat populations. *Molecular Ecology* 17:930–943.
Rhone, B., R. Vitalis, I. Goldringer, and I. Bonnin. 2010. Evolution of flowering time in experimental wheat populations: A comprehensive approach to detect genetic signatures of natural selection. *Evolution* 64:2110–2125.
Rhymer, J. M., and D. S. Simberloff. 1996. Extinction by hybridization and introgression. *Annual Review of Ecology and Systematics* 27:83–109.
Rice, L. B. 2009. The clinical consequences of antimicrobial resistance. *Current Opinion in Microbiology* 12:476–481.
Rice, W. R., and U. Friberg. 2009. A graphical approach to lineage selection between clonals and sexuals. Pages 75–97 in I. Schön, K. Martens, and P. van Dijk, eds., *Lost sex: The evolutionary biology of parthenogenesis*. Springer, New York.
Rice, W. R., S. Gavrilets, and U. Friberg. 2009. Sexually antagonistic chromosomal cuckoos. *Biology Letters* 5:686–688.

Rich, K. A., J. N. Thompson, and C. C. Fernandez. 2008. Diverse historical processes shape deep phylogeographic divergence in a pollinating seed parasite. *Molecular Ecology* 17:2430–2448.

Richards, T. A, G. Leonard, and D. M. Soanes. 2011. Gene transfer into the fungi. *Fungal Biology Reviews* 25:98–110.

Ricklefs, R. E. 2010. Evolutionary diversification, coevolution between populations and their antagonists, and the filling of niche space. *Proceedings of the National Academy of Sciences USA* 107:1265–1272.

Ricklefs, R., and E. Bermingham. 2008. The West Indies as a laboratory of biogeography and evolution. *Philosophical Transactions of the Royal Society B* 363:2393–2413.

Rico-Gray, V., and P. S. Oliveira. 2007. *The ecology and evolution of ant–plant interactions*. University of Chicago Press, Chicago.

Ridenhour, B. J., E. D. J. Brodie, and E. D. I. Brodie. 2007. Patterns of genetic differentiation in *Thamnophis* and *Taricha* from the Pacific Northwest. *Journal of Biogeography* 34:724–735.

Ridenhour, B. J., and S. N. Nuismer. 2007. Polygenic traits and parasite local adaptation. *Evolution* 61:368–376.

Ridenour, W. M., J. M. Vivanco, Y. Feng, J.-i. Horiuchi, and R. M. Callaway. 2008. No evidence for trade-offs: *Centaurea* plants from America are better competitors and defenders. *Ecological Monographs* 78:369–386.

Rieseberg, L. H., M. A. Archer, and R. K. Wayne. 1999. Transgressive segregation, adaptation and speciation. *Heredity* 83:363–372.

Rieseberg, L. H., O. Raymond, D. M. Rosenthal, Z. Lai, K. Livingstone, T. Nakazato, J. L. Durphy, et al. 2003. Major ecological transitions in wild sunflowers facilitated by hybridization. *Science* 301:1211–1216.

Rintelen, T. v., A. B. Wilson, A. Meyer, and M. Glaubrecht. 2004. Escalation and trophic specialization drive adaptive radiation of freshwater gastropods in ancient lakes on Sulawesi, Indonesia. *Proceedings of the Royal Society B* 271: 2541–2549.

Riska, B. 1981. Morphological variation in the horseshoe crab *Limulus polyphemus*. *Evolution* 35:647–658.

Rissler, L. J., and W. H. Smith. 2010. Mapping amphibian contact zones and phylogeographical break hotspots across the United States. *Molecular Ecology* 19:5404–5416.

Ritchie, M. G. 2007. Sexual selection and speciation. *Annual Review of Ecology, Evolution, and Systematics* 38:79–102.

Robertson, K. A., and A. Monteiro. 2005. Female *Bicyclus anynana* butterflies choose males on the basis of their doral UV-reflective eyespot pupils. *Proceedings of the Royal Society B* 272:1541–1546.

Robinson, J., M. J. Waller, S. C. Fail, H. McWilliam, R. Lopez, P. Parham, and S. G. E. Marsh. 2009. The IMGT/HLA database. *Nucleic Acids Research* 37: D1013–1017.

Rodrigues, A., M. Bacci, U. G. Mueller, A. Ortiz, and F. C. Pagnocca. 2008. Micro-fungal "weeds" in the leafcutter ant symbiosis. *Microbial Ecology* 56: 604–614.

Rodrigues, A., U. G. Mueller, H. D. Ishak, M. Bacci, and F. C. Pagnocca. 2011. Ecology of microfungal communities in gardens of fungus-growing ants (Hymenoptera: Formicidae): A year-long survey of three species of attine ants in Central Texas. *FEMS Microbiology Ecology* 78:244–255.

Rodriguez, R. J., J. F. White, A. E. Arnold, and R. S. Redman. 2009. Fungal endophytes: Diversity and functional roles. *New Phytologist* 182:314–330.

Roeselers, G., E. K. Mittge, W. Z. Stephens, D. M. Parichy, C. M. Cavanaugh, K. Guillemin, and J. F. Rawls. 2011. Evidence for a core gut microbiota in the zebrafish. *ISME Journal* 5:1595–1608.

Rokyta, D. R., C. J. Beisel, P. Joyce, M. T. Ferris, C. L. Burch, and H. A. Wichman. 2008. Beneficial fitness effects are not exponential for two viruses. *Journal of Molecular Evolution* 67:368–376.

Rokyta, D. R., C. Burch, S. B. Caudle, and H. A. Wichman. 2006. Horizontal gene transfer and the evolution of microvirid coliphage genomes. *Journal of Bacteriology* 188:1134–1142.

Rolff, J., and S. E. Reynolds, eds. 2009. *Insect infection and immunity*. Oxford University Press, Oxford.

Rolshausen, G., G. Segelbacher, K. A. Hobson, and H. M. Schaefer. 2009. Contemporary evolution of reproductive isolation and phenotypic divergence in sympatry along a migratory divide. *Current Biology* 19:2097–2101.

Ronce, O., and M. Kirkpatrick. 2001. When sources become sinks: Migrational meltdown in heterogeneous habitats. *Evolution* 55:1520–1531.

Roopnarine, P. D., and K. D. Angielczyk. 2012. The evolutionary palaeoecology of species and the tragedy of the commons. *Biology Letters* 23:147–150.

Rose, M. R. 2005. The effects of evolution are local: Evidence from experimental evolution in *Drosophila*. *Integrative and Comparative Biology* 45:486–491.

Rosindell, J., S. P. Hubbell, and R. S. Etienne. 2011. The unified neutral theory of biodiversity and biogeography at age ten. *Trends in Ecology and Evolution* 26:340–348.

Roslin, T., S. Gripenberg, J.-P. Salminen, M. Karonen, R. O'Hara, K. Pihlaja, and P. Pulkkinen. 2006. Seeing the trees for the leaves—Oaks as mosaics for a host-specific moth. *Oikos* 113:106–120.

Ross, C. L., and R. G. Harrison. 2002. A fine-scale spatial analysis of the mosaic hybrid zone between *Gryllus firmus* and *Gryllus pennsylvanicus*. *Evolution* 56: 2296–2312.

Rosvall, M., I. B. Dodd, S. Krishna, and K. Sneppen. 2006. Network models of phage-bacteria coevolution. *Physical Review E* 74:066105.

Rothstein, S. I., M. A. Patten, and R. C. Fleischer. 2002. Phylogeny, specialization, and brood parasite–host coevolution: Some possible pitfalls of parsimony. *Behavioral Ecology* 13:1–10.

Roux, F., L. Gao, and J. Bergelson. 2010. Impact of initial pathogen density on resistance and tolerance in a polymorphic disease resistance gene system in *Arabidopsis thaliana*. *Genetics* 185:283–291.

Rowland, J. M., and D. J. Emlen. 2009. Two thresholds, three male forms result in facultative male trimorphism in beetles. *Science* 323:773–776.

Roy, K., G. Hunt, D. Jablonski, A. Z. Krug, and J. W. Valentine. 2009. A macroevolutionary perspective on species range limits. *Proceedings of the Royal Society B* 276:1485–1493.

Ruano, F., S. Devers, O. Sanllorente, C. Errard, A. Tinaut, and A. Lenoir. 2011. A geographical mosaic of coevolution in a slave-making host–parasite system. *Journal of Evolutionary Biology* 24:1071–1079.

Rudgers, J. A., M. E. Afkhami, M. A. Rúa, A. J. Davitt, S. Hammer, and V. M. Huguet. 2009. A fungus among us: Broad patterns of endophyte distribution in the grasses. *Ecology* 90:1531–1539.

Rudgers, J. A., and S. Y. Strauss. 2004. A selection mosaic in the facultative mutualism between ants and wild cotton. *Proceedings of the Royal Society Biological Sciences Series B* 271:2481–2488.

Rudwick, M. J. S. 1970. The strategy of Lyell's *Principles of Geology*. *Isis* 61:4–33.

Ruegg, K. 2008. Genetic, morphological, and ecological characterization of a hybrid zone that spans a migratory divide. *Evolution* 62:452–466.

Rundle, H. D. 2002. A test of ecologically dependent postmating isolation between sympatric sticklebacks. *Evolution* 56:322–329.

Rundle, H. D., and M. C. Whitlock. 2001. A genetic interpretation of ecologically dependent isolation. *Evolution* 55:198–201.

Ruse, M. 1993. Evolution and progress. *Trends in Ecology and Evolution* 8:55–59.

Russell, J. E., and R. Stouthamer. 2011. The genetics and evolution of obligate reproductive parasitism in *Trichogramma pretiosum* infected with parthenogenesis-inducing *Wolbachia*. *Heredity* 106:58–67.

Saastamoinen, M., and I. Hanski. 2008. Genotypic and environmental effects on flight activity and oviposition in the Glanville fritillary butterfly. *American Naturalist* 171:701–712.

Sabeti, P. C., P. Varilly, B. Fry, J. Lohmueller, E. Hostetter, C. Cotsapas, X. Xie, et al. 2007. Genome-wide detection and characterization of positive selection in human populations. *Nature* 449:913–918.

Saccheri, I. J., F. Rousset, P. C. Watts, P. M. Brakefield, and L. M. Cook. 2008. Selection and gene flow on a diminishing cline of melanic peppered moths. *Proceedings of the National Academy of Sciences USA* 105:16212–16217.

Saenko, S. V., P. M. Brakefield, and P. Beldade. 2010. Single locus affects embryonic segment polarity and multiple aspects of an adult evolutionary novelty. *BMC Biology* 8:111.

Saikkonen, K., P. R. Wäli, and M. Helander. 2010. Genetic compatibility determines endophyte-grass combinations. *PLoS ONE* 5:e11395.

Sakai, A. K., F. W. Allendorf, J. S. Holt, D. M. Lodge, J. Molofsky, K. A. With, S. Baughman, et al. 2001. The population biology of invasive species. *Annual Review of Ecology and Systematics* 32:305–332.

Salamini, F., H. Özkan, A. Brandolini, R. Schäfer-Pregl, and W. Martin. 2002. Genetics and geography of wild cereal domestication in the Near East. *Nature Reviews Genetics* 3:429–444.

Salathé, M., R. M. May, and S. Bonhoeffer. 2005. The evolution of network topology by selective removal. *Journal of the Royal Society Interface* 2:533–536.

Salzburger, W. 2009. The interaction of sexually and naturally selected traits in the adaptive radiations of cichlid fishes. *Molecular Ecology* 18:169–185.

Sanford, E., and M. W. Kelley. 2011. Local adaptation in marine invertebrates. *Annual Review of Marine Science* 3:509–535.

Sanford, E., M. S. Roth, G. C. Johns, J. P. Wares, and G. N. Somero. 2003. Local selection and latitudinal variation in a marine predator–prey interaction. *Science* 300:1135–1137.

Sanford, E., and D. J. Worth. 2010. Local adaptation along a continuous coastline: Prey recruitment drives differentiation in a predatory snail. *Ecology* 91:891–901.

Sanger, F., G. M. Air, B. G. Barrell, N. L. Brown, A. R. Coulson, C. A. Fiddes, C. A. Hutchinson, et al. 1977. Nucleotide sequence of bacteriophage phi X174 DNA. *Nature* 264:687–695.

Sanjuan, R., A. Moya, and S. F. Elena. 2004. The distribution of fitness effects caused by single-nucleotide substitutions in an RNA virus. *Proceedings of the National Academy of Sciences USA* 101:8396–8401.

Sasu, M. A., K. L. Wall, and A. G. Stephenson. 2010. Antimicrobial nectar inhibits a florally transmitted pathogen of a wild *Cucurbita pepo* (Cucurbitaceae). *American Journal of Botany* 97:1025–1030.

Savile, D. 1975. Evolution and biogeography of Saxifragaceae with guidance from their rust parasites. *Annals of the Missouri Botanical Garden* 62:354–361.

Scanlan, D. J., M. Ostrowski, S. Mazard, A. Dufresne, L. Garczarek, W. R. Hess, A. F. Post, et al. 2009. Ecological genomics of marine picocyanobacteria. *Microbiology and Molecular Biology Reviews* 73:249–299.

Scanlan, P. D., A. R. Hall, L. D. C. Lopez-Pascua, and A. Buckling. 2011. Genetic basis of infectivity evolution in a bacteriophage. *Molecular Ecology* 20:981–989.

Schaedler, M., R. Brandl, and A. Kempel. 2010. Host plant genotype determines bottom-up effects in an aphid-parasitoid-predator system. *Entomologia Experimentalis et Applicata* 135:162–169.

Schardl, C. L., and K. D. Craven. 2003. Interspecific hybridization in plant-associated fungi and oomycetes: A review. *Molecular Ecology* 12:2861–2873.

Schatz, B., A. Geoffroy, B. Dainat, J.-M. Bessière, B. Buatois, M. Hossaert-McKey, and M.-A. Selosse. 2010. A case study of modified interactions with symbionts in a hybrid Mediterranean orchid. *American Journal of Botany* 97:1278–1288.

Scheiner, S. M. 2002. Selection experiments and the study of phenotypic plasticity. *Journal of Evolutionary Biology* 15:889–898.
Scheiner, S. M., and R. F. Lyman. 1991. The genetics of phenotypic plasticity. II. Response to selection. *Journal of Evolutionary Biology* 4:23–50.
Schemske, D. W. 2010. Adaptation and the origin of species. *American Naturalist* 176 (suppl. 1): S4–S25.
Schemske, D. W., and P. Bierzychudek. 2001. Evolution of flower color in the desert annual *Linanthus parryae*: Wright revisited. *Evolution* 55:1269–1282.
Schemske, D. W., and H. D. Bradshaw, Jr. 1999. Pollinator preference and the evolution of floral traits in monkeyflowers (*Mimulus*). *Proceedings of the National Academy of Sciences USA* 96:11910–11915.
Schiestl, F. P. 2010. The evolution of floral sent and insect chemical communication. *Ecology Letters* 13:643–656.
Schluter, D. 1988. Estimating the form of natural selection on a quantitative trait. *Evolution* 42:849–861.
———. 1994. Experimental evidence that competition promotes divergence in adaptive radiation. *Science* 266:798–801.
———. 1996. Ecological speciation in postglacial fishes. *Philosophical Transactions of the Royal Society of London B* 351:807–814.
———. 2000. *The ecology of adaptive radiation*. Oxford University Press, Oxford.
———. 2009. Evidence for ecological speciation and its alternative. *Science* 323:737–741.
Schluter, D., and G. L. Conte. 2009. Genetics and ecological speciation. *Proceedings of the National Academy of Sciences of the United States of America* 106:9955–9962.
Schluter, D., K. B. Marchinko, R. D. H. Barrett, and S. M. Rogers. 2010. Natural selection and the genetics of adaptation in threespine stickleback. *Philosophical Transactions of the Royal Society B* 365:2479–2486.
Schmid-Hempel, P. 2008. *Evolutionary parasitology: The integrated study of infections, immunology, ecology, and genetics*. Oxford University Press, Oxford.
Schmitz, O. J. 2010. *Resolving ecosystem complexity*. Princeton University Press, Princeton, NJ.
Schneider, D. J., and A. Collmer. 2010. Studying plant–pathogen interactions in the genomics era: Beyond molecular Koch's postulates to systems biology. *Annual Review of Phytopathology* 48:457–479.
Schnitzler, J., T. G. Barraclough, J. S. Boatwright, P. Goldblatt, J. C. Manning, M. P. Powell, T. Rebelo, et al. 2011. Causes of plant diversification in the cape biodiversity hotspot of South Africa. *Systematic Biology* 60:343–357.
Schoebel, C. N., C. Tellenbach, P. Spaak, and J. Wolinska. 2011. Temperature effects on parasite prevalence in a natural hybrid complex. *Biology Letters* 7:108–111.
Schoener, T. W. 2011. The newest synthesis: Understanding the interplay of evolutionary and ecological dynamics. *Science* 331:426–429.

Scholl, E., J. Thorne, and J. McCarter. 2003. Horizontally transferred genes in plant-parasitic nematodes: A high-throughput genomic approach. *Genome Biology* 4:R39.

Schön, I., G. Rossetti, and K. Martens. 2009. Darwinulid ostracods: Ancient asexual scandals or scandalous gossip? Pages 217–240 in I. Schön, K. Martens, and P. van Dijk, eds., *Lost sex: The evolutionary biology of parthenogensis*. Springer, New York.

Schulte, R. D., C. Makus, B. Hasert, N. K. Michiels, and H. Schulenburg. 2010. Multiple reciprocal adaptations and rapid genetic change upon experimental coevolution of an animal host and its microbial parasite. *Proceedings of the National Academy of Sciences USA* 107:7359–7364.

Schurko, A. M., M. Neiman, and J. M. Logsdon, Jr. 2009. Signs of sex: What we know and how we know it. *Trends in Ecology and Evolution* 24:208–217.

Schwander, T., L. Henry, and B. J. Crespi. 2011. Molecular evidence for ancient asexuality in *Timema* stick insects. *Current Biology* 21:1129–1134.

Schweitzer, J. A., M. D. Madritch, J. K. Bailey, C. J. LeRoy, D. G. Fischer, B. J. Rehill, R. L. Lindroth, et al. 2008. From genes to ecosystems: The genetic basis of condensed tannins and their role in nutrient regulation in a *Populus* model system. *Ecosystems* 11:1005–1020.

Scott, J. J., D.-C. Oh, M. C. Yuceer, K. D. Klepzig, J. Clardy, and D. R. Currie. 2008. Bacterial protection of beetle–fungus mutualism. *Science* 322:63.

Scriber, J. M. 2011. Impacts of climate warming on hybrid zone movement: Geographically diffuse and biologically porous "species borders." *Insect Science* 18:121–159.

Scriber, J. M., G. J. Ording, and R. J. Mercader. 2008. Introgression and parapatric speciation in a hybrid zone. Pages 69–87 in K. J. Tilmon, ed., *Specialization, speciation, and radiation: The evolutionary biology of herbivorous insects*. University of California Press, Berkeley.

Seehausen, O. 2004. Hybridization and adaptive radiation. *Trends in Ecology and Evolution* 19:198–207.

Segraves, K. A. 2010. Branching out with coevolutionary trees. *Evolution: Education and Outreach* 3:62–70.

Segraves, K. A., D. M. Althoff, and O. Pellmyr. 2005. Limiting cheaters in mutualism: Evidence from hybridization between mutualist and cheater yucca moths. *Proceedings of the Royal Society of London B* 272:2195–2201.

Segraves, K. A., and O. Pellmyr. 2004. Testing the out-of-Florida hypothesis on the origin of cheating in the yucca-yucca moth mutualism. *Evolution* 58:2266–2279.

Segraves, K. A., and J. N. Thompson. 1999. Plant polyploidy and pollination: Floral traits and insect visits to diploid and tetraploid *Heuchera grossulariifolia*. *Evolution* 53:1114–1127.

Segraves, K. A., J. N. Thompson, P. S. Soltis, and D. E. Soltis. 1999. Multiple origins of polyploidy and the geographic structure of *Heuchera grossulariifolia*. *Molecular Ecology* 8:253–262.

Serbus, L. R., C. Casper-Lindley, F. Landmann, and W. Sullivan. 2008. The genetics and cell biology of *Wolbachia*–host interactions. *Annual Review of Genetics* 42:683–707.

Shapiro, B., A. Prieto, H. F. Marín, B. A. González, M. T. P. Gilbert, and E. Willersiev. 2009. The Late Pleistocene distribution of vicuñas (*Vicugna vicugna*) and the "extinction" of the gracile llama ("*Lama gracilis*"): New molecular data. *Quaternary Science Reviews* 28:1369–1373.

Sharon, I., A. Alperovitch, F. Rohwer, M. Haynes, F. Glaser, N. Atamna-Ismaeel, R. Y. Pinter, et al. 2009. Photosystem I gene cassettes are present in marine virus genomes. *Nature* 461:258–262.

Shaw, R. G., and C. J. Geyer. 2010. Inferring fitness landscapes. *Evolution* 64: 2510–2520.

Shaw, R. G., C. J. Geyer, S. Wagenius, H. H. Hangelbroek, and J. R. Etterson. 2008. Unifying life-history analyses for inference of fitness and population growth. 172:E35–E47.

Shefferson, R. P., D. L. Taylor, M. Weiss, S. Garnica, M. K. McCormick, S. Adams, H. M. Gray, et al. 2007. The evolutionary history of mycorrhizal specificity among lady slipper's orchids. *Evolution* 61:1380–1390.

Sher, D., J. W. Thompson, N. Kashtan, L. Croal, and S. W. Chisholm. 2011. Response of *Prochlorococcus* ecotypes to co-culture with diverse marine bacteria. *ISME Journal* 5:1125–1132.

Sherborne, A. L., M. D. Thom, S. Paterson, F. Jury, W. E. R. Ollier, P. Stockley, R. J. Beynon, et al. 2007. The genetic basis of inbreeding avoidance in house mice. *Current Biology* 17:2061–2066.

Sherman, C. D. H., and D. J. Ayre. 2008. Fine-scale adaptation in a clonal sea anemone. *Evolution* 62:1373–1380.

Shields, J. L., J. W. Heath, and D. D. Heath. 2010. Marine landscape shapes hybrid zone in a broadcast spawning bivalve: Introgression and genetic structure in Canadian west coast Mytilus. *Marine Ecology Progress Series* 399:211–223.

Shizuka, D., and B. E. Lyon. 2011. Hosts improve the reliability of chick recognition by delaying the hatching of brood parasitic eggs. *Current Biology* 21:515–519.

Shuster, S. 2011. Differences in relative fitness among alternative mating tactics might be more apparent than real. *Journal of Animal Ecology* 80:905–907.

Shuster, S. M., and M. J. Wade. 1991. Equal mating success among male reproductive strategies in a marine isopod. *Nature* 350:608–610.

———. 2003. *Mating systems and strategies*. Princeton University Press, NJ.

Sibly, M. W., C. Winstanley, S. A. C. Godfrey, S. B. Levy, and R. W. Jackson. 2011. *Pseudomonas* genomes: Diverse and adaptable. *FEMS Microbiology Reviews* 35:652–680.

Sicard, D., and J.-L. Legras. 2011. Bread, beer and wine: Yeast domestication in the *Saccharomyces sensu* stricto complex. *Comptes Rendus Biologies* 334:229–236.

Sielezniew, M., M. Wlostowski, and I. Dziekanska. 2010. *Myrmica schencki* (Hymenoptera: Formicidae) as the primary host of *Phengaris* (*Maculinea*) *arion* (Lepidoptera: Lycaenidae) at heathlands in eastern Poland. *Sociobiology* 55:95–106.

Siepielski, A. M., and C. W. Benkman. 2004. Interactions among moths, crossbills, squirrels, and lodgepole pine in a geographic selection mosaic. *Evolution* 58:95–101.

———. 2007. Convergent patterns in the selection mosaic for two North American bird-dispersed pines. *Ecological Monographs* 77:203–220.

———. 2010. Conflicting selection from an antagonist and a mutualist enhances phenotypic variation in a plant. *Evolution* 64:1120–1128.

Siepielski, A. M., J. D. DiBattista, and S. M. Carlson. 2009. It's about time: The temporal dynamics of phenotypic selection in the wild. *Ecology Letters* 12:1261–1276.

Siepielski, A. M., J. D. DiBattista, J. A. Evans, and S. M. Carlson. 2011a. Differences in the temporal dynamics of phenotypic selection among fitness components in the wild. *Proceedings of the Royal Society B* 278:1572–1580.

Siepielski, A. M., K.-L. Hung, E. E. B. Bein, and M. A. McPeek. 2010. Experimental evidence for neutral community dynamics governing an insect assemblage. *Ecology* 91:847–857.

Siepielski, A. M., A. N. Mertens, B. L. Wilkinson, and M. A. McPeek. 2011b. Signature of ecological partitioning in the maintenance of damselfly diversity. *Journal of Animal Ecology* 80:1163–1173.

Silander, O. K., D. Weinreich, K. M. Wright, K. J. O'Keefe, C. U. Rang, P. E. Turner, and L. Chao. 2005. Widespread genetic exchange among terrestrial bacteriophages. *Proceedings of the National Academy of Sciences USA* 102:19009–19014.

Silby, M. W., A. M. Cerdeno-Tarraga, G. S. Vernikos, S. R. Giddens, R. W. Jackson, G. M. Preston, X.-X. Zhang, et al. 2009. Genomic and genetic analyses of diversity and plant interactions of *Pseudomonas fluorescens*. *Genome Biology* 10:R51.

Silvertown, J., L. Cook, R. Cameron, M. Dodd, K. McConway, J. Worthington, P. Skelton, et al. 2011. Citizen science reveals unexpected continental-scale evolutionary change in a model organism. *PLoS ONE* 6:e18927.

Simms, E. L., and D. L. Taylor. 2002. Partner choice in nitrogen-fixation mutualisms of legumes and rhizobia. *Integrative and Comparative Biology* 42:369–380.

Simões, P., M. R. Rose, A. Duarte, R. Gonçalves, and M. Matos. 2006. Evolutionary domestication in *Drosophila subobscura*. *Journal of Evolutionary Biology* 20:758–766.

Simpson, G. G. 1944. *Tempo and mode in evolution*. Columbia University Press, New York.

———. 1953. *The major features of evolution*. Columbia University Press, New York.

Sinervo, B., and R. Calsbeek. 2006. The developmental, physiological, neural, and genetical causes and consequences of frequency-dependent selection in the wild. *Annual Review of Ecology, Evolution, and Systematics* 37:581–610.

Sinervo, B., and C. M. Lively. 1996. The rock-paper-scissors game and the evolution of alternative male strategies. *Nature* 380:240–243.

Sinervo, B., E. Svensson, and T. Comendant. 2000. Density cycles and an offspring quantity and quality game driven by natural selection. *Nature* 406:14427–14432.

Singer, M. C., C. D. Thomas, and C. Parmesan. 1993. Rapid human-induced evolution of insect–host associations. *Nature* 366:681–683.

Singer, M. C., B. Wee, S. Hawkins, and M. Butcher. 2008. Rapid natural and anthropogenic diet evolution: Three examples from checkerspot butterflies. Pages 311–324 in K. J. Tilmon, ed., *Specialization, speciation, and radiation: The evolutionary biology of herbivorous insects*. University of California Press, Berkeley.

Sinsheimer, R. L. 1959. A single-stranded deoxyribonucleic acid from bacteriophage φX174. *Journal of Molecular Biology* 1:43–53.

Siol, M., S. I. Wright, and S. C. H. Barrett. 2010. The population genomics of plant adaptation. *New Phytologist* 188:313–332.

Six, D. L., and M. J. Wingfield. 2011. The role of phytopathogenicity in bark beetle–fungus symbioses: A challenge to the classic paradigm. *Annual Review of Entomology* 56:255–272.

Slack, E., S. Hapfelmeier, B. Stecher, Y. Velykoredko, M. Stoel, M. A. E. Lawson, M. B. Geuking, et al. 2009. Innate and adaptive immunity cooperate flexibly to maintain host-microbiota mutualism. *Science* 325:617–620.

Slattery, M., H. N. Kamel, S. Ankisetty, D. J. Gochfeld, C. A. Hoover, and R. W. Thacker. 2008. Hybrid vigor in a tropical Pacific soft-coral community. *Ecological Monographs* 78:423–443.

Sletvold, N., and J. Ågren. 2010. Pollinator-mediated selection on floral display and spur length in the orchid *Gymnadenia conopsea*. *International Journal of Plant Sciences* 171:999–1009.

Slove, J., and N. Janz. 2011. The relationship between diet breadth and geographic range size in the butterfly subfamily Nymphalinae: A study of global scale. *PLoS ONE* 6:e16057.

Smith, C. C. 1970. The coevolution of pine squirrels (*Tamiasciurus*) and conifers. *Ecological Monographs* 40:349–371.

Smith, C. I., C. S. Drummond, W. Godsoe, J. B. Yoder, and O. Pellmyr. 2009. Host specificity and reproductive success of yucca moths (*Tegeticula* spp. Lepidoptera: Prodoxidae) mirror patterns of gene flow between host plant varieties of the Joshua tree (*Yucca brevifolia*: Agavaceae). *Molecular Ecology* 18:5218–5229.

Smith, C. I., O. Pellmyr, D. M. Althoff, M. Balcázar-Lara, J. Leebens-Mack, and K. A. Segraves. 2008. Pattern and timing of diversification in *Yucca* (Agavaceae): Specialized pollination does not escalate rates of diversification. *Proceedings of the Royal Society B* 275:249–258.

Smith, D. C., and A. E. Douglas. 1987. *The biology of symbiosis*. Edward Arnold, London.
Smith, D. S., J. K. Bailey, S. M. Shuster, and T. G. Whitham. 2011. A geographic mosaic of trophic interactions and selection: Trees, aphids and birds. *Journal of Evolutionary Biology* 24:422–429.
Smith, J. W., and C. W. Benkman. 2007. A coevolutionary arms race causes ecological speciation in crossbills. *American Naturalist* 269:455–465.
Snell-Rood, E. C., J. D. Van Dyken, T. Cruickshank, M. J. Wade, and A. P. Moczek. 2010. Toward a population genetic framework of developmental evolution: The costs, limits, and consequences of phenotypic plasticity. *BioEssays* 32:71–81.
Sniegowski, P. D., P. J. Gerrish, and R. E. Lenski. 1997. Evolution of high mutation rates in experimental populations of *Escherichia coli*. *Nature* 387:703–705.
Snyder, J. C., and M. J. Young. 2011. Advances in understanding archaea–virus interactions in controlled and natural environments. *Current Opinion in Microbiology* 14:497–503.
Snyder, M. A., and Y. B. Linhart. 1998. Subspecific selectivity by a mammalian herbivore: Geographic differentiation of interactions between two taxa of *Sciurus aberti* and *Pinus ponderosa*. *Evolutionary Ecology* 12:755–765.
Sobel, J. M., G. F. Chen, L. R. Watt, and D. W. Schemske. 2010. The biology of speciation. *Evolution* 64:295–315.
Soderberg, R. J., and O. G. Berg. 2011. Kick-starting the ratchet: The fate of mutators in an asexual population. *Genetics* 187:1129–1137.
Soler, J. J., D. Martín-Gálvez, J. G. Martínez, M. Soler, D. Canestrari, J. M. Abad-Gómez, and A. P. Møller. 2011. Evolution of tolerance by magpies to brood parasitism by great spotted cuckoos. *Proceedings of the Royal Society B* 278:2047–2052.
Soler, M., C. Ruiz-Castellano, M. del Carmen Fernandez-Pinos, A. Rosler, J. Ontanilla, and T. Perez-Contreras. 2011. House sparrows selectively eject parasitic conspecific eggs and incur very low rejection costs. *Behavioral Ecology and Sociobiology* 65:1997–2005.
Soltis, D. E. 1986. Intergeneric hybridization between *Conimitella williamsii* and *Mitella stauropetala* (Saxifragaceae). *Systematic Botany* 11:293–297.
———. 1988. Karyotypes of *Bensoniella*, *Conimitella*, *Lithophragma*, and *Mitella*, and relationships in Saxifrageae (Saxifragaceae). *Systematic Botany* 13:64–72.
———. 2007. Saxifragaceae. Pages 418–435 in K. Kubitzki, ed., *The families and genera of vascular plants*. Springer, Berlin, Heidelberg.
Soltis, D. E., and B. A. Boehm. 1984a. Chromosomal and flavonoid chemical confirmation of intergeneric hybridization between *Tolmiea* and *Tellima* (Saxifragaceae). *Canadian Journal of Botany* 63:1309–1312.
———. 1984b. Karyology and flavonoid chemistry of the disjunct species of *Tiarella* (Saxifragaceae). *Systematic Botany* 9:441–447.

Soltis, D. E., and L. Hufford. 2002. Ovary position diversity in Saxifragaceae: Clarifying the homology of epigyny. *International Journal of Plant Sciences* 163:277–293.

Soltis, D., and R. Kuzoff. 1995. Discordance between nuclear and chloroplast phylogenies in the *Heuchera* group (Saxifragaceae). *Evolution* 49:727–742.

Soltis, D., R. Kuzoff, M. Mort, M. Zanis, M. Fishbein, L. Hufford, J. Koontz, et al. 2001. Elucidating deep-level phylogenetic relationships in Saxifragaceae using sequences for six chloroplastic and nuclear DNA regions. *Annals of the Missouri Botanical Garden* 88:669–693.

Soltis, D. E., D. R. Morgan, A. Grable, P. S. Soltis, and R. Kuzoff. 1993. Molecular systematics of *Saxifragaceae sensu stricto*. *American Journal of Botany* 80:1056–1081.

Soltis, D. E., and L. H. Rieseberg. 1986. Autopolyploidy in *Tolmiea menziesii* (Saxifragaceae): Genetic insights from enzyme electrophoresis. *American Journal of Botany* 73:310–318.

Soltis, D. E., and P. S. Soltis. 1989. Genetic consequences of autopolyploidy in *Tolmiea* (Saxifragaceae). *Evolution* 43:586–594.

———. 1999. Polyploidy: Recurrent formation and genome evolution. *Trends in Ecology and Evolution* 14:348–352.

Soltis, D. E, P. S. Soltis, T. G. Collier, and M. L. Edgerton. 1991. Chloroplast DNA variation within and among genera of the *Heuchera* group (Saxifragaceae): Evidence for chloroplast transfer and paraphyly. *American Journal of Botany* 78:1091–1112.

Soltis, D. E., P. S. Soltis, P. K. Endress, and M. W. Chase. 2005. *Phylogeny and evolution of angiosperms*. Sinauer Associates, Sunderland, MA.

Soltis, D. E., P. S. Soltis, J. C. Pires, A. Kovarik, J. A. Tate, and E. Mavrodiev. 2004. Recent and recurrent polyploidy in *Tragopogon* (Asteraceae): Cytogenetic, genomic and genetic comparisons. *Biological Journal of the Linnean Society* 82:485–501.

Soltis, D. E., P. S. Soltis, D. W. Schemske, J. F. Hancock, J. N. Thompson, B. C. Husband, and W. S. Judd. 2007. Autopolyploidy in angiosperms: Have we grossly underestimated the number of species? *Taxon* 56:13–30.

Soltis, P. S., G. M. Plunkett, S. J. Novak, and D. E. Soltis. 1995. Genetic variation in *Tragopogon* species: Additional origins of the allotetraploids *T. mirus* and *T. miscellus* (Compositae). *American Journal of Botany* 82:1329–1341.

Soltis, P. S., and D. E. Soltis. 2009. The role of hybridization in plant speciation. *Annual Review of Plant Biology* 60:561–588.

Sorenson, M., and R. Payne. 2002. Molecular genetic perspectives on avian brood parasitism. *Integrative and Comparative Biology* 42:388–400.

Sotka, E. E. 2005. Local adaptation in host use in marine invertebrates. *Ecology Letters* 8:448–459.

Sotka, E. E., J. P. Wares, and M. E. Hay. 2003. Geographic and genetic variation in feeding preferences for chemically defended seaweeds. *Evolution* 57:2262–2276.

Spanu, P. D., J. C. Abbott, J. Amselem, T. A. Burgis, D. M. Soanes, K. Stüber, E. Ver Loren van Themaat, et al. 2010. Genome expansion and gene loss in powdery mildew fungi reveal tradeoffs in extreme parasitism. *Science* 330:1543–1546.

Städler, T., and L. F. Delph. 2002. Ancient mitochondrial haplotypes and evidence for intragenic recombination in a gynodioecious plant. *Proceedings of the National Academy of Sciences USA* 99:11730–11735.

Stanek, M. T., T. F. Cooper, and R. E. Lenski. 2009. Identification and dynamics of a beneficial mutation in a long-term evolution experiment with *Escherichia coli*. *BMC Evolutionary Biology* 9:302.

Stanton, M. L., and T. M. Palmer. 2011. The high cost of mutualism: Effects of four species of East African ant symbionts on their myrmecophyte host tree. *Ecology* 92:1073–1082.

Stebbins, G. L. 1950. *Variation and evolution in plants*. Columbia University Press, New York.

Steele, J. A., P. D. Countway, L. Xia, P. D. Vigil, J. M. Beman, D. Y. Kim, C.-E. T. Chow, et al. 2011. Marine bacterial, archaeal and protistan association networks reveal ecological linkages. *ISME Journal* 5:1414–1425.

Stemshorn, K. C., F. A. Reed, A. W. Nolte, and D. Tautz. 2011. Rapid formation of distinct hybrid lineages after secondary contact of two fish species (*Cottus* sp.). *Molecular Ecology* 20:1475–1491.

Stenberg, P., and A. Saura. 2009. Cytology of asexual animals. Pages 63–74 in I. Schön, K. Martens, and P. van Dijk, eds., *Lost sex: The evolutionary biology of parthenogenesis*. Springer, New York.

Stewart, C. R., S. R. Casjens, S. G. Cresawn, J. M. Houtz, A. L. Smith, M. E. Ford, C. L. Peebles, et al. 2009. The genome of *Bacillus subtilis* bacteriophage SPO1. *Journal of Molecular Biology* 388:48–70.

Stinchcombe, J. R. 2005. Measuring natural selection on proportional traits: Comparisons of three types of selection estimates for resistance and susceptibility to herbivore damage. *Evolutionary Ecology* 19:363–373.

Stinchcombe, J. R., and H. E. Hoekstra. 2007. Combining population genomics and quantitative genetics: Finding the genes underlying ecologically important traits. *Heredity* 100:158–170.

Stinchcombe, J. R., and M. D. Rausher. 2002. The evolution of tolerance to deer herbivory: Modifications caused by the abundance of insect herbivores. *Proceedings of the Royal Society of London B* 269:1241–1246.

Stinchcombe, J. R., C. Weinig, K. D. Heath, M. T. Brock, and J. Schmitt. 2009. Polymorphic genes of major effect: Consequences for variation, selection and evolution in *Arabidopsis thaliana*. *Genetics* 182:911–922.

Stoks, R., and M. A. McPeek. 2006. A tale of two diversifications. *American Naturalist* 168:S50–S72.

Stoletzki, N., and A. Eyre-Walker. 2011. The positive correlation between dN/dS and dS in mammals is due to runs of adjacent substitutions. *Molecular Biology and Evolution* 28:1371–1380.

Stouthamer, R., J. E. Russell, F. Vavre, and L. Nunney. 2010. Intragenomic conflict in populations infected by parthenogenesis inducing *Wolbachia* ends with irreversible loss of sexual reproduction. *BMC Evolutionary Biology* 10:229.

Strassmann, J. E., O. M. Gilbert, and D. C. Queller. 2011. Kin discrimination and cooperation in microbes. *Annual Review of Microbiology* 65:349–367.

Strauss, S. Y., J. A. Lau, T. W. Schoener, and P. Tiffin. 2008. Evolution in ecological field experiments: Implications for effect size. *Ecology Letters* 11:199–207.

Strauss, S. Y., H. Sahli, and J. K. Conner. 2005. Toward a more trait-centered approach to diffuse (co)evolution. *New Phytologist* 165:81–90.

Studier, F. W., P. Daegelen, R. E. Lenski, S. Maslov, and J. F. Kim. 2009. Understanding the differences between genome sequences of *Escherichia coli* B strains REL606 and BL21(DE3) and comparison of the *E. coli* B and K-12 genomes. *Journal of Molecular Biology* 394:653–680.

Sturmbauer, C., W. Salzburger, N. Duftner, R. Schelly, and S. Koblmueller. 2010. Evolutionary history of the Lake Tanganyika cichlid tribe Lamprologini (Teleostei: Perciformes) derived from mitochondrial and nuclear DNA data. *Molecular Phylogenetics and Evolution* 57:266–284.

Suarez, A. V., and N. D. Tsutsui. 2008. The evolutionary consequences of biological invasions. *Molecular Ecology* 17:351–360.

Summers, R. W., R. J. G. Dawson, and R. E. Phillips. 2007. Assortative mating and patterns of inheritance indicate that the three crossbill taxa in Scotland are species. *Journal of Avian Biology* 38:153–162.

———. 2010. Temporal variation in breeding and cone size selection by three species of crossbills *Loxia* spp. in a native Scots pinewood. *Journal of Avian Biology* 41:219–228.

Sutherland, C. J., H. Babiker, M. J. Mackinnon, L. Ranford-Cartwright, and B. B. E. Sayed. 2011. Rational deployment of antimalarial drugs in Africa: Should first-line combination drugs be reserved for paediatric malaria cases? *Parasitology* 138:1459–1468.

Suttle, C. A. 2005. Viruses in the sea. *Nature* 437:356–361.

Suzuki, Y., and M. Nei. 2002. Origin and evolution of influenza virus hemagglutin genes. *Molecular Biology and Evolution* 19:501–509.

Svara, F., and D. J. Rankin. 2011. The evolution of plasmid-carried antibiotic resistance. *BMC Evolutionary Biology* 11:130.

Svensson, E. I., J. Abbott, and R. Hardling. 2005. Female polymorphism, frequency dependence, and rapid evolutionary dynamics in natural populations. *American Naturalist* 165:567–576.

Svensson, G. P, M. O. Hickman, Jr., S. Bartram, W. Boland, O. Pellmyr, and R. A. Raguso. 2005. Chemistry and geographic variation of floral scent in *Yucca filamentosa* (Agavaceae). *American Journal of Botany* 92:1624.

Svensson, G. P., O. Pellmyr, and R. A. Raguso. 2011. Pollinator attraction to volatiles from virgin and pollinated host flowers in a yucca/moth obligate mutualism. *Oikos* 120: 577–1583.

Swenson, N. G. 2010. Mapping the suturing of a continental biota. *Molecular Ecology* 19:5324–5327.

Swenson, N. G., and D. J. Howard. 2005. Clustering of contact zones, hybrid zones, and phylogeographic breaks in North America. *American Naturalist* 166:581–591.

Symonds, V. V., P. S. Soltis, and D. E. Soltis. 2010. Dynamics of polyploid formation in *Trapopogon* (Asteraceae): Recurrent formation, gene flow, and population structure. *Evolution* 64:1984–2003.

Tabashnik, B. E., A. J. Gassmann, D. W. Crowder, and Y. Carriére. 2008. Insect resistance to *Bt* crops: Evidence versus theory. *Nature Biotechnology* 26:199–202.

Tabata, J., Y. Hattori, H. Sakamoto, F. Yukuhiro, T. Fujii, S. Kugimiya, A. Mochizuki, et al. 2011. Male killing and incomplete inheritance of a novel spiroplasma in the moth *Ostrinia zaguliaevi*. *Microbial Ecology* 61:254–263.

Tack, A. J. M., and T. Roslin. 2010. Overrun by the neighbors: Landscape context affects strength and sign of local adaptation. *Ecology* 91:2253–2260.

Takiya, D. M., P. L. Tran, C. H. Dietrich, and N. A. Moran. 2006. Co-cladogenesis spanning three phyla: Leafhoppers (Insects: Hemiptera: Cicadellidae) and their dual bacteria symbionts. *Molecular Ecology* 15:4175–4191.

Tamas, I., L. Klasson, B. Canbäck, A. K. Näslund, A.-S. Eriksson, J. J. Wernegreen, J. P. Sandström, et al. 2002. 50 million years of genomic stasis in endosymbiotic bacteria. *Science* 296:2376–2379.

Tanaka, S., T. Nishida, and N. Ohsaki. 2007. Sequential rapid adaptation of indigenous parasitoid wasps to the invasive butterfly *Pieris brassicae*. *Evolution* 61:1791–1802.

Tanksley, S. D. 1993. Mapping polygenes. *Annual Review of Genetics* 27:205–233.

Tatum, E. L. 1945. X-ray induced mutant strains of *Escherichia coli*. *Proceedings of the National Academy of Sciences USA* 31:215–219.

Taylor, L. H., S. M. Latham, and M. E. J. Woolhouse. 2001. Risk factors for human disease emergence. *Philosophical Transactions of the Royal Society B* 356:983–989.

Taylor, R. L. 1965. The genus *Lithophragma* (Saxifragaceae). *University of California Publications in Botany* 37:1–89.

Tedersoo, L., T. W. May, and M. E. Smith. 2010. Ectomycorrhizal lifestyle in fungi: Global diversity, distribution, and evolution of phylogenetic lineages. *Mycorrhiza* 20:217–263.

Teixeira, L., A. Ferreira, and M. Ashburner. 2008. The bacterial symbiont *Wolbachia* induces resistance to RNA viral infections in *Drosophila melanogaster*. *PLoS Biology* 6:e2.

Teixeira, S., K. Foerster, and G. Bernasconi. 2009. Evidence for inbreeding depression and post-pollination selection against inbreeding in the dioecious plant *Silene latifolia*. *Heredity* 102:101–112.

Tellier, A., and J. K. M. Brown. 2007. Polymorphism in multilocus host parasite coevolutionary interactions. *Genetics* 177:1777–1790.

Temeles, E. J., and W. J. Kress. 2003. Adaptation in a plant–hummingbird association. *Science* 300:630–633.
———. 2010. Mate choice and mate competition by a tropical hummingbird at a floral resource. *Proceedings of the Royal Society B* 277:1607–1613.
Temeles, E. J., C. R. Koulouris, S. E. Sander, and W. J. Kress. 2009. Effect of flower shape and size on foraging performance and trade-offs in a tropical hummingbird. *Ecology* 90:1147–1161.
Teotónio, H., I. M. Chelo, M. Bradić, M. R. Rose, and A. D. Long. 2009. Experimental evolution reveals natural selection on standing genetic variation. *Nature Genetics* 41:251–257.
Texier, C., C. Vidau, B. Viguès, H. El Alaoui, and F. Delbac. 2010. Microsporidia: A model for minimal parasite–host interactions. *Current Opinion in Microbiology* 13:443–449.
Thayer, C. W. 1979. Biological bulldozers and the evolution of marine benthic communities. *Science* 203:458–461.
Thébault, E., and C. Fontaine. 2008. Does asymmetric specialization differ between mutualistic and trophic networks? *Oikos* 117:555–563.
———. 2010. Stability of ecological communities and the architecture of mutualistic and trophic networks. Science 329:853–856.
Thomas, J. A., D. J. Simcox, and R. T. Clarke. 2009. Successful conservation of a threatened *Maculinea* butterfly. *Science* 325:80–83.
Thompson, J. D., and R. Lumaret. 1992. The evolutionary dynamics of polyploid plants: Origins, establishment and persistence. *Trends in Ecology and Evolution* 7:302–307.
Thompson, J. N. 1978. Within-patch structure and dynamics in *Pastinaca sativa* and resource availability to a specialized herbivore. *Ecology* 59:443–448.
———. 1982. *Interaction and coevolution*. John Wiley and Sons, New York.
———. 1987. Symbiont-induced speciation. *Biological Journal of the Linnean Society* 32:385–393.
———. 1988a. Variation in preference and specificity in monophagous and oligophagous swallowtail butterflies. *Evolution* 42:118–128.
———. 1988b. Evolutionary genetics of oviposition preference in swallowtail butterflies. *Evolution* 42:1223–1234.
———. 1993. Preference hierarchies and the origin of geographic specialization in host use in swallowtail butterflies. *Evolution* 47:1585–1594.
———. 1994. *The coevolutionary process*. University of Chicago Press, Chicago.
———. 1995. The origins of host shifts in swallowtail butterflies versus other insects. Pages 195–203 in J. M. Scriber, Y. Tsubaki, and R. C. Lederhouse, eds., *Swallowtail Butterflies: Their Ecology and Evolutionary Biology*. Scientific Publishers, Gainesville, FL.
———. 1997. Evaluating the dynamics of coevolution among geographically structured populations. *Ecology* 78:1619–1623.

———. 1998. Rapid evolution as an ecological process. *Trends in Ecology and Evolution* 13:329–332.

———. 2005. *The geographic mosaic of coevolution*. University of Chicago Press, Chicago.

———. 2008a. Coevolution, cryptic speciation, and the persistence of plant–insect interactions. Pages 216–224 in K. J. Tilmon, ed., *Specialization, speciation, and radiation: The evolutionary biology of herbivorous insects*. University of California Press, Berkeley.

———. 2008b. Use the word evolution. *Evolution: Outreach and Education* 1:42–43.

———. 2009a. The coevolving web of life. *American Naturalist* 173:125–140.

———. 2009b. Which ecologically important traits are most likely to evolve rapidly? *Oikos* 118:1–3.

———. 2010. The adaptive radiation of coevolving prodoxid moths and their host plants: *Greya* moths and yucca moths. Pages 228–245 in P. R. Grant and B. R. Grant, eds., *In search of the causes of evolution: From field observations to mechanisms*. Princeton University Press, Princeton, NJ.

Thompson, J. N., and J. J. Burdon. 1992. Gene-for-gene coevolution between plants and parasites. *Nature* 360:121–125.

Thompson, J. N., and B. M. Cunningham. 2002. Geographic structure and dynamics of coevolutionary selection. *Nature* 417:735—738.

Thompson, J. N., B. M. Cunningham, K. A. Segraves, D. M. Althoff, and D. Wagner. 1997. Plant polyploidy and insect/plant interactions. *American Naturalist* 150:730–743.

Thompson, J. N., and C. C. Fernandez. 2006. Temporal dynamics of antagonism and mutualism in a geographically variable plant–insect interaction. *Ecology* 87:103–112.

Thompson, J. N., A.-L. Laine, and J. F. Thompson. 2010. Retention of mutualism in a geographically diverging interaction. *Ecology Letters* 13:1368–1377.

Thompson, J. N., and K. Merg. 2008. Evolution of polyploidy and diversification of plant-pollinator interactions. *Ecology* 89:2197–2206.

Thompson, J. N., S. L. Nuismer, and R. Gomulkiewicz. 2002. Coevolution and maladaptation. *Integrative and Comparative Biology* 42:381–387.

Thompson, J. N., S. L. Nuismer, and K. Merg. 2004. Plant polyploidy and the evolutionary ecology of plant/animal interactions. *Biological Journal of the Linnean Society* 82:511–519.

Thompson, J. N., and O. Pellmyr. 1992. Mutualism with pollinating seed parasites amid co-pollinators: Constraints on specialization. *Ecology* 73: 1780–1791.

Thompson, J. N., and K. A. Rich. 2011. Range edges and the molecular divergence of *Greya* moth populations. *Journal of Biogeography* 38:551–563.

Thompson, J. N., W. Wehling, and R. Podolsky. 1990. Evolutionary genetics of host use in swallowtail butterflies. *Nature* 344:148–150.

Thompson, J. N., and M. F. Willson. 1979. Evolution of temperate fruit/bird interactions: Phenological strategies. *Evolution* 33:973–982.

Thompson, L. R., Q. Zeng, L. Kelly, K. H. Huang, A. U. Singer, J. Stubbe, and S. W. Chisholm. 2011. Phage auxiliary metabolic genes and the redirection of cyanobacterial host carbon metabolism. *Proceedings of the National Academy of Sciences USA* 108:E757–E764.

Thorpe, A. S., and R. M. Callaway. 2011. Biogeographic differences in the effects of *Centaurea stoebe* on the soil nitrogen cycle: Novel weapons and soil microbes. *Biological Invasions* 13:1435–1445.

Thrall, P. H., and J. J. Burdon. 2002. Evolution of gene-for-gene systems in metapopulations: The effect of spatial scale of host and pathogen dispersal. *Plant Pathology* 51:169–184.

———. 2003. Evolution of virulence in a plant host–pathogen metapopulation. *Science* 299:1735–1737.

Thrall, P. H., J. J. Burdon, and J. D. Bever. 2002. Local adaptation in the *Linum marginale–Melampsora lini* host–pathogen interaction. *Evolution* 56:1340–1351.

Thrall, P. H., M. E. Hochberg, J. J. Burdon, and J. D. Bever. 2007. Coevolution of symbiotic mutualists and parasites in a community context. *Trends in Ecology and Evolution* 22:120–126.

Tian, D., M. B. Traw, J. Q. Chen, M. Kreitman, and J. Bergelson. 2003. Fitness costs of R-gene-mediated resistance in *Arabidopsis thaliana*. *Nature* 423:74–76.

Tilman, D. 2011. Diversification, biotic interchange, and the universal trade-off hypothesis. *American Naturalist* 178:355–371.

Tiritilli, M. E., and J. N. Thompson. 1988. Variation in swallowtail/plant interactions: Host selection and the shapes of survivorship curves. *Oikos* 53:153–160.

Toft, C., and S. G. E. Andersson. 2010. Evolutionary microbial genomics: Insights into bacterial host adaptation. *Nature Reviews Genetics* 11:465–475.

Toju, H. 2007. Interpopulation variation in predator foraging behaviour promotes the evolutionary divergence of prey. *Journal of Evolutionary Biology* 20:1544–1553.

———. 2008. Fine-scale local adaptation of weevil mouthpart length and *Camellia* pericarp thickness: Altitudinal gradient of a putative arms race. *Evolution* 62:1086–1102.

———. 2009. Natural selection drives the fine-scale divergence of a coevolutionary arms race involving a long-mouthed weevil and its obligate host plant. *BMC Evolutionary Biology* 9:273.

Toju, H., and T. Sota. 2006a. Imbalance of predator and prey armament: Geographic clines in phenotypic interface and natural selection. *American Naturalist* 167:105–117.

———. 2006b. Phylogeography and the geographic cline in the armament of a seed-predatory weevil: Effects of historical events vs. natural selection from the host plant. *Molecular Ecology* 15:4161–4173.

———. 2009. Do arms races punctuate evolutionary stasis? Unified insights from phylogeny, phylogeography and microevolutionary processes. *Molecular Ecology* 18:3940–3954.

Toju, H., S. Ueno, F. Taniguchi, and T. Sota. 2011. Metapopulation structure of a seed-predator weevil and its host plant in arms race coevolution. *Evolution* 65:1707–1722.

Tokuda, G., and H. Watanabe. 2007. Hidden cellulases in termites: Revision of an old hypothesis. *Biology Letters* 3:336–339.

Tomback, D. 1982. Dispersal of whitebark pine seeds by Clark's nutcracker: A mutualism hypothesis. *Journal of Animal Ecology* 51:451–467.

Tonsor, S. J., and S. M. Scheiner. 2007. Plastic trait integration across a CO_2 gradient in *Arabidopsis thaliana*. *American Naturalist* 169:E119–E140.

Tooker, J. F., J. R. Rohr, W. G. Abrahamson, and C. M. De Moraes. 2008. Gall insects can avoid and alter indirect plant defenses. *New Phytologist* 178:657–671.

Toräng, P., J. Ehrlén, and J. Ågren. 2008. Mutualists and antagonists mediate frequency-dependent selection on floral display. *Ecology* 89:1564–1572.

Touzet, P., and L. F. Delph. 2009. The effect of breeding system on polymorphism in mitochondrial genes of *Silene*. *Genetics* 181:631–644.

Trivers, R. L. 1974. Parent–offspring conflict. *Integrative and Comparative Biology* 14:249–264.

Trotter, M. V., and H. G. Spencer. 2009. Complex dynamics occur in a single-locus, multiallelic model of general frequency-dependent selection. *Theoretical Population Biology* 76:292–298.

Troyer, K. 1984. Microbes, herbivory, and the evolution of social behavior. *Journal of Theoretical Biology* 106:157–169.

Turelli, M., and N. H. Barton. 1994. Genetic and statistical analysis of strong selection on polygenic traits: What, me normal? *Genetics* 138:913–941.

———. 2006. Will population bottlenecks and multilocus epistasis increase additive genetic variance? *Evolution* 60:1763–1776.

Turelli, M., and A. A. Hoffmann. 1991. Rapid spread of an inherited incompatibility factor in California *Drosophila*. *Nature* 353:440–442.

———. 1995. Cytoplasmic incompatibility in *Drosophila simulans*: Dynamics and parameter estimates from natural populations. *Genetics* 140:1319–1338.

Turnbaugh, P. J., R. E. Ley, M. Hamady, C. M. Fraser-Liggett, R. Knight, and J. I. Gordon. 2007. The human microbiome project. *Nature* 449:804–810.

Turner, P. E., N. M. Morales, B. W. Alto, and S. K. Remold. 2010. Role of evolved host breadth in the initial emergence of an RNA virus. *Evolution* 64:3273–3286.

Udall, J. A., and J. F. Wendel. 2006. Polyploidy and crop improvement. *Crop Science* 46:S3–S14.

Urban, M. C., and D. K. Skelly. 2006. Evolving metacommunities: Toward an evolutionary perspective on metacommunities. *Ecology* 87:1616–1626.

Usher, K. M., B. Bergman, and J. A. Raven. 2007. Exploring cyanobacterial mutualisms. *Annual Review of Ecology, Evolution, and Systematics* 38:255–273.

Utsumi, S., Y. Ando, T. P. Craig, and T. Ohgushi. 2011. Plant genotypic diversity increases population size of a herbivorous insect. *Proceedings of the Royal Society B* 278:3108–3115.

Uyeda, J. C., T. F. Hansen, S. J. Arnold, and J. Pienaar. 2011. The million-year wait for macroevolutionary bursts. *Proceedings of the National Academy of Sciences USA* 108:15908–15913.

Valiente-Banuet, A., A. V. Rumebe, M. Verdú, and R. M. Callaway. 2006. Modern Quaternary plant lineages promote diversity through facilitation of ancient Tertiary lineages. *Proceedings of the National Academy of Sciences USA* 103:16812–16817.

Valiente-Banuet, A., and M. Verdú. 2007. Assembly through facilitation can increase the phylogenetic diversity of plant communities. *Ecology Letters* 10:1029–1036.

Vallin, A., S. Jakobsson, and C. Wiklund. 2007. "An eye for an eye"—On the generality of the intimidating quality of eyespots in a butterfly and a hawkmoth. *Behavioral Ecology and Sociobiology* 61:1419–1424.

Vamosi, S. M., and D. Schluter. 2004. Character shifts in the defensive armor of sympatric sticklebacks. *Evolution* 58:376–385.

Van Bocxlaer, B., D. V. Damme, and C. S. Feibel. 2008. Gradual versus punctuated equilibrium evolution in the Turkana Basin molluscs: Evolutionary events or biological invasions? *Evolution* 62:511–520.

Van den Abbeele, P., T. Van de Wiele, W. Verstraete, and S. Possemiers. 2011. The host selects mucosal and luminal associations of coevolved gut microorganisms: A novel concept. *FEMS Microbiology Reviews* 35:681–704.

Van Der Putten, W. 2009. A multitrophic perspective on functioning and evolution of facilitation in plant communities. *Journal of Ecology* 97:1131–1138.

Van Doorn, G. S., and U. Dieckmann. 2006. The long-term evolution of multilocus traits under frequency-dependent disruptive selection. *Evolution* 60:2226–2238.

Van Noorden, R. 2010. Demand for malaria drug soars. *Nature* 466:672–673.

Van Valen, L. 1973. A new evolutionary law. *Evolutionary Theory* 1:1–30.

Van Zandt, P. A., and S. Mopper. 1998. A meta-analysis of adaptive deme formation in phytophagous insect populations. *American Naturalist* 152:595–604.

Vander Wall, S. B. 2002. Masting in animal-dispersed pine facilitates seed dispersal. *Ecology* 83:3508–3516.

———. 2008. On the relative contributions of wind vs. animals to seed dispersal of four Sierra Nevada pines. *Ecology* 89:1837–1849.

van't Hof, A. E., N. Edmonds, M. Dalikova, F. Marec, and I. J. Saccheri. 2011. Industrial melanism in British peppered moths has a singular and recent mutational origin. *Science* 332:958–960.

Vanthournout, B., J. Swaegers, and F. Hendrickx. 2011. Spiders do not escape reproductive manipulations by *Wolbachia*. *BMC Evolutionary Biology* 11:15.

Vásquez, D. P., C. J. Melián, N. M. Williams, N. Blüthgen, B. R. Krasnov, and R. Poulin. 2007. Species abundance and asymmetric interaction strength in ecological networks. *Oikos* 116:1120–1127.

Vautrin, E., and F. Vavre. 2009. Interactions between vertically transmitted symbionts: Cooperation or conflict? *Trends in Microbiology* 17:95–99.

Venditti, C., and M. Pagel. 2010. Speciation as an active force in promoting genetic evolution. *Trends in Ecology and Evolution* 25:14–20.

Venn, A. A., J. E. Loram, and A. E. Douglas. 2008. Photosynthetic symbioses in animals. *Journal of Experimental Botany* 59:1069–1080.

Venner, S., C. Feschotte, and C. Biemont. 2009. Dynamics of transposable elements: Towards a community ecology of the genome. *Trends in Genetics* 25:317–323.

Vercken, E., J. Clobert, and B. Sinervo. 2010. Frequency-dependent reproductive success in female common lizards: A real-life hawk-dove-bully game? *Oecologia* 162:49–58.

Verdú, M., P. Rey, J. Alcantara, and G. Siles. 2009. Phylogenetic signatures of facilitation and competition in successional communities. *Journal of Ecology* 97:1171–1180.

Verdú, M., and A. Valiente-Banuet. 2008. The nested assembly of plant facilitation networks prevents species extinctions. *American Naturalist* 172:751–760.

Vermeer, K. M. C. A., M. Dicke, and P. W. Jong. 2010. The potential of a population genomics approach to analyse geographic mosaics of plant–insect coevolution. *Evolutionary Ecology* 25:977–992.

Vermeij, G. J. 1977. The Mesozoic marine revolution: Evidence from snails, predators, and grazers. *Paleobiology* 3:245–258.

———. 1987. *Evolution and escalation*. Princeton University Press, Princeton, NJ.

———. 1994. The evolutionary interaction among species: Selection, escalation, and coevolution. *Annual Review of Ecology and Systematics* 25:219–236.

———. 2002a. Characters in context: Molluscan shells and the forces that mold them. *Paleobiology* 28:41–54.

———. 2002b. Evolution in the consumer age: Predators and the history of life. Pages 375–393 in M. Kowalewski and P. H. Kelley, eds., *The fossil record of predation*. Paleontological Society Papers, vol. 8. New Haven, CT.

Vermeij, G. J., and R. K. Grosberg. 2010. The great divergence: When did diversity on land exceed that in the sea? *Integrative and Comparative Biology* 50:675–682.

Via, S. 2009. Natural selection in action during speciation. *Proceedings of the National Academy of Sciences USA* 106:9939–9946.

Via, S., and D. J. Hawthorne. 2002. The genetic architecture of ecological specialization: Correlated gene effects on host use and habitat choice in pea aphids. *American Naturalist* 159:S76–S88.

Vignieri, S. N., J. G. Larson, and H. E. Hoekstra. 2010. The selective advantage of crypsis in mice. *Evolution* 64:2153–2158.

Vijendravarma, R. K., A. R. Kraaijeveld, and H. C. J. Godfray. 2009. Experimental evolution shows *Drosophila melanogaster* resistance to a microsporidian pathogen has fitness costs. *Evolution* 63:104–114.

Vines, T. H., and D. Schluter. 2006. Strong assortative mating between allopatric sticklebacks as a by-product of adaptation to different environments. *Proceedings of the Royal Society B* 273:911–916.

Vogwill, T., A. Fenton, A. Buckling, M. E. Hochberg, and M. A. Brockhurst. 2009. Source populations act as coevolutionary pacemakers in experimental selection mosaics containing hotspots and coldspots. *American Naturalist* 173:E171–E176.

Volkov, I., J. R. Banavar, S. P. Hubbell, and A. Maritan. 2009. Inferring species interactions in tropical forests. *Proceedings of the National Academy of Sciences USA* 106:13854–13859.

Vos, M., and X. Didelot. 2009. A comparison of homologous recombination rates in bacteria and archaea. *ISME Journal* 3:199–208.

Vrba, E. S. 2005. Mass turnover and heterochrony events in response to physical change. *Paleobiology* 31:157.

Vrijenhoek, R. C., and E. D. Parker, Jr. 2009. Geographical parthenogenesis: General purpose and frozen niche variation. Pages 9–131 in I. Schön, K. Martens, and P. van Dijk, eds., *Lost sex: The evolutionary biology of parthenogenesis*. Springer, New York.

Wade, M. J. 2007. The co-evolutionary genetics of ecological communities. *Nature Reviews Genetics* 8:185–195.

Wade, M. J., and C. Goodnight. 2006. Cyto-nuclear epistasis: Two-locus random genetic drift in hermaphroditic and dioecious species. *Evolution* 60:643–659.

Wade, M. J., and J. R. Griesemer. 1998. Population heritability: Empirical studies of evolution in metapopulations. *American Naturalist* 151:135–147.

Wade, M. J., and D. E. McCauley. 2005. Paternal leakage sustains the cytoplasmic polymorphism underlying gynodioecy but remains invasible by nuclear restorers. *American Naturalist* 166:592–602.

Wagner, A. 2005. *Robustness and evolvability in living systems*. Princeton University Press, Princeton, NJ.

Wagner, G. P., and J. Zhang. 2011. The pleiotropic structure of the genotype-phenotype map: The evolvability of complex organisms. *Nature Reviews Genetics* 12:204–213.

Wain, L. V., J. A. L. Armour, and M. D. Tobin. 2009. Genomic copy number variation, human health, and disease. *Lancet* 374:340–350.

Wäli, P. R., J. U. Ahlholm, M. Helander, and K. Saikkonen. 2007. Occurrence and genetic structure of the systemic grass endophyte *Epichloe festucae* in fine fescue populations. *Microbial Ecology* 53:20–29.

Walker, J. A., and M. A. Bell. 2000. Net evolutionary trajectory of body shape evolution within a microgeographic radiation of threespine sticklebacks (*Gasterosteus aculeatus*). *Journal of Zoology* 252:293–302.

Walls, E. A., J. Berkson, and S. A. Smith. 2002. The horseshoe crab, *Limulus polyphemus*: 200 million years of existence, 100 years of study. *Reviews in Fisheries Science* 10:39–73.

Walsh, M. R., and D. M. Post. 2011. Interpopulation variation in a fish predator drives evolutionary divergence in prey in lakes. *Proceedings of the Royal Society B* 278:2628–2637.

Walsh, M. R., and D. N. Reznick. 2011. Experimentally induced life-history evolution in a killifish in response to the introduction of guppies. *Evolution* 65:1021–1036.

Wang, J., L. Zhang, J. Li, A. Lawton-Rauh, and D. Tian. 2011. Unusual signatures of highly adaptable R-loci in closely-related *Arabidopsis* species. *Gene* 482:24–33.

Wang, X., H. Wang, J. Wang, R. Sun, J. Wu, S. Liu, Y. Bai, et al. 2011. The genome of the mesopolyploid crop species *Brassica rapa*. *Nature Genetics* 43:1035–1039.

Waser, N. M., and J. Ollerton. 2006. *Plant–pollinator interactions: From specialization to generalization*. University of Chicago Press, Chicago.

Waters, C., and B. L. Bassler. 2005. Quorum sensing: Cell-to-cell communication in bacteria. *Annual Review of Cell and Developmental Biology* 21:319–346.

Watt, W. B. 1977. Adaptation at specific loci. I. Natural selection on phosphoglucose isomerase of *Colias* butterflies: Biochemical and population aspects. *Genetics* 87:177–194.

Watt, W.B., and R. Cassin. 1983. Adaptation at specific loci. III. Field behavior and survivorship differences among *Colias* PGI genotypes are predictable from in vitro biochemistry. *Genetics* 103:725–739.

Watt, W. B., C. W. Wheat, E. H. Meyer, and J.-F. Martin. 2003. Adaptation at specific loci. VII. Natural selection, dispersal and the diversity of molecular-functional variation patterns among butterfly species complexes (*Colias*: Lepidoptera, Pieridae). *Molecular Ecology* 12:1265–1275.

Webb, C. O., D. D. Ackerly, and S. W. Kembel. 2008. Phylocom: Software for the analysis of community phylogenetic structure and character evolution. *Bioinformatics* 24:2098–2100.

Webb, C. O., D. D. Ackerly, M. A. McPeek, and M. J. Donoghue. 2002. Phylogenies and community ecology. *Annual Review of Ecology and Systematics* 33:475–505.

Weeks, A. R., M. Turelli, W. R. Harcombe, K. T. Reynolds, and A. A. Hoffman. 2007. From parasite to mutualist: Rapid evolution of *Wolbachia* in natural populations of *Drosophila*. *PLoS Biology* 5:997–1005.

Wehling, W. F., and J. N. Thompson. 1997. Evolutionary conservatism of oviposition preference in a widespread polyphagous insect herbivore, *Papilio zelicaon*. *Oecologia* 111:209–215.

Weiblen, G. D., and B. G. Brehm. 1996. Reproductive strategies and barriers to hybridization between *Tellima grandiflora* and *Tolmiea menziesii* (Saxifragaceae). *American Journal of Botany* 83:910–918.

Weiner, J. 1995. *The beak of the finch*. Vintage, New York.

Weir, J. T., E. Bermingham, and D. Schluter. 2009. The great American biotic interchange in birds. *Proceedings of the National Academy of Sciences USA* 106: 21737–21742.

Weir, J. T., and D. Schluter. 2007. The latitudinal gradient in recent speciation and extinction rates of birds and mammals. *Science* 315:1574–1576.

Weis, V. M. 2008. Cellular mechanisms of Cnidarian bleaching: Stress causes the collapse of symbiosis. *Journal of Experimental Biology* 211:3059–3066.

Welbergen, J. A., and N. B. Davies. 2009. Strategic variation in mobbing as a front line of defense against brood parasitism. *Current Biology* 19:235–240.

Welch, D. B. M., C. Ricci, and M. Meselson. 2009. Bdelloid rotifers: Progress in understanding the success of an evolutionary scandal. Pages 259–279 in I. Schön, K. Martens, and P. Dijk, eds., *Lost sex: The evolutionary biology of parthenogenesis*. Springer, New York.

Wellems, T. E., K. Hayton, and R. M. Fairhurst. 2009. The impact of malaria parasitism: From corpuscles to communities. *Journal of Clinical Investigation* 119:2496–2505.

Werren, J. H. 1997. *Wolbachia* and speciation. Pages 245–260 in D. Howard and S. H. Berlocher, eds., *Endless forms: Species and speciation*. Oxford University Press, Oxford.

———. 2011. Selfish genetic elements, genetic conflict, and evolutionary innovation. *Proceedings of the National Academy of Sciences USA* 108:10863–10870.

Werren, J. H,, L. Baldo, and M. E. Clark. 2008. *Wolbachia*: Master manipulators of invertebrate biology. *Nature Reviews Microbiology* 6:741–751.

Werren, J. H., and J. Jaenike. 1995. *Wolbachia* and cytoplasmic incompatibility in mycophagous *Drosophila* and their relatives. *Heredity* 75:320–326.

Werren, J. H., U. Nur, and C.-I. Wu. 1988. Selfish genetic elements. *Trends in Ecology and Evolution* 11:297–302.

West, D. A. 2003. *Fritz Müller: A naturalist in Brazil*. Pocahontas Press, Blacksburg, VA.

West, S. A. 2009. *Sex allocation*. Princeton University Press, Princeton, NJ.

West, S. A., S. P. Diggle, A. Buckling, A. Gardner, and A. S. Griffin. 2007. The social lives of microbes. *Annual Review of Ecology Evolution and Systematics* 38:53–77.

West-Eberhard, M. J. 1983. Sexual selection, social competition and speciation. *Quarterly Review of Biology* 58:155–183.

———. 2005. Developmental plasticity and the origin of species differences. *Proceedings of the National Academy of Sciences USA* 102 (suppl. 1): 6543–6549.

Westley, P. A. H. 2011. What invasive species reveal about the rate and form of contemporary phenotypic change in nature. *American Naturalist* 177:496–509.

Whalon, M., D. Mota-Sanchez, R. M. Hollingworth, and L. Duynslager. 2011. Arthropod pesticide resistance database; http://www.pesticideresistance.org/.

Wheat, C. W., H. W. Fescemyer, J. Kvist, E. Tas, J. Cristobal Vera, M. J. Frilander, I. Hanski, et al. 2011. Functional genomics of life history variation in a butterfly metapopulation. *Molecular Ecology* 20:1813–1828.

Wheat, C. W., C. R. Haag, J. H. Marden, I. Hanski, and M. J. Frilander. 2010. Nucleotide polymorphism at a gene (*Pgi*) under balancing selection in a butterfly metapopulation. *Molecular Biology and Evolution* 27:267–281.

Whitham, T. G. 1981. Individual trees as heterogeneous environments: Adaptation to herbivory or epigenetic noise? Pages 9–27 in R. F. Denno and H. Dingle, eds., *Insect life history patterns: Habitat and geographic variation*. Springer-Verlag, New York.

———. 1989. Plant hybrid zones as sinks for pests. *Science* 244:1490–1493.

Whitham, T. G., J. K. Bailey, J. A. Schweitzer, S. M. Shuster, R. K. Bangert, C. J. LeRoy, E. Lonsdorf, et al. 2006. A framework for community and ecosystem genetics: From genes to ecosystems. *Nature Reviews Genetics* 7:510–523.

Whitham, T. G., G. D. Martinsen, K. D. Floate, H. S. Dungey, B. M. Potts, and P. Keim. 1999. Plant hybrid zones affect biodiversity: Tools for a genetic-based understanding of community structure. *Ecology* 80:416–428.

Whitman, W., D. Coleman, and W. Wiebe. 1998. Prokaryotes: The unseen majority. *Proceedings of the National Academy of Sciences USA* 95:6578–6583.

Whitney, K. D., and C. A. Gabler. 2008. Rapid evolution in introduced species, "invasive traits" and recipient communities: Challenges for predicting invasive potential *Diversity and Distributions* 14:569–580.

Wichman, H. A., M. R. Badgett, L. A. Scott, C. M. Boulianne, and J. J. Bull. 1999. Different trajectories of parallel evolution during viral adaptation. *Science* 285:422–424.

Wichman, H., and C. Brown. 2010. Experimental evolution of viruses: Microviridae as a model system. *Philosophical Transactions of the Royal Society B* 365:2495–2501.

Wichman, H. A., J. Millstein, and J. J. Bull. 2005. Adaptive molecular evolution for 13,000 phage generations: A possible arms race. *Genetics* 170:19–31.

Wichman, H. A., L. A. Scott, C. D. Yarber, and J. J. Bull. 2000. Experimental evolution recapitulates natural evolution. *Philosophical Transactions of the Royal Society B* 355:1677–1684.

Wiedenbeck, J., and F. M. Cohan. 2011. Origins of bacterial diversity through horizontal genetic transfer and adaptation to new ecological niches. FEMS *Microbiology Reviews* 35:957–976.

Wiens, J. J., D. D. Ackerly, A. P. Allen, B. L. Anacker, L. B. Buckley, H. V. Cornell, E. I. Damschen, et al. 2010. Niche conservatism as an emerging principle in ecology and conservation biology. *Ecology Letters* 13:1310–1324.

Wiklund, C. 1981. Generalist vs. specialist oviposition behaviour in *Papilio machaon* (Lepidoptera) and functional aspects on the hierarchy of oviposition preferences. *Oikos* 36:163–170.

Wiklund, C., and M. Friberg. 2008. Enemy-free space and habitat-specific host specialization in a butterfly. *Oecologia* 157:287–294.

———. 2009. The evolutionary ecology of generalization: Among-year variation in host plant use and offspring survival in a butterfly. *Ecology* 90:3406–3417.

Wilczek, A. M., L. T. Burghardt, A. R. Cobb, M. D. Cooper, S. M. Welch, and J. Schmitt. 2010. Genetic and physiological bases for phenological responses to current and predicted climates. *Philosophical Transactions of the Royal Society B* 365:3129–3147.

Wilczek, A. M., J. L. Roe, M. C. Knapp, M. D. Cooper, C. Lopez-Gallego, L. J. Martin, C. D. Muir, et al. 2009. Effects of genetic perturbation on seasonal life history plasticity. *Science* 323:930–934.

Wilf, P., C. C. Labandeira, K. R. Johnson, and B. Ellis. 2006. Decoupled plant and insect diversity after the end-Cretaceous extinction. *Science* 313:1112–1115.

Williams, B. L., C. T. Hanifin, E. D. Brodie, and E. D. Brodie III. 2010. Tetrodotoxin affects survival probability of rough-skinned newts (*Taricha granulosa*) faced with TTX-resistant garter snake predators (*Thamnophis sirtalis*). *Chemoecology* 20:285–290.

Williams, G. C. 1975. *Sex and evolution*. Princeton University Press, Princeton, NJ.

Williams, P., K. Winzer, W. C. Chan, and M. Camara. 2007. Look who's talking: Communication and quorum sensing in the bacterial world. *Philosophical Transactions of the Royal Society B* 362:1119–1134.

Williamson, S. H., M. J. Hubisz, A. G. Clark, B. A. Payseur, C. D. Bustamante, and R. Nielsen. 2007. Localizing recent adaptive evolution in the human genome. *PLoS Genetics* 3:e90.

Willis, B. L., M. J. H. van Oppen, D. J. Miller, S. V. Vollmer, and D. J. Ayre. 2006. The role of hybridization in the evolution of reef corals. *Annual Review of Ecology Evolution and Systematics* 37:489–517.

Willson, M. F., and C. Whelan. 1993. Variation of dispersal phenology in a bird-dispersed shrub, *Cornus drummondii*. *Ecological Monographs* 63:151–172.

Wilson, A. B., K. Noack-Kunnmann, and A. Meyer. 2000. Incipient speciation in sympatric Nicaraguan crater lake cichlid fishes: Sexual selection versus ecological diversification. *Proceedings of the Royal Society B* 267:2133–2141.

Wilson, C. G., and P. W. Sherman. 2010. Anciently asexual bdelloid rotifers escape lethal fungal parasites by drying up and blowing away. *Science* 327:574–576.

Wilson, E. O. 1971. *The insect societies*. Harvard University Press, Cambridge, MA.

Wilson, H. B., M. P., Hassell, and H. C. J. Godfray. 1996. Host–parasitoid food webs: Dynamics, persistence, and invasion. *American Naturalist* 148:787–806.

Wing, S. L., F. Herrera, C. A. Jaramillo, C. Gómez-Navarro, P. Wilf, and C. C. Labandeira. 2009. Late Paleocene fossils from the Cerrejon Formation, Colombia, are the earliest record of Neotropical rainforest. *Proceedings of the National Academy of Sciences USA* 106:18627–18632.

Wolf, P. G., P. S. Soltis, and D. E. Soltis. 1989. Tetrasomic inheritance and chromosome pairing behaviour in the naturally occurring autotetraploid *Heuchera grossulariifolia* (Saxifragaceae). *Genome* 32:655–659.

Wommack, K. E., and R. R. Colwell. 2000. Virioplankton: Viruses in aquatic ecosystems. *Microbiology and Molecular Biology Reviews* 64:69–114.

Wood, T. E., N. Takebayashi, M. S. Barker, I. Mayrose, P. B. Greenspoon, and L. H. Rieseberg. 2009. The frequency of polyploid speciation in vascular plants. *Proceedings of the National Academy of Sciences USA* 106:13875–13879.

Woods, R. J., J. E. Barrick, T. F. Cooper, U. Shrestha, M. R. Kauth, and R. E. Lenski. 2011. Second-order selection for evolvability in a large *Escherichia coli* population. *Science* 331:1433–1436.

Wright, S. 1931. Evolution in Mendelian populations. *Genetics* 16:97–159.

———. 1932. The roles of mutation, inbreeding, crossbreeding and selection in evolution. *Proceedings of the 6th International Congress of Genetics* 1:356–366.

———. 1948. On the roles of directed and random changes in gene frequency in the genetics of populations. *Evolution* 2:279–294.

Wu, C.-I. 2001. The genic view of the process of speciation. *Journal of Evolutionary Biology* 14:851–864.

Wu, J., and I. T. Baldwin. 2010. New insights into plant responses to the attack from insect herbivores. *Annual Review of Genetics* 44:1–24.

Xavier, J. B. 2011. Social interaction in synthetic and natural microbial communities. *Molecular Systems Biology* 7:483.

Yang, M.-C., C. A. Chen, H.-L. Hsieh, and C.-P. Chen. 2007. Population subdivision of the tri-spine horseshoe crab, *Tachypleus tridentatus*, in Taiwan Strait. *Zoological Science* 24:219–224.

Yeaman, S., and M. C. Whitlock. 2011. The genetic architecture of adaptation under migration–selection balance. *Evolution* 65:1897–1911.

Yoshida, S., S. Maruyama, H. Nozaki, and K. Shirasu. 2010. Horizontal gene transfer by the parasitic plant *Striga hermonthica*. *Science* 328:1128.

Young, T. P., and M. L. Stanton. 2003. Effects of natural and simulated herbivory on spine lengths of *Acacia drepanolobium* in Kenya. *Oikos* 101:171–179.

Yuan, X., S. Xiao, and T. N. Taylor. 2005. Lichen-like symbioses 600 mya. *Science* 308:1017–1020.

Yutin, N., K. S. Makarova, S. L. Mekhedov, Y. I. Wolf, and E. V. Koonin. 2008. The deep archaeal roots of eukaryotes. *Molecular Biology and Evolution* 25:1619–1630.

Zangerl, A. R., and M. R. Berenbaum. 2003. Phenotypic matching in wild parsnip and parsnip webworms: Causes and consequences. *Evolution* 57:806–815.

———. 2005. Increase in toxicity of an invasive weed after reassociation within its coevolved herbivore. *Proceedings of the National Academy of Sciences USA* 102: 15529–15532.

Zangerl, A. R., M. C. Stanley, and M. R. Berenbaum. 2008. Selection for chemical trait remixing in an invasive weed after reassociation with a coevolved species. *Proceedings of the National Academy of Sciences USA* 105:4547–4552.

Zbinden, M., C. R. Haag, and D. Ebert. 2008. Experimental evolution of field populations of *Daphnia magna* in response to parasite treatment. *Journal of Evolutionary Biology* 21:1068–1078.

Zhaxybayeva, O., and W. F. Doolittle. 2011. Lateral gene transfer. *Current Biology* 21:R243.

Zhaxybayeva, O., K. S. Swithers, P. Lapierre, G. P. Fournier, D. M. Bickhart, R. T. DeBoy, K. E. Nelson, et al. 2009. On the chimeric nature, thermophilic origin, and phylogenetic placement of the Thermotogales. *Proceedings of the National Academy of Sciences USA* 106:5865–5870.

Zimmer, C. 2008. *Microcosm: E. coli and the new science of life*. Pantheon, New York.

Zinger, L., L. A. Amaral-Zettler, J. A. Fuhrman, M. C. Horner-Devine, S. M. Huse, D. B. M. Welch, J. B. H. Martiny, et al. 2011. Global patterns of bacterial Beta-diversity in seafloor and seawater ecosystems. *PLoS ONE* 6:e24570.

Index